VOLUME ONE HUNDRED AND SEVENTY

CURRENT TOPICS IN DEVELOPMENTAL BIOLOGY

Erythropoiesis/Hematopoiesis

CURRENT TOPICS IN DEVELOPMENTAL BIOLOGY

"A meeting-ground for critical review and discussion of developmental processes"
A.A. Moscona and Alberto Monroy (Volume 1, 1966)

VOLUME ONE HUNDRED AND SEVENTY

CURRENT TOPICS IN DEVELOPMENTAL BIOLOGY

Erythropoiesis/Hematopoiesis

Edited by

JAMES J. BIEKER

Mount Sinai School of Medicine,
Department of Cell, Developmental, and Regenerative Biology,
New York, NY, United States

ELSEVIER

ACADEMIC PRESS
An imprint of Elsevier

Academic Press is an imprint of Elsevier
125 London Wall, London, EC2Y 5AS, United Kingdom
50 Hampshire Street, 5th Floor, Cambridge, MA 02139, United States

First edition 2026

ISBN: 978-0-443-43128-9
ISSN: 0070-2153

For information on all Academic Press publications
visit our website at https://www.elsevier.com/books-and-journals

Publisher: Zoe Kruze
Editorial Project Manager: Devwart Chauhan
Production Project Manager: Bharatwaj Varatharajan
Cover Designer: Muthukumaran Thangaraj

Typeset by MPS Limited, India

Contents

Contributors

Sarah E. Adams
Department of Biological, Geological, and Environmental Sciences; Center for Gene Regulation in Health and Disease, at Cleveland State University, Cleveland, OH, United States

Divya Beri
Laboratory of Blood-Borne Parasites, Lindsley F. Kimball Research Institute, New York Blood Center, Rye, NY, United States

James J. Bieker
Department of Stem Cell Biology & Regenerative Medicine, Mount Sinai School of Medicine, New York, NY, United States

Yi-Tian Dai
Laboratory of Blood-Borne Parasites, Lindsley F. Kimball Research Institute, New York Blood Center, Rye, NY, United States

James Davies
Oxford National Institute of Health Research Biomedical Research Centre, University of Oxford, Oxford, United Kingdom

Nicholas Denny
MRC Molecular Haematology Unit, MRC Weatherall Institute of Molecular Medicine, Radcliffe Department of Medicine, University of Oxford, Oxford, United Kingdom

Anita R. Dhara
Department of Biological, Geological, and Environmental Sciences; Center for Gene Regulation in Health and Disease, at Cleveland State University, Cleveland, OH, United States

Yesenia Earth
Division of Hematology, Department of Medicine, School of Medicine, Washington University in St. Louis, St. Louis, MO, United States

Merlin Nithya Gnanapragasam
Department of Biological, Geological, and Environmental Sciences; Center for Gene Regulation in Health and Disease, at Cleveland State University; Center for RNA Science and Therapeutics, Case Western Reserve University, Cleveland, OH, United States

Jill A. Henrich
Division of Hematology and Oncology, Department of Medicine; Biomedical Graduate Studies, University of Pennsylvania Perelman School of Medicine, Philadelphia, PA, United States

Muhammad Ishfaq
Division of Classical Hematology, Department of Medicine, Pittsburgh Heart Lung and Blood Vascular Medicine Institute, University of Pittsburgh, Pittsburgh, PA, United States

Peng Ji
Department of Pathology, Feinberg School of Medicine, Northwestern University, Chicago, IL, United States

Yoon-A Kang
Division of Hematology, Department of Medicine, School of Medicine, Washington University in St. Louis, St. Louis, MO, United States

Hsiang-Ying Lee
State Key Laboratory of Membrane Biology, Peking-Tsinghua Center for Life Sciences, School of Life Sciences, Peking University, Beijing, P.R. China

In-Won Lee
Division of Hematology, Department of Medicine, School of Medicine, Washington University in St. Louis, St. Louis, MO, United States

Hangpeng Li
MRC Molecular Haematology Unit, MRC Weatherall Institute of Molecular Medicine, Radcliffe Department of Medicine, University of Oxford, Oxford, United Kingdom

Cheryl A. Lobo
Laboratory of Blood-Borne Parasites, Lindsley F. Kimball Research Institute, New York Blood Center, Rye, NY, United States

Sumayea B. Mahmood Meem
Department of Biological, Geological, and Environmental Sciences; Center for RNA Science and Therapeutics, Case Western Reserve University, Cleveland, OH, United States

Kaustav Mukherjee
Department of Stem Cell Biology & Regenerative Medicine, Mount Sinai School of Medicine, New York, NY, United States

Vikram R. Paralkar
Division of Hematology and Oncology, Department of Medicine; Department of Cell and Developmental Biology; Abramson Family Cancer Research Institute; Penn Epigenetics Institute, University of Pennsylvania Perelman School of Medicine, Philadelphia, PA, United States

Parina N. Patel
Department of Biological, Geological, and Environmental Sciences; Center for Gene Regulation in Health and Disease, at Cleveland State University, Cleveland, OH, United States

Mahesh Ramamoorthy
Center for Gene Regulation in Health and Disease, at Cleveland State University, Cleveland, OH, United States

Kehan Ren
Department of Medicine, Renaissance School of Medicine, Stony Brook University, Stony Brook, NY, United States

Marilis Rodriguez
Laboratory of Blood-Borne Parasites, Lindsley F. Kimball Research Institute, New York Blood Center, Rye, NY, United States

Eleanor I. Sams
Division of Hematology and Oncology, Department of Medicine; Biomedical Graduate Studies, University of Pennsylvania Perelman School of Medicine, Philadelphia, PA, United States

Raja Sundari Meenakshi Sundaram
Center for Gene Regulation in Health and Disease, at Cleveland State University, Cleveland, OH, United States

Pratap Vydyam
Laboratory of Blood-Borne Parasites, Lindsley F. Kimball Research Institute, New York Blood Center, Rye, NY, United States

Pan Wang
Department of Pathology, Feinberg School of Medicine, Northwestern University, Chicago, IL, United States

Rachael A. White
Department of Biological, Geological, and Environmental Sciences; Center for RNA Science and Therapeutics, Case Western Reserve University, Cleveland, OH, United States

Yvette Y. Yien
Division of Classical Hematology, Department of Medicine, Pittsburgh Heart Lung and Blood Vascular Medicine Institute, University of Pittsburgh, Pittsburgh, PA, United States

Yiyang Zhang
State Key Laboratory of Membrane Biology, Peking-Tsinghua Center for Life Sciences, School of Life Sciences, Peking University, Beijing, P.R. China

Shicong Zhu
State Key Laboratory of Membrane Biology, Peking-Tsinghua Center for Life Sciences, School of Life Sciences, Peking University, Beijing, P.R. China

CHAPTER ONE

Ribosomal RNA transcription in normal and leukemic hematopoiesis

Jill A. Henrich[a,b,1], Eleanor I. Sams[a,b,1], and Vikram R. Paralkar[a,c,d,e,*]

[a]Division of Hematology and Oncology, Department of Medicine, University of Pennsylvania Perelman School of Medicine, Philadelphia, PA, United States
[b]Biomedical Graduate Studies, University of Pennsylvania Perelman School of Medicine, Philadelphia, PA, United States
[c]Department of Cell and Developmental Biology, University of Pennsylvania Perelman School of Medicine, Philadelphia, PA, United States
[d]Abramson Family Cancer Research Institute, University of Pennsylvania Perelman School of Medicine, Philadelphia, PA, United States
[e]Penn Epigenetics Institute, University of Pennsylvania Perelman School of Medicine, Philadelphia, PA, United States
*Corresponding author. e-mail address: Vikram.Paralkar@pennmedicine.upenn.edu

Contents

[1] These authors contributed equally.

Current Topics in Developmental Biology, Volume 170
ISSN 0070-2153, https://doi.org/10.1016/bs.ctdb.2026.06.008

Abstract

Hematopoiesis is governed by precisely coordinated transcriptional programs that balance hematopoietic stem cell (HSC) self-renewal with the ability to differentiate into multiple blood lineages. During normal hematopoiesis, cells maintain tight control over the level of ribosomal RNA (rRNA) transcription by RNA Polymerase I (Pol I) and over mature rRNA abundance (ribosomal subunits). Although the factors governing RNA Polymerase II (Pol II)-mediated gene expression and their roles in cell fate determination are well characterized, far less is known about the regulation of rRNA transcription and ribosome abundance. Little is known also about how ribosome numbers contribute to hematopoietic cell fate. This review summarizes current literature on the mechanisms by which hematopoietic cell types fine-tune rRNA transcription and ribosomal subunit abundance throughout cell fate trajectories, and how dysregulation of these processes contributes to acute myeloid leukemia (AML).

1. Introduction – Hematopoiesis and ribosome biogenesis

Hematopoiesis is the process by which the diverse cell types of the blood system are generated through a hierarchical program of cell fate decisions. Hematopoietic stem cells (HSCs) balance self-renewal and differentiation to produce multipotent progenitors and lineage-committed progenitors, which then go on to sustain day-to-day mature blood cell production in homeostatic or stress conditions (Olson et al., 2020; Cedar & Bergman, 2011; Liggett & Sankaran, 2020; Rieger & Schroeder, 2012). Tight regulation of this hierarchy is critical for normal hematopoietic development and function, and its disruption underlies a range of hematologic malignancies, including acute myeloid leukemia (AML). AML is a highly aggressive malignancy characterized by the aberrant self-renewal, impaired differentiation, and uncontrolled proliferation of immature myeloid cells. Both normal hematopoietic development and leukemic transformation are known to be regulated in part through transcriptional programs mediated by RNA Polymerase II (Pol II). In the normal hematopoietic system, these programs are orchestrated by hierarchies of lineage-specific transcription factors that coordinate Pol II activity to establish and maintain cell-type-specific gene expression programs (Orkin & Zon, 2008). These Pol II transcriptional networks are well established as key determinants of hematopoietic cell fate decisions and contributors to leukemogenesis.

Less appreciated, however, is the regulation of ribosomal RNA (rRNA) transcription by RNA Polymerase I (Pol I) and its contribution to hematopoietic cell behavior. This is despite the fact that rRNA transcription accounts for a large fraction of all cellular transcription. Ribosome biogenesis is one of the most energy-intensive processes in proliferating cells, with ribosome biogenesis factors constituting a substantial part of the cellular proteome. Despite this considerable investment, rRNA transcription and ribosome biogenesis remain understudied in the context of cell-type-specific transcriptional regulation.

In mammalian cells, rRNA is transcribed in the nucleolus by Pol I from a subset of hundreds of ribosomal DNA (rDNA) repeats (Fig. 1). The nascent 47S precursor rRNA (pre-rRNA) undergoes a highly coordinated series of processing steps to generate the mature 18S, 5.8S, and 28S rRNAs, which assemble with the 5S rRNA (transcribed by RNA Pol III in the nucleoplasm) and 79 ribosomal proteins to form the mature 40S and 60S ribosomal subunits (Paralkar, 2024; Baßler & Hurt, 2019). The mature rRNAs form the structural and catalytic core of the ribosome, highlighting their central role in ribosome function.

The nucleolus itself is a dynamic, membraneless nuclear compartment organized into three subdomains: the central fibrillar center (FC), the surrounding dense fibrillar component (DFC), and the outer granular component (GC). rDNA repeats are localized to the FC, with rRNA transcription occurring at the FC-DFC interface. As pre-rRNA processing and ribosomal subunit assembly proceed, pre-ribosomal particles transit outward through the DFC and GC before being exported to the cytoplasm for final maturation (Quinodoz et al., 2025). Notably, leukemic myeloblasts have long been recognized to contain enlarged and numerous nucleoli (Smetana et al., 1997; Hein et al., 2013), a feature suggestive of elevated rRNA transcription and ribosome biogenesis.

In this review, we synthesize current knowledge of rRNA transcription levels and mature rRNA (ribosome subunit) abundance across the normal hematopoietic hierarchy and in AML. We discuss the molecular steps of rRNA transcription and ribosome assembly, highlighting potential modes of regulation that may modulate Pol I activity and downstream ribosome biogenesis in a cell-type-specific manner. Our discussion seeks to dissect the emerging view that rRNA transcription and ribosome biogenesis constitute a regulated, cell-type-specific feature of the hematopoietic hierarchy.

1.1 Scope and limitations of this review

Several important areas of ribosome biogenesis biology fall outside the scope of this review. Foundational and emerging insights into Pol I, rDNA, rRNA, and ribosome biology have been derived largely from non-mammalian

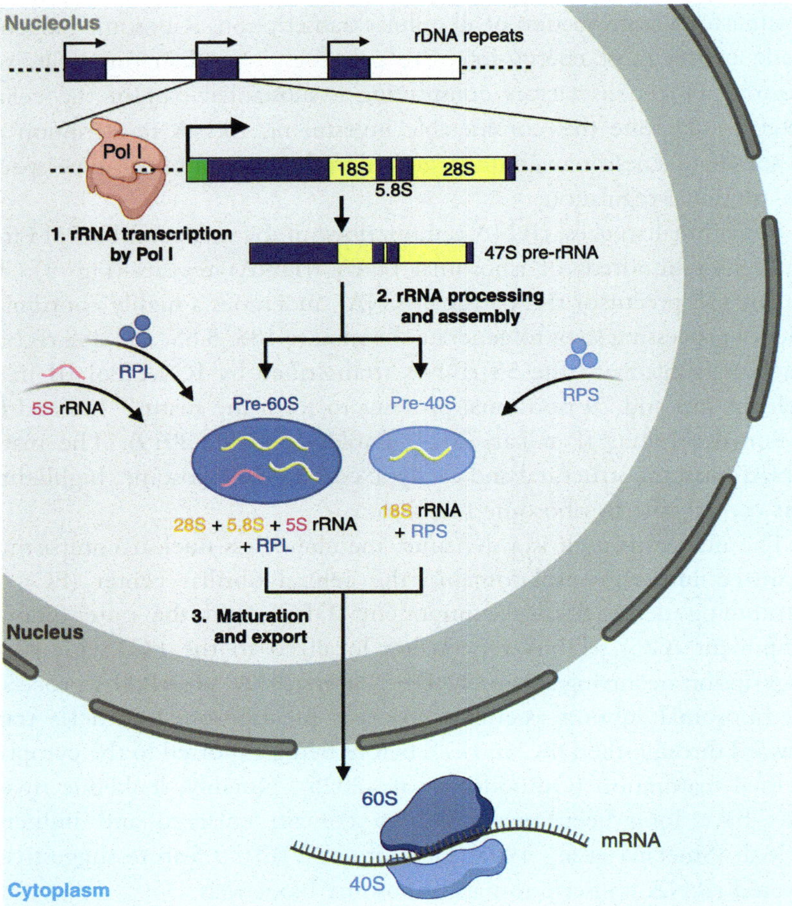

Fig. 1 Overview of ribosome biogenesis. Schematic of the steps of ribosome biogenesis, which begins with (1) transcription of rRNA by Pol I, followed by (2) rRNA processing and assembly, and (3) maturation of ribosomal subunits and export to the cytoplasm. Pol I, RNA Polymerase I; RPS, small subunit ribosomal proteins; RPL, large subunit ribosomal proteins.

systems, ranging from bacteria and yeast to amphibian and fish oocytes. While these studies are instrumental in shaping our current understanding and are briefly touched upon in relevant sections, readers are referred to comprehensive reviews for detailed discussion of them (Woolford & Susan, 2013; Henras et al., 2008; Dörner et al., 2023; Shajani et al., 2011). In addition, rRNA transcription and ribosome biogenesis have been studied in non-hematopoietic mammalian tissues and contexts, including circadian regulation in the liver (Jouffe et al., 2013; Sinturel et al., 2017), dynamic regulation in the brain (Li et al., 2025), early embryonic development (Falcon et al., 2022; Woolnough et al., 2016), and diverse cancer contexts (Ruggero & Pandolfi, 2003; Montanaro et al., 2008; Pelletier et al., 2018). Although we reference selected findings from these areas when mechanistically informative, we do not subject them to comprehensive discussion.

Additionally, it is now well-appreciated that ribosomes are not identical, whether within a single cell or across tissues within a single organism. Ribosome-to-ribosome variations arise from rDNA sequence variants, rRNA modifications, or variable ribosomal protein composition, paralogs, or post-translational modifications. While we briefly address qualitative differences in ribosome composition, our discussion is not comprehensive, and readers are directed to recent reviews on ribosome heterogeneity and specialization (Ferretti & Karbstein, 2019; Miller et al., 2023). In addition, we do not focus on translational control or ribosome function (which occurs after ribosome subunits have completed their full maturation), and instead refer readers to dedicated reviews on translational regulation (Brito et al., 2024).

Finally, although we briefly address the hematopoietic disorder Diamond-Blackfan Anemia, which is caused by a combination of p53 stabilization and reduced ribosome abundance arising from mutations in ribosomal proteins and other biogenesis factors, we do not discuss it in detail, and instead refer readers to more comprehensive reviews (Liu & Karlsson, 2024; Mills & Green, 2017; Narla & Ebert, 2010; Farley-Barnes et al., 2019).

2. rRNA transcription is highly dynamic in normal and leukemic hematopoiesis

Using a FISH-Flow assay that enables direct quantification of nascent rRNA (surrogate for transcription rate) and mature rRNA (ribosome subunit) in rare and heterogeneous populations (Antony et al., 2022, 2023), we recently established a detailed quantitative profile of their abundance across the normal mouse hematopoietic tree (Sams et al., 2025). Consistent with previous studies (Hayashi et al., 2014; Jarzebowski et al., 2018), we found that both rRNA transcription level and ribosome subunit abundance are highly dynamic and cell-type-specific, with 9-fold variation in rRNA transcription and 11-fold variation in ribosome subunit abundance across cell types (Fig. 2A). Within this variation, rRNA levels are intermediate in HSCs, increase to their highest levels in lineage-committed progenitors, and progressively decline during terminal differentiation. The magnitude and dynamics of these changes are lineage-specific, with the erythroid lineage exhibiting a broader dynamic range than maturing neutrophils or monocytes, while bone marrow-derived B cell and T cell populations generally maintain low to intermediate rRNA levels. Importantly, correlation between rRNA transcription and ribosome subunit abundance does not scale linearly, indicating that ribosome number is not determined solely by rRNA transcriptional output and likely involves additional levels of regulation (Fig. 2B). Extending these observations beyond normal hematopoiesis, we formally tested the longstanding assumption that ribosome biogenesis is upregulated in AML and found that both rRNA transcription and ribosome subunit abundance are significantly elevated in an AML mouse model driven by homozygous germline *Flt3* internal tandem

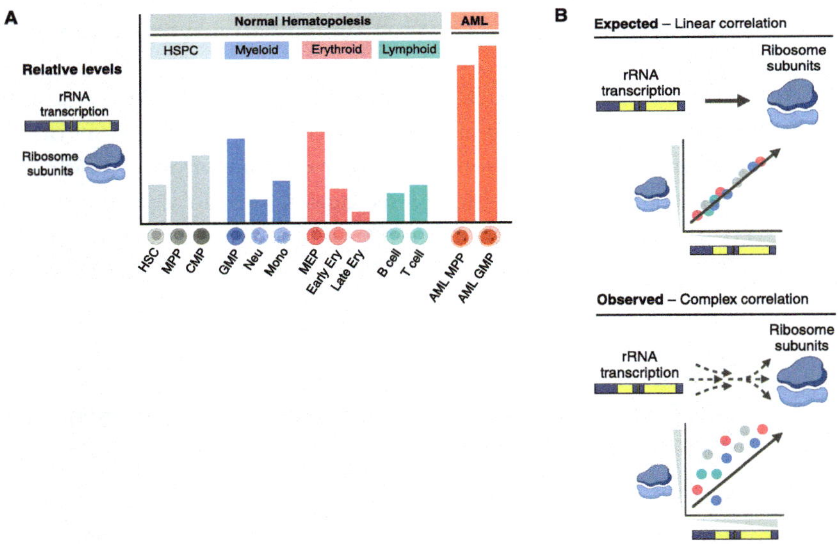

Fig. 2 rRNA transcription level and ribosome abundance in normal and leukemic hematopoiesis (A). Schematic of trends in rRNA levels (both nascent and mature) across different lineages of normal hematopoiesis and in AML. (B) Expected (top) and observed (bottom) models of the correlation between rRNA transcription level and ribosome subunit abundance across the normal hematopoietic system. HSC, hematopoietic stem cell; MPP, multipotent progenitor; CMP, common myeloid progenitor; GMP, granulocyte-monocyte progenitor; Neu, neutrophil; Mono, monocyte; MEP, megakaryocyte-erythroid progenitor; Early Ery, early erythroblast; Late Ery, late erythroblast.

duplication (*Flt3*-ITD) and hematopoietic *Dnmt3a* knockout (Fig. 2A). Whether increased rRNA levels in AML reflects AML-specific mechanisms, such as aberrant oncoprotein activity, versus heightened activity of canonical ribosome biogenesis pathways (e.g., increased expression of Pol I), remains unresolved.

This tight regulation of rRNA transcription and ribosome abundance across the hematopoietic hierarchy raises a fundamental question of why distinct cell types require different ribosome numbers. Hematopoietic cells vary widely in properties that are closely linked to ribosome function, including global protein synthesis rates, selective translation programs, cell size, and proliferation dynamics. Global translation levels are tightly regulated across hematopoiesis, and their disruption can impair cell function, particularly in HSCs, where altered translation compromises self-renewal and reconstitution capacity (Signer et al., 2014, 2016; Lv et al., 2021; Spevak et al., 2020; Cai et al., 2015; Khajuria et al., 2018). Plasma cells, which are specialized for high levels of antibody production, exhibit an extensive rough endoplasmic reticulum (Shaffer et al., 2004), indicative of a high ribosome content, although, to our knowledge, ribosome abundance in these cells has not been directly quantified or compared with other hematopoietic cells. Notably, however, global protein synthesis rates do not consistently correlate with ribosome abundance across hematopoietic cell types (Sams et al., 2025; Signer et al., 2014, 2016). Some populations, most prominently HSCs, maintain relatively high ribosome levels despite low translational output. Similarly, naive T cells harbor substantial pools of inactive free ribosome subunits that are rapidly engaged in translation only upon activation (Wolf et al., 2020).

Ribosome abundance may also influence gene-specific translational control, consistent with the Ribosome Concentration Hypothesis (Mills & Green, 2017), which proposes that ribosome levels directly shape gene-specific translation output based on intrinsic features of individual mRNA transcripts (e.g., translation initiation rate). Indeed, in support of this model, hematopoietic cells exhibit distinct translational landscapes (Wang et al., 2025), potentially influenced in part by their distinct ribosome numbers. Additionally, perturbations in ribosome biogenesis have been shown to promote differentiation in myeloid progenitor cell lines (Hayashi et al., 2014), and it is possible that altered ribosome availability contributes to this effect by reshaping cell-type-specific translational landscapes.

Beyond translation, ribosome levels are also linked to cell growth and proliferation. While the relationship between ribosome abundance and cell size across hematopoietic cell types remains poorly defined, increased ribosome numbers would be expected to support cell growth. Likewise, higher ribosome levels may support increased proliferative capacity; however, as with translation, ribosome abundance does not consistently scale with cell cycle dynamics across hematopoiesis (Sams et al., 2025). Neutrophils and polychromatic erythroblasts contain equal numbers of ribosomes despite neutrophils being markedly less proliferative; moreover, although polychromatic erythroblasts are equally as proliferative as their pro-erythroblast precursors, they contain nearly 4-fold fewer ribosomes (Sams et al., 2025). Together, these observations indicate that translation, cell size, and proliferative capacity are part of a highly complex functional demand that likely shapes cell-type-specific ribosome numbers.

The importance of precise regulation of ribosome biogenesis in hematopoietic development and maintenance is underscored by the hematopoietic disorders Diamond-Blackfan Anemia and Shwachman-Diamond Syndrome, which are characterized in part by bone marrow failure and an increased risk of hematologic malignancies, including AML. These disorders belong to a broader class of inherited diseases termed ribosomopathies, which are caused by mutations in ribosomal proteins or ribosome processing and assembly factors that result in reduced ribosome production or activity (Mills & Green, 2017; Narla & Ebert, 2010; Farley-Barnes et al., 2019). Collectively, these observations highlight ribosome abundance as a tightly regulated and functionally relevant feature of both normal and dysregulated hematopoiesis, motivating deeper investigation into the mechanisms that establish and maintain cell-type-specific ribosome numbers.

3. Steps of rRNA transcription

The hundreds of near-identical copies of rDNA in the genome are arranged in tandem arrays in nucleolar organization regions (NORs) across 5 chromosomes in humans and mice. In humans, NORs exist on the short arms (p-arms) of acrocentric chromosomes 13, 14, 15, 21, and 22 (Henderson et al., 1972). Each rDNA repeat consists of a ~13.4 kb transcribed region encoding the 47S pre-rRNA, followed by a ~30 kb intergenic spacer (IGS) (Fig. 3). Each repeat contains two promoters, an upstream spacer promoter and downstream 47S promoter spaced ~450 bp apart in humans and ~1750 bp apart in mice, separated by "enhancer repeats," which are periodic

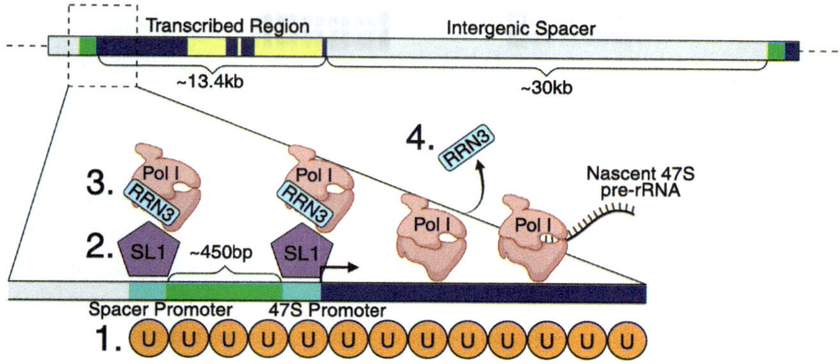

Fig. 3 Steps of rRNA transcription by Pol I. 1. UBTF binds across the promoter and entire transcribed region. 2. SL1, a five-subunit TATA-binding protein complex, binds to the spacer promoter and promoter. 3. Pol I bound by its initiation factor RRN3 binds to the rDNA promoter to form the pre-initiation complex. 4. RRN3 detaches and Pol I elongates and transcribes 47S nascent pre-rRNA.

sequences of ~120–140 bp each (George et al., 2023). Pol I initiates transcription at the downstream 47S promoter. Early studies in Xenopus laevis demonstrated that the spacer promoter and enhancer repeats are required for maximal rRNA transcription from the 47S promoter (De Winter & Moss, 1986; Caudy & Pikaard, 2002). It is reasonable to presume that they similarly promote rRNA transcription in mammals, though the precise mechanisms through which they act remain unclear.

rRNA is transcribed through a multi-step process that requires coordinated assembly of multiple core transcription factors at rDNA repeats (Fig. 3). At any given time, a subset of rDNA repeats are activated through binding of the High Mobility Group (HMG)-box protein UBTF, which binds to the promoters, the enhancer repeats, and throughout the entire 13.4 kb transcribed region, promoting complete removal of nucleosomes from the length of the transcribed region (Sanij et al., 2008). UBTF is a key architectural protein for activating rRNA transcription, and was first identified by its ability to bind rDNA promoters and enhancer repeats (Bell et al., 1988; Pikaard et al., 1989). Through its multiple HMG box domains, UBTF is thought to bend rDNA and form a nucleosome-like structure that is critical for maintaining the transcribed region of rDNA in a transcriptionally permissive state (Putnam et al., 1994; Stefanovsky et al., 2001).

Once bound to rDNA, UBTF then recruits SL1, a five subunit TATA-binding protein complex, to the rDNA promoters in a phosphorylation-dependent manner, where it interacts cooperatively with SL1 to stabilize promoter binding (Tuan et al., 1999; Bell et al., 1988). The 13-subunit Pol I, bound by its initiation factor RRN3, is then recruited to the spacer promoter and 47S promoter. Pol I requires RRN3 to bind rDNA, as RRN3 acts as a bridge between Pol I and SL1, thereby facilitating promoter recognition (Miller et al., 2001; Yuan et al., 2002). Together, UBTF, SL1, and Pol I-RRN3, assemble at both rDNA promoters to form the pre-initiation complex (PIC). Once the Pol I-RRN3 complex is loaded and transcription of 47S pre-rRNA is initiated from the 47S promoter, RRN3 detaches and Pol I elongates and transcribes the 47S pre-rRNA. It currently remains debated whether Pol I undergoes promoter-proximal pausing analogous to Pol II (Core & Adelman, 2019). Termination of rRNA transcription is mediated by Transcription Termination Factor 1 (TTF-1), which binds to terminator elements located at the 3′ end of the transcribed region and promotes proper release of Pol I (Grummt et al., 1986; Evers et al., 1995).

4. Modes of rRNA transcriptional regulation

In the following sections, we present **Conceptual Framework #1**, outlining the key determinants of cellular rRNA transcriptional capacity which thereby influence cell-type-specific transcriptional output (Fig. 4). We organize this framework into four major regulatory layers: (1) Opening and closing of rDNA repeats, (2) Regulation of Pol I activity at open rDNA loci, (3) Regulation of the expression of core rDNA transcription machinery, and (4) Regulation of cellular nucleoside triphosphate (NTP) availability. Within each section, we discuss the molecular mechanisms governing these processes and the regulatory layers that collectively modulate rRNA transcriptional output.

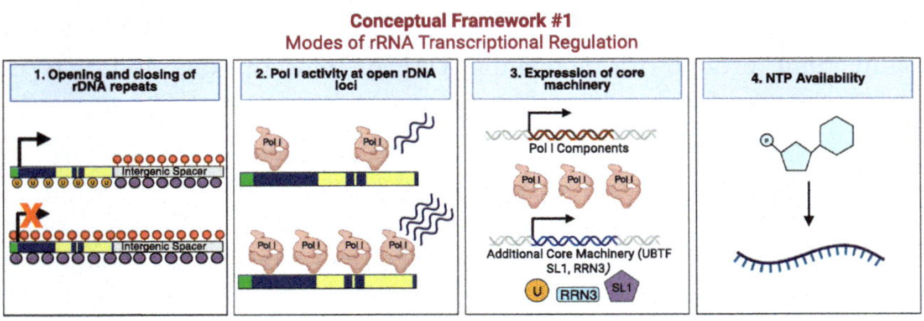

Fig. 4 Conceptual framework #1. Visual outline of conceptual framework #1: Modes of rRNA transcriptional regulation.

4.1 A brief note on MYC and p53

4.1.1 MYC

It is well established that the critical developmental and oncogenic transcription factor MYC functions as a master transcriptional regulator coordinating the activity of all three RNA polymerases (Campbell & White, 2014). MYC is known to influence each of the regulatory layers outlined in this conceptual framework. To date, the extent to which variations in rRNA transcription are fully attributable to MYC, and which specific aspect of its activity represents its most critical point of control, remains unclear. Rather than dedicating a separate section to MYC, we integrate its established roles into each designated section of the framework, discussing its contributions within the context of the specific regulatory processes described below.

4.1.2 p53

p53 is a central tumor suppressor and a key downstream effector of diverse cellular stress responses, including DNA damage, oncogene activation, metabolic stress, and nucleolar stress. Under homeostatic conditions, the E3 ubiquitin ligase MDM2 binds to p53 and mediates its ubiquitination and proteasomal degradation, maintaining low basal p53 levels. Perturbation of Pol I transcription or ribosome biogenesis induces nucleolar stress and leads to stabilization of p53 (Xiao et al., 2026; Bywater et al., 2012). A well-characterized mechanism underlying this response involves the ribosomal proteins RPL5 and RPL11, which, together with 5S rRNA, form a complex that binds to and inhibits MDM2. This interaction prevents MDM2-mediated ubiquitination and degradation of p53, resulting in rapid p53 stabilization (Bursać et al., 2012; Donati et al., 2013; Sloan et al., 2013). Recent studies also report that free RPL22 protein binds *MDM4* mRNA and induces exon skipping, leading to reduced full length MDM4 protein and therefore p53 stabilization (Weinstein et al., 2024; Jansen et al., 2024). Accumulation of p53 activates transcriptional programs governing cell cycle arrest, apoptosis, and genome stability (Wang & Attardi, 2022). Given that p53 activation is a consequence of impaired ribosome biogenesis, and given that there is only limited evidence indicating that it directly represses rRNA transcription (Zhai & Comai, 2000), we do not discuss p53 in detail in this review. For comprehensive information of nucleolar stress signaling and p53 regulation, readers are referred to prior reviews (Woods et al., 2015; Lindström et al., 2022).

4.2 Open versus closed ribosomal DNA copies

4.2.1 The discovery of distinct rDNA chromatin states

Electron microscopy imaging of chromatin spreads from non-mammalian vertebrates, fruit flies, and yeast first demonstrated the presence of two distinct populations of rDNA repeats: actively transcribed repeats loaded with Pol I and depleted of nucleosomes, and inactive repeats bound by nucleosomes (Miller & Beatty, 1969; McKnight & Miller, 1976; Rattner et al., 1982). Later studies confirmed this finding across diverse eukaryotic species using psoralen crosslinking, a technique in which an intercalating agent preferentially crosslinks nucleosome-free DNA upon UV irradiation, allowing for nucleosome-bound (inactive) and nucleosome-depleted (active) rDNA copies to be distinguished by electrophoretic mobility (Sogo et al., 1984; Dammann et al., 1993; Conconi et al., 1989). Application of psoralen crosslinking to the murine erythroleukemia (MEL) cell line demonstrated that roughly half of its rDNA copies are active (Conconi et al., 1989).

These studies established a foundational model in which rDNA repeats exist either in a fully open and transcriptionally active state, with the entire ~13.4-kb-long stretch from the promoter to the transcription termination site depleted of nucleosomes and loaded with Pol I, or in a closed, nucleosome-bound and transcriptionally inactive state (Fig. 5). Of note, the IGS remains consistently nucleosome-bound, regardless of the transcriptional status of the adjacent transcribed region (Brock & Bird, 1997).

4.2.2 Open copies of rDNA

Open rDNA copies constitute the transcriptionally competent fraction and are defined by complete loss of nucleosomes and extensive UBTF occupancy across the promoter and entire transcribed region (Conconi et al., 1989; Sanij et al., 2008) (Fig. 5). Mechanistically, UBTF displaces histones throughout the transcribed region and functions to both maintain rDNA in an open transcriptionally permissive state and facilitate the recruitment of the basal Pol I transcriptional machinery (Sanij et al., 2008; Hamdane et al., 2014).

Given that open rDNA copies can be identified by MNase accessibility of the transcribed region, we recently quantified rDNA accessibility across hematopoietic cell types by mapping human and mouse ATAC-seq datasets to rDNA (Sams et al., 2025). We found that the fraction of accessible rDNA copies varies across cell types, and notably, closely parallels cell-type-specific patterns of rRNA transcription. Specifically, rDNA accessibility peaks as HSCs differentiate into myeloid and erythroid progenitors and then

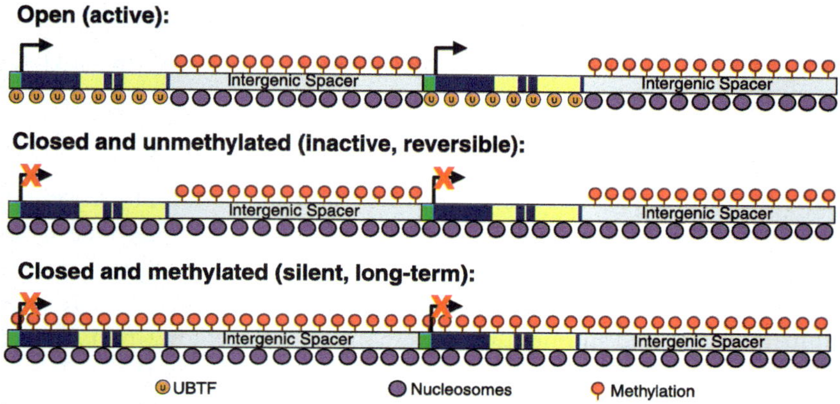

Fig. 5 Chromatin states of rDNA. Active rDNA repeats are devoid of nucleosomes, unmethylated across the transcribed region, and are bound by UBTF. Closed and unmethylated (inactive and reversible) repeats are nucleosome-bound but also unmethylated across the transcribed region. Closed and methylated (silent and long-term) repeats are both nucleosome-bound and methylated throughout the transcribed region. The IGS is always nucleosome-bound and methylated in all copies, irrespective of the status of the flanking transcribed regions.

progressively decreases during terminal myeloid and erythroid maturation. We also identified increased rDNA accessibility in an *Npm1c*-mut, *Flt3*-ITD AML mouse model, as well as in immature versus mature cells within human AML patient samples. Together, these findings support a model in which dynamic control of rDNA repeat opening and closing contributes to differences in rRNA transcription across normal and leukemic hematopoiesis.

Notably, cells with a high fraction of accessible rDNA repeats from both the K562 cell line (CML in blast crisis) and primary human hematopoietic stem and progenitor cells exhibit distinct global chromatin landscapes enriched for specific transcription factor motifs, including factors with myeloid-specific dependencies (e.g., SP1) as well as universally essential regulators (e.g., NRF1) (Sams et al., 2025). Whether the transcriptional programs associated with these factors drive increased rDNA accessibility or instead arise as a consequence of increased accessibility remains to be determined. Additionally, in a variety of hematopoietic and other cell types, nucleosomes immediately upstream of the rDNA Spacer Promoter display canonical active marks including H3K9ac, H3K4me2, H3K4me3, and H3K27ac (Antony et al., 2022; Herdman et al., 2017; Moss et al., 2019). While these marks are striking, their precise roles in regulating Pol I activity remain to be elucidated. Importantly, although rDNA opening provides one layer of rRNA transcriptional regulation, rRNA synthesis can also be modulated at already open rDNA copies through control of Pol I abundance or activity, as we detail later in this review.

4.2.3 Closed and unmethylated copies of rDNA

Closed rDNA copies are characterized by absence of UBTF occupancy, exclusion of additional Pol I pre-initiation complex components, and complete nucleosome-wrapping. There are two distinct epigenetic states of closed rDNA, one that corresponds to a reversible (poised) inactive state (unmethylated), and another that corresponds to a likely long-term silenced state (methylated) (Fig. 5). Notably, the IGS is fully methylated in all copies, irrespective of the methylation status of the flanking transcribed regions.

Depletion of UBTF alone is sufficient to close rDNA repeats and prevent Pol I pre-initiation complex formation, independent of CpG methylation status (Sanij et al., 2008; Hamdane et al., 2014; Herdman et al., 2017). This closed yet unmethylated configuration of rDNA corresponds to a transcriptionally inactive but reversible state, and restoration of UBTF re-opens repeats and rescues active rDNA copy number (Poortinga et al., 2004; Sanij et al., 2008; Theophanous et al., 2023). Consistent with the cell-type-specific patterns of rDNA accessibility observed across hematopoiesis (Sams et al. 2025), these findings suggest that UBTF-dependent inactivation provides a dynamic mechanism to modulate rRNA transcription, potentially contributing to different rDNA transcription levels throughout hematopoiesis (Sanij et al., 2008; Sams et al., 2025).

4.2.4 Closed and methylated copies of rDNA

Methylated, transcriptionally silent rDNA is characterized by extensive CpG methylation across the rDNA promoter and transcribed region, in addition to complete nucleosome-wrapping (Santoro et al., 2002; Hori et al., 2021; Potapova et al., 2025) (Fig. 5).

Pharmacologic inhibition of DNA methylation provides functional evidence of its rDNA silencing role. In NIH-3T3 fibroblasts, treatment with the hypomethylating agent 5-aza-2′-deoxycytidine reduces rDNA promoter methylation and increases rRNA transcription (Santoro & Grummt, 2001). Similar effects are observed across multiple human myelomonocytic leukemia cell lines (Raval et al., 2012). More recent studies have shown that prolonged treatment of human cells with the DNMT1 inhibitor GSK-3484862 results in global hypomethylation of rDNA, specifically restoring UBTF binding and transcriptional activity to previously silent rDNA repeats that could be distinguished using rRNA sequence variants (Potapova et al., 2025). Collectively, these findings support a model in which DNA methylation stabilizes the silent state of rDNA repeats, and its removal is required for UBTF binding and subsequent Pol I PIC formation. Although the precise mechanisms governing rDNA silencing through methylation remain under investigation, the nucleolar remodeling complex (NoRC) has been implicated as a potential mediator through its ability to recruit DNA methyltransferases to rDNA promoters (Santoro et al., 2002). It is worth considering that DNA methyltransferase inhibitors, which are standard of care therapy in myeloid malignancies, may theoretically change the balance of methylated and unmethylated rDNA copies.

To date, constitutive rDNA silencing seems to represent a highly stable epigenetic state that can be inherited independent of cellular identity (Potapova et al., 2025). In line with this, our recent work found that the percentage of methylated rDNA copies seems to be fairly constant (15–25 %) across normal mouse and human hematopoietic cell types, as well as in mouse AML, despite highly variable levels of rRNA transcription (Sams et al., 2025). This indicates that dynamic changes in rRNA transcription within an organ system are likely not driven by differences in the numbers of unmethylated rDNA copies.

4.2.5 Regulation of UBTF expression in hematopoiesis

Mechanisms that regulate cellular UBTF abundance are thought to ultimately determine the balance between open and closed copies of rDNA. Studies in the HL-60 promyelocytic cell line show that terminal differentiation is accompanied by ~90 % reduction in expression of the *UBTF* gene and its protein levels, as well as its occupancy at rDNA, leading to a reduced fraction of active rDNA repeats and reduced rRNA transcription (Sanij et al., 2008).

Transcription of *UBTF* is likely controlled by multiple transcription factors, but a primary known regulator is the Myc-Max-Mad network, with MYC activating and MAD1 repressing *UBTF* transcription. In agreement with our finding of higher rRNA transcription levels in undifferentiated versus differentiated myeloid cells (Sams et al., 2025), MYC binding is enriched at the *UBTF* promoter in undifferentiated granulocytes, while MAD1 binding is enriched at the *UBTF* promoter in differentiated granulocytes (Poortinga et al., 2004). MYC overexpression in differentiated granulocytes increases both rRNA transcription and *UBTF* mRNA, demonstrating that MYC promotes rRNA transcription in part by increasing cellular UBTF levels (Poortinga et al., 2004).

Additionally, UBTF function is regulated by post-translational modifications (PTMs), particularly phosphorylation. Phosphorylation of UBTF at T117 and T201 impairs its ability to bind and bend rDNA, thus decreasing its occupancy and activation at rDNA (Sanij et al., 2008). While this directly affects UBTF occupancy, it is important to note that multiple components of the rDNA core transcriptional machinery have important regulatory PTMs that modulate their activity, which will be discussed further in the following section of Conceptual Framework #1.

4.2.6 UBTF-tandem duplications (UBTF-TD)

Tandem duplications in the *UBTF* gene (UBTF-TD) were recently discovered as drivers of pediatric and adult AML (Umeda et al., 2022; Duployez et al., 2023). These somatic mutations are always heterozygous (such that one normal copy of *UBTF* is always retained) and involve either in-frame insertions on the 3′ end of exon 13 of UBTF (Internal tandem duplication, ITD) or in-frame duplication of exon 13 (partial tandem duplication, PTD) (Umeda et al., 2022). Studies thus far seem to indicate that UBTF-TD neither compromises nor enhances its action at rDNA but instead confers on it the novel ability to aberrantly occupy loci of the *HOXA/B* genes and *MEIS1* and drive their hyperexpression (Umeda et al., 2022; Barajas et al., 2024). This oncogenic ability is dependent on the Menin complex and is impaired by treatment with small molecule inhibitors of the Menin-KMT2A interaction. This mechanism unifies the pathogenic mechanism of UBTF-TD with NPM1c mutation, KMT2A-rearrangement, and NUP98-rearrangement, all of which converge on Menin-driven expression of a myeloid oncogenic gene network (Faber et al., 2009; Heikamp et al., 2022; Uckelmann et al., 2023). Notably, though eventual upregulation of rRNA transcription may be part of the downstream program of many oncogenic drivers, the direct action of UBTF-TD appears to be exerted through an aberrant Pol II transcriptional program, not Pol I.

4.2.7 rDNA copy number

The total number of rDNA copies varies widely between species, and between individuals of the same species. Estimated rDNA copy number calculations for white British participants from ~500,000 people in the UK Biobank whole-genome sequencing dataset showed that most individuals harbor between 200 and 350 rDNA repeats in their diploid genome, with only a few outliers having under 150 or over 450 copies (Wang & Lemos, 2017; Rodriguez-Algarra et al., 2024). There likely also exist ancestry-related variations in rDNA copy numbers, though larger sequencing studies will be required to resolve these differences (Parks et al., 2018). More recently, comparative genomic analyses across humans and great apes revealed that each individual (whether human or ape) has a unique rDNA landscape, characterized by distinct copy numbers, sizes of arrays on each of the chromosomes that harbor rDNA, and active versus inactive rDNA variants (Potapova et al., 2025; Solar et al., 2025; Yoo et al., 2025). Of relevance to the hematopoietic system, germline rDNA copy number in the UK Biobank was found to be positively associated with peripheral blood neutrophil counts and negatively associated with lymphocyte counts (Rodriguez-Algarra et al., 2024). Given the complexity of mechanisms responsible for fine-tuning blood cell counts, this association could have a diversity of biological explanations, however, the obvious question such work raises is whether higher germline copy numbers can lead to higher absolute numbers of active rDNA copies and potentially higher rRNA transcription in some or all cell types.

4.3 Regulation of Pol I activity at open and active rDNA

At open, transcriptionally competent rDNA repeats, the rate of rRNA transcription can be fine-tuned through the regulation of Pol I activity. As stated in **Steps of rRNA transcription** (Fig. 3), productive transcription requires efficient assembly of the PIC, which requires coordinated interaction between UBTF, SL1, RRN3, and Pol I. Thus, any process that impairs or promotes the formation of the PIC or productive elongation ultimately determines Pol I occupancy and productive transcription of rRNA.

Dynamic regulation of rRNA transcription and CDK-dependent phosphorylation across the cell cycle provides a physiological example of this regulatory logic. rRNA transcription is low in early G1, increases throughout G1, and remains elevated in S and G2 phases (Fig. 6A). Continued rRNA transcription throughout most of the cell cycle allows cells to double their ribosome content in preparation for division (Fig. 6B). During mitosis, rRNA transcription is shut off by eviction of Pol I from rDNA despite UBTF remaining bound to rDNA as a mitotic bookmark, which preserves rDNA accessibility and enables transcriptionally competent rDNA copies to be maintained across cell divisions (Conconi et al., 1989; Grob et al., 2014). This silencing is driven by phosphorylation-dependent inactivation of SL1,

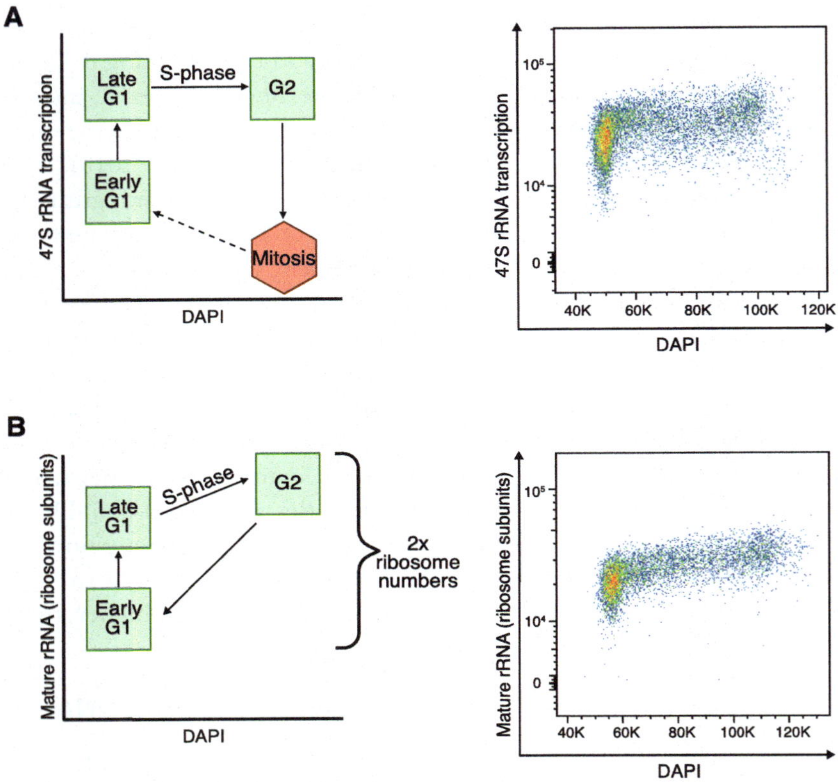

Fig. 6 rRNA transcription and mature rRNA levels throughout the cell cycle. (A). rRNA transcription increases throughout G1, remains high during S-phase and G2, and shuts off during mitosis. Note: Complete mitotic shutoff of nascent rRNA transcription is inferred from reports of Pol I eviction from mitotic chromosomes, but isn't consistently observed as a distinct "M" cell population on FISH-Flow. Further investigation with FISH-Flow on mitosis-enriched cells is warranted to validate this model. (B). Mature rRNA increases throughout G1, S-phase, and G2, in order to double ribosome numbers prior to mitosis. After mitosis, ribosome numbers are distributed to daughter cells.

preventing PIC formation. As an example, Cdc2/Cyclin B-mediated phosphorylation of SL1 subunits during mitosis disrupts SL1-UBTF interaction (Heix et al., 1998; Kihm et al., 1998). As cells re-enter G1, activation of rRNA transcription requires phosphorylation of UBTF at Ser484 by G1-specific CDK/Cyclin complexes (Voit et al., 1999; Voit & Grummt, 2001). During S-phase, Ser484 is no longer phosphorylated, and UBTF is preferentially phosphorylated at Ser388 by CDK2/Cyclin E or CDK2/Cyclin A, which persists throughout G2 and is required for UBTF interaction with Pol I, promoting transcriptional activation (Voit & Grummt, 2001).

In hematopoietic populations, there is a diversity of cell cycle kinetics and doubling times across hematopoietic cell types, and we have found that cell cycle distribution does not cleanly correlate with cell type specific rates of rRNA transcription (Sams et al., 2025). Instead, the dynamic regulation of Pol I activity through PTMs likely extends beyond cell cycle effects alone. These effects may be mediated by signaling pathways and alternative kinases that target PIC components, allowing fine-tuning of rRNA transcription in a cell-type-specific manner.

Each member of the core machinery – UBTF, SL1, RRN3 and Pol I – has multiple phosphorylation sites and is modulated by PTMs (Lawrence et al., 2026). RRN3 serves a central regulatory role because its phosphorylation state can activate or inhibit the formation of Pol I-RRN3 complex, which is critical for Pol I to bind to rDNA promoters (Tuan et al., 1999). Growth factor, nutrient, and stress-response pathways converge on RRN3 to modulate Pol I occupancy and processivity to fine-tune rRNA transcription in accordance with cell demands (Sharifi & Bierhoff, 2018; Lawrence et al., 2026). As a conceptual example, we have summarized the 8 known phosphorylation sites on RRN3 and the signaling pathways that regulate them (Table 1, Fig. 7). With the exception of Ser199 and Thr200, which lie at a critical binding interface between RRN3 and Pol I, most of the mapped phosphorylation sites lie within unstructured regions (Misiaszek et al., 2021). Though we classify these modifications as activating, inactivating, or dynamically regulated, the molecular mechanisms by which most of these sites exert their effects remain unknown, as do their precise timings of phosphorylation and dephosphorylation. It is indeed possible that they are all dynamically regulated during the Pol I transcription cycle.

4.3.1 Potential shared regulation of Pol I and Pol II

Pol I occupancy at rDNA reflects a dynamic cycle of promoter binding, initiation, release, and elongation. As mentioned previously, it remains debated whether Pol I undergoes promoter proximal pausing similar to Pol II. In support of shared regulators between Pol I and Pol II, transcriptional CDKs have recently been implicated in Pol I control. Notably, CDK9, the catalytic subunit of the super elongation complex that phosphorylates the Pol II C-terminal domain, was recently shown to promote rRNA transcription. Treatment with flavopiridol, a CDK9 inhibitor, results in

Table 1 RRN3 Phosphorylation Sites.

RRN3 Phosphorylation site	Placed by	Removed by	Activating, Inactivating, or Dynamic	Overview	References
Ser44	Cdk2/cyclin E	PP2A	Activating	Phosphorylated by Cdk2/cyclin E. mTOR signaling downregulates PP2A to prevent phosphatase activity	Mayer et al. (2004)
Ser170	Casein Kinase 2 (CK2)	FCP1	Dynamic	Dynamically placed and removed during the Pol I transcriptional cycle	Bierhoff et al. (2008)
Ser172	Casein Kinase 2 (CK2)	FCP1	Dynamic	Dynamically placed and removed during the Pol I transcriptional cycle	Bierhoff et al. (2008)
Ser199	AMPK	mTOR signaling	Inactivating	Nutrient-deprivation induced phosphorylation by AMPK signaling. Lies with Thr200 at the interface with POLR1F (subunit of Pol I), and likely weakens interaction between RRN3 and Pol I	Mayer et al. (2004), Lv et al. (2025)
Thr200	JNK2	Unspecified	Inactivating	Stress induced phosphorylation by JNK2. Lies with Ser199 at the interface with POLR1F (subunit of Pol I), and likely weakens interaction between RRN3 and Pol I	Mayer et al. (2005)
Ser633	ERK	Unspecified	Activating	Mitogen mediated activation leads to phosphorylation of Ser649 by RSK and subsequent phosphorylation of Ser633 by ERK	Zhao et al. (2003)
Ser635	AMPK	Unspecified	Inactivating	Nutrient-deprivation induced phosphorylation by AMPK signaling. Impairs association of the Pol I-RRN3 complex with SL1	Hoppe et al. (2009)
Ser649	RSK	Unspecified	Activating	See Ser633 above	Zhao et al. (2003)

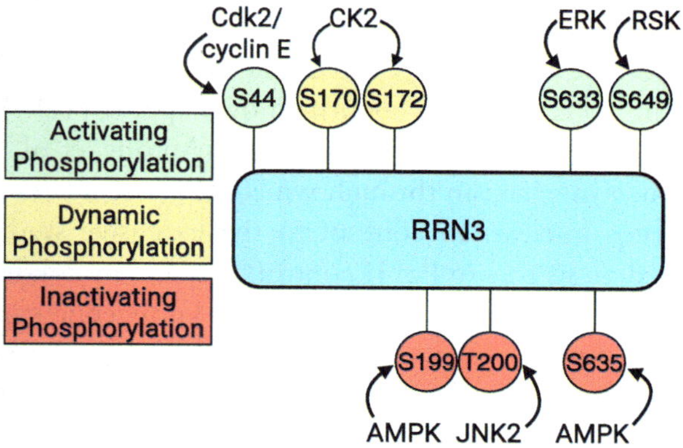

Fig. 7 RRN3 phosphorylation. Known activating, inactivating, and dynamic phosphorylation sites on RRN3, and the pathway and/or kinase known to deposit the mark.

complete loss of nucleolar structural integrity within an hour, strongly pointing to a direct role in promoting the activity either of Pol I or of some other component of the core rDNA machinery (Potapova et al., 2023).

4.3.2 Regulation of Pol I activity by lineage-specific TFs

One possible mechanism for cell-type-specific control of rRNA transcription is through the direct actions of lineage-specific transcription factors (TFs) at rDNA loci. As mentioned below, a handful of groups studying specific TFs have stumbled upon the binding of their TFs-of-interest to rDNA. However, because rDNA sequences were historically not included in genome assemblies used for high-throughput mapping, the potential binding of most TFs and chromatin factors at rDNA remained unexplored. Until recently, only a few publications had taken an unbiased look at rDNA-TF occupancy (Zentner et al., 2011, 2014). We recently published custom genomes optimized for human and mouse rDNA mapping (George et al., 2023), and re-mapped ~2200 publicly available ChIP-Seq datasets for 250 TFs and chromatin proteins (Antony et al., 2022). We revealed that numerous hematopoietic and oncogenic transcription factors bind to rDNA at previously unrecognized canonical motif sequences, raising the possibility that they act combinatorially to fine-tune cell-type-specific rates of rRNA transcription. We address salient TFs here, but refer readers to a past review in which we dissected TF-rDNA regulation in detail, and included a discussion of important considerations when distinguishing direct versus indirect roles in Pol I regulation (Paralkar, 2024).

MYC has been shown to bind its canonical E-box motifs in human rDNA and promote occupancy of SL1 and Pol I at the rDNA promoter (Grandori et al., 2005). We too confirmed overlapping peaks of MYC and MAX at rDNA in human and mouse cells (Antony et al., 2022). The precise molecular mechanism through which MYC regulates Pol I activity however remains unclear. In contrast to the extensive studies that have been performed on its role in Pol II control, very little work has been done on its direct effects on Pol I, creating a major knowledge gap.

The RUNX family of transcription factors has multiple binding motifs throughout human, mouse, and rat rDNA loci (Young et al., 2007). In particular, RUNX1, which plays critical roles in normal hematopoiesis and which is mutated in leukemia, binds to rDNA and may boost HSC ribosome numbers through a combination of direct and indirect mechanisms involving upregulation of a broad ribosome biogenesis gene program (Cai et al., 2015). Notably, RUNX2 is reported to repress rRNA transcription in non-hematopoietic cells (Young et al., 2007). Given these opposing reports for these two family members, further work will be required to disentangle their precise contributions in hematopoietic and leukemic contexts where both are simultaneously expressed. Notably, RUNX proteins associate with rDNA during mitosis, and their striking co-localization with UBTF in mitotic spreads can be visualized by immunofluorescence, warranting investigation into their roles as potential mitotic bookmarkers (Young et al., 2007). We observed that RUNX1, RUNX2, and RUNX3 all show broad patterns of occupancy across the entire transcribed region of rDNA (Antony et al., 2022), indicating that their recruitment to rDNA may not be through specific DNA motifs, but more likely through interactions with UBTF itself.

CEBPA likely plays a direct role in promoting Pol I activity. We identified the CEBP family of TFs as the largest set of TF ChIP-Seq tracks with rDNA peaks, and found that multiple family members showed a consistent sharp peak on rDNA at a conserved, canonical motif sequence within the transcribed region of human and mouse rDNA (Antony et al., 2022). Since CEBPA is the master TF of the myeloid lineage, and also a specific dependency in AML, we used a degron system to determine the immediate effects of its loss on rRNA transcription. After extensively validating the rDNA occupancy of CEBPA, and demonstrating that it binds actively transcribed rDNA repeats, we showed that its degradation results in a rapid decrease (within hours) in Pol I and RRN3 occupancy on rDNA, followed by a reduction in rRNA transcription. Notably, this effect occurs without a change in the cellular levels of any of the core rDNA machinery proteins (Antony et al., 2022).

CTCF and all cohesin complex subunits show a prominent peak immediately upstream of the rDNA spacer promoter in both human and mouse cells (Antony et al., 2022; Herdman et al., 2017). This conserved occupancy, and the location of this peak at the exact site where the nucleosome-depleted stretch of rDNA commences, is strongly indicative of an insulation or boundary-like function. However, this potential role has not been explored in detail, and, to date, studies of CTCF at rDNA have relied on the effects of its knockdown (van de Nobelen et al., 2010; Huang et al., 2013), which are limited by the inability to uncouple direct roles at rDNA from indirect effects arising from its extensive genomewide roles.

4.4 Expression of Pol I transcriptional machinery

Another critical determinant of rRNA transcription is the abundance of the core Pol I transcriptional machinery, as reduced levels of any of its components can impair pre-initiation complex (PIC) formation. As discussed above for UBTF, expression of individual PIC components can influence rRNA transcriptional output. Because each factor is encoded by a Pol II-transcribed gene, Pol I activity can be constrained by their expression levels, which vary across cell types, during differentiation, and in disease states. Indeed, we recently identified that expression of the core Pol I transcriptional machinery is downregulated during differentiation of myeloid progenitors to neutrophils, consistent with reduced rRNA transcription in maturing neutrophils (Sams et al., 2025). These expression changes may be coordinated in part by MYC, which has been shown to directly regulate expression of a broad network of Pol I components (*POLR1B* and *POLR1E*) and associated factors (*UBTF, RRN3,* and the SL1 subunit *TAF1C*) in myeloid cells. Reduced level of MYC during granulocytic differentiation leads to reduced expression of these components and concordant decrease in rRNA transcription (Poortinga et al., 2011). More recently, acute MYC degradation coupled with nascent RNA sequencing (SLAM-seq) demonstrated that MYC loss rapidly decreases transcription of numerous genes encoding Pol I machinery components (Muhar et al., 2018).

More broadly, overexpression of core Pol I transcriptional machinery appears to be a recurrent feature across multiple cancer types (Hwang & Denicourt, 2024). Datasets from The Cancer Genome Atlas show that genes encoding *RRN3,* multiple Pol I subunits, and the SL1 subunit *TAF1A* are consistently overexpressed in a wide array of tumors (Rossetti et al., 2016). Increased *UBTF* expression has also been documented in several cancers and is associated with enhanced rRNA transcription and proliferative capacity (Huang et al., 2002; Rajput et al., 2015). Collectively, these data suggest that upregulation of core Pol I transcription machinery represents a central mechanism by which rRNA transcription is elevated in cancer.

These findings further support the idea that differences in rRNA transcription throughout hematopoiesis and its upregulation in AML can, at least in part, be attributed to differential expression of the Pol I transcriptional machinery. Careful quantification will be required to determine the relative contributions of direct regulation of Pol I activity versus indirect regulation via Pol II-dependent expression of its components.

4.5 Cellular NTP availability

A final important modulator of rRNA transcriptional output is the cellular availability of NTPs, the building blocks of rRNA. rRNA comprises approximately 80% of total cellular RNA and is transcribed at exceptionally high levels. Although the precise fraction of total transcription devoted to rRNA is context-dependent and has not been formally quantified across mammalian cell types, rRNA synthesis accounts for at least ~35% (and likely more) of total cellular transcription (Moss et al., 2007). Additionally, the nascent rRNA transcript is ~13.4 kb in length, imposing a substantial demand for NTPs to sustain continuous synthesis.

While the cellular level of NTPs can activate nutrient- and energy-sensing pathways such as mTOR and AMPK, both of which contribute to the regulation of rRNA transcription, there is also evidence for more direct need for NTP availability (Hoxhaj et al., 2017; Hoppe et al., 2009). Pharmacologic and genetic perturbations that limit both purine and pyrimidine pools impair rRNA synthesis across multiple cell types and tumor models. This effect likely reflects a direct limitation of the nucleotides required to sustain nascent rRNA transcription.

4.5.1 Purine availability affects rRNA transcription

Purine nucleotides (ATP and GTP) make up two of the four nucleotide building blocks of RNA. Purines are produced through two main pathways: either *de novo* synthesis (in which new purines are created) or salvage (in which free bases are reused and converted back into nucleotides), both of which occur in the cytosol (Villa et al., 2019).

It was shown in the 1970s that subjecting mammalian cells to amino acid starvation results in immediate decrease of cellular ATP and GTP levels and inhibition of rRNA synthesis, which can be reversed by replenishing the cellular purine pool (Grummt & Grummt, 1976). More recently, inhibitors of IMPDH1 and IMPDH2, key rate limiting enzymes in the *de novo* guanine nucleotide biosynthesis pathway (Villa et al., 2019), have been shown to rapidly decrease rRNA transcription across multiple systems including glioblastoma cells, and K562 and CCRF-CEM (T-ALL) leukemia cell lines (Kofuji et al., 2019; Huang et al., 2008). Notably, in glioblastoma, the effects of IMPDH inhibition occur without changes in PI3K, mTOR, or ERK/RSK signaling (Kofuji et al., 2019).

In AML, venetoclax (Ven), a BCL-2 inhibitor used in clinical practice, suppresses oxidative phosphorylation (OXPHOS) independent of its primary BCL-2 inhibitory activity (Roca-Portoles et al., 2020). Ven-resistant AML cells adapt to OXPHOS inhibition by upregulating *de novo* guanine nucleotide biosynthesis, which enhances ribosome biogenesis through increased rRNA transcription. This increase in turn suppresses activation of the impaired ribosome biogenesis checkpoint and prevents p53 stabilization (Kawano et al., 2026).

4.5.2 Pyrimidine availability affects rRNA transcription

Pyrimidine nucleotides (UTP and CTP) constitute the other two nucleotide building blocks of RNA. Pyrimidine synthesis occurs across the cytosol and mitochondrial membrane, and, like purine synthesis, involves *de novo* synthesis or the salvage pathway (Villa et al., 2019). DHODH and UMPS are two critical enzymes involved in the *de novo* pyrimidine synthesis pathway (Villa et al., 2019). Across multiple cancers, including glioblastoma, colorectal carcinoma, and breast cancer, DHODH inhibition reduces rRNA transcription (Lafita-Navarro et al., 2020; Hubackova et al., 2020).

In both human and mouse AML models, DHODH inhibition promotes myeloid differentiation, resulting in reduced leukemic burden and improved survival. These effects are reversed by uridine supplementation, indicating that loss of UMP synthesis is the primary driver of differentiation (Sykes et al., 2016). How reduced pyrimidine availability translates into altered cell fate in AML remains to be fully elucidated, yet given the dependence of rRNA transcription on pyrimidine availability, suppression of ribosome biogenesis is a plausible downstream mediator. Given that progenitor cells exhibit high rRNA transcription levels that decline during hematopoietic differentiation, it is plausible that suppression of rRNA synthesis actively facilitates differentiation rather than merely accompanying it, thereby contributing to this phenotype in leukemic cells.

Collectively, these findings underscore the critical importance of NTP availability in sustaining rRNA transcription, and reduced nucleotide pools impair rRNA synthesis both directly (limiting the substrates required for nascent transcript elongation), and indirectly (nucleotide-sensing mechanisms that attenuate Pol I activity).

4.5.3 Transcriptional regulation of NTP synthesis in mammalian cells

MYC and ATF3 are central transcriptional regulators of NTP availability in mammalian cells. Genome-wide mapping of MYC binding sites in human B cell lymphoma, together with gene expression studies in B cell lymphoma, MYC-ER rat fibroblasts, and murine hepatocellular carcinoma, demonstrate that MYC directly regulates the majority of *de novo* nucleotide biosynthetic genes, including key enzymes such as *PPAT*, *DHODH*, *IMPDH1*, and *IMPDH2* (Liu et al., 2008). More recently, MYC has been shown to directly control transcription of all six enzymes in the *de novo* purine synthesis pathway, highlighting its role in maintaining nucleotide biosynthesis (Muhar et al., 2018). In AML, ATF3 directly activates genes involved in serine and one-carbon metabolism required for purine biosynthesis, while also upregulating key enzymes in the *de novo* purine and pyrimidine synthesis pathways (Di et al., 2021). By controlling the capacity of cells to generate NTPs, these transcriptional regulators determine the nucleotide supply available for rRNA transcription.

4.6 Summary

rRNA transcription can be regulated through multiple different modalities. At the chromatin level, rDNA repeats can either be active, reversibly inactive and unmethylated, or silent and methylated. At active rDNA repeats, rRNA synthesis can be fine-tuned by regulating PIC assembly, with multiple post-translational modifications of UBTF, SL1, RRN3, and Pol I modulating transcriptional initiation and elongation. The abundance of members of the core rDNA machinery themselves can be controlled by expression of the genes that encode them. Finally, cellular NTP availability can directly limit rRNA transcription, further linking ribosome biogenesis to metabolic state. It is likely that contributions of each of these factors underlie the variation in rRNA transcription level across the hematopoietic system.

5. Production of mature ribosomes from nascent rRNA

Beyond regulation of rRNA transcription itself, multiple downstream processes shape the abundance and composition of mature rRNAs and ribosome subunits, thereby influencing both the quantity and quality of the cellular ribosome pool. Notably, we identified that the correlation between rRNA transcription level and mature rRNA (ribosome subunit) abundance is non-linear and cell-type-specific across the hematopoietic system (Fig. 2B), and we also found a prominent and unexpected excess of 28S rRNA (60S subunit) in all cell types, ranging from 20 % excess in B cells to 60 %–90 % excess across the normal myeloid lineage (Sams et al., 2025). These trends indicate that additional regulatory steps downstream of rRNA transcription play a dynamic role in controlling ribosome subunit abundance. Accordingly, in this section we present **Conceptual Framework #2**, which focuses on downstream regulatory processes influencing ribosome subunit abundance (Fig. 8). We organize this framework into four principal regulatory pathways: (1) rRNA processing and assembly, (2) rRNA modifications and ribosome heterogeneity, (3) Cell cycling dynamics, and (4) rRNA degradation and ribosome turnover. For each pathway, we describe the underlying molecular mechanisms and discuss how their activity may influence cell-type-specific ribosome subunit abundance across the hematopoietic system.

5.1 rRNA processing and assembly

The nascent 47S pre-rRNA transcript must be processed into the mature 18S, 5.8S, and 28S rRNAs, which are folded and assembled with ribosomal proteins to generate the 40S (small subunit; SSU) and 60S (large subunit; LSU) ribosomal subunits. In mammalian cells, the 47S rRNA co-transcriptionally

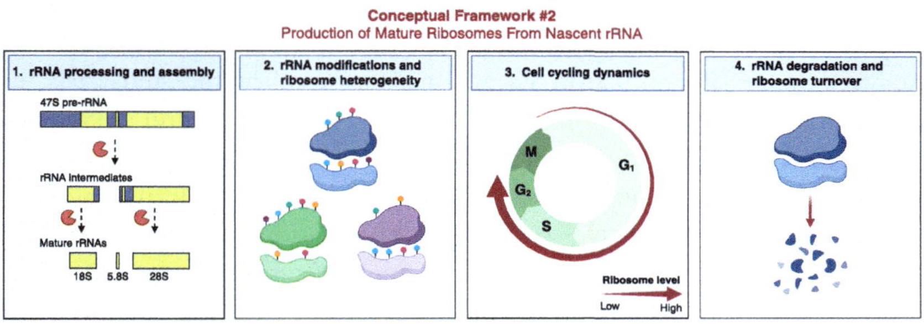

Fig. 8 Conceptual framework #2. Visual outline of conceptual framework #2: Production of mature ribosomes from nascent rRNA.

folds with ribosomal proteins (RP; RPS and RPL), the 5S rRNA (transcribed in the nucleoplasm by RNA Pol III), and a host of ribosome biogenesis factors to form the 90S pre-ribosome, which then splits into the pre–40S (containing 18S rRNA) and pre-60S (containing 28S, 5.8S, and 5S rRNAs) subunits (Baßler & Hurt, 2019; Aubert et al., 2018; Henras et al., 2015; Thomson et al., 2013; Dörner et al., 2023). The majority of rRNA processing and subunit assembly occurs in the nucleolus, with a limited number of late maturation steps taking place in the nucleoplasm and cytoplasm to eventually form mature 40S and 60S subunits. Since ribosome assembly occurs co-transcriptionally, and rRNAs are folded into pre-ribosomal particles from the earliest stages of transcription, there is no evidence that mature rRNAs exist at any point during their lifespan in a "free" or unbound state. In total, in addition to the 4 mature rRNAs and 79 ribosomal proteins, ribosome biogenesis requires the coordinated activity of over 200 assembly factors and approximately 80 small nucleolar RNAs (snoRNAs) (Baßler & Hurt, 2019).

Processing of the 47S pre-rRNA into rRNA intermediates and ultimately into mature rRNAs involves a highly complex and ordered series of endonucleolytic and exonucleolytic cleavage and trimming events that progressively remove the 5′ and 3′ external transcribed spacers (5′ETS and 3′ETS) and the internal transcribed spacers (ITS1 and ITS2) flanking the mature rRNA sequences. While alternative processing pathways have been described, the predominant pathway involves rapid cleavages in the 5′ETS within minutes of transcription initiation (Quinodoz et al. 2025), followed by cleavage within ITS1 to separate the 90S pre-ribosome into pre-40S and pre-60S intermediates. These early events are followed by a series of complex processing, folding, and assembly steps that together require around 2 h (Quinodoz et al., 2025).

Whether rRNA processing and assembly kinetics are actively regulated in a cell-type-specific manner to tune ribosome biogenesis rates remains an open question. Emerging evidence suggests that rRNA processing dynamics can vary across differentiation states. In mouse embryonic stem cells, differentiation is associated with accumulation of the 32S and 36S rRNA intermediates relative to nascent pre-rRNA, consistent with altered processing kinetics (McCann et al., 2020). Similarly, studies of human neuronal differentiation models have reported accumulation of the 30S rRNA intermediate in differentiated neurons relative to their progenitors (Pan et al., 2025). In the hematopoietic system, further support for the functional relevance of rRNA processing dynamics comes from Diamond-Blackfan Anemia, a ribosomopathy classically caused by ribosome haploinsufficiency arising from mutations in ribosomal proteins but also linked to mutations in rRNA processing factors, most notably TSR2 (Gripp et al., 2014; Khajuria et al., 2018). Loss of TSR2 in vitro results in reduced levels of mature 18S rRNA, accumulation of its immediate precursor 18SE rRNA, decreased ribosome abundance, and impaired erythroid differentiation (Khajuria et al., 2018). Together, these studies suggest that rRNA processing and assembly may be modulated in a cell-type-specific manner, or, at minimum, may show inherent variations within certain lineages. Ongoing advances in high-resolution spatiotemporal imaging (Quinodoz et al., 2025; Pan et al., 2025) and sequencing (Pastore et al., 2025) approaches to map rRNA processing intermediates will help resolve rRNA processing dynamics and their influence on cell-type-specific ribosome biogenesis rates.

5.2 rRNA modifications and ribosome heterogeneity

During ribosome assembly, rRNA processing is tightly coordinated with RP incorporation and extensive rRNA modification. Mammalian rRNAs are highly modified, with approximately 2% of rRNA nucleotides bearing modifications (over 220 annotated sites) (Taoka et al., 2018; Sloan et al., 2017; Gay et al., 2022; Natchiar et al., 2017; Lafontaine, 2015). The two most abundant rRNA modification types are 2′-O-methylation (2′-O-Me), catalyzed by Fibrillarin and guided by C/D box snoRNAs, and pseudouridylation (Ψ), catalyzed by Dyskerin and guided by H/ACA box snoRNAs. Additional modifications include nucleotide base methylations (e.g., m^6A) and acetylation (e.g., ac^4C). rRNA modifications are enriched at critical functional regions of the ribosome, including the peptidyl transferase center, decoding center, and intersubunit bridges, where they contribute to ribosome structure and stability as well as translation fidelity, kinetics, and specificity (Sloan et al., 2017; Gay et al., 2022; Ferretti & Karbstein, 2019; Natchiar et al., 2017; Taoka et al., 2018; Lafontaine, 2015).

Importantly, rRNA modification patterns are heterogeneous within the ribosome pool in a single cell and vary across cell types and physiological contexts. This heterogeneity provides a major source of ribosome diversity and enables functional specialization by modulating ribosome dynamics and selective translational output (Ferretti & Karbstein, 2019; Shi et al., 2017; Milenkovic & Novoa, 2025). In addition to rRNA modifications, ribosome heterogeneity can also arise from variable RP incorporation and from post-translational modifications of RPs (reviewed in (Miller et al., 2023; Ferretti & Karbstein, 2019)).

There is growing evidence that rRNA modifications are regulated in cell-type-specific and context-specific manners, including in the hematopoietic system. In normal hematopoiesis, different cell types exhibit distinct snoRNA expression profiles (Wang et al., 2025; Rabany et al., 2026; Warner et al., 2018) and rRNA modification landscapes (Rabany et al., 2026; Wang et al., 2025; Zhou et al., 2023). Proper regulation of these programs is required to maintain cell identity and functional capacity, particularly in hematopoietic stem cells (Wang et al., 2025; Rabany et al., 2026). Loss of the SNORD113–114 cluster of snoRNAs, enriched specifically in long-term HSCs, results in loss of methylation at target rRNA modification sites, ultimately leading to impaired HSC reconstitution capacity (Wang et al., 2025). Beyond normal hematopoiesis, dysregulation of snoRNA expression and rRNA modification has been documented in leukemic contexts. Specific snoRNA expression patterns (Warner et al., 2018; Valleron et al., 2012; Pauli et al., 2020) and rRNA modification signatures (Zhou et al., 2017, 2023), such as SNORD35A (Zhou et al. 2017) and SNORD42A (Pauli et al. 2020), are enriched in leukemia patient samples and are associated with increased leukemic blast cell counts and leukemic stem cell frequency. These leukemia-associated snoRNAs and rRNA modifications have been shown to be essential for leukemic cell proliferation, self-renewal, and leukemogenic capacity in both in vitro and in vivo models (Zhou et al., 2023, 2017; Pauli et al., 2020). Knockdown of SNORD35A in transplanted leukemic cells significantly delays leukemogenesis (Zhou et al., 2017), and knockdown of SNORD42A impairs the colony-forming capacity of four different leukemia cell lines (Pauli et al., 2020). Caution should be exercised however in interpreting studies of snoRNA manipulation, since many have functions beyond merely directing rRNA modifications, and have been reported to be required for proper rRNA folding and ribosome assembly, as well as in non-nucleolar roles such as mRNA splicing and miRNA biology (Huang et al., 2022).

It is also worth noting that most rRNA modifications are deposited co-transcriptionally or during early rRNA processing and folding, coincident with pre-ribosomal particle assembly (Sloan et al., 2017; Aubert et al., 2018; Quinodoz et al., 2025; Pastore et al., 2025). This close coordination between rRNA modification, processing, and assembly suggests that qualitative ribosome features, such as rRNA modification profiles or RP composition, may contribute not only to ribosome function but also to ribosome production dynamics. In doing so, these features may shape both the quantity and quality of the ribosome pool in a cell-type-specific manner. Emerging approaches such as Nanopore Direct RNA Sequencing now enable high-resolution, comprehensive profiling of rRNA modifications states in nascent and maturing rRNAs, offering new opportunities to resolve how rRNA modification heterogeneity influences cell-type-specific ribosome production (Milenkovic et al., 2025; Pastore et al., 2025).

5.3 Cell cycling dynamics

As described above in the context of regulation of Pol I occupancy, ribosome biogenesis is tightly coupled to cell cycle progression, as proliferating cells must increase their biosynthetic capacity to support growth and ensure that daughter cells inherit sufficient ribosome pools. Ribosome abundance is therefore determined not only by rRNA transcription and assembly, but also by dilution through cell division. Thus, in principle, cells with identical rRNA transcription rates could maintain different steady-state ribosome levels if they divide at different rates, with faster cycling cells experiencing more rapid dilution of their ribosome pools.

While this model provides one possible explanation for why nascent rRNA synthesis and mature ribosome abundance may become uncoupled across hematopoietic cell types, since different populations display a wide range of proliferation rates (Sams et al., 2025; Signer et al., 2016; Eastman et al., 2020), our recent work indicates that differences in cell cycle dynamics alone are insufficient to account for observed variation in ribosome abundance. In the hematopoietic system, rRNA transcription levels and mature rRNA (ribosome subunit) abundance do not trivially correlate with proliferation dynamics (Sams et al., 2025). Cell types with similar rRNA transcription levels and cycling behaviors can exhibit markedly different levels of mature rRNAs (e.g., classical and intermediate monocytes), while other cell types with comparable mature rRNA abundance and proliferation dynamics can display substantial differences in rRNA transcription levels (e.g., MEPs and MPP2/3/4s). In addition, HSCs, which are largely in G0 and do not divide for weeks at a time (Wilson et al., 2008), surprisingly exhibit an intermediate level of rRNA transcription rather than the very low levels one might expect with quiescence. Together, these findings demonstrate that dilution of ribosomes through cell cycling represents only one component of a more complex regulatory framework that shapes ribosome abundance.

5.4 rRNA degradation and ribosome turnover

rRNA and ribosome degradation represent an important but poorly explored avenue of regulation of ribosome abundance. Very few studies have directly investigated ribosome subunit half-lives and the mechanisms governing ribosome turnover under normal, unstressed conditions. Available evidence from mammalian cells indicates that mature rRNAs, and therefore ribosomes themselves, are relatively long-lived, with reported half-lives typically on the order of days (Menzies & Gold, 1972; Menzies et al., 1972; An & Harper, 2018; Yang & Karbstein, 2024). Early metabolic labeling studies tracking ribosome turnover in different rat tissues revealed tissue-to-tissue variability in half-lives, ranging from 5.9 days in the liver to 9 days in the lung (Menzies et al., 1972).

While still an evolving area of investigation, particularly in mammalian systems, there is a substantial literature on ribosome degradation pathways induced by biogenesis errors, ribosome damage or dysfunction, and cellular stress (see reviews (Yang & Karbstein, 2024; Lafontaine, 2010; Kazibwe et al., 2019; An & Harper, 2020)). Early ribosome biogenesis quality control involves nucleolar surveillance pathways that monitor rRNA processing and assembly and target aberrant subunit intermediates for degradation through the nuclear RNA exosome before they achieve full maturity (Shcherbik et al. 2010; Wang & Pestov, 2011; Sudo et al. 2016). However, these degradation processes have largely been studied in aberrant, manipulated, or stressed cells, and there is limited work on their baseline activity in homeostatic tissues.

Once fully matured, ribosome subunit degradation is largely coupled to surveillance of translation dynamics. Ribosomes that encounter persistent translation defects are resolved through ribosome quality control pathways that monitor stalled or collided ribosomes and can trigger downstream activation of nonfunctional rRNA decay (NRD) to eliminate defective ribosomal subunits (LaRiviere et al., 2006; Cole et al., 2009; Coria et al., 2025). Translation defects that engage NRD include ribosome stalling or collisions that can arise from functional impairments in rRNAs, including sequence mutations in the decoding or peptidyl transferase centers, aberrant rRNA modifications, or accumulated oxidative damage (LaRiviere et al., 2006; Cole et al., 2009; Coria et al., 2025; Huang et al., 2025). In yeast, NRD pathways are well established and are mechanistically distinct for the 40S and 60S ribosomal subunits (Cole et al., 2009; Lafontaine, 2010). In mammalian systems, the 18S rRNA/40S subunit NRD pathway has only recently begun to be mechanistically defined, and the pathway governing 28S rRNA/60S subunit NRD remains poorly understood. Recent work on NRD in mammalian cells has shown that defective 40S subunits are selectively targeted for disassembly through ubiquitination of ribosomal proteins, resulting in partial subunit remodeling and exposure of the 18S rRNA for degradation by cytoplasmic RNA exosome machinery (Coria et al., 2025; Huang et al., 2025; Ford et al., 2025). This process is coordinated through a regulatory axis involving GCN2, RNF10, and RIOK3 (Coria et al., 2025; Huang et al., 2025; Garzia et al., 2021; Ford et al., 2025). This regulatory pathway can also target individual 40S subunits for degradation in response to nutrient deprivation or other forms of cellular stress (Huang et al., 2025; Ford et al., 2025).

Beyond translation-coupled surveillance of individual ribosomes, stress conditions such as nutrient deprivation have historically been shown to induce bulk ribosome degradation through a selective autophagy pathway termed "ribophagy" (Kraft et al., 2008). Although still an evolving area of investigation in mammalian systems, NUFIP1 and RPL12 have been identified as ribophagy "receptors" that promote bulk lysosomal degradation of ribosomes under stress conditions (Wyant et al., 2018; Chen et al., 2025). The extent to which cellular stress preferentially activates targeted NRD of individual ribosomal subunits versus bulk ribophagy remains unresolved, and may depend on specific stress contexts (An et al., 2020).

It is possible that differential engagement of ribosome surveillance and degradation pathways may contribute to cell-type-specific regulation of ribosome abundance in the hematopoietic system. Such regulation could arise from variable propensities to generate defective ribosomes, accumulate damaged ribosomes, or from differences in the activity of key degradation pathway components, including GCN2, RNF10, and RIOK3 in NRD, or NUFIP1 and RPL12 in ribophagy. In addition, differential activity of subunit-specific surveillance pathways, either during subunit maturation in the nucleolus or once fully mature in the cytoplasm, may contribute to the mismatches in 40S and 60S subunit numbers found across hematopoietic cell types, the physiological function and implications of which are currently unknown. These emerging studies provide a framework for considering ribosome degradation as an additional layer through which steady-state ribosome subunit numbers may be fine-tuned in a cell-type-specific manner.

5.5 Summary

Taken together, ribosome abundance is regulated at multiple levels downstream of rRNA transcription, including coordinated control of rRNA processing, modification, and assembly, cell cycle-dependent dilution, and ribosome surveillance and degradation pathways. Differential engagement of these processes provides a mechanism by which distinct hematopoietic cell types can establish and maintain cell-type-specific ribosome numbers.

6. Therapeutic inhibition of rRNA transcription

Given the central role of rRNA transcription and ribosome function in normal and malignant cell biology, rRNA transcription has emerged as an attractive target in both experimental and therapeutic contexts. A limited number of small molecule inhibitors are commonly used to suppress rRNA transcription, including the classical agents Actinomycin D (ActD), CX-5461, and BMH-21, as well as newly emerging inhibitors PMR-116 and NSH76.

The classical rRNA transcription inhibitors ActD, CX-5461, and BMH-21 all function as DNA intercalators. DNA intercalators are planar molecules that insert between adjacent DNA base pairs, inducing local DNA unwinding and helix elongation (Mukherjee & Sasikala, 2013). Because of the high GC content of rDNA, intercalators often exhibit preferential and pronounced effects on rRNA transcription. However, DNA intercalation can also disrupt chromatin structure, induce or stabilize Z-DNA, and elicit replication fork stalling, DNA damage, or broader genomic instability and transcriptional dysregulation across the genome (Espinoza et al. 2024). Consequently, the cellular effects of these agents are not restricted to inhibition of rRNA transcription, and they are known to produce immediate S-phase and G2/M cell cycle accumulation that is better explained by defects in DNA replication and mitosis (Quin et al., 2016; Makhale et al., 2021; Hein et al., 2017; Negi & Brown, 2015). Careful interpretation of the effects of these drugs is therefore required even under conditions where overt DNA damage is not detected. Given their widespread use in basic and translational research, we briefly summarize the mechanisms and limitations of ActD, CX-5461, and BMH-21 in this section, and we additionally highlight PMR-116 and NSH76 as emerging inhibitors of rRNA transcription that may offer improved specificity.

Actinomycin D (ActD) is an antibiotic that has long been used as an RNA polymerase inhibitor, exhibiting preferential inhibition of Pol I at low doses and broader inhibition of Pol II and Pol III at higher concentrations (Perry & Kelley, 1970; Espinoza et al., 2024; Ferreira et al., 2020). In addition to suppressing rRNA transcription, ActD interferes with normal topoisomerase activity, induces DNA damage (Trask & Muller, 1988), and can bind and stabilize G-quadruplex structures (Hudson et al., 2009; Kang & Park, 2009). Accordingly, while ActD has been widely used in basic and translational studies as well as clinical applications, including in hematopoietic settings (Damaskou et al., 2025; Hayashi et al., 2014), the extent to which observed responses are due to rRNA transcription alone remains difficult to resolve.

CX-5461 was originally identified in a screen for inhibitors of rRNA transcription (Drygin et al., 2011; Haddach et al., 2012), and has been shown to robustly suppress rRNA transcription across cell types and tissue systems in vitro and in vivo. These properties have ultimately led to its use in clinical settings as a chemotherapeutic agent in various malignancies, including AML (Khot et al., 2019). In addition to its role as a DNA intercalator, CX-5461 also inhibits rRNA transcription by interfering with the activity of the rDNA transcription preinitiation complex (PIC) (Mars et al., 2020). However, CX-5461 has also been shown to induce DNA damage (Quin et al., 2016; Sanij et al., 2020; Mars et al., 2020; Xu et al., 2017; Cameron et al., 2024) and act as a topoisomerase poison (Li et al., 2018; Bruno et al., 2020; Yan et al., 2021; Pan et al., 2021; Bossaert et al., 2021). As such, CX-5461 cannot be considered a selective inhibitor of rRNA transcription, as its cellular effects likely reflect the combined consequences of Pol I inhibition and broader genotoxic stress (see reviews (Ferreira et al., 2020; Zisi et al., 2022)).

BMH-21 was originally identified in a screen for compounds that activate p53 (Peltonen et al., 2010) and was subsequently shown to robustly suppress rRNA transcription. Although BMH-21 is also a DNA intercalator, it does not elicit the canonical DNA damage response or stabilization of topoisomerase cleavage complexes associated with DNA strand breaks that are observed with ActD and CX-5461 (Colis et al., 2014). Instead, BMH-21 inhibits Pol I transcription at multiple stages, including initiation, promoter escape, and elongation (Jacobs et al., 2022; Peltonen et al., 2014; Wei et al., 2018), and interestingly, also induces proteasomal degradation of Pol I (Peltonen et al., 2014; Wei et al., 2018; Espinoza et al., 2024). BMH-21 has demonstrated efficacy in reducing proliferation of AML cell lines (Hein et al., 2017) and in reducing rRNA transcription in cultured primary normal hematopoietic cells (Jarzebowski et al., 2018). However, recent work indicates that BMH-21 can also promote proteasomal degradation of RNA Pol II, drive formation of Z-DNA, and trap topoisomerase on chromatin without inducing overt DNA damage

(Espinoza et al., 2024). These findings underscore the need for additional studies to more comprehensively dissect the specificity of BMH-21 as a Pol I inhibitor.

PMR-116 was recently identified in a screen designed to identify inhibitors of rRNA transcription that lack the off-target effects associated with classical DNA intercalating agents (Ferreira et al., 2025). PMR-116 reduces rRNA transcription by inducing Pol I stalling at rDNA enhancers and promoters, thereby preventing promoter clearance and transcription, although the precise molecular mechanism underlying this stalling remains unclear. PMR-116 exerts these effects in the absence of global DNA damage, and early studies demonstrate efficacy of PMR-116 in multiple in vivo models of hematological malignancies, including AML, multiple myeloma, and B cell lymphoma, as well as in various solid tumors. Notably, PMR-116 synergizes with several hematologic therapeutics in AML cell line models, including the BCL-2 inhibitor venetoclax.

NSH76 was recently identified in a screen for small molecules targeting the Pol I-specific initiation factor RRN3 (Sarkar et al., 2025). NSH76 directly binds and inhibits RRN3, leading to reduced occupancy of PIC components at the rDNA promoter and decreased rRNA transcription. Importantly, these effects occur in the absence of detectable DNA damage, large scale mutagenesis, or chromosomal instability. Through inhibition of Pol I, NSH76 sensitizes chemotherapy-resistant cell lines and impairs migratory and invasive behavior. While further work is required to define the full spectrum of its cellular effects and improve its potency and pharmacological properties, the specific targeting of the Pol I PIC positions NSH76 as one of the only Pol I inhibitors with a known selective mechanism.

7. Conclusion

Regulation of rRNA transcription by Pol I has historically been overlooked in most studies of cell-type-specific and disease-specific transcriptional control. However, recent work highlights rRNA transcription and ribosome biogenesis as highly dynamic processes with clear influences on cell behavior. We recently found that rRNA transcription levels and ribosome subunit abundance are cell-type-specific across the normal hematopoietic system and are aberrantly elevated in AML. Our findings that rRNA levels are uncoupled from global protein synthesis and cell cycle dynamics, together with the non-linear relationships between rRNA transcription and ribosome abundance as well as unequal subunit stoichiometry, indicate that ribosome biogenesis involves multiple layers of complex regulation that are adapted in a cell-type-specific manner. Integrating these findings with the regulatory pathways discussed in this review will help resolve how and why ribosome biogenesis is fine-tuned across hematopoietic cell types and dysregulated in disease.

Acknowledgements

We thank members of the Paralkar Lab for valuable scientific discussions. Figures were created in BioRender (https://BioRender.com/eoo02q7) and Adobe Illustrator.

Funding

V.R.P. is supported by National Institutes of Health (NIH) grant R35 GM138035 (National Institute of General Medical Sciences [NIGMS]) and the University of Pennsylvania University Research Foundation [URF] Grants Program. We thank the Greg Wolf Fund, and Meghneel Gore and Anjali Dave, for generous philanthropic funding in support of this work. E.I.S. and J.A.H. are Cell and Molecular Biology students through the Biomedical Graduate Studies program at the University of Pennsylvania. E.I.S. is supported by the PERFORM-KUH U2C/TL1 grant DK136784 (National Institute of Diabetes and Digestive and Kidney Diseases [NIDDK]). J.A.H. is supported by NIH grant T32 HL007439 (National Heart, Lung, and Blood Institute [NHLBI]). Research on this project was supported by the University of Pennsylvania Abramson Cancer Center Support Grant, supported by National Cancer Institute (NCI) grant P30 CA016520.

References

An, H., & Harper, J. W. (2018). Systematic analysis of ribophagy in human cells reveals bystander flux during selective autophagy. *Nature Cell Biology, 20*(2), 135–143.

An, H., & Harper, J. W. (2020). Ribosome abundance control via the ubiquitin-proteasome system and autophagy. *Journal of Molecular Biology, 432*(1), 170–184.

An, H., Alban, O., Maria, K., Joao, A. P., & Wade, H. J. (2020). Systematic quantitative analysis of ribosome inventory during nutrient stress. *Nature, 583*(7815), 303–309.

Antony, C., George, S. S., Blum, J., et al. (2022). Control of ribosomal RNA synthesis by hematopoietic transcription factors. *Molecular Cell, 82*(20), 3826–3839.e9.

Antony, C., Somers, P., Gray, E. M., Pimkin, M., & Paralkar, V. R. (2023). FISH-flow to quantify nascent and mature ribosomal RNA in mouse and human cells. *STAR Protocols, 4*(3), 102463.

Aubert, M., O'Donohue, M.-F., Lebaron, S., & Gleizes, P.-E. (2018). Pre-ribosomal RNA processing in human cells: From mechanisms to congenital diseases. *Biomolecules, 8*(4), https://doi.org/10.3390/biom8040123.

Barajas, Milad Rasouli, J. M., Umeda, M., et al. (2024). Acute myeloid leukemias with UBTF tandem duplications are sensitive to menin inhibitors. *Blood, 143*(7), 619–630.

Baßler, J., & Hurt, E. (2019). Eukaryotic ribosome assembly. *Annual Review of Biochemistry, 88*(June), 281–306.

Bell, S. P., Learned, R. M., Jantzen, H. M., & Tjian, R. (1988). Functional cooperativity between transcription factors UBF1 and SL1 mediates human ribosomal RNA synthesis. *Science (New York, N.Y.), 241*, 1192–1197.

Bierhoff, H., Dundr, M., Michels, A. A., & Grummt, I. (2008). Phosphorylation by casein kinase 2 facilitates rRNA gene transcription by promoting dissociation of TIF-IA from elongating RNA Polymerase I. *Molecular and Cellular Biology, 28*(16), 4988–4998.

Bossaert, M., Pipier, A., Riou, J.-F., et al. (2021). Transcription-associated topoisomerase 2α (TOP2A) activity is a major effector of cytotoxicity induced by G-quadruplex ligands. *eLife, 10*(June), https://doi.org/10.7554/eLife.65184.

Brito, Q., Jailson, Irene, D.-L., & Ramakrishnan, V. (2024). The molecular basis of translation initiation and its regulation in eukaryotes. *Nature Reviews. Molecular Cell Biology, 25*(3), 168–186.

Brock, G. J., & Bird, A. (1997). Mosaic methylation of the repeat unit of the human ribosomal RNA genes. *Human Molecular Genetics, 6*(3), 451–456.

Bruno, P. M., Lu, M., Dennis, K. A., et al. (2020). The primary mechanism of cytotoxicity of the chemotherapeutic agent CX-5461 Is topoisomerase II poisoning. *Proceedings of the National Academy of Sciences of the United States of America, 117*(8), 4053–4060.

Bursać, S., Brdovčak, M. C., Pfannkuchen, M., et al. (2012). Mutual protection of ribosomal proteins L5 and L11 from degradation is essential for p53 activation upon ribosomal biogenesis stress. *Proceedings of the National Academy of Sciences of the United States of America, 109*(50), 20467–20472.

Bywater, M. J., Poortinga, G., Sanij, E., et al. (2012). Inhibition of RNA Polymerase I as a therapeutic strategy to promote cancer-specific activation of p53. *Cancer Cell, 22*(1), 51–65.

Cai, X., Gao, L., Teng, L., et al. (2015). Runx1 deficiency decreases ribosome biogenesis and confers stress resistance to hematopoietic stem and progenitor cells. *Cell Stem Cell, 17*(2), 165–177.

Cameron, D. P., Sornkom, J., Alsahafi, S., et al. (2024). CX-5461 preferentially induces Top2α-dependent DNA breaks at ribosomal DNA loci. *Biomedicines, 12*(7), https://doi.org/10.3390/biomedicines12071514.

Campbell, K. J., & White, R. J. (2014). MYC regulation of cell growth through control of transcription by RNA Polymerases I and III. *Cold Spring Harbor Perspectives in Medicine, 4*(5), https://doi.org/10.1101/cshperspect.a018408.

Caudy, A. A., & Pikaard, C. S. (2002). Xenopus ribosomal RNA gene intergenic spacer elements conferring transcriptional enhancement and nucleolar dominance-like competition in oocytes. *The Journal of Biological Chemistry, 277*(35), 31577–31584.

Cedar, H., & Bergman, Y. (2011). Epigenetics of haematopoietic cell development. *Nature Reviews. Immunology, 11*(7), 478–488.

Chen, Y., Hu, J., Zhao, P., et al. (2025). Rpl12 Is a conserved ribophagy receptor. *Nature Cell Biology, 27*(3), 477–492.

Cole, S. E., LaRiviere, F. J., Merrikh, C. N., & Moore, M. J. (2009). A convergence of rRNA and mRNA quality control pathways revealed by mechanistic analysis of nonfunctional rRNA decay. *Molecular Cell, 34*(4), 440–450.

Colis, L., Peltonen, K., Sirajuddin, P., et al. (2014). DNA intercalator BMH-21 inhibits RNA Polymerase I independent of DNA damage response. *Oncotarget, 5*(12), 4361–4369.

Conconi, A., Widmer, R. M., Koller, T., & Sogo, J. M. (1989). Two different chromatin structures coexist in ribosomal RNA genes throughout the cell cycle. *Cell, 57*(5), 753–761.

Core, L., & Adelman, K. (2019). Promoter-proximal pausing of RNA Polymerase II: A nexus of gene regulation. *Genes & Development, 33*(15-16), 960–982.

Coria, A. R., Shah, A., Shafieinouri, M., et al. (2025). The integrated stress response regulates 18S nonfunctional rRNA decay in mammals. *Molecular Cell, 85*(4), 787–801.e8.

Damaskou, A., Wilson, R., Gozdecka, M., et al. (2025). Posttranscriptional depletion of ribosome biogenesis factors engenders therapeutic vulnerabilities in NPM1-mutant AML. *Blood, 146*(10), 1239–1252.

Dammann, R., Lucchini, R., Koller, T., & Sogo, J. M. (1993). Chromatin structures and transcription of rDNA in yeast saccharomyces cerevisiae. *Nucleic Acids Research, 21*(10), 2331–2338.

De Winter, R. F., & Moss, T. (1986). Spacer promoters are essential for efficient enhancement of X. Laevis ribosomal transcription. *Cell, 44*(2), 313–318.

Di, M., Martinez, D. E., Kanefsky, J. S., et al. (2021). ATF3 coordinates serine and nucleotide metabolism to drive cell cycle progression in acute myeloid leukemia. *Molecular Cell, 81*(13), 2752–2764.e6.

Donati, G., Peddigari, S., Mercer, C. A., & Thomas, G. (2013). 5S Ribosomal RNA is an essential component of a nascent ribosomal precursor complex that regulates the Hdm2-p53 checkpoint. *Cell Reports, 4*(1), 87–98.

Dörner, K., Ruggeri, C., Zemp, I., & Kutay, U. (2023). Ribosome biogenesis factors-from names to functions. *The EMBO Journal, 42*(7), e112699.

Drygin, D., Lin, A., Bliesath, J., et al. (2011). Targeting RNA Polymerase I with an oral small molecule CX-5461 inhibits ribosomal RNA synthesis and solid tumor growth. *Cancer Research, 71*(4), 1418–1430.

Duployez, N., Vasseur, L., Kim, R., et al. (2023). UBTF tandem duplications define a distinct subtype of adult de novo acute myeloid leukemia. *Leukemia: Official Journal of the Leukemia Society of America, Leukemia Research Fund, U. K, 37*(6), 1245–1253.

Eastman, A. E., Chen, X., Hu, X., et al. (2020). Resolving cell cycle speed in one snapshot with a live-cell fluorescent reporter. *Cell Reports, 31*(12), 107804.

Espinoza, J. A., Kanellis, D. C., Saproo, S., et al. (2024). Chromatin damage generated by DNA intercalators leads to degradation of RNA Polymerase II. *Nucleic Acids Research, 52*(8), 4151–4166.

Evers, R., Smid, A., Rudloff, U., Lottspeich, F., & Grummt, I. (1995). Different domains of the murine RNA Polymerase I-specific termination factor mTTF-I serve distinct functions in transcription termination. *The EMBO Journal, 14*(6), 1248–1256.

Faber, J., Krivtsov, A. V., Stubbs, M. C., et al. (2009). HOXA9 Is required for survival in human MLL-rearranged acute leukemias. *Blood, 113*(11), 2375–2385.

Falcon, K. T., Watt, K. E. N., Dash, S., et al. (2022). Dynamic regulation and requirement for ribosomal RNA transcription during mammalian development. *Proceedings of the National Academy of Sciences of the United States of America, 119*(31), e2116974119.

Farley-Barnes, K. I., Ogawa, L. M., & Baserga, S. J. (2019). Ribosomopathies: Old concepts, new controversies. *Trends in Genetics: TIG, 35*(10), 754–767.

Ferreira, R., Hannan, K. M., Panov, K., et al. (2025). The novel RNA Polymerase I transcription inhibitor PMR-116 exploits a critical therapeutic vulnerability in a broad-spectrum of high *MYC* malignancies. *bioRxiv*. https://doi.org/10.1101/2025.04.19.649466 April 21.

Ferreira, R., Schneekloth Jr, J. S., Panov, K. I., Hannan, K. M., & Hannan, R. D. (2020). Targeting the RNA Polymerase I transcription for cancer therapy comes of age. *Cells, 9*(2), https://doi.org/10.3390/cells9020266.

Ferretti, M. B., & Karbstein, K. (2019). Does functional specialization of ribosomes really exist? *RNA (New York, N.Y.),* 521–538.

Ford, P. W., Garshott, D. M., Narasimhan, M., et al. (2025). RNF10 and RIOK3 facilitate 40S ribosomal subunit degradation upon 60S biogenesis disruption or amino acid starvation. *Cell Reports, 44*(3), 115371.

Garzia, A., Meyer, C., & Tuschl, T. (2021). The E3 ubiquitin ligase RNF10 modifies 40S ribosomal subunits of ribosomes compromised in translation. *Cell Reports, 36*(5), 109468.

Gay, D. M., Lund, A. H., & Jansson, M. D. (2022). Translational control through ribosome heterogeneity and functional specialization. *Trends in Biochemical Sciences, 47*(1), 66–81.

George, S. S., Pimkin, M., & Paralkar, V. R. (2023). Construction and validation of customized genomes for human and mouse ribosomal DNA mapping. *The Journal of Biological Chemistry, 299*(6), 104766.

Grandori, C., Gomez-Roman, N., Felton-Edkins, Z. A., et al. (2005). C-Myc binds to human ribosomal DNA and stimulates transcription of rRNA genes by RNA Polymerase I. *Nature Cell Biology,* 7(3), 311–318.

Gripp, K. W., Curry, C., Olney, A. H., et al. (2014). Diamond-blackfan anemia with mandibulofacial dystostosis is heterogeneous, including the novel DBA genes TSR2 and RPS28. *American Journal of Medical Genetics. Part A, 164*(9), 2240–2249.

Grob, A., Colleran, C., & McStay, B. (2014). Construction of synthetic nucleoli in human cells reveals how a major functional nuclear domain is formed and propagated through cell division. *Genes & Development, 28*(3), 220–230.

Grummt, I., & Grummt, F. (1976). Control of nucleolar RNA synthesis by the intracellular pool sizes of ATP and GTP. *Cell,* 7(3), 447–453.

Grummt, I., Kuhn, A., Bartsch, I., & Rosenbauer, H. (1986). A transcription terminator located upstream of the mouse rDNA initiation site affects rRNA synthesis. *Cell, 47*(6), 901–911.

Haddach, M., Schwaebe, M. K., Michaux, J., et al. (2012). Discovery of CX-5461, the first direct and selective inhibitor of RNA Polymerase I, for cancer therapeutics. *ACS Medicinal Chemistry Letters, 3*(7), 602–606.

Hamdane, N., Stefanovsky, V. Y., Tremblay, M. G., et al. (2014). Conditional inactivation of upstream binding factor reveals its epigenetic functions and the existence of a somatic nucleolar precursor body. *PLoS Genetics, 10*(8), e1004505.

Hayashi, Y., Kuroda, T., Kishimoto, H., Wang, C., Iwama, A., & Kimura, K. (2014). Downregulation of rRNA transcription triggers cell differentiation. *PLoS One, 9*(5), e98586.

Heikamp, E. B., Henrich, J. A., Perner, F., et al. (2022). The Menin-MLL1 interaction is a molecular dependency in NUP98-rearranged AML. *Blood, 139*(6), 894–906.

Hein, N., Cameron, D. P., Hannan, K. M., et al. (2017). Inhibition of Pol I transcription treats murine and human AML by targeting the leukemia-initiating cell population. *Blood, 129*(21), 2882–2895.

Hein, N., Hannan, K. M., George, A. J., Sanij, E., & Hannan, R. D. (2013). The nucleolus: An emerging target for cancer therapy. *Trends in Molecular Medicine, 19*(11), 643–654.

Heix, J., Vente, A., Voit, R., Budde, A., Michaelidis, T. M., & Grummt, I. (1998). Mitotic silencing of human rRNA synthesis: Inactivation of the promoter selectivity factor SL1 by cdc2/cyclin B-mediated phosphorylation. *The EMBO Journal, 17*(24), 7373–7381.

Henderson, A. S., Warburton, D., & Atwood, K. C. (1972). Location of ribosomal DNA in the human chromosome complement. *Proceedings of the National Academy of Sciences of the United States of America, 69*(11), 3394–3398.

Henras, A. K., Soudet, J., Gérus, M., et al. (2008). The post-transcriptional steps of eukaryotic ribosome biogenesis. *Cellular and Molecular Life Sciences: CMLS, 65*(15), 2334–2359.

Henras, A. K., Plisson-Chastang, C., O'Donohue, M.-F., Chakraborty, A., & Gleizes, P.-E. (2015). An overview of pre-ribosomal RNA processing in eukaryotes. *Wiley Interdisciplinary Reviews. RNA, 6*(2), 225–242.

Herdman, C., Mars, J.-C., Stefanovsky, V. Y., et al. (2017). A unique enhancer boundary complex on the mouse ribosomal RNA genes persists after loss of Rrn3 or UBF and the inactivation of RNA Polymerase I transcription. *PLoS Genetics, 13*(7), e1006899.

Hoppe, S., Bierhoff, H., Cado, I., et al. (2009). AMP-activated protein kinase adapts rRNA synthesis to cellular energy supply. *Proceedings of the National Academy of Sciences of the United States of America, 106*(42), 17781–17786.

Hori, Y., Shimamoto, A., & Kobayashi, T. (2021). The human ribosomal DNA array is composed of highly homogenized tandem clusters. *Genome Research, 31*(11), 1971–1982.

Hoxhaj, G., Hughes-Hallett, J., Timson, R. C., et al. (2017). The mTORC1 signaling network senses changes in cellular purine nucleotide levels. *Cell Reports, 21*(5), 1331–1346.

Huang, K., Jia, J., Wu, C., et al. (2013). Ribosomal RNA gene transcription mediated by the master genome regulator protein CCCTC-binding factor (CTCF) Is negatively regulated by the condensin complex. *The Journal of Biological Chemistry, 288*(36), 26067–26077.

Huang, M., Ji, Y., Zhang, K. I. Y., & Mitchell, B. (2008). Guanine nucleotide depletion inhibits pre-ribosomal RNA synthesis and causes nucleolar disruption. *Leukemia Research, 32*(1), 131–141.

Huang, R., Wu, T., Xu, L., Liu, A., Ji, Y., & Hu, G. (2002). Upstream binding factor up-regulated in hepatocellular carcinoma is related to the survival and cisplatin-sensitivity of cancer cells. *FASEB Journal: Official Publication of the Federation of American Societies for Experimental Biology, 16*(3), 293–301.

Huang, Z.-H., Du, Y.-P., Wen, J.-T., Lu, B.-F., & Zhao, Y. (2022). snoRNAs: Functions and mechanisms in biological processes, and roles in tumor pathophysiology. *Cell Death Discovery, 8*(1), 259.

Huang, Z., Diehl, F. F., Wang, M., et al. (2025). RIOK3 mediates the degradation of 40S ribosomes. *Molecular Cell, 85*(4), 802–814.e12.

Hubackova, S., Davidova, E., Boukalova, S., et al. (2020). Replication and ribosomal stress induced by targeting pyrimidine synthesis and cellular checkpoints suppress p53-deficient tumors. *Cell Death & Disease, 11*(2), 110.

Hudson, J. S., Brooks, S. C., & Graves, D. E. (2009). Interactions of actinomycin D with human telomeric G-quadruplex DNA. *Biochemistry, 48*(21), 4440–4447.

Hwang, S.-P., & Denicourt, C. (2024). The impact of ribosome biogenesis in cancer: From proliferation to metastasis. *NAR Cancer, 6*(2), zcae017.

Jacobs, R. Q., Huffines, A. K., Laiho, M., & Schneider, D. A. (2022). The small-molecule BMH-21 directly inhibits transcription elongation and DNA occupancy of RNA Polymerase I in vivo and in vitro. *The Journal of Biological Chemistry, 298*(1), 101450.

Jansen, J., Bohnsack, K. E., Böhlken-Fascher, S., Bohnsack, M. T., & Dobbelstein, M. (2024). The ribosomal protein L22 binds the MDM4 Pre-mRNA and promotes exon skipping to activate p53 upon nucleolar stress. *Cell Reports, 43*(8), 114610.

Jarzebowski, L., Bouteiller, M. L., Coqueran, S., et al. (2018). *Mouse adult hematopoietic stem cells actively synthesize ribosomal RNA. RNA (New York, N.Y.), 24*, 1803–1812.

Jouffe, C., Cretenet, G., Symul, L., et al. (2013). The circadian clock coordinates ribosome biogenesis. *PLoS Biology, 11*(1), e1001455.

Kang, H.-J., & Park, H.-J. (2009). Novel molecular mechanism for actinomycin D activity as an oncogenic promoter G-quadruplex binder. *Biochemistry, 48*(31), 7392–7398.

Kawano, Riichiro Ikeda, G., Ishihara, D., et al. (2026). Guanine nucleotides drive ribosome biogenesis and glycolytic reprogramming in acute myeloid leukemia stem cells. *Blood, 147*(7), 768–782.

Kazibwe, Z., Liu, A.-Y., MacIntosh, G. C., & Bassham, D. C. (2019). The ins and outs of autophagic ribosome turnover. *Cells, 8*(12), https://doi.org/10.3390/cells8121603.

Khajuria, Mathias Munschauer, R. K., Ulirsch, J. C., et al. (2018). Ribosome levels selectively regulate translation and lineage commitment in human hematopoiesis. *Cell, 173*(1), 90–103.e19.

Khot, A., Brajanovski, N., Cameron, D. P., et al. (2019). First-in-human RNA Polymerase I transcription inhibitor CX-5461 in patients with advanced hematologic cancers: Results of a phase I dose-escalation study. *Cancer Discovery, 9*(8), 1036–1049.

Kihm, A. J., Hershey, J. C., Haystead, T. A., Madsen, C. S., & Owens, G. K. (1998). Phosphorylation of the rRNA transcription factor upstream binding factor promotes its association with TATA binding protein. *Proceedings of the National Academy of Sciences of the United States of America, 95*(25), 14816–14820.

Kofuji, S., Hirayama, A., Eberhardt, A. O., et al. (2019). IMP dehydrogenase-2 drives aberrant nucleolar activity and promotes tumorigenesis in glioblastoma. *Nature Cell Biology, 21*(8), 1003–1014.

Kraft, C., Deplazes, A., Sohrmann, M., & Peter, M. (2008). Mature ribosomes are selectively degraded upon starvation by an autophagy pathway requiring the Ubp3p/Bre5p ubiquitin protease. *Nature Cell Biology, 10*(5), 602–610.

Lafita-Navarro, Carmen, M., Venkateswaran, N., Kilgore, J. A., et al. (2020). Inhibition of the de novo pyrimidine biosynthesis pathway limits ribosomal RNA transcription causing nucleolar stress in glioblastoma cells. *PLoS Genetics, 16*(11), e1009117.

Lafontaine, D. L. J. (2010). A 'Garbage Can' for ribosomes: How eukaryotes degrade their ribosomes. *Trends in Biochemical Sciences, 35*(5), 267–277.

Lafontaine, D. L. J. (2015). Noncoding RNAs in eukaryotic ribosome biogenesis and function. *Nature Structural & Molecular Biology, 22*(1), 11–19.

LaRiviere, F. J., Cole, S. E., Ferullo, D. J., & Moore, M. J. (2006). A late-acting quality control process for mature eukaryotic rRNAs. *Molecular Cell, 24*(4), 619–626.

Lawrence, I. R., Sutton, E. C., Bhaskar, S., & Baserga, S. J. (2026). Signaling to make human ribosomes: Connections between the cytoplasm and the nucleolus. *Molecular Cell, 86*(3), 469–490.

Liggett, L. A., & Sankaran, V. G. (2020). Unraveling hematopoiesis through the lens of genomics. *Cell, 182*(6), 1384–1400.

Li, G., Shen, J., Cao, J., et al. (2018). Alternative splicing of human telomerase reverse transcriptase in gliomas and its modulation mediated by CX-5461. *Journal of Experimental & Clinical Cancer Research: CR, 37*(1), 78.

Lindström, M. S., Bartek, J., & Maya-Mendoza, A. (2022). p53 at the crossroad of DNA replication and ribosome biogenesis stress pathways. *Cell Death and Differentiation, 29*(5), 972–982.

Liu, Y., & Karlsson, S. (2024). Perspectives of current understanding and therapeutics of diamond-blackfan anemia. *Leukemia: Official Journal of the Leukemia Society of America, Leukemia Research Fund, U. K, 38*(1), 1–9.

Liu, Y.-C., Li, F., Handler, J., et al. (2008). Global regulation of nucleotide biosynthetic genes by c-Myc. *PLoS One, 3*(7), e2722.

Li, W., Zhang, J., & Liu, Q. (2025). Kin17 promotes rDNA transcription, ribosomal biogenesis, and cortical lamination. *EMBO Reports, 26*(17), 4283–4311.

Lv, J., Wang, S., Liu, T., et al. (2025). Nutrient stress diverts RRN3 from rRNA transcription to alternative polyadenylation of autophagy mRNAs in ovarian cancer. *Cell Death & Disease, 16*(1), 849.

Lv, K., Gong, C., Antony, C., et al. (2021). HectD1 controls hematopoietic stem cell regeneration by coordinating ribosome assembly and protein synthesis. *Cell Stem Cell, 28*(7), 1275–1290.e9.

Makhale, A., Nanayakkara, D., Raninga, P., Khanna, K. K., & Kalimutho, M. (2021). CX-5461 enhances the efficacy of APR-246 via induction of DNA damage and replication stress in triple-negative breast cancer. *International Journal of Molecular Sciences, 22*(11), https://doi.org/10.3390/ijms22115782.

Mars, J.-C., Tremblay, M. G., Valere, M., et al. (2020). The chemotherapeutic agent CX-5461 irreversibly blocks RNA Polymerase I initiation and promoter release to cause nucleolar disruption, DNA damage and cell inviability. *NAR Cancer, 2*(4), zcaa032.

Mayer, C., Bierhoff, H., & Grummt, I. (2005). The nucleolus as a stress sensor: JNK2 inactivates the transcription factor TIF-IA and down-regulates rRNA synthesis. *Genes & Development, 19*(8), 933–941.

Mayer, C., Zhao, J., Yuan, X., & Grummt, I. (2004). mTOR-dependent activation of the transcription factor TIF-IA Links rRNA synthesis to nutrient availability. *Genes & Development, 18*(4), 423–434.
McCann, K. L., Kavari, S. L., Burkholder, A. B., Phillips, B. T., & Hall, T. M. T. (2020). H/ACA snoRNA levels are regulated during stem cell differentiation. *Nucleic Acids Research, 48*(15), 8686–8703.
McKnight, S. L., & Miller, O. L., Jr (1976). Ultrastructural patterns of RNA synthesis during early embryogenesis of drosophila melanogaster. *Cell, 8*(2), 305–319.
Menzies, R. A., & Gold, P. H. (1972). The apparent turnover of mitochondria, ribosomes and sRNA of the brain in young adult and aged rats. *Journal of Neurochemistry, 19*(7), 1671–1683.
Menzies, R. A., Mishra, R. K., & Gold, P. H. (1972). The turnover of ribosomes and soluble RNA in a variety of tissues of young adult and aged rats. *Mechanisms of Ageing and Development, 1*(January), 117–132.
Milenkovic, I., Cruciani, S., Llovera, L., et al. (2025). Epitranscriptomic rRNA fingerprinting reveals tissue-of-origin and tumor-specific signatures. *Molecular Cell, 85*(1), 177–190.e7.
Milenkovic, I., & Novoa, E. M. (2025). Dynamic rRNA modifications as a source of ribosome heterogeneity. *Trends in Cell Biology, 35*(7), 604–614.
Miller, G., Panov, K. I., Friedrich, J. K., Trinkle-Mulcahy, L., Lamond, A. I., & Zomerdijk, J. C. (2001). hRRN3 Is essential in the SL1-mediated recruitment of RNA Polymerase I to rRNA gene promoters. *The EMBO Journal, 20*(6), 1373–1382.
Miller, O. L., J., & Beatty, B. R. (1969). *Visualization of nucleolar genes. Science (New York, N.Y.), 164*, 955–957.
Miller, S. C., MacDonald, C. C., Kellogg, M. K., Karamysheva, Z. N., & Karamyshev, A. L. (2023). Specialized ribosomes in health and disease. *International Journal of Molecular Sciences, 24*(7), https://doi.org/10.3390/ijms24076334.
Mills, E. W., & Green, R. (2017). Ribosomopathies: There's strength in numbers. *Science (New York, N. Y.), 358*(6363), https://doi.org/10.1126/science.aan2755.
Misiaszek, A. D., Girbig, M., Grötsch, H., et al. (2021). Cryo-EM structures of human RNA Polymerase I. *Nature Structural & Molecular Biology, 28*(12), 997–1008.
Montanaro, L., Treré, D., & Derenzini, M. (2008). Nucleolus, ribosomes, and cancer. *The American Journal of Pathology, 173*(2), 301–310.
Moss, T., Langlois, F., Gagnon-Kugler, T., & Stefanovsky, V. (2007). A housekeeper with power of attorney: The rRNA genes in ribosome biogenesis. *Cellular and Molecular Life Sciences: CMLS, 64*(1), 29–49.
Moss, T., Mars, J.-C., Tremblay, M. G., & Sabourin-Felix, M. (2019). The chromatin landscape of the ribosomal RNA genes in mouse and human. *Chromosome Research: An International Journal on the Molecular, Supramolecular and Evolutionary Aspects of Chromosome Biology, 27*(1-2), 31–40.
Muhar, M., Ebert, A., Neumann, T., et al. (2018). SLAM-Seq defines direct gene-regulatory functions of the BRD4-MYC axis. *Science (New York, N.Y.), 360*, 800–805.
Mukherjee, A., & Sasikala, W. D. (2013). Drug-DNA intercalation: From discovery to the molecular mechanism. *Advances in Protein Chemistry and Structural Biology, 92*, 1–62.
Narla, A., & Ebert, B. L. (2010). Ribosomopathies: Human disorders of ribosome dysfunction. *Blood, 115*(16), 3196–3205.
Natchiar, S. K., Myasnikov, A. G., Kratzat, H., Hazemann, I., & Klaholz, B. P. (2017). Visualization of chemical modifications in the human 80S ribosome structure. *Nature, 551*(7681), 472–477.
Negi, S. S., & Brown, P. (2015). rRNA synthesis inhibitor, CX-5461, activates ATM/ATR pathway in acute lymphoblastic leukemia, arrests cells in G2 phase and induces apoptosis. *Oncotarget, 6*(20), 18094–18104.

Nobelen, S. V. D., Rosa-Garrido, M., Leers, J., et al. (2010). CTCF regulates the local epigenetic state of ribosomal DNA repeats. *Epigenetics & Chromatin, 3*(1), 19.
Olson, O. C., Kang, Y.-A., & Passegué, E. (2020). Normal hematopoiesis is a balancing act of self-renewal and regeneration. *Cold Spring Harbor Perspectives in Medicine, 10*(12), https://doi.org/10.1101/cshperspect.a035519.
Orkin, S. H., & Zon, L. I. (2008). Hematopoiesis: An evolving paradigm for stem cell biology. *Cell, 132*(4), 631–644.
Pan, M., Wright, W. C., Chapple, R. H., et al. (2021). The chemotherapeutic CX-5461 primarily targets TOP2B and exhibits selective activity in high-risk neuroblastoma. *Nature Communications, 12*(1), 6468.
Pan, Y.-H., Shan, L., Zhang, Y.-Y., et al. (2025). Pre-rRNA spatial distribution and functional organization of the nucleolus. *Nature, 646*(8083), 227–235.
Paralkar, V. R. (2024). Transcription factor regulation of ribosomal RNA in hematopoiesis. *Current Opinion in Hematology, 31*(4), 199–206.
Parks, M. M., Kurylo, C. M., Dass, R. A., et al. (2018). Variant ribosomal RNA alleles are conserved and exhibit tissue-specific expression. *Science Advances, 4*(2), eaao0665.
Pastore, S., Wacheul, L., Lehmann, L., et al. (2025). Mapping human Pre-rRNA processing and modification at single nucleotide resolution using long read nanopore sequencing. *bioRxiv*. https://doi.org/10.1101/2025.03.01.640970 March 3.
Pauli, C., Liu, Y., Rohde, C., et al. (2020). Site-specific methylation of 18S ribosomal RNA by SNORD42A Is required for acute myeloid leukemia cell proliferation. *Blood, 135*(23), 2059–2070.
Pelletier, J., Thomas, G., & Volarević, S. (2018). Ribosome biogenesis in cancer: New players and therapeutic avenues. *Nature Reviews. Cancer, 18*(1), 51–63.
Peltonen, K., Colis, L., Liu, H., et al. (2010). Identification of novel p53 pathway activating small-molecule compounds reveals unexpected similarities with known therapeutic agents. *PLoS One, 5*(9), e12996.
Peltonen, K., Colis, L., Liu, H., et al. (2014). A targeting modality for destruction of RNA Polymerase I that possesses anticancer activity. *Cancer Cell, 25*(1), 77–90.
Perry, R. P., & Kelley, D. E. (1970). Inhibition of RNA synthesis by actinomycin D: Characteristic dose-response of different RNA species. *Journal of Cellular Physiology, 76*(2), 127–139.
Pikaard, C. S., McStay, B., Schultz, M. C., Bell, S. P., & Reeder, R. H. (1989). The xenopus ribosomal gene enhancers bind an essential Polymerase I transcription factor, xUBF. *Genes & Development, 3*(11), 1779–1788.
Poortinga, G., Hannan, K. M., Snelling, H., et al. (2004). MAD1 and c-MYC regulate UBF and rDNA transcription during granulocyte differentiation. *The EMBO Journal, 23*(16), 3325–3335.
Poortinga, G., Wall, M., Sanij, E., et al. (2011). C-MYC coordinately regulates ribosomal gene chromatin remodeling and Pol I availability during granulocyte differentiation. *Nucleic Acids Research, 39*(8), 3267–3281.
Potapova, T. A., Kostos, P., McKinney, S., et al. (2025). Chromosome-specific epigenetic control and transmission of ribosomal DNA arrays in hominidae genomes. *Cell Genomics, 5*(12), 101031.
Potapova, T. A., Unruh, J. R., Conkright-Fincham, J., et al. (2023). Distinct states of nucleolar stress induced by anticancer drugs. *eLife, 12*(December), https://doi.org/10.7554/eLife.88799.
Putnam, C. D., Copenhaver, G. P., Denton, M. L., & Pikaard, C. S. (1994). The RNA Polymerase I transactivator upstream binding factor requires its dimerization domain and high-mobility-group (HMG) box 1 to bend, wrap, and positively supercoil enhancer DNA. *Molecular and Cellular Biology, 14*(10), 6476–6488.

Quin, J., Chan, K. T., Devlin, J. R., et al. (2016). Inhibition of RNA Polymerase I transcription initiation by CX-5461 activates non-canonical ATM/ATR signaling. *Oncotarget, 7*(31), 49800–49818.

Quinodoz, S. A., Jiang, L., Abu-Alfa, A. A., et al. (2025). Mapping and engineering RNA-Driven architecture of the multiphase nucleolus. *Nature, 644*(8076), 557–566.

Rabany, O., Dror, S. B., Arafat, M., et al. (2026). Dynamic rRNA methylation regulates translation in the hematopoietic system and is essential for stem cell fitness. *Blood, 147*(5), 520–533.

Rajput, P., Kumar Shukla, S., & Kumar, V. (2015). The HBx oncoprotein of hepatitis B virus potentiates cell transformation by inducing c-Myc-dependent expression of the RNA Polymerase I transcription factor UBF. *Virology Journal, 12*(April), 62.

Rattner, J. B., Saunders, C., Davie, J. R., & Hamkalo, B. A. (1982). Ultrastructural organization of yeast chromatin. *The Journal of Cell Biology, 93*(1), 217–222.

Raval, A., Sridhar, K. J., Patel, S., Turnbull, B. B., Greenberg, P. L., & Mitchell, B. S. (2012). Reduced rRNA expression and increased rDNA promoter methylation in CD34+ cells of patients with myelodysplastic syndromes. *Blood, 120*(24), 4812–4818.

Rieger, M. A., & Schroeder, T. (2012). Hematopoiesis. *Cold Spring Harbor Perspectives in Biology, 4*(12), https://doi.org/10.1101/cshperspect.a008250.

Roca-Portoles, A., Rodriguez-Blanco, G., Sumpton, D., et al. (2020). Venetoclax causes metabolic reprogramming independent of BCL-2 inhibition. *Cell Death & Disease, 11*(8), 616.

Rodriguez-Algarra, F., Evans, D. M., & Rakyan, V. K. (2024). Ribosomal DNA copy number variation associates with hematological profiles and renal function in the UK biobank. *Cell Genomics, 4*(6), 100562.

Rossetti, S., Wierzbicki, A. J., & Sacchi, N. (2016). Mammary epithelial morphogenesis and early breast cancer. Evidence of involvement of basal components of the RNA Polymerase I transcription machinery. *Cell Cycle (Georgetown, Tex.), 15*, 2515–2526.

Ruggero, D., & Pandolfi, P. P. (2003). Does the ribosome translate cancer? *Nature Reviews. Cancer, 3*(3), 179–192.

Sams, E. I., Feist, V. K., Gray, E. M., et al. (2025). Dynamics of ribosomal RNA transcription and abundance in normal and leukemic hematopoiesis. *bioRxiv*. https://doi.org/10.1101/2025.08.01.668217 August 1.

Sanij, E., Hannan, K. M., Xuan, J., et al. (2020). CX-5461 activates the DNA damage response and demonstrates therapeutic efficacy in high-grade serous ovarian cancer. *Nature Communications, 11*(1), 2641.

Sanij, E., Poortinga, G., Sharkey, K., et al. (2008). UBF levels determine the number of active ribosomal RNA genes in mammals. *The Journal of Cell Biology, 183*(7), 1259–1274.

Santoro, R., Li, J., & Grummt, I. (2002). The nucleolar remodeling complex NoRC mediates heterochromatin formation and silencing of ribosomal gene transcription. *Nature Genetics, 32*(3), 393–396.

Santoro, R., & Grummt, I. (2001). Molecular mechanisms mediating methylation-dependent silencing of ribosomal gene transcription. *Molecular Cell, 8*(3), 719–725.

Sarkar, S. S., Sharma, M., Karmakar, A., et al. (2025). NSH76: A selective inhibitor of RRN3 and RNA Polymerase I transcription with potential for cancer therapy. *Journal of Translational Medicine, 23*(1), 1131.

Shaffer, A. L., Shapiro-Shelef, M., Iwakoshi, N. N., et al. (2004). XBP1, downstream of Blimp-1, expands the secretory apparatus and other organelles, and increases protein synthesis in plasma cell differentiation. *Immunity, 21*(1), 81–93.

Shajani, Z., Sykes, M. T., & Williamson, J. R. (2011). Assembly of bacterial ribosomes. *Annual Review of Biochemistry, 80*, 501–526.

Sharifi, S., & Bierhoff, H. (2018). Regulation of RNA Polymerase I transcription in development, disease, and aging. *Annual Review of Biochemistry, 87*(June), 51–73.

Shcherbik, N., Wang, M., Lapik, Y. R., Srivastava, L., & Pestov, D. G. (2010). Polyadenylation and degradation of incomplete RNA Polymerase I transcripts in mammalian cells. *EMBO Reports, 11*(2), 106–111.

Shi, Z., Fujii, K., Kovary, K. M., et al. (2017). Heterogeneous ribosomes preferentially translate distinct subpools of mRNAs genome-wide. *Molecular Cell, 67*(1), 71–83.e7.

Signer, R. A. J., Magee, J. A., Salic, A., & Morrison, S. J. (2014). Haematopoietic stem cells require a highly regulated protein synthesis rate. *Nature, 509*(7498), 49–54.

Signer, R. A. J., Qi, L., Zhao, Z., et al. (2016). The rate of protein synthesis in hematopoietic stem cells is limited partly by 4E-BPs. *Genes & Development, 30*(15), 1698–1703.

Sinturel, F., Gerber, A., Mauvoisin, D., et al. (2017). Diurnal oscillations in liver mass and cell size accompany ribosome assembly cycles. *Cell, 169*(4), 651–663.e14.

Sloan, K. E., Bohnsack, M. T., & Watkins, N. J. (2013). The 5S RNP couples p53 homeostasis to ribosome biogenesis and nucleolar stress. *Cell Reports, 5*(1), 237–247.

Sloan, K. E., Warda, A. S., Sharma, S., Entian, K.-D., Lafontaine, D. L. J., & Bohnsack, M. T. (2017). Tuning the ribosome: The influence of rRNA modification on eukaryotic ribosome biogenesis and function. *RNA Biology, 14*(9), 1138–1152.

Smetana, K., Šubrtová, H., Jirásková, I., & Rosa, L. (1997). A further study on the incidence of nucleoli in myeloblasts of patients suffering from acute myeloid leukemia. *Hematology (Amsterdam, Netherlands), 2*(5), 379–385.

Sogo, J. M., Ness, P. J., Widmer, R. M., Parish, R. W., & Koller, T. (1984). Psoralen-crosslinking of DNA as a probe for the structure of active nucleolar chromatin. *Journal of Molecular Biology, 178*(4), 897–919.

Solar, S. J., Hebbar, P., Leonardo, de Lima, G., et al. (2025). Origin and evolution of acrocentric chromosomes in human and great apes. *bioRxiv*. https://doi.org/10.64898/2025.12.22.696095 December 23.

Spevak, C. C., Elias, H. K., Kannan, L., et al. (2020). Hematopoietic stem and progenitor cells exhibit stage-specific translational programs via mTOR- and CDK1-dependent mechanisms. *Cell Stem Cell, 26*(5), 755–765.e7.

Stefanovsky, V. Y., Pelletier, G., Bazett-Jones, D. P., Crane-Robinson, C., & Moss, T. (2001). DNA looping in the RNA Polymerase I enhancesome is the result of non-cooperative in-phase bending by two UBF molecules. *Nucleic Acids Research, 29*(15), 3241–3247.

Sudo, H., Nozaki, A., Uno, H., Ishida, Y.-I., & Nagahama, M. (2016). Interaction properties of human TRAMP-like proteins and their role in pre-rRNA 5′ETS turnover. *FEBS Letters, 590*(17), 2963–2972.

Sykes, D. B., Kfoury, Y. S., Mercier, F. E., et al. (2016). Inhibition of dihydroorotate dehydrogenase overcomes differentiation blockade in acute myeloid leukemia. *Cell, 167*(1), 171–186.e15.

Taoka, M., Nobe, Y., Yamaki, Y., et al. (2018). Landscape of the complete RNA chemical modifications in the human 80S ribosome. *Nucleic Acids Research, 46*(18), 9289–9298.

Theophanous, A., Christodoulou, A., Mattheou, C., Sibai, D. S., Moss, T., & Santama, N. (2023). Transcription factor UBF depletion in mouse cells results in downregulation of both downstream and upstream elements of the rRNA transcription network. *The Journal of Biological Chemistry, 299*(10), 105203.

Thomson, E., Ferreira-Cerca, S., & Hurt, E. (2013). Eukaryotic ribosome biogenesis at a glance. *Journal of Cell Science, 126*(Pt 21), 4815–4821.

Trask, D. K., & Muller, M. T. (1988). Stabilization of type i topoisomerase-DNA covalent complexes by actinomycin D. *Proceedings of the National Academy of Sciences of the United States of America, 85*(5), 1417–1421.

Tuan, J. C., Zhai, W., & Comai, L. (1999). Recruitment of TATA-binding protein-TAFI complex SL1 to the human ribosomal DNA promoter is mediated by the carboxy-terminal activation domain of upstream binding factor (UBF) and is regulated by UBF phosphorylation. *Molecular and Cellular Biology, 19*(4), 2872–2879.

Uckelmann, H. J., Haarer, E. L., Takeda, R., et al. (2023). Mutant NPM1 directly regulates oncogenic transcription in acute myeloid leukemia. *Cancer Discovery, 13*(3), 746–765.

Umeda, M., Ma, J., Huang, B. J., et al. (2022). Integrated genomic analysis identifies UBTF tandem duplications as a recurrent lesion in pediatric acute myeloid leukemia. *Blood Cancer Discovery, 3*(3), 194–207.

Valleron, W., Laprevotte, E., Gautier, E.-F., et al. (2012). Specific small nucleolar RNA expression profiles in acute leukemia. *Leukemia: Official Journal of the Leukemia Society of America, Leukemia Research Fund, U. K, 26*(9), 2052–2060.

Villa, E., Ali, E. S., Sahu, U., & Ben-Sahra, I. (2019). Cancer cells tune the signaling pathways to empower de novo synthesis of nucleotides. *Cancers, 11*(5), https://doi.org/10.3390/cancers11050688.

Voit, R., & Grummt, I. (2001). Phosphorylation of UBF at serine 388 Is required for interaction with RNA Polymerase I and activation of rDNA transcription. *Proceedings of the National Academy of Sciences of the United States of America, 98*(24), 13631–13636.

Voit, R., Hoffmann, M., & Grummt, I. (1999). Phosphorylation by G1-specific Cdk-cyclin complexes activates the nucleolar transcription factor UBF. *The EMBO Journal, 18*(7), 1891–1899.

Wang, H., Zhang, Z., Han, C., et al. (2025). SNORD113-114 cluster maintains haematopoietic stem cell self-renewal via orchestrating the translation machinery. *Nature Cell Biology, 27*(2), 246–261.

Wang, M., & Lemos, B. (2017). Ribosomal DNA copy number amplification and loss in human cancers is linked to tumor genetic context, nucleolus activity, and proliferation. *PLoS Genetics, 13*(9), e1006994.

Wang, M., & Attardi, L. D. (2022). A balancing Act: p53 activity from tumor suppression to pathology and therapeutic implications. *Annual Review of Pathology, 17*(January), 205–226.

Wang, M., & Pestov, D. G. (2011). 5′-End surveillance by Xrn2 acts as a shared mechanism for mammalian Pre-rRNA maturation and decay. *Nucleic Acids Research, 39*(5), 1811–1822.

Warner, W. A., Spencer, D. H., Trissal, M., et al. (2018). Expression profiling of snoRNAs in normal hematopoiesis and AML. *Blood Advances, 2*(2), 151–163.

Weinstein, H. N. W., Hu, K., Fish, L., et al. (2024). RPL22 Is a tumor suppressor in MSI-High cancers and a splicing regulator of MDM4. *Cell Reports, 43*(8), 114622.

Wei, T., Najmi, S. M., Liu, H., et al. (2018). Small-molecule targeting of RNA Polymerase I activates a conserved transcription elongation checkpoint. *Cell Reports, 23*(2), 404–414.

Wilson, A., Laurenti, E., Oser, G., et al. (2008). Hematopoietic stem cells reversibly switch from dormancy to self-renewal during homeostasis and repair. *Cell, 135*(6), 1118–1129.

Wolf, T., Jin, W., Zoppi, G., et al. (2020). Dynamics in protein translation sustaining T cell preparedness. *Nature Immunology, 21*(8), 927–937.

Woods, S. J., Hannan, K. M., Pearson, R. B., & Hannan, R. D. (2015). The nucleolus as a fundamental regulator of the p53 response and a new target for cancer therapy. *Biochimica et Biophysica Acta, 1849*(7), 821–829.

Woolford, J. L., Jr, & Susan, J. B. (2013). Ribosome biogenesis in the yeast saccharomyces cerevisiae. *Genetics, 195*(3), 643–681.

Woolnough, J. L., Atwood, B. L., Liu, Z., Zhao, R., & Giles, K. E. (2016). The regulation of rRNA gene transcription during directed differentiation of human embryonic stem cells. *PLoS One, 11*(6), e0157276.

Wyant, G. A., Abu-Remaileh, M., Frenkel, E. M., et al. (2018). *NUFIP1 Is a ribosome receptor for starvation-induced ribophagy. Science (New York, N.Y.), 360*, 751–758.

Xiao, Y., Pan, J., Fan, J., et al. (2026). RNA helicase DDX21 sustains hematopoietic stem cell homeostasis by preventing nucleolar stress. *Cell Investigation, 2*(1), 100060.

Xu, H., Antonio, M. D., McKinney, S., et al. (2017). CX-5461 Is a DNA G-quadruplex stabilizer with selective lethality in BRCA1/2 deficient tumours. *Nature Communications, 8*(February), 14432.

Yang, Y.-M., & Karbstein, K. (2024). Ribosome assembly and repair. *Annual Review of Cell and Developmental Biology, 40*(1), 241–264.

Yan, S., Xuan, J., Brajanovski, N., et al. (2021). The RNA Polymerase I transcription inhibitor CX-5461 cooperates with Topoisomerase 1 inhibition by enhancing the DNA damage response in homologous recombination-proficient high-grade serous ovarian cancer. *British Journal of Cancer, 124*(3), 616–627.

Yoo, D., Rhie, A., Hebbar, P., et al. (2025). Complete sequencing of ape genomes. *Nature, 641*(8062), 401–418.

Young, D. W., Hassan, M. Q., Pratap, J., et al. (2007). Mitotic occupancy and lineage-specific transcriptional control of rRNA genes by Runx2. *Nature, 445*(7126), 442–446.

Yuan, X., Zhao, J., Zentgraf, H., Hoffmann-Rohrer, U., & Grummt, I. (2002). Multiple Interactions between RNA Polymerase I, TIF-IA and TAF(I) subunits regulate pre-initiation complex assembly at the ribosomal gene promoter. *EMBO Reports, 3*(11), 1082–1087.

Zentner, G. E., Balow, S. A., & Scacheri, P. C. (2014). *Genomic characterization of the mouse ribosomal DNA locus. G3 (Bethesda, Md.), 4*, 243–254.

Zentner, G. E., Saiakhova, A., Manaenkov, P., Adams, M. D., & Scacheri, P. C. (2011). Integrative genomic analysis of human ribosomal DNA. *Nucleic Acids Research, 39*(12), 4949–4960.

Zhai, W., & Comai, L. (2000). Repression of RNA Polymerase I transcription by the tumor suppressor p53. *Molecular and Cellular Biology, 20*(16), 5930–5938.

Zhao, J., Yuan, X., Frödin, M., & Grummt, I. (2003). ERK-dependent phosphorylation of the transcription initiation factor TIF-IA Is required for RNA Polymerase I transcription and cell growth. *Molecular Cell, 11*(2), 405–413.

Zhou, F., Aroua, N., Liu, Y., et al. (2023). A dynamic rRNA ribomethylome drives stemness in acute myeloid leukemia. *Cancer Discovery, 13*(2), 332–347.

Zhou, F., Liu, Y., Rohde, C., et al. (2017). AML1-ETO requires enhanced C/D box snoRNA/RNP formation to induce self-renewal and leukaemia. *Nature Cell Biology, 19*(7), 844–855.

Zisi, A., Bartek, J., & Lindström, M. S. (2022). Targeting ribosome biogenesis in cancer: Lessons learned and way forward. *Cancers, 14*(9), https://doi.org/10.3390/cancers14092126.

CHAPTER TWO

Regulation of erythroid development by iron metabolism

Muhammad Ishfaq and Yvette Y. Yien*
Division of Classical Hematology, Department of Medicine, Pittsburgh Heart Lung and Blood Vascular Medicine Institute, University of Pittsburgh, Pittsburgh, PA, United States
*Corresponding author. e-mail address: yieny@pitt.edu

Contents

Abstract

Erythropoiesis is a highly complex process involving a highly coordinated set of cellular processes, culminating with extrusion of the nucleus resulting in formation of a mature erythrocyte. The main function of erythroid cells is to synthesize hemoglobin, which utilizes most of the iron in the body. As such, erythropoiesis is a key driver of systemic iron metabolism. Iron metabolism also plays a central role in erythropoietic development and is essential for processes such as proliferation, cytoskeletal remodeling and nuclear condensation that are required for erythropoietic development. As such, defects in iron metabolism cause a wide range of developmental defects in vertebrate models of hematopoiesis. Recent studies have uncovered a treasure trove of mechanisms by which iron sensing and utilization interacts with the erythroid developmental process. This chapter surveys the mechanisms by which iron metabolism facilitates terminal erythroid differentiation and suggests areas for further investigation based on studies of iron-dependent pathways in non-erythroid cells.

Current Topics in Developmental Biology, Volume 170
ISSN 0070-2153, https://doi.org/10.1016/bs.ctdb.2026.06.005

1. Introduction

Iron plays a fundamental role in biology because of its redox potential which allows it to readily acquire either the Fe^{2+} or Fe oxidation states. In the form of iron-containing cofactors, this property supports ubiquitous cellular processes including oxidative phosphorylation, redox reactions, and oxygen transport (Acín-Pérez et al., 2008; Federico et al., 2022; Ruskoski & Boal, 2021; Socolovsky, 2024; Vallières et al., 2024). Yet, the same redox activity that makes iron indispensable also makes it toxic: unbound or mismanaged iron catalyzes the formation of reactive oxygen species (ROS) through Fenton chemistry, leading to oxidative injury and cellular damage. To balance these properties, organisms have evolved sophisticated mechanisms that maintain tight control over iron absorption, storage, and utilization. Iron metabolism pathways provide important developmental and metabolic cues to all cells in the body (Barrientos et al., 2015; Braymer et al., 2024; Golub et al. 2006a, 2006b, 2007; Hoffbrand et al., 1976; Matak et al., 2016; Puig et al., 2017; Xavier-Ferrucio et al., 2019; Xu et al., 2015) and are central to the function and development of erythroid cells, whose main function is to synthesize heme and hemoglobin for oxygen transport.

Disruption of iron metabolism genes cause broadly heterogenous erythroid phenotypes (Brownlie et al., 1998; Donovan et al., 2000, 2005; Perfetto et al., 2025a, 2025b; Shaw et al., 2006; Troadec et al., 2011; Wingert et al., 2005; Yien et al., 2014), indicating that iron metabolism participates in multiple distinct processes during erythroid differentiation. There is also extensive crosstalk between systemic iron metabolism and erythroid cell development. Because developing erythroid cells utilize the vast majority of the body's iron, erythropoietic disorders also severely disrupt systemic iron metabolism, causing iron restriction in non-erythroid tissues or iron overload (An et al., 2023; Castro-Mollo et al., 2021; Castruccio Castracani et al., 2025; Kautz et al., 2014; Kremyanskaya et al., 2024; Levy & Ginzburg, 2025). While the centrality of iron in erythroid cell biology and the crosstalk between erythropoiesis and systemic iron metabolism are well established, most studies on erythroid iron have focused on its role in hemoglobinization and its disruption by systemic iron deficiency. In this chapter, we are therefore focusing on the less understood questions on iron metabolism regulates basic cellular processes, other than hemoglobin production, that are essential to erythroid development.

2. Erythroid iron metabolism

Erythropoiesis is the process by which hematopoietic stem cells (HSCs) differentiate into mature erythrocytes via the common myeloid progenitors (CMPs) and megakaryocyte–erythroid progenitors (MEPs) before committing to the erythroid lineage. Committed erythroid cells become progressively more differentiated in a series of events requiring proliferation, changes in cell morphology and cytoskeletal composition, and hemoglobin formation, typically occurring in the late erythroblast populations. In the terminal steps of erythroid maturation, erythroid cells undergo cell cycle arrest, nuclear condensation and finally, enucleation (An et al., 2014; Hu et al., 2013; Koury et al., 2002; Liu et al., 2013). Each erythroid progenitor population is defined by the expression of specific cell surface markers and cellular processes (An & Mohandas, 2011; Hu et al., 2013; Liu et al., 2013) (summarized in Fig. 1). Each of these differentiation steps, from erythroid specification to terminal differentiation, is intricately co-regulated with iron metabolism.

Adult humans make approximately 200 billion erythrocytes each day, requiring 20–25 mg of iron per day. Developing erythroid cells are a major driver of systemic iron homeostasis (Kautz & Nemeth, 2014). To handle this demand, organisms have evolved complex regulatory networks enabling sensing of erythropoietic demand and iron status. Organisms have also evolved parallel mechanisms for erythroid cells to acquire, traffic and process the massive quantities of iron required for terminal differentiation and hemoglobin production. Iron metabolism proteins are tightly incorporated into physical complexes or biochemical pathways that facilitate life-essential processes in erythroid cells.

Iron metabolism is tightly integrated with erythropoietic activity through the hormone erythroferrone (ERFE), which links EPO signaling to hepatic hepcidin (HAMP) regulation (Kautz et al., 2014, 2015). The primary role of ERFE is to suppress hepatic HAMP production, which increases iron efflux from enterocytes and macrophages by stabilizing ferroportin (Kautz et al., 2015). Increased erythropoietic demand, signaled by EPO or hypoxia, stimulates erythroblast ERFE production, which in turn suppresses hepcidin, increasing iron efflux from enterocytes, hepatocytes and macrophages via FPN1 to meet the demands of hemoglobin synthesis (Kautz et al., 2015; Paul et al., 2021). Disruption of this axis results in elevated hepcidin and functional iron deficiency (Fig. 2).

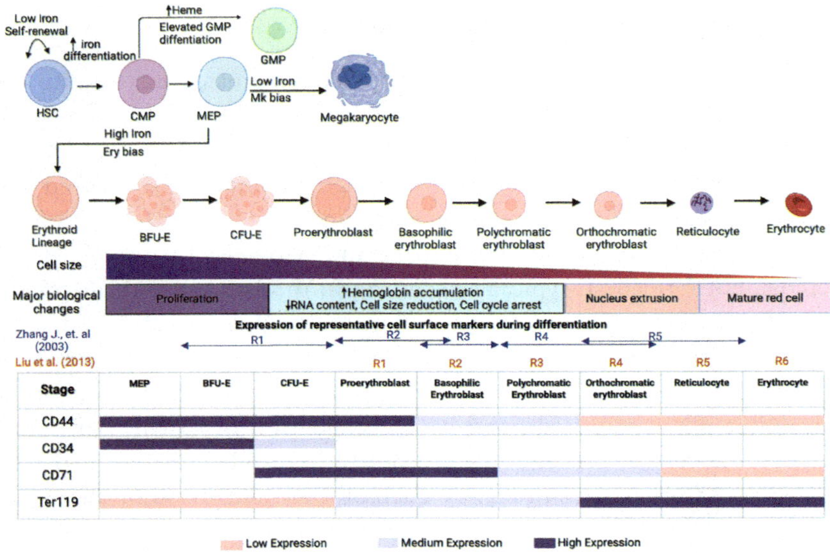

Fig. 1 Overview of erythropoiesis and regulation of erythroid cell fate by iron status. Erythroid cell fate is governed by iron metabolism at multiple points during hematopoietic differentiation. Labile iron levels contribute to self-renewal vs. differentiation fate decisions in the HSC population, while heme levels contribute to CMP differentiation down the MEP vs. granulocytic fate. Within the MEP population, high iron favors erythroid cell fate. During erythroid terminal differentiation, cells undergo a regulated series of cell proliferations, culminating in arrest and nuclear extrusion. During the differentiation process, erythroid cells synthesize large quantities of hemoglobin and undergo morphological changes that are essential for formation of a mature red cell. Erythroid cells at different stages of maturation are morphologically distinct and express specific cell surface markers, enabling one to distinguish specific erythroid populations (Refs. Liu et al., 2013; Zhang et al., 2003).

The EPO-ERFE-HAMP axis is highly responsive to hypoxia, linking erythropoietic drive with increased iron mobilization. Under hypoxic conditions, HIF1 and 2 activate transcription of EPO (Haase, 2010), which in turn, stimulate proliferation, survival and terminal differentiation of early erythroid progenitors. This increases production of ERFE and decreased *Hamp* expression (Liu et al., 2012), driving increased erythroid iron utilization. Hypoxia-inducible factor 2α (HIF2α) enhances intestinal iron absorption and erythropoietic drive by increasing duodenal DMT1 expression under hypoxic conditions, integrating oxygen sensing, iron uptake, and erythroid output (Mastrogiannaki et al., 2013). The ability of hypoxia to increase erythropoietic drive and iron mobilization has recently been leveraged in developing hypoxia inducible factor prolyl hydroxylase

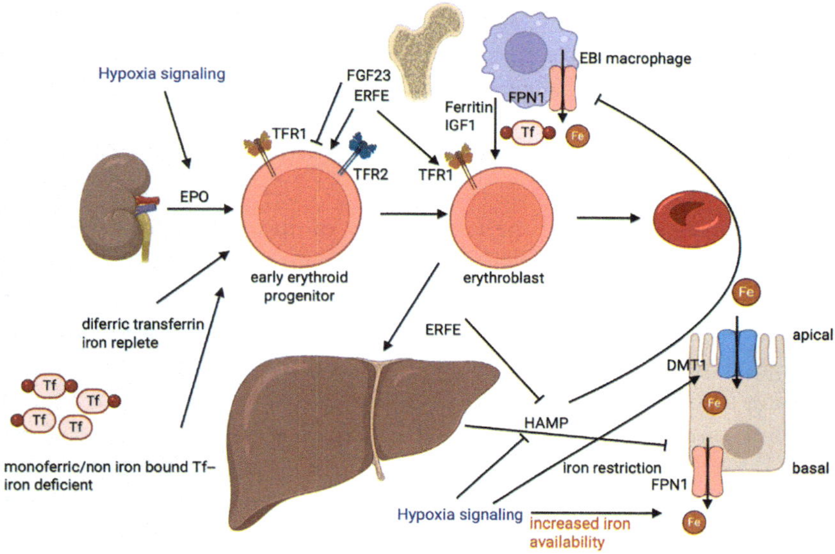

Fig. 2 Systemic iron homeostasis and sensing. Erythroid differentiation is highly responsive to physiological cues that link differentiation stimuli (EPO) with serum iron levels (monoferric or diferric transferrin saturation), hepatic iron sensing and iron acquisition from the gut, as well as iron and growth factor signaling in the bone and bone marrow microenvironment. Hypoxia signaling for instance, increases expression of DMT1 and expression of FPN1 via suppression of hepatic HAMP to increase iron availability concurrent with upregulation of EPO to increase erythropoietic drive. Osteogenic cells and erythroblastic macrophages are also essential sources of growth factors (IGF1, FGF23) and ERFE which signal iron availability to the developing erythroblast; erythroblastic macrophages also supply transferrin bound iron and ferritin to developing erythroid cells.

inhibitor (HIF-PHI) drugs to treat anemias of sickle cell disease and chronic kidney disease (Haider et al., 2023; Hsieh et al., 2007; Nakai et al., 2023; Sharma et al., 2026) (summarized in Fig. 2).

Erythroid maturation is supported by erythroblastic islands (EBIs), specialized bone marrow niches composed of a central macrophage surrounded by developing erythroblasts. Central macrophages supply bioavailable iron through the ferroportin transporter , clear extruded nuclei during enucleation, and provide cytokines such as IGF1 critical for erythroblast survival and terminal differentiation (Li et al., 2019; Manwani & Bieker, 2008). Disruption of macrophage–erythroblast interactions impair maturation and contributes to ineffective erythropoiesis in disorders like β-thalassemia. EBIs therefore exemplify the integration of systemic regulators, including EPO and iron, with local microenvironmental cues to sustain efficient red cell production. In this section, we will discuss the interplay between erythroid lineage commitment and terminal differentiation with iron metabolism with an emphasis on erythroid-specific iron acquisition and processing.

2.1 Overview of iron acquisition and intracellular iron metabolism

Dietary iron is absorbed in the lumenal side of duodenal enterocytes. Non-heme Fe^{3+} is first reduced to Fe^{2+} by duodenal cytochrome b (Dcytb) and imported into enterocytes by divalent metal transporter 1 (DMT1). Within the enterocytes, iron is either stored in ferritin or exported into circulation through ferroportin (FPN1), where it is oxidized by hephaestin to Fe^{3+}, allowing its binding to transferrin (Semenza, 1999). Heme iron is acquired via heme carrier protein 1 (HCP1) (Le Blanc et al., 2012) and degraded by heme oxygenase 1 (HO1) to release Fe^{2+} (An & Mohandas, 2011; Koury et al., 2002). Transport of iron through FPN1 is tightly regulated by hepcidin (HAMP), a peptide hormone that is secreted by hepatocytes. HAMP binds to ferroportin, causing its internalization and degradation, thereby reducing dietary iron uptake or release of iron from storage organs (Billesbølle et al., 2020; Ramey et al., 2010). HAMP expression is increased when organisms are iron loaded or experiencing inflammation (Camaschella et al., 2022). Conversely, its expression is decreased during hypoxia (Haase, 2010) or during increased erythropoietic demand. This occurs by production of erythroferrone by erythroblasts (Kautz et al., 2014) or by the hepatic cytokine FGL1 (Sardo et al., 2024), ensuring that iron supply matches physiological needs. These mechanisms are summarized in Fig. 2.

Most cellular iron is acquired via transferrin binding. Transferrin binds to two Fe^{3+} ions and binds to transferrin receptor 1 (TfR1), expressed on the cell surface. The Tf-TfR1 complex is internalized in an acidic endosome where Fe^{3+} is reduced to Fe^{2+} by STEAP3 (Ohgami et al., 2005). The mechanisms by which iron is exported from the endosome are still hotly debated. One hypothesis proposes that endosomal iron is transported to the mitochondria by direct physical contact (Hamdi et al., 2016; Sheftel et al., 2007). Another hypothesis describes a mechanism by which iron is exported from the endosome to the cytoplasm by DMT1 (Levy et al., 1999), stored in cytosolic ferritin, and subsequently released for heme synthesis by proteolytic degradation of the ferritin "cage" (Patel et al., 2021; Ryu et al., 2017). During the transition from early erythroblast to terminal differentiation, erythroid cells employ specialized mechanisms that upregulate iron acquisition to meet the requirements for hemoglobinization.

Mitochondria are the site where iron-containing cofactors are synthesized, and mitochondrial iron processing is required for incorporation of iron into cytosolic proteins (Shepherd et al., 2023). The vast majority of iron in erythroid cells is used to make heme. Heme synthesis is initiated by synthesis of aminolevulinic acid (ALA) by condensation of glycine and succinyl Co-A, catalyzed by the aminolevulinate synthase (ALAS) enzymes. Here, glycine is imported into the mitochondrial matrix by SLC25A38 (Fernández-Murray et al., 2016; Lunetti et al., 2016) where it is used for ALA synthesis, while succinyl Co-A may be obtained from the Kreb's cycle or via glutamine transmination (Burch et al., 2018). Erythroid cells mainly rely on ALAS2 which is under the control of the IRE/IRP system (reviewed here (Iwai, 2019); Wilkinson & Pantopoulos, 2014; Zhang et al., 2014) and transcriptionally activated by erythroid transcription factors, KLF1 and GATA1. ALA is exported into the cytosol and converted into tetrapyrrole intermediates via a series of enzymatic reactions. In the terminal steps, coproporphyrinogen III (CPgenIII) is transported through the mitochondrial outer membrane, where it is converted to protoporphyrinogen IX (PPgenIX) via CPOX in the intermembrane space. PPgenIX is transported through the mitochondrial inner membrane into the mitochondrial matrix via a process that requires the transmembrane protein TMEM14C (Yien et al., 2014), where it is finally converted to protoporphyrin IX (PPIX) and heme via the PPOX and FECH enzymes, which reside in a physical complex bound to the mitochondrial inner membrane (Medlock et al., 2015). This terminal process is also co-regulated with mitochondrial iron import, which occurs via the mitoferrin transporters (MFRN1 and MFRN2). Erythroid mitochondrial transport largely occurs through erythroid-enriched MFRN1 transporter and requires FAM210B for maximum efficiency (Baumann et al., 2025; Christenson et al., 2018; Shaw et al., 2006; Yien et al., 2018). In differentiating erythroid cells, FAM210B is also required for maximal FECH activity (Yien et al., 2018). Heme export from the mitochondria is thought to occur through the FLVCR1B transporter (Chiabrando et al., 2012). The distribution of heme from the mitochondria to other subcellular compartments and hemoproteins is dependent on heme chaperones TANGO2 (Jayaram et al., 2025; Sun et al., 2022), GAPDH (Jayaram et al., 2025) and PGRMC1 (Piel et al., 2016; Piel Iii et al., 2026) (Fig. 3).

Although most iron in the erythroblast is used for heme synthesis, Fe-S clusters are essential cofactors for DNA replication, mitosis and oxidative phosphorylation. Synthesis of Fe-S clusters is highly complex and more completely reviewed here (Maio & Rouault, 2020). Within the mitochondrial

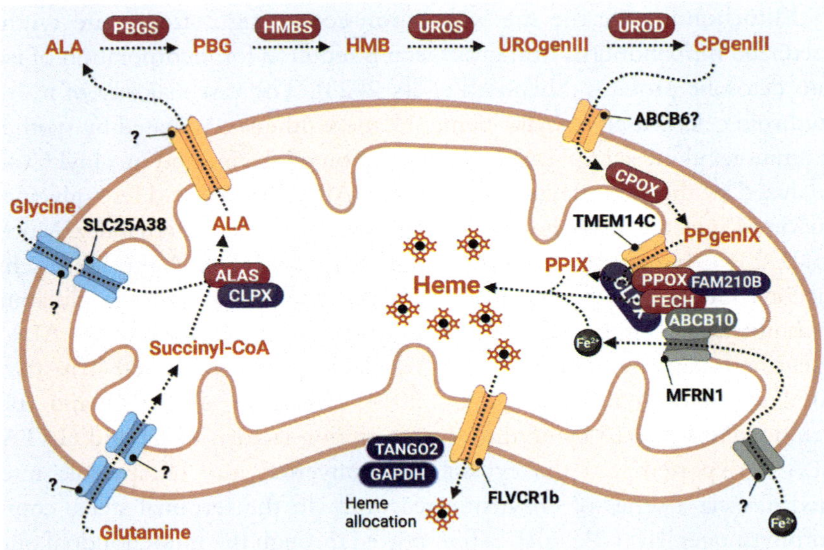

Fig. 3 Schematic of heme metabolism and its regulation. Transporters of heme intermediates are major regulators of heme synthesis, many of which remain unknown. Within the mitochondria, heme synthesis enzymes are tightly regulated in the context of metabolomes where interacting proteins such as CLPX regulate their stability and activity. At the terminal steps of heme synthesis, transporters such as MFRN1 and TMEM14C, along with accessory proteins such as FAM210B and ABCB10, are in a physical complex with the heme synthetic enzymes PPOX and FECH enabling efficient transfer of substrates from one enzyme to the next, minimizing deleterious redox reactions. Heme is then exported into the cytosol by FLVCR1B and incorporated into cytosolic apo-proteins with the help of the TANGO2 and GAPDH chaperones.

matrix, sulfur is removed from cysteine and delivered by NFS1 to ISCU, the mitochondrial Fe-S assembly protein. [2Fe-2S] clusters are assembled on ISCU when iron is provided. Chaperone proteins then deliver [2Fe−2S] clusters to recipient proteins. Alternatively, [2Fe−2S] proteins are transferred to secondary carriers such as GLRX5, which is essential for erythroid heme synthesis, which then transfer them to recipient proteins (Wingert et al., 2005). Mitochondrial Fe-S clusters may then be exported to the cytosol via ABCB7 (Braymer et al., 2024; Cavadini et al., 2007; Qi et al., 2014). Other Fe-S biogenesis factors may acquire [2Fe−2S] clusters from ISCU to assemble [4Fe−4S] clusters, which are the most abundant species in Fe-S binding proteins. Erythroid synthesis is highly dependent on the Fe-S machinery. ALAS2 is regulated by an iron response element (IRE) in its 5′UTR. When iron is limiting, the iron response element binding protein (IRP2) binds to this

IRE, suppressing its translation. Under iron replete conditions, IRP2 is degraded by by FXBL5 (Vashisht et al., 2009); independently of FXBL5, IRP2 also binds to Fe-S, precluding binding to its IRE (Terzi et al., 2021). The release of IRP2 from the ALAS2 IRE subsequently translation of ALAS2 when erythroid cells are iron replete. In this manner, ALAS2 expression is coupled with cellular iron availability and Fe-S synthesis. More directly, mammalian ALAD and FECH are [4Fe−4S] (Liu et al., 2020) and [2Fe−2S]-binding (Crooks et al., 2010; Ferreira et al., 1994; Schneider-Yin et al., 2000) heme synthesis enzymes respectively, underscoring the intricate coregulation of heme and Fe-S cluster synthesis. Mutations in *GLRX5* have caused defects in ALAS2 and FECH function in human patients (Daher et al., 2019). More broadly, the cytosol and nucleus also contain their own Fe-S assembly machinery, and it is thought that Fe-S clusters are incorporated into their apoproteins in the subcellular compartment where they reside within the cell (reviewed here (Maio & Rouault, 2020); Braymer & Lill, 2017). However, prior to assembly in cytosolic and nuclear Fe-S clusters, iron must still be processed within the mitochondria (Shepherd et al., 2023).

Within the cell, iron is stored in cytosolic or mitochondrial ferritin. Cytosolic ferritin is composed of heavy (H) and light (L) chains, and serves as the main intracellular iron storage complex, capable of sequestering approximately 4500 iron atoms in a process requiring PCBP1 and 2 for iron delivery (Ryu et al., 2017). During iron deficiency, ferritinophagy—a selective autophagic process mediated by nuclear receptor coactivator 4 (NCOA4)—transports ferritin to lysosomes, where Fe^{2+} is released for cellular use (Dowdle et al., 2014; Patel et al., 2021; Ryu et al., 2017). Factors that determine whether iron enters the Fe-S or heme synthesis pathways, or is sequestered in ferritin, are still completely unknown but given the highly cell-specific nature of iron requirements, likely to be critical for cellular function.

2.2 Iron regulatory mechanisms in erythropoiesis

MEPs give rise to early erythroblasts, first the "burst forming units-erythroid" (BFU-Es) and then "colony forming units-erythroid" (CFU-Es). These erythroblasts progressively mature through a series of developmental stages that are marked by expression of specific cell surface markers (Liu et al., 2013; Wong et al., 2011) (Fig. 1). Expression of CD71 or TFR1 is markedly upregulated during the BFU-E to CFU-E (R1 to R2) transition, denoting a significant increase in iron demand. During the CFU-E (TER119 negative; R2 in Fig. 1) to the late erythroblast transition

(TER119 positive; R3), expression of globin, heme synthesis, erythroid membrane protein and erythroid transcription factor encoding genes are significantly upregulated (Chen et al., 2013; Chung et al., 2015; Wong et al., 2011; Yien et al., 2014, 2018; Zhang et al., 2003). Many of these genes are under the transcriptional control of the erythropoietic regulators, KLF1 and GATA1 (Barbarani et al., 2019; Bieker & Philipsen, 2024; Doty et al., 2019; El Hoss et al., 2026; Gnanapragasam & Bieker, 2017; Liao & Bresnick, 2022, 2024). These developmental transitions are intricately intertwined with cellular iron acquisition and utilization. In this section, we discuss the developmental regulation of iron metabolism during erythroid differentiation.

2.2.1 Erythroid iron acquisition and transport

Erythroid-specific mechanisms that couple iron acquisition and utilization with the needs of the developing erythrocyte become increasingly important in the CFU-E to late erythroblast transition. Genes coding for proteins such as SNX3, which facilitate the recycling of TFR1 receptors to the cell surface, increasing the rate of iron acquisition, are transcriptionally upregulated at this stage (Chen et al., 2013). At the same time, iron acquisition must be kept in check. Excessively rapid endosomal recycling accelerates erythroid development and increases the proportion of iron that is trafficked to the heme synthesis pathway in the absence of normal developmental stimuli. This results in erythroid cells that exhibit markers of premature senescence and membrane fragility (Liu et al., 2026). The "brake" in erythroid iron acquisition and heme synthesis is TMEM187, a Golgi transmembrane protein (Liu et al., 2026). These data indicate that iron itself triggers the processes that facilitate erythroid development (beyond regulation of hemoglobin synthesis), and the kinetics of iron acquisition must be tightly controlled to couple iron import with the induction of developmental pathways that are required for proper erythroid maturation. In the absence of the TMEM187 "brake", there is increased entry of iron into the heme synthesis pathway, indicating the existence of mechanisms that preferentially direct iron into heme synthesis at the expense of Fe-S cluster synthesis during erythroid differentiation.

Mitochondrial iron import primarily occurs via MFRN1, an inner membrane protein which is predominantly expressed in erythroid cells and transports Fe^{2+} into the mitochondrial matrix for heme and Fe-S cluster synthesis (Christenson et al., 2018; Shaw et al., 2006). Within the inner mitochondrial membrane, MFRN1 assembles with the ATP-binding cassette transporter ABCB10 and FECH which both stabilizes MFRN1 and enables efficient transfer of iron to heme synthesis during terminal erythropoiesis (Chen et al. 2009a, 2010). Another FECH and PPOX binding protein, FAM210B, is required for maximal mitochondrial iron import in concert with MFRN1 during terminal erythropoiesis, particularly during erythropoietic stress (Perfetto et al., 2025b; Yien et al., 2018).

2.2.2 Mechanisms of iron sensing during erythropoiesis

Erythropoiesis is a resource-intensive process that can compete with the requirements of non-erythroid cells when the body is iron restricted. Organisms have therefore evolved mechanisms by which early erythroblasts can sense iron status in their environment, avoiding committing resources to terminal erythroid differentiation during iron deficiency, hence decreasing the iron demand on the body and preventing ineffective erythropoiesis.

Erythroid cells sense iron saturation of dimeric transferrin, where monoferric transferrin signals iron deficiency and alters serum EPO levels, via a mechanism which discriminates according to directionality of the lobes (Parrow et al., 2019). In particular, lack of iron binding on the N-lobe of transferrin exacerbates the iron-restricted response (relative to an unoccupied C-lobe) and decreases EPO responsiveness of the erythroblast (Parrow et al., 2019). Although most erythroid iron uptake occurs via TFR1, TFR2 plays a role in trafficking transferrin to the lysosome, erythroid iron sensing and facilitating interactions between the lysosome and mitochondria (Khalil et al., 2017a). While TFR2 is better understood for its role in extracellular iron sensing, it plays an important role in regulating surface expression of EPOR in erythroid cells during iron deficiency. Erythroid TFR2 expression is destabilized during iron deficiency. Iron deficiency decreases aconitase activity and isocitrate levels, which in turn destabilizes the expression of the receptor trafficking regulator Scribble by targeting the TFR2-Scribble trafficking vesicles to lysozomes. This prevents Scribble from localizing to EPOR-containing transport vesicles and trafficking them to the cell surface. When iron is replete, Scribble is allowed to facilitate EPOR presentation at the cell surface, allowing proliferation and differentiation of the early erythroblast (Khalil et al., 2017b) (Fig. 4).

Erythropoietin (EPO)-driven survival, proliferation and maturation prior to hemoglobin synthesis is an important developmental checkpoint for early erythroid cells. EPO responsiveness in early erythroblasts requires adequate

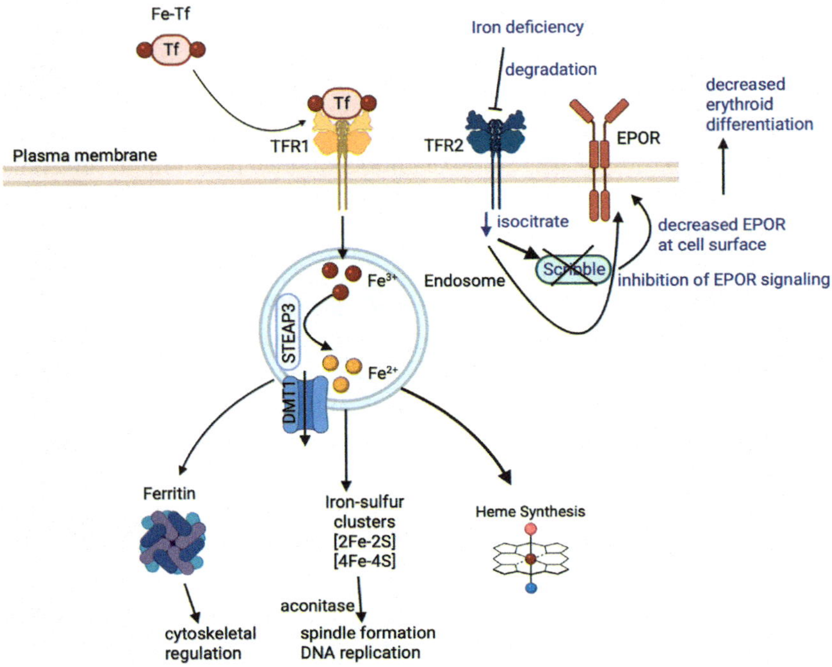

Fig. 4 Iron acquisition and sensing. Erythroid cells acquire the vast majority of their iron via a TFR1 mediated mechanism which endocytoses Fe-bound transferrin. TFR2 plays a secondary role in iron acquisition. During iron deficiency, which decreases isocitrate levels, TFR2, in concert with Scribble, an EPO receptor trafficking protein, are targeted for lysosomal degradation, when then decreases EPO receptor presentation at the cell surface. This causes a resultant decrease in EPO receptor signaling and decreased erythroid proliferation and survival, linking iron deficiency with sensitivity to erythroid differentiation stimuli. Iron that is acquired via TFR1 is reduced in an acidified endosome via STEAP proteins and is incorporated into ferritin, iron-sulfur clusters or heme, all of which play important roles in erythroid differentiation.

iron uptake, which is sensed by the iron responsive proteins 1 and 2 (IRP1 and 2). When iron is replete, IRP1 functions as a cytosolic aconitase, but converts to an RNA binding protein during iron deficiency that regulates RNA stability and translation of proteins involved in iron response. Aconitase enzymatic activity and isocitrate levels therefore function as key signals for iron sufficiency in the developing erythroblast. While isocitrate itself did not affect intracellular iron, decreased aconitase levels in iron deficient erythroblasts promoted PKCα/β hyperactivation, which was associated with inhibition of EPO signaling activity. Addition of isocitrate in iron-deficient erythroblasts decreased PKCα/β hyperactivation, which increased EPO sensitivity, heme synthesis and cytoskeletal stability and bypassed an important iron restriction signal (Bullock et al., 2010; Goldfarb et al., 2021) (Summarized in Fig. 4).

Lastly, EPO regulates erythroid mitochondrial iron utilization and sensing via its phosphorylation of the heme synthesis enzyme FECH (Chung et al., 2017) and activating transcription of the mitochondrial protein FAM210B which facilitates maximal mitochondrial iron import during terminal erythroid differentiation (Perfetto et al., 2025b; Yien et al., 2018). In the absence of FAM210B, systemic iron sensing is dysregulated, and organisms are less able to respond to erythropoietic stress (Perfetto et al., 2025b). Further studies on FAM210B are likely to shed light on the functional interaction between mitochondrial status and systemic iron homeostasis during high erythropoietic demand.

2.2.3 Heme regulation during erythropoiesis

ALAS2 is the erythroid-specific, rate-limiting enzyme that catalyzes the committed step of heme synthesis. Its mRNA contains a 5′ iron-responsive element (IRE) that permits translational repression by iron regulatory proteins IRP1 and IRP2 during iron deficiency, limiting heme synthesis during erythroid iron restriction (Cotter et al., 1994; Iwai, 2019; Wilkinson & Pantopoulos, 2014; Zhang et al., 2014). The ALAS enzymes require a covalently attached pyridoxal-5′-phosphate (PLP) cofactor (Astner et al., 2005; Bishop et al., 1990; Hunter & Ferreira, 2011). Import of mitochondrial PLP occurs through SLC25A38 (Pena et al., 2025), which may indicate a role for SLC25A38 in PLP delivery to ALAS2 and control of ALAS2 in addition to its role in mitochondrial glycine import. Mitochondrial heme inhibits ALAS2 activity, establishing a heme-dependent checkpoint that prevents porphyrin accumulation and resultant oxidative damage (Chitrakar et al., 2026). ALAS2 stability is regulated by the CLPXP proteolytic complex in a heme dependent manner (Cottle et al., 2024), but presumably with a higher heme threshold than the non-erythroid ALAS1, which is degraded by the LON and CLPXP proteolytic complexes when heme is in excess (Kubota et al., 2016). The erythroid-specific ALAS2 is generally subject to mechanisms that promote its expression under iron replete conditions, consistent with its function in an iron-rich erythroid cell. In contrast, the non-erythroid ALAS1 is degraded when heme levels are high as a negative feedback mechanism.

FECH, the terminal enzyme of the heme synthesis pathway, resides on the inner mitochondrial membrane and catalyzes the insertion of Fe^{2+} into protoporphyrin IX to form heme. Structurally, FECH functions as a homodimer, with each subunit containing a [2Fe–2S] cluster that is essential for protein stability, activity (Crooks et al., 2010; Dailey et al., 1994, 2007) and pH sensing (Shah et al., 2012). The incorporation of the FECH [2Fe−2S] cluster is mediated by the Fe-S assembly protein, GLRX5 (Daher et al., 2019; Ye et al., 2010). The CLPX protein unfoldase plays a role in promoting FECH activity and iron utilization through a mechanism that is still unclear (Rondelli et al., 2021). During erythroid differentiation, FECH relies on several post-transcriptional regulation mechanisms and protein-protein interactions for its maximal function. The RNA binding protein, RBM38, promotes *Fech* mRNA splicing and stability, increasing translation of FECH protein; loss of RBM38 caused both microcytic anemia and protoporphyria (Xie et al., 2025). During terminal erythropoiesis and EPO stimulation (Bullock et al., 2010; Khalil et al., 2017b), FECH is phosphorylated by PKA, which increases its activity and the erythroid cell's capacity for heme synthesis (Chung et al., 2017). FECH function requires intact mitochondria structure and its interactions with the mitochondria contact side and cristae organizing system are essential for its activity (Dietz et al., 2021). FECH interacts with MFRN1 (Chen et al. 2009a, 2010) and FAM210B (Yien et al., 2018), a mitochondrial transmembrane protein in erythroid cells, and these interactions are likely to preferentially direct iron to the heme synthesis pathway during terminal differentiation.

3. Regulation of erythroid differentiation processes by iron metabolism

3.1 Erythroid specification

Iron metabolism is essential for definitive hematopoietic cell fate and lineage determination. In HSCs, high levels of labile iron limit regenerative capacity, reprogram HSC metabolism (Sighinolfi et al., 2026) and bias HSCs toward a megakaryocytic fate (Kao et al., 2024). Conversely, iron deficiency enhances HSC self-renewal at the expense of differentiation (Zhang et al., 2022). Notably, in vitro differentiation of *Tfr1* conditional knockout, iron deficient HSCs, was restored by addition of hemin (Wang et al., 2020a), indicating that conversion to heme-dependent cofactors is essential for iron to regulate HSC cell fate.

HSCs respond to infection by promoting myelopoiesis at the expense of erythropoiesis, and part of this response may result from iron restriction that occurs during infection (Zaretsky et al., 2014). When HSCs and common myeloid progenitors (CMPs) are exposed to infectious stimuli, there is reduced BACH1 and 2 binding to myeloid genes, increasing the proportion of myeloid progenitor cells. This finding was also recapitulated by treatment with exogenous heme (Kato et al., 2018) which functions a pro-inflammatory mediator and is a damage-associated molecular pattern (DAMP) which activates the innate immune system during infection or tissue damage (Martins & Knapp, 2018). Hence in the HSC or CMP context, exposure to excess heme triggers an inflammatory response, biasing development towards a pro-myeloid fate at the expense of erythropoiesis.

At the MEP stage, systemic iron deficiency biases differentiation to the megakaryocyte lineage at the expense of erythroid differentiation (Xavier-Ferrucio et al., 2019). Iron-deficient MEPs sense iron deficiency via TFR2; this in turn decreases ERK phosphorylation and increases megakaryocyte commitment (Xavier-Ferrucio et al., 2019). This is often observed within the clinic, where iron deficiency is associated with thrombocytosis and increased risk of venous thromboembolism (Hung et al., 2025; Talamo et al., 2025). (Summarized in Fig. 1).

3.2 Terminal differentiation

Beyond hemoglobin production, defects in iron metabolism cause highly variable clinical symptoms and phenotypes in model organisms. Most homozygous mutants with iron metabolism defects suffer from embryonic lethality and experience impaired differentiation during the early to late erythroblast transition (Troadec et al., 2011; Yien et al., 2014). Iron metabolism mutant erythroid cells tend to have morphological defects and enlarged nuclei. Systemic iron deficiency causes microcytic anemia in patients (reviewed here (Brugnara, 2003)). This is phenocopied in model organisms with genetic defects in iron acquisition, such as mutations in *Tfr1, Dmt1* and *Fpn1* (Donovan et al., 2000, 2005; Fleming et al., 1997; Wingert et al., 2004). Defects in mitochondrial iron or porphyrin transport on the other hand cause macrocytosis, as observed in genetic mutants of *Tmem14c* (Yien et al., 2014), *Mfrn1* (Perfetto et al., 2025a; Shaw et al., 2006; Troadec et al., 2011), *Alas2* (Brownlie et al., 1998). Other phenotypes where there are differences across mutants include nuclear condensation (Brownlie et al., 1998; Donovan et al., 2002; Fleming et al., 1997; Perfetto et al., 2025a; Yien et al., 2014) erythroid cell number (Perfetto et al., 2025b) and cell cycle (Fu & Richardson, 2007; Patel et al., 2021; Perfetto et al., 2022). In this section, we will review the cellular processes that are associated with terminal erythroid differentiation and outline the associated iron-related regulatory processes.

3.2.1 Cytoskeletal maintenance

During terminal erythropoiesis, progressively differentiated populations of erythroblasts are morphologically distinct due to developmentally regulated expression of erythroid membrane and cytoskeletal proteins (Chen et al., 2009b). As previously mentioned, defects in erythroid iron metabolism also causes dysregulation of erythroid morphology, indicating a poorly understood role for iron in maintenance of the erythroid cytoskeleton. In chickens and mammalian models, the ferritin heavy chain (FTH1) is a microtubule stabilizing protein in erythroid cells (Feick et al., 1991; Goldfarb et al., 2021). Iron restriction in erythroid cells induced centrosomal disorganization and disruption of the microtubule cytoskeleton within 18 h, prior to other developmental phenotypes. This phenotype was not apparent in non-erythroid cells (Goldfarb et al., 2021; Koury et al., 1989). Genetic knockdown of FTH1, or its degradation in vivo during iron deficiency, also caused dysregulation of the erythroid microtubule cytoskeleton, which was reversed when erythroid cells were treated with ferritin inducers. In contrast, overexpression of FTH1 and ferritin light chain (FTL) promoted erythoid differentiation and restored microtubule structure in iron restricted progenitors (Goldfarb et al., 2021). These studies were among the first to mechanistically connect iron metabolism with erythroid cytoskeletal regulation (Fig. 4).

We similarly observed morphological defects in our studies of mitochondrial iron deficient (*Mfrn1* mutant) and systemic iron deficient (*Fpn1*) erythroid cells that recapitulated studies from other labs (Perfetto et al., 2025a; Prevalence, 2023; Shaw et al., 2006). Using the zebrafish as an animal model, we sought to distinguish the roles of mitochondrial vs. cellular iron import in erythroid development. *Mfrn1* mutant erythroid cells were macrocytic, had large nuclei, and had variable shapes, unlike wild-type erythroid cells which were elliptical with condensed nuclei. In contrast, *Fpn1* mutant erythroid cells were microcytic. Treatment with supplemental iron partially reversed the morphological defects, again indicating that iron played a role in regulation of cellular morphology (Perfetto et al., 2025a). *Actin* genes were among the most upregulated in *Mfrn1* mutant erythroid cells, suggesting that cytoskeletal structure was one of the pathways dependent on mitochondrial iron. While labile cytosolic iron may be incorporated into cytoskeletal regulating proteins, mitochondrial processing is required for iron cofactor production and synthesis of intermediates that enable iron to be incorporated in cytosolic apoproteins (Shepherd et al., 2023). Further studies that identify the cofactor synthesis pathways and target proteins will go far to unravel the functional interactions between iron and cytoskeletal regulation. These will likely include transcriptional mechanisms, as heme plays a major role in regulating the erythroid transcriptome; loss of heme causes defects in expression of genes encoding erythroid membrane proteins, including SLC4A1, TUBB2A and TUBB2B (Liao & Bresnick, 2022).

3.2.2 Cell cycle regulation

Cell cycle entry is a central event in erythroid lineage determination and terminal differentiation and is highly dependent on adequate iron availability (Fu & Richardson, 2007; Le & Richardson, 2004; Nurtjahja-Tjendraputra et al., 2007). During lineage specification, a larger increase in cell cycle speed in MEPs is associated with erythroid fate, while cell cycle inhibition in MEPs results increases megakaryocyte specification (Lu et al., 2018). Erythroid cells are highly proliferative prior terminal differentiation and enucleation (reviewed here (Socolovsky, 2024); Gnanapragasam & Bieker, 2017; Hidalgo et al., 2021) and their entry into terminal differentiation is closely coordinated with the speed of their G1 and S phases (Socolovsky, 2024).

Cell division and DNA replication are processes that are highly dependent on iron containing proteins (Puig et al., 2017). The rate limiting step of DNA replication in the S phase is the synthesis of ribonucleotides by the enzyme ribonucleotide reductase, which is an iron binding enzyme whose activity is decreased during iron deficiency leading to inhibition of deoxyribonucleotide synthesis, preventing DNA synthesis (Hoffbrand et al., 1976; Lederman et al., 1984; McClarty et al., 1990; Puig et al., 2017; Ruskoski & Boal, 2021). Many proteins in the DNA synthesis pathway also require Fe-S clusters including DNA polymerases (Agnarelli et al., 2025; Federico et al., 2022; Kiktev et al., 2021) and DNA repair complexes (Lisova et al., 2022; Zhang et al., 2024a, 2024b) Adequate iron is also required for entry into G1 via regulation of cyclins and p21 (Fu & Richardson, 2007; Nurtjahja-Tjendraputra et al., 2007; Richardson et al., 2009). Given the high proliferative capacity of early erythroid progenitors, it is unsurprising that iron status functions as an early developmental and cell cycle checkpoint to curb inefficient erythroid iron use during iron deficiency. Indeed, our recent studies indicated that the erythroid developmental defect in *Mfrn1* mutant erythroid cells was primarily caused by cell cycle arrest that was reversed by iron supplementation, releasing them from their G2/M arrest (Perfetto et al., 2025b).

Because evidence from cell culture and in vitro studies predominantly focused on the role of iron in G1/S, and the importance of G1 commitment in cell cycle entry, we were surprised to find that *Mfrn1* deficient erythroid cells were arrested at the G2/M transition. These studies may reflect the fact that *Mfrn1* is upregulated during the transition to the late erythroblast stage, and hence *Mfrn1* deficiency did not affect the formation of early erythroid progenitors, which we also observed (Perfetto et al., 2025a). *Mfrn1* deficient erythroid cells had very enlarged nuclei, which reflected the fact that these cells had 4X DNA content because their DNA had replicated but the cells were not undergoing mitosis. Studies in non-erythroid cells have identified several iron-dependent proteins required for completion of mitosis and cytokinesis. The mitotic cytokinesin KIF4A requires Fe-S cluster binding for correct localization to the spindle mid-body during anaphase. In its absence, or in a KIF4A mutant that is lacking Fe-S binding activity, cells are arrested at prometaphase and lacking in chromatin condensation required for metaphase (Ben-Shimon et al., 2018; Ito et al., 2010; Wandke et al., 2012; Wu & Chen, 2008). Fe-S incorporation into KIF4A requires MMS19, a component of the cytosolic Fe-S assembly (CIA) complex, which is broadly required for incorporation of Fe-S complexes into proteins involved in DNA replication and maintenance of genome integrity (Stehling et al., 2012). MMS19 is also required for activating the cyclin activating kinase activity of CDK7, enabling it to promote cell division (Nag et al., 2018). These roles of MMS19 and the importance of Fe-S assembly in cell division are evident in the phenotype of *Mms19* deficient cells which lack proper chromosome segregation, lack mitotic spindle assembly and orientation, and have decreased microtubule polymerization (Chippalkatti et al., 2020). These roles of MMS19 in promoting Fe-S assembly and mitotic spindle formation overlap with the requirement for ferritins in erythroid microtubule organization (Goldfarb et al., 2021). While these studies were conducted in the non-erythroid context, they suggest mechanisms by which Fe-S proteins may regulate the G2/M transition during terminal erythroid differentiation.

Transcriptomic studies have also identified a role for heme in expression of cytokinesis genes such as *Kif2a* and cell cycle regulators such as *Cdkn2b* (Liao et al., 2020). The G2/M arrest in *Mfrn1*-mutant erythroid cells indicates that even cytosolic Fe-S clusters or iron cofactors require mitochondrial iron processing for incorporation into apoproteins that regulate microtubule polymerization and spindle formation. Indeed, genes involved in Fe-S assembly are among those that are most upregulated in *Mfrn1*-mutant erythroid cells (Perfetto et al., 2025a).

3.2.3 Transcription regulation

Erythroid-specific transcription and iron metabolism are highly coregulated. Most erythroid iron metabolism genes are under the transcriptional control of KLF1 and GATA1, master regulators of the erythroid transcriptome. Conversely, iron and heme are essential components of transcription factors and histone modification enzymes that regulate erythroid transcription (Liao & Bresnick, 2022, 2024; Ogawa et al., 2001; Shi et al., 2004; Tsukada et al., 2006; Zenke-Kawasaki et al., 2007). The first erythroid transcription factor known to be directly regulated by heme is the BACH1 transcriptional repressor which serves as a transcriptional sensor of heme. In low cellular heme concentrations, BACH1 constitutively binds to its target genes, including α and *β-globin*. Upon heme binding, BACH1 is released from its binding sites, exported from the nucleus and degraded, de-repressing its target genes (Ogawa et al., 2001; Suzuki et al., 2004; Tahara et al., 2004a, 2004b). BACH1 also represses genes that promote iron export and iron recycling via heme degradation such as *Fpn1* and *Ho1* which are derepressed by heme, providing a negative feedback loop (Marro et al., 2010; Miyazaki et al., 2010).

Independent of BACH1, heme promotes its own transcriptional program in differentiating erythroid cells in concert with the master erythroid regulator, GATA1 (Doty et al., 2019; Tanimura et al. 2016, 2018) to meet the unique needs of the erythroid cell for hemoglobin synthesis. Heme itself promotes expression of *Gata1* mRNA (Doty et al., 2019) and then regulates GATA1 activated genes by increasing chromatin accessibility of GATA1 transcriptional targets which included the globin genes, and genes that regulate protein stability and cytoskeletal structure such as *Slc4a1* (Liao et al., 2020). Early in erythroid development, heme increases transcription of ribosomal RNAs and heme synthesis genes (Doty et al., 2019). Unexpectedly, heme facilitated both the dynamic transcription activation and downregulation of zinc transporters in concert with requirements of the developing erythroid cell. Regulation of zinc transport by heme caused a steep increase in intracellular zinc during initial erythroid development which confers cell survival. Heme/GATA1 subsequently increase the expression of a zinc exporter, with a concomitant decrease of the zinc importer during terminal differentiation, decreasing intracellular zinc as differentiation progresses. As abnormally elevated intracellular zinc levels accelerate erythroid differentiation, the regulated decrease of zinc by heme/GATA1 appears to keep erythroid differentiation in check to avoid subsequent problems such as early senescence during accelerated differentiaton (Liao et al., 2020; Liu et al., 2026; Tanimura et al., 2018). As KLF1 and GATA1 are zinc finger transcription factors, control of erythroid zinc levels may also be a feedback mechanism by which crosstalk between erythroid transcription and heme regulation occurs.

Later in erythropoiesis, heme accumulation downregulates *Gata1* gene expression (Doty et al., 2019) and protein levels by displacing it from its binding to HSP70, allowing degradation of GATA1 by caspase-3 (Gastou et al., 2017; Ribeil et al., 2007). Downregulation of GATA1 when there is high intracellular heme is accompanied by decreased expression of mitotic spindle genes, contributing to the normal termination of cell division during the last stages of erythropoiesis (Doty et al., 2019). These results are consistent with our in vivo data demonstrating that erythroid cells which are deficient in mitochondrial iron and heme express aberrantly high levels of *Gata1* mRNA, which fail to terminally differentiate and remain macrocytic (Kobayashi & Yamamoto, 2007; Perfetto et al., 2025a; Whyatt et al., 2000). Of note, the response of GATA1 to heme depends on endogenously synthesized heme, a signal for iron repletion and effective erythropoiesis, rather than extracellular heme, which is a pro-inflammatory signal (Doty et al., 2019).

3.2.4 Translational regulation

During erythroid development, it is essential that synthesis of globin subunits is coordinated with heme synthesis to avoid accumulation of unfolded proteins or excess heme, both of which are toxic to the cell. Coordination between heme availability and globin translation is primarily carried out by eIF2α kinase, which is primarily expressed in immature erythroblasts and reticulocytes (HRI). When erythroid cells are heme replete, heme binding by HRI promotes disulfide formation within itself, deactivating its kinase activity (Chen et al., 1989). During heme deficiency, non-heme-bound HRI is autophosphorylated on multiple sites, allowing it to dimerize and become autoactivated (Bauer et al., 2001). In its active form, HRI then phosphorylates the α-subunit of eIF2, inhibiting translational initiation and further rounds of translation. Thus, HRI prevents aggregation of globin chains when erythroid cells are heme deficient. HRI deficiency exacerbates the phenotypes of erythropoietic protoporphyria, iron deficiency and β-thalassemia, causing excessive globin synthesis and accumulation of protein aggregates (Han et al., 2001, 2005).

HRI more broadly regulates translation initiation and elongation by its translational inhibition of ribosomal protein synthesis during heme and iron deficiency, and integrates heme availability with stress response pathways in erythroid cells. The mTORC1 (mammalian target of rapamycin complex 1)/4E-BP (eukaryotic translation initiation factor 4E binding protein) pathway is an important mechanism for sensing nutrient levels, particularly bioavailability of neutral essential amino acids and iron. EPO stimulation of erythroid cells activates the mTORC/4E-BP pathway to increase protein synthesis (Chung et al., 2015; Knight et al., 2014). During iron deficiency, HRI mediated signaling represses the mTORC1 pathway and cyclin D3 expression in late erythroblasts, mitigating ineffective erythropoiesis. In summary, HRI-mediated heme sensing and signaling broadly coordinates heme availability with translation and proliferation during erythroid differentiation, mitigating proteostatic stress and ineffective erythropoiesis (Zhang et al., 2018).

4. Interplay between iron homeostasis and the erythropoietic niche

Macrophages play important roles regulating iron homeostasis in their resident tissues by serving as transient iron reservoirs or scavengers (Martins et al., 2026; Winn et al., 2020). This is especially the case for erythroid cells which develop within erythroblastic islands (EBI), consisting of a central macrophage ringed by developing erythroblasts (Manwani & Bieker, 2008). Specific markers for EBI macrophages are still being identified, but they are currently characterized by their expression of F4/80, CD169, SPI1, KLF1 and EPOR (Chow et al., 2013; Zhang et al., 2021; Invalid citation, no date; Romano et al., 2022). EBI macrophages support erythropoiesis by providing local exposure to growth factors such as IGF1 (Miyagawa et al., 2000) and ferritin or transferrin-bound iron which facilitates hemoglobin synthesis terminal erythropoiesis (Li et al., 2019; Leimberg et al., 2008). Upon completion of terminal differentiation, EBI macrophages phagocytose extruded erythroblast nuclei (Manwani & Bieker, 2008; Li et al., 2019). Proximity to EBI macrophages support erythroblast proliferation and development (Rhodes et al., 2008; Ramos et al., 2013; Liao et al., 2018), while disruption of some, but not all, erythroid-macrophage interactions disrupts enucleation and terminal maturation, causing perinatal death with severe anemia (Soni et al., 2006; Lee et al., 2006).

Monocyte differentiation into iron-regulating macrophages is tightly controlled by heme. During hemolysis, exogenous heme binds to monocyte and macrophage BACH1, de-repressing the *SpiC* gene which activates differentiation of red pulp spleen macrophages that recycle iron from senescent red cells, and bone marrow macrophages (Haldar et al., 2014; Kohyama et al., 2009). Excess heme biases macrophages toward the M1, pro-inflammatory phenotype (Vinchi et al., 2016). This is in contrast to EBI macrophages, which tend to express both M1 and M2 phenotypic markers, with M2 macrophages tending to be anti-inflammatory, tissue resident macrophages that support repair and homeostasis (Romano et al., 2022). The control of erythropoiesis in response to infection tends to be mediated through EBI macrophages which express inflammatory cytokine receptors, and sequester iron in response to increased hepcidin levels, instead of releasing it to the developing erythron during infection (Soares & Weiss, 2015). In summary, macrophages integrate iron homeostatic and immune signals from the extracellular environment including heme, hepcidin and cytokine signaling which in turn generate an erythropoietic response within the developmental niche (Romano et al., 2022).

Adult erythropoiesis occurs in the bone marrow, and recent studies are discovering that osteogenic cells also play an important role in iron metabolism during erythroid development. Osteoblasts secrete ERFE, which regulates bone modeling via its role in mediating BMP signaling (Arezes et al., 2018; Wang et al., 2020b). Osteoblast-derived ERFE also plays a critical role in suppressing hepatic hepcidin production by sequestering BMPs (Castro-Mollo et al., 2021; Arezes et al., 2018), which in turn increases iron supply during increased erythropoietic demand (Na-Phatthalung et al., 2026). Paradoxically, osteoblastic ERFE decreases red cell count and terminal erythroid maturation during stress erythropoiesis and expression of erythropoietic genes such as *Epor* and *Gata1*. In contrast, ERFE derived from hematopoietic lineages did not have this function. These effects of osteoblastic ERFE occurred by BMP sequestration and suggest that osteoblastic ERFE plays a role in regulating the erythropoietic niche to respond to erythropoietic stress and iron status (Na-Phatthalung et al., 2026).

Osteogenic cells also secrete fibroblast growth factor 23 (FGF23), which acts on the kidney to induce urinary phosphate excretion and decrease phosphate reabsorption into circulation when phosphate levels are high. Independent of phosphate regulation, FGF23 also regulates erythropoiesis and iron homeostasis (Park et al., 2024). During iron deficiency, both osteoblasts/osteocytes and erythroid cells secrete FGF23. Treatment with EPO improved erythroid iron utilization and decreased serum FGF23, implicating increased FGF23 levels in erythroid iron restriction and suggesting functional crosstalk between phosphate and iron metabolism pathways (Noonan et al., 2020). In β-thalassemia and myelodysplastic syndromes, FGF23 levels are elevated causing bone defects, but systemic inhibition of FGF23 was not only able to normalize bone mineralization, but also significantly improved HSC function and erythropoiesis (Aprile et al., 2023; Weidner et al., 2020). This could be caused by additive effects of osteoblast/osteocyte FGF23, whose roles are in phosphate regulation, and also the effect of excessive FGF23 secretion from erythroid cells, which inhibits erythropoiesis (Courbon et al., 2026). Currently, the effects of FGF23 and osteoblast-derived ERFE on erythroid development and iron homeostasis suggest that there is extensive crosstalk between bone cell populations and the erythropoietic niche microenvironment within the bone marrow.

5. Outstanding questions and conclusion

In recent years, the field has made tremendous strides in understanding how iron is systemically regulated, particularly with the discovery of the *Hamp* and *Erfe* iron regulatory hormones, which have led to novel therapeutics for hematologic diseases (Kremyanskaya et al., 2024; Langer & Esrick, 2021; Manolova et al., 2019; Leal et al., 2026; Nai et al., 2025). We have also made progress in identifying novel transporters of iron, heme and heme intermediates, which continue to be a rich area of inquiry. Lastly, identifying novel characteristics of the erythroblastic niche and identifying the roles of niche cells in iron homeostasis and erythroid development has blossomed with the advent of technologies like scRNAseq that make it possible to unveil the phenotypic and metabolic heterogeneity of tissue-specific macrophages and other cell types. While major progress understanding erythroid development and metabolism within a cell-autonomous context has been made using immortalized and primary erythroid cell culture, the next step is to investigate these developmental pathways in the context of the erythroblastic niche. Being able to identify and pharmacologically target dysregulated erythropoietic niche cells will open up another frontier in gene therapies and design of therapeutic strategies for iron and hematologic diseases (Castruccio Castracani et al., 2025; Ramos et al., 2013; Borges et al., 2025).

Currently, there are several major questions pertaining to the interaction of iron and erythroid development that remain to be addressed. With a better understanding of the mechanisms of mitochondrial iron import, can we determine how iron fate in erythroid and other cell types is determined? Iron-containing proteins are diverse and require iron in different forms, requiring mechanisms to selectively traffic iron to cofactor synthesis pathways. Currently, these iron trafficking mechanisms are unidentified; however, the existence of heme metabolomes suggests that physical interactions between iron transport proteins and cofactor synthesis enzyme pathways play a major role in trafficking iron toward specific cofactors. Detailed protein-protein interaction studies and quantitation of iron cofactors as they are synthesized are essential to elucidate this issue.

The delivery of iron-containing cofactors to their target apoproteins is an area that has made tremendous progress in recent years, particularly with the identification of assembly proteins that deliver Fe-S clusters to specific targets (Maio & Rouault, 2020). Recently, there have also been significant advances in understanding heme delivery to apoproteins from the Stuehr lab which identified nitric oxide and GAPDH as factors contributing to mitochondrial heme export and heme loading onto apoproteins (Jayaram et al., 2025; Das Sinha et al., 2025; Dai & Stuehr, 2025; Dai et al., 2022). While these studies were largely carried out in mammalian cell culture, they now make it possible to develop these techniques for use in primary cells and animal models that more closely recapitulate the erythroid developmental context.

We envision that our understanding of how iron facilitates erythroid development will exponentially increase with the continuing development of technical and conceptual advances. Advances in vivo imaging, cell culture, genomics and genetic manipulation of animal models have all contributed to understanding erythroid development in all its exquisite detail. Combined with the contributions of basic biochemistry in our understanding of iron metabolism, this knowledge has given rise to novel therapeutic strategies for hemoglobinopathies, porphyrias and myelodysplastic syndromes that leverage our ability to correct iron mishandling and thereby decrease the disease burden in patients. With advances in basic science and increasing collaborations between clinicians and basic scientists, the future is bright for translating these findings into individualized therapeutic strategies for patients.

Acknowledgements

This book chapter was supported in part by National Institutes of Health grant R01 DK137107 and an award from the American Porphyria Experts Consortium (YYY).

References

Acín-Pérez, R., Fernández-Silva, P., Peleato, M. L., Pérez-Martos, A., & Enriquez, J. A. (2008). Respiratory active mitochondrial supercomplexes. *Molecular Cell, 32*(4), 529–539. https://doi.org/10.1016/j.molcel.2008.10.021.

Agnarelli, A., Buckley-Benbow, L., Ozgencil, M., et al. (2025). The genetic and biochemical basis of human leading strand synthesis. *Nature Communications, 17*(1), 412. https://doi.org/10.1038/s41467-025-67107-7.

An, W., Feola, M., Levy, M., et al. (2023). Iron chelation improves ineffective erythropoiesis and iron overload in myelodysplastic syndrome mice. *Elife, 12*. https://doi.org/10.7554/eLife.83103.

An, X., & Mohandas, N. (2011). Erythroblastic islands, terminal erythroid differentiation and reticulocyte maturation. *International Journal of Hematology, 93*(2), 139–143. https://doi.org/10.1007/s12185-011-0779-x.

An, X., Schulz, V. P., Li, J., et al. (2014). Global transcriptome analyses of human and murine terminal erythroid differentiation. *Blood, 123*(22), 3466–3477. https://doi.org/10.1182/blood-2014-01-548305.

Aprile, A., Raggi, L., Bolamperti, S., et al. (2023). Inhibition of FGF23 is a therapeutic strategy to target hematopoietic stem cell niche defects in β-thalassemia. *Science Translational Medicine, 15*(698), eabq3679. https://doi.org/10.1126/scitranslmed.abq3679.

Arezes, J., Foy, N., McHugh, K., et al. (2018). Erythroferrone inhibits the induction of hepcidin by BMP6. *Blood, 132*(14), 1473–1477. https://doi.org/10.1182/blood-2018-06-857995.

Astner, I., Schulze, J. O., van den Heuvel, J., Jahn, D., Schubert, W. D., & Heinz, D. W. (2005). Crystal structure of 5-aminolevulinate synthase, the first enzyme of heme biosynthesis, and its link to XLSA in humans. *The EMBO Journal, 24*(18), 3166–3177. https://doi.org/10.1038/sj.emboj.7600792.

Barbarani, G., Fugazza, C., Strouboulis, J., & Ronchi, A. E. (2019). The pleiotropic effects of GATA1 and KLF1 in physiological erythropoiesis and in dyserythropoietic disorders. *Frontiers in Physiology, 10*, 91. https://doi.org/10.3389/fphys.2019.00091.

Barrientos, T., Laothamatas, I., Koves, T. R., et al. (2015). Metabolic catastrophe in mice lacking transferrin receptor in muscle. *EBioMedicine, 2*(11), 1705–1717. https://doi.org/10.1016/j.ebiom.2015.09.041.

Bauer, B. N., Rafie-Kolpin, M., Lu, L., Han, A., & Chen, J.-J. (2001). Multiple autophosphorylation is essential for the formation of the active and stable homodimer of heme-regulated eIF2α kinase. *Biochemistry, 40*(38), 11543–11551. https://doi.org/10.1021/bi010983s.

Baumann, N., Wagener, R. J., Javed, A., et al. (2025). Regional differences in progenitor metabolism shape brain growth during development. *Cell, 188*(13), 3567–3582.e20. https://doi.org/10.1016/j.cell.2025.04.003.

Ben-Shimon, L., Paul, V. D., David-Kadoch, G., et al. (2018). Fe-S cluster coordination of the chromokinesin KIF4A alters its subcellular localization during mitosis. *Journal of Cell Science, 131*(12), https://doi.org/10.1242/jcs.211433.

Bieker, J. J., & Philipsen, S. (2024). Erythroid Krüppel-Like Factor (KLF1): A surprisingly versatile regulator of erythroid differentiation. *Advances in Experimental Medicine and Biology, 1459*, 217–242. https://doi.org/10.1007/978-3-031-62731-6_10.

Billesbølle, C. B., Azumaya, C. M., Kretsch, R. C., et al. (2020). Structure of hepcidin-bound ferroportin reveals iron homeostatic mechanisms. *Nature, 586*(7831), 807–811. https://doi.org/10.1038/s41586-020-2668-z.

Bishop, D. F., Henderson, A. S., & Astrin, K. H. (1990). Human delta-aminolevulinate synthase: assignment of the housekeeping gene to 3p21 and the erythroid-specific gene to the X chromosome. *Genomics,* 7(2), 207–214. https://doi.org/10.1016/0888-7543(90)90542-3.

Borges, M. D., de Souza, M. A., Paes, I. F., et al. (2025). Monocyte-red blood cell crosstalk supports clearance and heme metabolism in sickle cell anemia. *Frontiers in Immunology, 16*, 1699306. https://doi.org/10.3389/fimmu.2025.1699306.

Braymer, J. J., & Lill, R. (2017). Iron–sulfur cluster biogenesis and trafficking in mitochondria. *Journal of Biological Chemistry, 292*(31), 12754–12763. https://doi.org/10.1074/jbc.R117.787101.

Braymer, J. J., Stehling, O., Stümpfig, M., et al. (2024). Requirements for the biogenesis of [2Fe-2S] proteins in the human and yeast cytosol. *Proceedings of the National Academy of Sciences of the United States of America, 121*(21), e2400740121. https://doi.org/10.1073/pnas.2400740121.

Brownlie, A., Donovan, A., Pratt, S. J., et al. (1998). Positional cloning of the zebrafish sauternes gene: A model for congenital sideroblastic anaemia. *Nature Genetics, 20*(3), 244–250. https://doi.org/10.1038/3049.

Brugnara, C. (2003). Iron deficiency and erythropoiesis: New diagnostic approaches. *Clinical Chemistry, 49*(10), 1573–1578. https://doi.org/10.1373/49.10.1573.

Bullock, G. C., Delehanty, L. L., Talbot, A.-L., et al. (2010). Iron control of erythroid development by a novel aconitase-associated regulatory pathway. *Blood, 116*(1), 97–108. https://doi.org/10.1182/blood-2009-10-251496.

Burch, J. S., Marcero, J. R., Maschek, J. A., et al. (2018). Glutamine via α-ketoglutarate dehydrogenase provides succinyl-CoA for heme synthesis during erythropoiesis. *Blood, 132*(10), 987–998. https://doi.org/10.1182/blood-2018-01-829036.

Camaschella, C., Pagani, A., Silvestri, L., & Nai, A. (2022). The mutual crosstalk between iron and erythropoiesis. *International Journal of Hematology, 116*(2), 182–191. https://doi.org/10.1007/s12185-022-03384-y.

Castro-Mollo, M., Gera, S., Ruiz-Martinez, M., et al. (2021). The hepcidin regulator erythroferrone is a new member of the erythropoiesis-iron-bone circuitry. *Elife, 10*. https://doi.org/10.7554/eLife.68217.

Castruccio Castracani, C., Breda, L., Papp, T. E., et al. (2025). An erythroid-specific lentiviral vector improves anemia and iron metabolism in a new model of XLSA. *Blood, 145*(1), 98–113. https://doi.org/10.1182/blood.2024025846.

Cavadini, P., Biasiotto, G., Poli, M., et al. (2007). RNA silencing of the mitochondrial ABCB7 transporter in HeLa cells causes an iron-deficient phenotype with mitochondrial iron overload. *Blood, 109*(8), 3552–3559. https://doi.org/10.1182/blood-2006-08-041632.

Chen, C., Garcia-Santos, D., Ishikawa, Y., et al. (2013). Snx3 regulates recycling of the transferrin receptor and iron assimilation. *Cell Metabolism, 17*(3), 343–352. https://doi.org/10.1016/j.cmet.2013.01.013.

Chen, J. J., Yang, J. M., Petryshyn, R., Kosower, N., & London, I. M. (1989). Disulfide bond formation in the regulation of eIF-2 α kinase by heme. *Journal of Biological Chemistry, 264*(16), 9559–9564. https://doi.org/10.1016/S0021-9258(18)60568-X.

Chen, K., Liu, J., Heck, S., Chasis, J. A., An, X., & Mohandas, N. (2009b). Resolving the distinct stages in erythroid differentiation based on dynamic changes in membrane protein expression during erythropoiesis. *Proceedings of the National Academy of Sciences, 106*(41), 17413–17418. https://doi.org/10.1073/pnas.0909296106.

Chen, W., Dailey, H. A., & Paw, B. H. (2010). Ferrochelatase forms an oligomeric complex with mitoferrin-1 and Abcb10 for erythroid heme biosynthesis. *Blood, 116*(4), 628–630. https://doi.org/10.1182/blood-2009-12-259614.

Chen, W., Paradkar, P. N., Li, L., et al. (2009a). Abcb10 physically interacts with mitoferrin-1 (Slc25a37) to enhance its stability and function in the erythroid mitochondria. *Proceedings of the National Academy of Sciences of the United States of America, 106*(38), 16263–16268. https://doi.org/10.1073/pnas.0904519106.

Chiabrando, D., Marro, S., Mercurio, S., et al. (2012). The mitochondrial heme exporter FLVCR1b mediates erythroid differentiation. *The Journal of Clinical Investigation, 122*(12), 4569–4579. https://doi.org/10.1172/jci62422.

Chippalkatti, R., Egger, B., & Suter, B. (2020). Mms19 promotes spindle microtubule assembly in Drosophila neural stem cells. *PLoS Genetics, 16*(11), e1008913. https://doi.org/10.1371/journal.pgen.1008913.

Chitrakar, I., Roberson, A. B., Ayres-Galhardo, P. H., & Brown, B. L. (2026). A reversible feedback mechanism regulating mitochondrial heme synthesis. *The Journal of Biological Chemistry, 302*(2), 111089. https://doi.org/10.1016/j.jbc.2025.111089.

Chow, A., Huggins, M., Ahmed, J., et al. (2013). CD169^{+} macrophages provide a niche promoting erythropoiesis under homeostasis and stress. *Nature Medicine, 19*(4), 429–436. https://doi.org/10.1038/nm.3057.

Christenson, E. T., Gallegos, A. S., & Banerjee, A. (2018). In vitro reconstitution, functional dissection, and mutational analysis of metal ion transport by mitoferrin-1. *The Journal of Biological Chemistry, 293*(10), 3819–3828. https://doi.org/10.1074/jbc.M117.817478.

Chung, J., Bauer, D. E., Ghamari, A., et al. (2015). The mTORC1/4E-BP pathway coordinates hemoglobin production with L-leucine availability. *Science Signaling, 8*(372), ra34. https://doi.org/10.1126/scisignal.aaa5903.

Chung, J., Wittig, J. G., Ghamari, A., et al. (2017). Erythropoietin signaling regulates heme biosynthesis. *Elife, 6*. https://doi.org/10.7554/eLife.24767.

Cotter, P. D., Rucknagel, D. L., & Bishop, D. F. (1994). X-linked sideroblastic anemia: Identification of the mutation in the erythroid-specific delta-aminolevulinate synthase gene (ALAS2) in the original family described by Cooley. *Blood, 84*(11), 3915–3924.

Cottle, T., Joh, L., Posner, C., DeCosta, A., & Kardon, J. R. (2024). An adaptor for feedback regulation of heme biosynthesis by the mitochondrial protease CLPXP. *bioRxiv*. https://doi.org/10.1101/2024.07.05.602318.

Courbon, G., Thomas, J. J., Duque, E. J., et al. (2026). Erythroid-produced intact FGF23 is a paracrine inhibitor of erythropoiesis. *Blood*. https://doi.org/10.1182/blood.2025030634.

Crooks, D. R., Ghosh, M. C., Haller, R. G., Tong, W. H., & Rouault, T. A. (2010). Posttranslational stability of the heme biosynthetic enzyme ferrochelatase is dependent on iron availability and intact iron-sulfur cluster assembly machinery. *Blood, 115*(4), 860–869. https://doi.org/10.1182/blood-2009-09-243105.

Daher, R., Mansouri, A., Martelli, A., et al. (2019). GLRX5 mutations impair heme biosynthetic enzymes ALA synthase 2 and ferrochelatase in Human congenital sideroblastic anemia. *Molecular Genetics and Metabolism, 128*(3), 342–351. https://doi.org/10.1016/j.ymgme.2018.12.012.

Dai, Y., Fleischhacker, A. S., Liu, L., et al. (2022). Heme delivery to heme oxygenase-2 involves glyceraldehyde-3-phosphate dehydrogenase. *Biological Chemistry, 403*(11-12), 1043–1053. https://doi.org/10.1515/hsz-2022-0230.

Dai, Y., & Stuehr, D. J. (2025). Heme delivery into soluble guanylyl cyclase requires a heme redox change and is regulated by NO and Hsp90 by distinct mechanisms. *The Journal of Biological Chemistry, 301*(3), 108315. https://doi.org/10.1016/j.jbc.2025.108315.

Dailey, H. A., Finnegan, M. G., & Johnson, M. K. (1994). Human ferrochelatase is an iron-sulfur protein. *Biochemistry, 33*(2), 403–407. https://doi.org/10.1021/bi00168a003.

Dailey, H. A., Wu, C. K., Horanyi, P., et al. (2007). Altered orientation of active site residues in variants of human ferrochelatase. Evidence for a hydrogen bond network involved in catalysis. *Biochemistry, 46*(27), 7973–7979. https://doi.org/10.1021/bi700151f.

Das Sinha, P., Islam, S., Biswas, P., & Stuehr, D. J. (2025). Nitric oxide regulates cytochrome P450 2D6 and 3A4 activity via concentration-dependent modulation of heme loading. *The Journal of Biological Chemistry, 301*(11), 110772. https://doi.org/10.1016/j.jbc.2025.110772.

Dietz, J. V., Willoughby, M. M., Piel, R. B., 3rd, et al. (2021). Mitochondrial contact site and cristae organizing system (MICOS) machinery supports heme biosynthesis by enabling optimal performance of ferrochelatase. *Redox Biology, 46*, 102125. https://doi.org/10.1016/j.redox.2021.102125.

Donovan, A., Brownlie, A., Dorschner, M. O., et al. (2002). The zebrafish mutant gene chardonnay (cdy) encodes divalent metal transporter 1 (DMT1). *Blood, 100*(13), 4655–4659. https://doi.org/10.1182/blood-2002-04-1169.

Donovan, A., Brownlie, A., Zhou, Y., et al. (2000). Positional cloning of zebrafish ferroportin1 identifies a conserved vertebrate iron exporter. *Nature, 403*(6771), 776–781. https://doi.org/10.1038/35001596.

Donovan, A., Lima, C. A., Pinkus, J. L., et al. (2005). The iron exporter ferroportin/Slc40a1 is essential for iron homeostasis. *Cell Metabolism, 1*(3), 191–200. https://doi.org/10.1016/j.cmet.2005.01.003.

Doty, R. T., Yan, X., Lausted, C., et al. (2019). Single-cell analyses demonstrate that a heme-GATA1 feedback loop regulates red cell differentiation. *Blood, 133*(5), 457–469. https://doi.org/10.1182/blood-2018-05-850412.

Dowdle, W. E., Nyfeler, B., Nagel, J., et al. (2014). Selective VPS34 inhibitor blocks autophagy and uncovers a role for NCOA4 in ferritin degradation and iron homeostasis in vivo. *Nature Cell Biology, 16*(11), 1069–1079. https://doi.org/10.1038/ncb3053.

El Hoss, S., Lizarralde-Iragorri, M. A., Maciel, T. T., & Hermine, O. (2026). Erythropoiesis in health and disease: Distinguishing defective and ineffective erythropoiesis. *Hemasphere, 10*(1), e70297. https://doi.org/10.1002/hem3.70297.

Federico, G., Carrillo, F., Dapporto, F., et al. (2022). NCOA4 links iron bioavailability to DNA metabolism. *Cell Reports, 40*(7), 111207. https://doi.org/10.1016/j.celrep.2022.111207.

Feick, P., Foisner, R., & Wiche, G. (1991). Immunolocalization and molecular properties of a high molecular weight microtubule-bundling protein (syncolin) from chicken erythrocytes. *The Journal of Cell Biology, 112*(4), 689–699. https://doi.org/10.1083/jcb.112.4.689.

Fernández-Murray, J. P., Prykhozhij, S. V., Dufay, J. N., et al. (2016). Glycine and folate ameliorate models of congenital sideroblastic anemia. *PLoS Genetics, 12*(1), e1005783. https://doi.org/10.1371/journal.pgen.1005783.

Ferreira, G. C., Franco, R., Lloyd, S. G., et al. (1994). Mammalian ferrochelatase, a new addition to the metalloenzyme family. *Journal of Biological Chemistry, 269*(10), 7062–7065. https://doi.org/10.1016/S0021-9258(17)37245-9.

Fleming, M. D., Trenor, C. C., Su, M. A., et al. (1997). Microcytic anaemia mice have a mutation in Nramp2, a candidate iron transporter gene. *Nature Genetics, 16*(4), 383–386. https://doi.org/10.1038/ng0897-383.

Fu, D., & Richardson, D. R. (2007). Iron chelation and regulation of the cell cycle: 2 mechanisms of posttranscriptional regulation of the universal cyclin-dependent kinase inhibitor p21CIP1/WAF1 by iron depletion. *Blood, 110*(2), 752–761. https://doi.org/10.1182/blood-2007-03-076737.

Gastou, M., Rio, S., Dussiot, M., et al. (2017). The severe phenotype of Diamond-Blackfan anemia is modulated by heat shock protein 70. *Blood Advances, 1*(22), 1959–1976. https://doi.org/10.1182/bloodadvances.2017008078.

Gnanapragasam, M. N., & Bieker, J. J. (2017). Orchestration of late events in erythropoiesis by KLF1/EKLF. *Current Opinion in Hematology, 24*(3), 183–190. https://doi.org/10.1097/moh.0000000000000327.

Goldfarb, A. N., Freeman, K. C., Sahu, R. K., et al. (2021). Iron control of erythroid microtubule cytoskeleton as a potential target in treatment of iron-restricted anemia. *Nature Communications, 12*(1), 1645. https://doi.org/10.1038/s41467-021-21938-2.

Golub, M. S., Hogrefe, C. E., Germann, S. L., Capitanio, J. P., & Lozoff, B. (2006a). Behavioral consequences of developmental iron deficiency in infant rhesus monkeys. *Neurotoxicology and Teratology, 28*(1), 3–17. https://doi.org/10.1016/j.ntt.2005.10.005.

Golub, M. S., Hogrefe, C. E., & Germann, S. L. (2007). Iron deprivation during fetal development changes the behavior of juvenile rhesus monkeys. *The Journal of Nutrition, 137*(4), 979–984. https://doi.org/10.1093/jn/137.4.979.

Golub, M. S., Hogrefe, C. E., Tarantal, A. F., et al. (2006b). Diet-induced iron deficiency anemia and pregnancy outcome in rhesus monkeys. *The American Journal of Clinical Nutrition, 83*(3), 647–656. https://doi.org/10.1093/ajcn.83.3.647.

Haase, V. H. (2010). Hypoxic regulation of erythropoiesis and iron metabolism. *American Journal of Physiology-Renal Physiology, 299*(1), F1–F13. https://doi.org/10.1152/ajprenal.00174.2010.

Haider, M. U., Furqan, M., & Mehmood, Q. (2023). Daprodustat: A potential game-changer in renal anemia therapy-A perspective. *Frontiers in Pharmacology, 14*, 1249492. https://doi.org/10.3389/fphar.2023.1249492.

Haldar, M., Kohyama, M., So, Alex Y.-L., et al. (2014). Heme-mediated SPI-C induction promotes monocyte differentiation into iron-recycling macrophages. *Cell, 156*(6), 1223–1234. https://doi.org/10.1016/j.cell.2014.01.069.

Hamdi, A., Roshan, T. M., Kahawita, T. M., Mason, A. B., Sheftel, A. D., & Ponka, P. (2016). Erythroid cell mitochondria receive endosomal iron by a "kiss-and-run" mechanism. *Biochimica et Biophysica Acta, 1863*(12), 2859–2867. https://doi.org/10.1016/j.bbamcr.2016.09.008.

Han, A.-P., Fleming, M. D., & Chen, J.-J. (2005). Heme-regulated eIF2α kinase modifies the phenotypic severity of murine models of erythropoietic protoporphyria and β-thalassemia. *The Journal of Clinical Investigation, 115*(6), 1562–1570. https://doi.org/10.1172/JCI24141.

Han, A. P., Yu, C., Lu, L., et al. (2001). Heme-regulated eIF2alpha kinase (HRI) is required for translational regulation and survival of erythroid precursors in iron deficiency. *The EMBO Journal, 20*(23), 6909–6918. https://doi.org/10.1093/emboj/20.23.6909.

Hidalgo, D., Bejder, J., Pop, R., et al. (2021). EpoR stimulates rapid cycling and larger red cells during mouse and human erythropoiesis. *Nature Communications, 12*(1), 7334. https://doi.org/10.1038/s41467-021-27562-4.

Hoffbrand, A. V., Ganeshaguru, K., Hooton, J. W., & Tattersall, M. H. (1976). Effect of iron deficiency and desferrioxamine on DNA synthesis in human cells. *British Journal of Haematology, 33*(4), 517–526. https://doi.org/10.1111/j.1365-2141.1976.tb03570.x.

Hsieh, M. M., Linde, N. S., Wynter, A., et al. (2007). HIF prolyl hydroxylase inhibition results in endogenous erythropoietin induction, erythrocytosis, and modest fetal hemoglobin expression in rhesus macaques. *Blood, 110*(6), 2140–2147. https://doi.org/10.1182/blood-2007-02-073254.

Hu, J. R. J., Xue, F., et al. (2013). Isolation and functional characterization of human erythroblasts at distinct stages: implications for understanding of normal and disordered erythropoiesis in vivo. *Blood, 121*(16), 3246–3253. https://doi.org/10.1182/blood-2013-01-476390.

Hung, K. C., Weng, H. L., Hsu, C. W., Lai, Y. C., & Chen, I. W. (2025). Association between iron deficiency anemia and risk of venous thromboembolism: a multi-institutional retrospective study. *Frontiers in Nutrition, 12*, 1660622. https://doi.org/10.3389/fnut.2025.1660622.

Hunter, G. A., & Ferreira, G. C. (2011). Molecular enzymology of 5-aminolevulinate synthase, the gatekeeper of heme biosynthesis. *Biochimica et Biophysica Acta, 1814*(11), 1467–1473. https://doi.org/10.1016/j.bbapap.2010.12.015.

Ito, S., Tan, L. J., Andoh, D., et al. (2010). MMXD, a TFIIH-independent XPD-MMS19 protein complex involved in chromosome segregation. *Molecular Cell, 39*(4), 632–640. https://doi.org/10.1016/j.molcel.2010.07.029.

Iwai, K. (2019). Regulation of cellular iron metabolism: Iron-dependent degradation of IRP by SCF(FBXL5) ubiquitin ligase. *Free Radical Biology & Medicine, 133*, 64–68. https://doi.org/10.1016/j.freeradbiomed.2018.09.011.

Jayaram, D. T., Sivaram, P., Biswas, P., Dai, Y., Sweeny, E. A., & Stuehr, D. J. (2025). Heme allocation in eukaryotic cells relies on mitochondrial heme export through FLVCR1b to cytosolic GAPDH. *Nature Communications, 16*(1), 7972. https://doi.org/10.1038/s41467-025-62819-2.

Kao, Y. R., Chen, J., Kumari, R., et al. (2024). An iron rheostat controls hematopoietic stem cell fate. *Cell Stem Cell, 31*(3), 378–397.e12. https://doi.org/10.1016/j.stem.2024.01.011.

Kato, H., Itoh-Nakadai, A., Matsumoto, M., et al. (2018). Infection perturbs Bach2- and Bach1-dependent erythroid lineage 'choice' to cause anemia. *Nature Immunology, 19*(10), 1059–1070. https://doi.org/10.1038/s41590-018-0202-3.

Kautz, L., Jung, G., Du, X., et al. (2015). Erythroferrone contributes to hepcidin suppression and iron overload in a mouse model of β-thalassemia. *Blood, 126*(17), 2031–2037. https://doi.org/10.1182/blood-2015-07-658419.

Kautz, L., Jung, G., Valore, E. V., Rivella, S., Nemeth, E., & Ganz, T. (2014). Identification of erythroferrone as an erythroid regulator of iron metabolism. *Nature Genetics, 46*(7), 678–684. https://doi.org/10.1038/ng.2996.

Kautz, L., & Nemeth, E. (2014). Molecular liaisons between erythropoiesis and iron metabolism. *Blood, 124*(4), 479–482. https://doi.org/10.1182/blood-2014-05-516252.

Khalil, S., Delehanty, L., Grado, S., et al. (2017b). Iron modulation of erythropoiesis is associated with Scribble-mediated control of the erythropoietin receptor. *Journal of Experimental Medicine, 215*(2), 661–679. https://doi.org/10.1084/jem.20170396.

Khalil, S., Holy, M., Grado, S., et al. (2017a). A specialized pathway for erythroid iron delivery through lysosomal trafficking of transferrin receptor 2. *Blood Advances, 1*(15), 1181–1194. https://doi.org/10.1182/bloodadvances.2016003772.

Kiktev, D. A., Dominska, M., Zhang, T., et al. (2021). The fidelity of DNA replication, particularly on GC-rich templates, is reduced by defects of the Fe-S cluster in DNA polymerase δ. *Nucleic Acids Research, 49*(10), 5623–5636. https://doi.org/10.1093/nar/gkab371.

Knight, Z. A., Schmidt, S. F., Birsoy, K., Tan, K., & Friedman, J. M. (2014). A critical role for mTORC1 in erythropoiesis and anemia. *eLife, 3*, e01913. https://doi.org/10.7554/eLife.01913.

Kobayashi, M., & Yamamoto, M. (2007). Regulation of GATA1 gene expression. *Journal of Biochemistry, 142*(1), 1–10. https://doi.org/10.1093/jb/mvm122.

Kohyama, M., Ise, W., Edelson, B. T., et al. (2009). Role for Spi-C in the development of red pulp macrophages and splenic iron homeostasis. *Nature, 457*(7227), 318–321. https://doi.org/10.1038/nature07472.

Koury, M. J., Sawyer, S. T., & Brandt, S. J. (2002). New insights into erythropoiesis. *Current Opinion in Hematology, 9*(2), 93–100. https://doi.org/10.1097/00062752-200203000-00002.

Koury, S. T., Koury, M. J., & Bondurant, M. C. (1989). Cytoskeletal distribution and function during the maturation and enucleation of mammalian erythroblasts. *The Journal of Cell Biology, 109*(6 Pt 1), 3005–3013. https://doi.org/10.1083/jcb.109.6.3005.

Kremyanskaya, M., Kuykendall, A. T., Pemmaraju, N., et al. (2024). Rusfertide, a hepcidin mimetic, for control of erythrocytosis in polycythemia vera. *The New England Journal of Medicine, 390*(8), 723–735. https://doi.org/10.1056/NEJMoa2308809.

Kubota, Y., Nomura, K., Katoh, Y., Yamashita, R., Kaneko, K., & Furuyama, K. (2016). Novel mechanisms for heme-dependent degradation of ALAS1 protein as a component of negative feedback regulation of heme biosynthesis. *The Journal of Biological Chemistry, 291*(39), 20516–20529. https://doi.org/10.1074/jbc.M116.719161.

Langer, A. L., & Esrick, E. B. (2021). β-Thalassemia: Evolving treatment options beyond transfusion and iron chelation. *Hematology/the Education Program of the American Society of Hematology, 2021*(1), 600–606. https://doi.org/10.1182/hematology.2021000313.

Le Blanc, S., Garrick, M. D., & Arredondo, M. (2012). Heme carrier protein 1 transports heme and is involved in heme-Fe metabolism. *American Journal of Physiology-Cell Physiology, 302*(12), C1780–C1785. https://doi.org/10.1152/ajpcell.00080.2012.

Le, N. T., & Richardson, D. R. (2004). Iron chelators with high antiproliferative activity up-regulate the expression of a growth inhibitory and metastasis suppressor gene: A link between iron metabolism and proliferation. *Blood, 104*(9), 2967–2975. https://doi.org/10.1182/blood-2004-05-1866.

Leal, S., Denardo, A., Van Echten, A., et al. (2026). Halofuginone suppresses hepcidin by a heparan sulfate-dependent mechanism to treat iron disorders in mice. *Blood Advances, 10*(9), 3020–3031. https://doi.org/10.1182/bloodadvances.2025017943.

Lederman, H. M., Cohen, A., Lee, J. W., Freedman, M. H., & Gelfand, E. W. (1984). Deferoxamine: A reversible S-phase inhibitor of human lymphocyte proliferation. *Blood, 64*(3), 748–753.

Lee, G., Lo, A., Short, S. A., et al. (2006). Targeted gene deletion demonstrates that the cell adhesion molecule ICAM-4 is critical for erythroblastic island formation. *Blood, 108*(6), 2064–2071. https://doi.org/10.1182/blood-2006-03-006759.

Leimberg, M. J., Prus, E., Konijn, A. M., & Fibach, E. (2008). Macrophages function as a ferritin iron source for cultured human erythroid precursors. *Journal of Cellular Biochemistry, 103*(4), 1211–1218. https://doi.org/10.1002/jcb.21499.

Levy, J. E., Jin, O., Fujiwara, Y., Kuo, F., & Andrews, N. C. (1999). Transferrin receptor is necessary for development of erythrocytes and the nervous system. *Nature Genetics, 21*(4), 396–399. https://doi.org/10.1038/7727.

Levy, M. V., & Ginzburg, Y. Z. (2025). Iron-Loading anemias. *Advances in Experimental Medicine and Biology, 1480*, 145–161. https://doi.org/10.1007/978-3-031-92033-2_11.

Li, W., Wang, Y., Zhao, H., et al. (2019). Identification and transcriptome analysis of erythroblastic island macrophages. *Blood, 134*(5), 480–491. https://doi.org/10.1182/blood.2019000430.

Liao, C., Prabhu, K. S., & Paulson, R. F. (2018). Monocyte-derived macrophages expand the murine stress erythropoietic niche during the recovery from anemia. *Blood, 132*(24), 2580–2593. https://doi.org/10.1182/blood-2018-06-856831.

Liao, R., & Bresnick, E. H. (2024). Endogenous small molecule effectors in GATA transcription factor mechanisms governing biological and pathological processes. *Experimental Hematology, 137*, 104252. https://doi.org/10.1016/j.exphem.2024.104252.

Liao, R., & Bresnick, E. H. (2022). Heme as a differentiation-regulatory transcriptional cofactor. *International Journal of Hematology, 116*(2), 174–181. https://doi.org/10.1007/s12185-022-03404-x.

Liao, R., Zheng, Y., Liu, X., et al. (2020). Discovering how heme controls genome function through heme-omics. *Cell Reports, 31*(13), 107832. https://doi.org/10.1016/j.celrep.2020.107832.

Lisova, A. E., Baranovskiy, A. G., Morstadt, L. M., Babayeva, N. D., Stepchenkova, E. I., & Tahirov, T. H. (2022). The iron-sulfur cluster is essential for DNA binding by human DNA polymerase ε. *Scientific Reports, 12*(1), 17436. https://doi.org/10.1038/s41598-022-21550-4.

Liu, G., Sil, D., Maio, N., et al. (2020). Heme biosynthesis depends on previously unrecognized acquisition of iron-sulfur cofactors in human amino-levulinic acid dehydratase. *Nature Communications, 11*(1), 6310. https://doi.org/10.1038/s41467-020-20145-9.

Liu, J., Zhang, J., Ginzburg, Y., et al. (2013). Quantitative analysis of murine terminal erythroid differentiation in vivo: Novel method to study normal and disordered erythropoiesis. *Blood, 121*(8), e43–e49. https://doi.org/10.1182/blood-2012-09-456079.

Liu, Q., Davidoff, O., Niss, K., & Haase, V. H. (2012). Hypoxia-inducible factor regulates hepcidin via erythropoietin-induced erythropoiesis. *The Journal of Clinical Investigation, 122*(12), 4635–4644. https://doi.org/10.1172/jci63924.

Liu, Y., Zhang, W., Cai, J., et al. (2026). TMEM187 is a novel modulator in the regulation of erythropoiesis. *Blood*. https://doi.org/10.1182/blood.2025031135.

Lu, Y. C., Sanada, C., Xavier-Ferrucio, J., et al. (2018). The molecular signature of megakaryocyte-erythroid progenitors reveals a role for the cell cycle in fate specification. *Cell Reports, 25*(8), 2083–2093.e4. https://doi.org/10.1016/j.celrep.2018.10.084.

Lunetti, P., Damiano, F., De Benedetto, G., et al. (2016). Characterization of human and yeast mitochondrial glycine carriers with implications for heme biosynthesis and anemia. *The Journal of Biological Chemistry, 291*(38), 19746–19759. https://doi.org/10.1074/jbc.M116.736876.

Maio, N., & Rouault, T. A. (2020). Outlining the complex pathway of mammalian Fe-S cluster biogenesis. *Trends in Biochemical Sciences, 45*(5), 411–426. https://doi.org/10.1016/j.tibs.2020.02.001.

Manolova, V., Nyffenegger, N., Flace, A., et al. (2019). Oral ferroportin inhibitor ameliorates ineffective erythropoiesis in a model of β-thalassemia. *The Journal of Clinical Investigation, 130*(1), 491–506. https://doi.org/10.1172/jci129382.

Manwani, D., & Bieker, J. J. (2008). The erythroblastic island. *Current Topics in Developmental Biology, 82*, 23–53. https://doi.org/10.1016/s0070-2153(07)00002-6.

Marro, S., Chiabrando, D., Messana, E., et al. (2010). Heme controls ferroportin1 (FPN1) transcription involving Bach1, Nrf2 and a MARE/ARE sequence motif at position -7007 of the FPN1 promoter. *Haematologica, 95*(8), 1261–1268. https://doi.org/10.3324/haematol.2009.020123.

Martins, R., Blankehaus, B., Braza, F., et al. (2026). Homeostatic control of energy metabolism by monocyte-derived macrophages. *The EMBO Journal, 45*(1), 106–150. https://doi.org/10.1038/s44318-025-00622-x.

Martins, R., & Knapp, S. (2018). Heme and hemolysis in innate immunity: Adding insult to injury. *Current Opinion in Immunology, 50*, 14–20. https://doi.org/10.1016/j.coi.2017.10.005.

Mastrogiannaki, M., Matak, P., & Peyssonnaux, C. (2013). The gut in iron homeostasis: Role of HIF-2 under normal and pathological conditions. *Blood, 122*(6), 885–892. https://doi.org/10.1182/blood-2012-11-427765.

Matak, P., Matak, A., Moustafa, S., et al. (2016). Disrupted iron homeostasis causes dopaminergic neurodegeneration in mice. *Proceedings of the National Academy of Sciences of the United States of America, 113*(13), 3428–3435. https://doi.org/10.1073/pnas.1519473113.

McClarty, G. A., Chan, A. K., Choy, B. K., & Wright, J. A. (1990). Increased ferritin gene expression is associated with increased ribonucleotide reductase gene expression and the establishment of hydroxyurea resistance in mammalian cells. *The Journal of Biological Chemistry, 265*(13), 7539–7547.

Medlock, A. E., Shiferaw, M. T., Marcero, J. R., et al. (2015). Identification of the mitochondrial heme metabolism complex. *PLoS One, 10*(8), e0135896. https://doi.org/10.1371/journal.pone.0135896.

Miyagawa, S., Kobayashi, M., Konishi, N., Sato, T., & Ueda, K. (2000). Insulin and insulin-like growth factor I support the proliferation of erythroid progenitor cells in bone marrow through the sharing of receptors. *British Journal of Haematology, 109*(3), 555–562. https://doi.org/10.1046/j.1365-2141.2000.02047.x.

Miyazaki, T., Kirino, Y., Takeno, M., et al. (2010). Expression of heme oxygenase-1 in human leukemic cells and its regulation by transcriptional repressor Bach1. *Cancer Science, 101*(6), 1409–1416. https://doi.org/10.1111/j.1349-7006.2010.01550.x.

Mukherjee, K., Xue, L., Planutis, A., Gnanapragasam, M. N., Chess, A., & Bieker, J. J. (2021). EKLF/KLF1 expression defines a unique macrophage subset during mouse erythropoiesis. *Elife*, e61070.

Nag, R. N., Niggli, S., Sousa-Guimarães, S., Vazquez-Pianzola, P., & Suter, B. (2018). Mms19 is a mitotic gene that permits Cdk7 to be fully active as a Cdk-activating kinase. *Development (Cambridge, England), 145*(2), https://doi.org/10.1242/dev.156802.

Nai, A., Silvestri, L., Asperti, M., & Vinchi, F. (2025). Current landscape of hepcidin therapeutics. *Advances in Experimental Medicine and Biology, 1480*, 399–418. https://doi.org/10.1007/978-3-031-92033-2_26.

Nakai, T., Iwamura, Y., Kato, K., et al. (2023). Drugs activating hypoxia-inducible factors correct erythropoiesis and hepcidin levels via renal EPO induction in mice. *Blood Advances*, 7(15), 3793–3805. https://doi.org/10.1182/bloodadvances.2023009798.

Na-Phatthalung, P., Caloen, Gv, Planoutene, M., et al. (2026). Erythroferrone derived from osteoblasts regulates stress erythropoiesis. *Proceedings of the National Academy of Sciences, 123*(18), e2537627123. https://doi.org/10.1073/pnas.2537627123.

Noonan, M. L., Clinkenbeard, E. L., Ni, P., et al. (2020). Erythropoietin and a hypoxia-inducible factor prolyl hydroxylase inhibitor (HIF-PHDi) lowers FGF23 in a model of chronic kidney disease (CKD). *Physiological Reports, 8*(11), e14434. https://doi.org/10.14814/phy2.14434.

Nurtjahja-Tjendraputra, E., Fu, D., Phang, J. M., & Richardson, D. R. (2007). Iron chelation regulates cyclin D1 expression via the proteasome: A link to iron deficiency-mediated growth suppression. *Blood, 109*(9), 4045–4054. https://doi.org/10.1182/blood-2006-10-047753.

Ogawa, K., Sun, J., Taketani, S., et al. (2001). Heme mediates derepression of Maf recognition element through direct binding to transcription repressor Bach1. *The EMBO Journal, 20*(11), 2835–2843.

Ohgami, R. S., Campagna, D. R., Greer, E. L., et al. (2005). Identification of a ferrireductase required for efficient transferrin-dependent iron uptake in erythroid cells. *Nature Genetics, 37*(11), 1264–1269. https://doi.org/10.1038/ng1658.

Park, M. Y., Agoro, R., Jankauskas, S. S., Le Henaff, C., & Sitara, D. (2024). Phosphorus-independent role of FGF23 in erythropoiesis and iron homeostasis. *PLoS One, 19*(12), e0315228. https://doi.org/10.1371/journal.pone.0315228.

Parrow, N. L., Li, Y., Feola, M., et al. (2019). Lobe specificity of iron binding to transferrin modulates murine erythropoiesis and iron homeostasis. *Blood, 134*(17), 1373–1384. https://doi.org/10.1182/blood.2018893099.

Patel, S. J., Protchenko, O., Shakoury-Elizeh, M., Baratz, E., Jadhav, S., & Philpott, C. C. (2021). The iron chaperone and nucleic acid-binding activities of poly(rC)-binding protein 1 are separable and independently essential. *Proceedings of the National Academy of Sciences of the United States of America, 118*(25), https://doi.org/10.1073/pnas.2104666118.

Paul, R., Elena, G., Stefania, R., et al. (2021). Induction of erythroferrone in healthy humans by micro-dose recombinant erythropoietin or high-altitude exposure. *Haematologica, 106*(2), 384–390. https://doi.org/10.3324/haematol.2019.233874.

Pena, I. A., Shi, J. S., Chang, S. M., et al. (2025). SLC25A38 is required for mitochondrial pyridoxal 5′-phosphate (PLP) accumulation. *Nature Communications, 16*(1), 978. https://doi.org/10.1038/s41467-025-56130-3.

Perfetto, M., Danoff, A., Ishfaq, M., et al. (2025a). Mitochondrial iron transport via MFRN1 is required for erythroid cell cycle progression. *Blood Advances*. https://doi.org/10.1182/bloodadvances.2024014887.

Perfetto, M., Danoff, A., Ishfaq, M., et al. (2022). Mitochondrial iron transport via MFRN1 is required for erythroid cell cycle progression. *bioRxiv, 2024*, 08.22.504833. https://doi.org/10.1101/2022.08.22.504833.

Perfetto, M., Ishfaq, M., Mohideen, A., et al. (2025b). FAM210B regulates iron homeostasis and sex-specific responses in stress erythropoiesis. *Experimental Hematology*, 104797. https://doi.org/10.1016/j.exphem.2025.104797.

Piel Iii, R. B., Obi, C. D., Viana, M. P., et al. (2026). Progesterone receptor membrane component 1 (PGRMC1) regulates Heme trafficking through mitochondria-ER junctions. *Journal of Inorganic Biochemistry, 275*, 113093. https://doi.org/10.1016/j.jinorgbio.2025.113093.

Piel, R. B., 3rd, Shiferaw, M. T., Vashisht, A. A., et al. (2016). A novel role for progesterone receptor membrane component 1 (PGRMC1): A partner and regulator of ferrochelatase. *Biochemistry, 55*(37), 5204–5217. https://doi.org/10.1021/acs.biochem.6b00756.

Prevalence (2023). years lived with disability, and trends in anaemia burden by severity and cause, 1990-2021: Findings from the Global Burden of Disease Study 2021. *The Lancet Haematology, 10*(9), e713–e734. https://doi.org/10.1016/s2352-3026(23)00160-6.

Puig, S., Ramos-Alonso, L., Romero, A. M., & Martínez-Pastor, M. T. (2017). The elemental role of iron in DNA synthesis and repair. *Metallomics, 9*(11), 1483–1500. https://doi.org/10.1039/c7mt00116a.

Qi, W., Li, J., & Cowan, J. A. (2014). A structural model for glutathione-complexed iron-sulfur cluster as a substrate for ABCB7-type transporters. *Chemical Communications (Camb), 50*(29), 3795–3798. https://doi.org/10.1039/c3cc48239a.

Ramey, G., Deschemin, J. C., Durel, B., Canonne-Hergaux, F., Nicolas, G., & Vaulont, S. (2010). Hepcidin targets ferroportin for degradation in hepatocytes. *Haematologica, 95*(3), 501–504. https://doi.org/10.3324/haematol.2009.014399.

Ramos, P., Casu, C., Gardenghi, S., et al. (2013). Macrophages support pathological erythropoiesis in polycythemia vera and β-thalassemia. *Nature Medicine, 19*(4), 437–445. https://doi.org/10.1038/nm.3126.

Rhodes, M. M., Kopsombut, P., Bondurant, M. C., Price, J. O., & Koury, M. J. (2008). Adherence to macrophages in erythroblastic islands enhances erythroblast proliferation and increases erythrocyte production by a different mechanism than erythropoietin. *Blood, 111*(3), 1700–1708. https://doi.org/10.1182/blood-2007-06-098178.

Ribeil, J. A., Zermati, Y., Vandekerckhove, J., et al. (2007). Hsp70 regulates erythropoiesis by preventing caspase-3-mediated cleavage of GATA-1. *Nature, 445*(7123), 102–105. https://doi.org/10.1038/nature05378.

Richardson, D. R., Kalinowski, D. S., Lau, S., Jansson, P. J., & Lovejoy, D. B. (2009). Cancer cell iron metabolism and the development of potent iron chelators as anti-tumour agents. *Biochimica et Biophysica Acta (BBA) - General Subjects, 1790*(7), 702–717. https://doi.org/10.1016/j.bbagen.2008.04.003.

Romano, L., Seu, K. G., Papoin, J., et al. (2022). Erythroblastic islands foster granulopoiesis in parallel to terminal erythropoiesis. *Blood, 140*(14), 1621–1634. https://doi.org/10.1182/blood.2022015724.

Rondelli, C. M., Perfetto, M., Danoff, A., et al. (2021). The ubiquitous mitochondrial protein unfoldase CLPX regulates erythroid heme synthesis by control of iron utilization and heme synthesis enzyme activation and turnover. *The Journal of Biological Chemistry, 297*(2), 100972. https://doi.org/10.1016/j.jbc.2021.100972.

Ruskoski, T. B., & Boal, A. K. (2021). The periodic table of ribonucleotide reductases. *The Journal of Biological Chemistry, 297*(4), 101137. https://doi.org/10.1016/j.jbc.2021.101137.

Ryu, M. S., Zhang, D., Protchenko, O., Shakoury-Elizeh, M., & Philpott, C. C. (2017). PCBP1 and NCOA4 regulate erythroid iron storage and heme biosynthesis. *The Journal of Clinical Investigation, 127*(5), 1786–1797. https://doi.org/10.1172/jci90519.

Sardo, U., Perrier, P., Cormier, K., et al. (2024). The hepatokine FGL1 regulates hepcidin and iron metabolism during anemia in mice by antagonizing BMP signaling. *Blood, 143*(13), 1282–1292. https://doi.org/10.1182/blood.2023022724.

Schneider-Yin, X., Gouya, L., Dorsey, M., Rüfenacht, U., Deybach, J. C., & Ferreira, G. C. (2000). Mutations in the iron-sulfur cluster ligands of the human ferrochelatase lead to erythropoietic protoporphyria. *Blood, 96*(4), 1545–1549.

Semenza, G. L. (1999). Regulation of mammalian O2 homeostasis by hypoxia-inducible factor 1. *Annual Review of Cell and Developmental Biology, 15*, 551–578. https://doi.org/10.1146/annurev.cellbio.15.1.551.

Shah, D. I., Takahashi-Makise, N., Cooney, J. D., et al. (2012). Mitochondrial Atpif1 regulates haem synthesis in developing erythroblasts. *Nature, 491*(7425), 608–612. https://doi.org/10.1038/nature11536.

Sharma, M., Ishfaq, M., Mondal, P., et al. (2026). Daprodustat augments red cell and gamma-globin production in models of sickle cell disease. *Blood Red Cells & Iron*. (In press).

Shaw, G. C., Cope, J. J., Li, L., et al. (2006). Mitoferrin is essential for erythroid iron assimilation. *Nature, 440*(7080), 96–100. https://doi.org/10.1038/nature04512.

Sheftel, A. D., Zhang, A. S., Brown, C., Shirihai, O. S., & Ponka, P. (2007). Direct interorganellar transfer of iron from endosome to mitochondrion. *Blood, 110*(1), 125–132. https://doi.org/10.1182/blood-2007-01-068148.

Shepherd, R. E., Kreinbrink, A. C., Njimoh, C. L., Vali, S. W., & Lindahl, P. A. (2023). Yeast mitochondria import aqueous Fe(II) and, when activated for iron-sulfur cluster assembly, export or release low-molecular-mass iron and also export iron that incorporates into cytosolic proteins. *Journal of the American Chemical Society, 145*(25), 13556–13569. https://doi.org/10.1021/jacs.2c13439.

Shi, Y., Lan, F., Matson, C., et al. (2004). Histone demethylation mediated by the nuclear amine oxidase homolog LSD1. *Cell, 119*(7), 941–953.

Sighinolfi, S., Cassina, L., Lidonnici, M. R., et al. (2026). Iron overload damages mitochondria and induces metabolic rewiring of hematopoietic stem cells towards glycolysis. *Blood*. https://doi.org/10.1182/blood.2025031552.

Soares, M. P., & Weiss, G. (2015). The Iron age of host-microbe interactions. *EMBO Reports, 16*(11), 1482–1500. https://doi.org/10.15252/embr.201540558.

Socolovsky, M. (2024). Pas de deux: The coordinated coupling of erythroid differentiation with the cell cycle. *Current Opinion in Hematology, 31*(3), 96–103. https://doi.org/10.1097/moh.0000000000000811.

Soni, S., Bala, S., Gwynn, B., Sahr, K. E., Peters, L. L., & Hanspal, M. (2006). Absence of erythroblast macrophage protein (Emp) leads to failure of erythroblast nuclear extrusion. *The Journal of Biological Chemistry, 281*(29), 20181–20189. https://doi.org/10.1074/jbc.M603226200.

Stehling, O., Vashisht, A. A., Mascarenhas, J., et al. (2012). MMS19 assembles iron-sulfur proteins required for DNA metabolism and genomic integrity. *Science (New York, N. Y.), 337*(6091), 195–199. https://doi.org/10.1126/science.1219723.

Sun, F., Zhao, Z., Willoughby, M. M., et al. (2022). HRG-9 homologues regulate haem trafficking from haem-enriched compartments. *Nature, 610*(7933), 768–774. https://doi.org/10.1038/s41586-022-05347-z.

Suzuki, H., Tashiro, S., Hira, S., et al. (2004). Heme regulates gene expression by triggering Crm1-dependent nuclear export of Bach1. *The EMBO Journal, 23*(13), 2544–2553. https://doi.org/10.1038/sj.emboj.7600248.

Tahara, T., Sun, J., Igarashi, K., & Taketani, S. (2004a). Heme-dependent up-regulation of the alpha-globin gene expression by transcriptional repressor Bach1 in erythroid cells. *Biochemical and Biophysical Research Communications, 324*(1), 77–85. https://doi.org/10.1016/j.bbrc.2004.09.022.
Tahara, T., Sun, J., Nakanishi, K., et al. (2004b). Heme positively regulates the expression of beta-globin at the locus control region via the transcriptional factor Bach1 in erythroid cells. *The Journal of Biological Chemistry, 279*(7), 5480–5487. https://doi.org/10.1074/jbc.M302733200.
Talamo, G., Oyeleye, O., Paudel, A., et al. (2025). Platelet levels before and after iron replacement therapy in patients with iron deficiency anemia. *Hematology (Amsterdam, Netherlands), 30*(1), 2458358. https://doi.org/10.1080/16078454.2025.2458358.
Tanimura, N., Liao, R., Wilson, G. M., et al. (2018). GATA/Heme multi-omics reveals a trace metal-dependent cellular differentiation mechanism. *Developmental Cell, 46*(5), 581–594.e4. https://doi.org/10.1016/j.devcel.2018.07.022.
Tanimura, N., Miller, E., Igarashi, K., et al. (2016). Mechanism governing heme synthesis reveals a GATA factor/heme circuit that controls differentiation. *EMBO Reports, 17*(2), 249–265. https://doi.org/10.15252/embr.201541465.
Terzi, E. M., Sviderskiy, V. O., Alvarez, S. W., Whiten, G. C., & Possemato, R. (2021). Iron-sulfur cluster deficiency can be sensed by IRP2 and regulates iron homeostasis and sensitivity to ferroptosis independent of IRP1 and FBXL5. *Science Advances,* 7(22), https://doi.org/10.1126/sciadv.abg4302.
Troadec, M. B., Warner, D., Wallace, J., et al. (2011). Targeted deletion of the mouse Mitoferrin1 gene: From anemia to protoporphyria. *Blood, 117*(20), 5494–5502. https://doi.org/10.1182/blood-2010-11-319483.
Tsukada, Y.-i, Fang, J., Erdjument-Bromage, H., et al. (2006). Histone demethylation by a family of JmjC domain-containing proteins. *Nature, 439*(7078), 811–816.
Vallières, C., Benoit, O., Guittet, O., et al. (2024). Iron-sulfur protein odyssey: Exploring their cluster functional versatility and challenging identification. *Metallomics, 16*(5), https://doi.org/10.1093/mtomcs/mfae025.
Vashisht, A. A., Zumbrennen, K. B., Huang, X., et al. (2009). Control of iron homeostasis by an iron-regulated ubiquitin ligase. *Science (New York, N. Y.), 326*(5953), 718–721. https://doi.org/10.1126/science.1176333.
Vinchi, F., Costa da Silva, M., Ingoglia, G., et al. (2016). Hemopexin therapy reverts heme-induced proinflammatory phenotypic switching of macrophages in a mouse model of sickle cell disease. *Blood, 127*(4), 473–486. https://doi.org/10.1182/blood-2015-08-663245.
Wandke, C., Barisic, M., Sigl, R., et al. (2012). Human chromokinesins promote chromosome congression and spindle microtubule dynamics during mitosis. *The Journal of Cell Biology, 198*(5), 847–863. https://doi.org/10.1083/jcb.201110060.
Wang, C.-Y., Xu, Y., Traeger, L., et al. (2020b). Erythroferrone lowers hepcidin by sequestering BMP2/6 heterodimer from binding to the BMP type I receptor ALK3. *Blood, 135*(6), 453–456. https://doi.org/10.1182/blood.2019002620.
Wang, S., He, X., Wu, Q., et al. (2020a). Transferrin receptor 1-mediated iron uptake plays an essential role in hematopoiesis. *Haematologica, 105*(8), 2071–2082. https://doi.org/10.3324/haematol.2019.224899.
Weidner, H., Baschant, U., Lademann, F., et al. (2020). Increased FGF-23 levels are linked to ineffective erythropoiesis and impaired bone mineralization in myelodysplastic syndromes. *JCI Insight, 5*(15), https://doi.org/10.1172/jci.insight.137062.
Whyatt, D., Lindeboom, F., Karis, A., et al. (2000). An intrinsic but cell-nonautonomous defect in GATA-1-overexpressing mouse erythroid cells. *Nature, 406*(6795), 519–524. https://doi.org/10.1038/35020086.

Wilkinson, N., & Pantopoulos, K. (2014). The IRP/IRE system in vivo: Insights from mouse models. *Frontiers in Pharmacology, 5*, 176. https://doi.org/10.3389/fphar.2014.00176.

Wingert, R. A., Brownlie, A., Galloway, J. L., et al. (2004). The chianti zebrafish mutant provides a model for erythroid-specific disruption of transferrin receptor 1. *Development (Cambridge, England), 131*(24), 6225–6235. https://doi.org/10.1242/dev.01540.

Wingert, R. A., Galloway, J. L., Barut, B., et al. (2005). Deficiency of glutaredoxin 5 reveals Fe-S clusters are required for vertebrate haem synthesis. *Nature, 436*(7053), 1035–1039. https://doi.org/10.1038/nature03887.

Winn, N. C., Volk, K. M., & Hasty, A. H. (2020). Regulation of tissue iron homeostasis: The macrophage "ferrostat". *JCI Insight, 5*(2), https://doi.org/10.1172/jci.insight.132964.

Wong, P., Hattangadi, S. M., Cheng, A. W., Frampton, G. M., Young, R. A., & Lodish, H. F. (2011). Gene induction and repression during terminal erythropoiesis are mediated by distinct epigenetic changes. *Blood, 118*(16), e128–e138. https://doi.org/10.1182/blood-2011-03-341404.

Wu, G., & Chen, P. L. (2008). Structural requirements of chromokinesin Kif4A for its proper function in mitosis. *Biochemical and Biophysical Research Communications, 372*(3), 454–458. https://doi.org/10.1016/j.bbrc.2008.05.065.

Xavier-Ferrucio, J., Scanlon, V., Li, X., et al. (2019). Low iron promotes megakaryocytic commitment of megakaryocytic-erythroid progenitors in humans and mice. *Blood, 134*(18), 1547–1557. https://doi.org/10.1182/blood.2019002039.

Xie, X., Zou, A., Zhang, L., et al. (2025). Rbm38 deficiency impairs erythroid heme biosynthesis and induces porphyria via reduced ferrochelatase expression. *Blood, 146*(25), 3106–3123. https://doi.org/10.1182/blood.2025028783.

Xu, W., Barrientos, T., Mao, L., Rockman, H. A., Sauve, A. A., & Andrews, N. C. (2015). Lethal cardiomyopathy in mice lacking transferrin receptor in the heart. *Cell Reports, 13*(3), 533–545. https://doi.org/10.1016/j.celrep.2015.09.023.

Ye, H., Jeong, S. Y., Ghosh, M. C., et al. (2010). Glutaredoxin 5 deficiency causes sideroblastic anemia by specifically impairing heme biosynthesis and depleting cytosolic iron in human erythroblasts. *The Journal of Clinical Investigation, 120*(5), 1749–1761. https://doi.org/10.1172/jci40372.

Yien, Y. Y., Robledo, R. F., Schultz, I. J., et al. (2014). TMEM14C is required for erythroid mitochondrial heme metabolism. *The Journal of Clinical Investigation, 124*(10), 4294–4304. https://doi.org/10.1172/jci76979.

Yien, Y. Y., Shi, J., Chen, C., et al. (2018). FAM210B is an erythropoietin target and regulates erythroid heme synthesis by controlling mitochondrial iron import and ferrochelatase activity. *The Journal of Biological Chemistry, 293*(51), 19797–19811. https://doi.org/10.1074/jbc.RA118.002742.

Zaretsky, A. G., Engiles, J. B., & Hunter, C. A. (2014). Infection-Induced changes in hematopoiesis. *The Journal of Immunology, 192*(1), 27–33. https://doi.org/10.4049/jimmunol.1302061.

Zenke-Kawasaki, Y., Dohi, Y., Katoh, Y., et al. (2007). Heme induces ubiquitination and degradation of the transcription factor Bach1. *Molecular and Cellular Biology, 27*(19), 6962–6971. https://doi.org/10.1128/mcb.02415-06.

Zhang, D., Gao, X., Li, H., et al. (2022). The microbiota regulates hematopoietic stem cell fate decisions by controlling iron availability in bone marrow. *Cell Stem Cell, 29*(2), 232–247.e7. https://doi.org/10.1016/j.stem.2021.12.009.

Zhang, D. L., Ghosh, M. C., & Rouault, T. A. (2014). The physiological functions of iron regulatory proteins in iron homeostasis - an update. *Frontiers in Pharmacology, 5*, 124. https://doi.org/10.3389/fphar.2014.00124.

Zhang, H., Wang, S., Liu, D., et al. (2021). EpoR-tdTomato-Cre mice enable identification of EpoR expression in subsets of tissue macrophages and hematopoietic cells. *Blood, 138*(20), 1986–1997. https://doi.org/10.1182/blood.2021011410.

Zhang, J., Socolovsky, M., Gross, A. W., & Lodish, H. F. (2003). Role of Ras signaling in erythroid differentiation of mouse fetal liver cells: Functional analysis by a flow cytometry-based novel culture system. *Blood, 102*(12), 3938–3946. https://doi.org/10.1182/blood-2003-05-1479.

Zhang, Q., Chen, C., Zou, X., et al. (2024a). Iron promotes ovarian cancer malignancy and advances platinum resistance by enhancing DNA repair via FTH1/FTL/POLQ/RAD51 axis. *Cell Death and Disease, 15*(5), 329. https://doi.org/10.1038/s41419-024-06688-5.

Zhang, S., Macias-Garcia, A., Velazquez, J., Paltrinieri, E., Kaufman, R. J., & Chen, J.-J. (2018). HRI coordinates translation by eIF2αP and mTORC1 to mitigate ineffective erythropoiesis in mice during iron deficiency. *Blood, 131*(4), 450–461. https://doi.org/10.1182/blood-2017-08-799908.

Zhang, T., Liu, Q., Chen, Q., & Wu, H. (2024b). Iron regulatory protein two facilitates ferritinophagy and DNA damage/repair through guiding ATG9A trafficking. *The Journal of Biological Chemistry, 300*(10), 107767. https://doi.org/10.1016/j.jbc.2024.107767.

CHAPTER THREE

Mediator at the Helm: Coordinating transcription and biomolecular condensates in hematopoiesis

Shicong Zhu[1], Yiyang Zhang[1], and Hsiang-Ying Lee*
State Key Laboratory of Membrane Biology, Peking-Tsinghua Center for Life Sciences, School of Life Sciences, Peking University, Beijing, P.R. China
*Corresponding author. e-mail address: slee@pku.edu.cn

Contents

[1] These authors contributed equally to this work.

Current Topics in Developmental Biology, Volume 170
ISSN 0070-2153, https://doi.org/10.1016/bs.ctdb.2026.06.002

Abstract

Hematopoiesis relies on precisely coordinated transcriptional programs that balance stem cell self-renewal, lineage commitment, and terminal differentiation. Central to this regulation is the Mediator complex, a large multi-subunit transcriptional co-regulator that integrates signals from transcription factors and chromatin regulators to control RNA polymerase II (Pol II) activity. The dynamic and modular composition of Mediator enables context-dependent transcriptional outputs, while individual subunits can exert specialized regulatory functions during hematopoietic lineage specification, thereby contributing to cell-fate-specific transcriptional outputs. Recent advances further reveal that transcriptional regulation is shaped by the spatial organization of regulatory machinery with biomolecular condensates formed through liquid-liquid phase separation (LLPS), particularly at super-enhancers. In this emerging framework, Mediator functions not only as a transcriptional integrator but also as a key coordinator of transcriptional machinery within condensates at cell-fate-related gene loci. In this chapter, we summarize how distinct Mediator subunits confer specific modes of transcriptional regulation and discuss how the interplay between Mediator and phase-separated condensates shapes transcriptional control during hematopoiesis. We highlight how specific subunits, including MED1 and MED26, participate in distinct regulatory modes in erythropoiesis, spanning super-enhancer-driven transcriptional activation, progenitor expansion, and condensate-associated mechanisms that influence Pol II pausing and global transcription repression during terminal differentiation. Together, these findings support a model in which Mediator integrates transcriptional regulation with nuclear organization through condensate-mediated mechanisms, providing a conceptual framework for understanding hematopoietic cell fate decisions and transcriptional dysregulation in hematological diseases.

1. The Mediator complex as a dynamic regulator of transcription

1.1 Structure, modular organization, and transcriptional function of Mediator

The Mediator complex is an evolutionarily conserved transcriptional co-regulator that coordinates communication between sequence-specific transcription factors and the RNA polymerase II (Pol II) transcription machinery from yeast to human (Kang et al., 2002; Malik et al., 2004; Nayak & Taatjes, 2022; Richter et al., 2022; Soutourina, 2018; Zhang et al., 2005). Structurally, Mediator is organized into four modules: the Head, Middle, Tail, and the CDK8 kinase module, also referred to as the Mediator kinase module (MKM) (Fig. 1A).

The Head, Middle, and Tail modules together constitute the core Mediator complex (cMED), whereas the kinase module associates with cMED in a reversible manner. Functionally, cMED promotes basal transcription by facilitating Pol II recruitment to TFIID-bound promoters, stabilizing assembly of the pre-initiation complex (PIC), stimulating phosphorylation of the Pol II C-terminal domain (CTD), and promoting

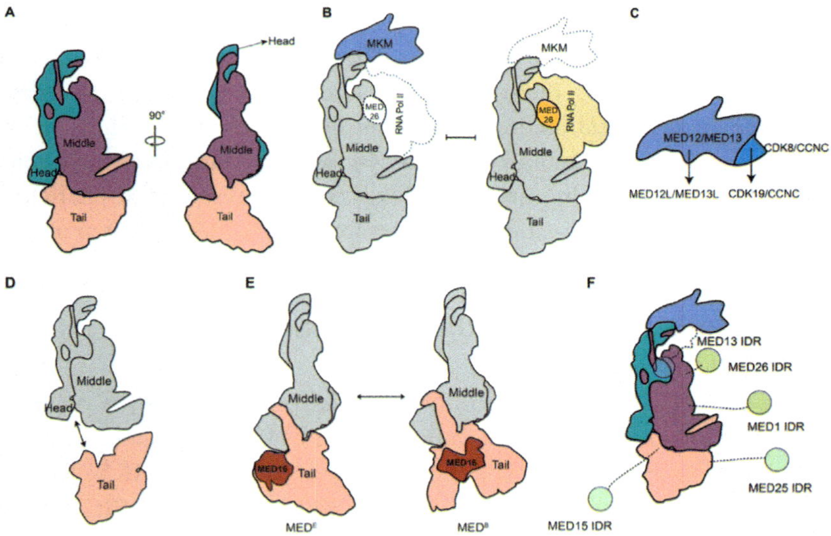

Fig. 1 Structural organization and alternative compositional states of the Mediator complex. (A). Schematic representation of the core Mediator complex (cMED), consisting of the Head module (shown in blue), Middle module (shown in violet) and Tail module (shown in pink). (B). Model illustrating the mutually exclusive association of the Mediator kinase module (MKM) and RNA polymerase II (Pol II) with the Mediator complex. (MKM is shown in blue; Pol II and MED26 are shown in light yellow and yellow, respectively) (Osman et al., 2021; Sato et al., 2004; Takahashi et al., 2011; Tsai et al., 2013; Zhu et al., 2024). (C). Paralogous subunits within the kinase module. In vertebrates, MED12, MED13, and CDK8 might be replaced by their paralogs MED12L, MED13L, and CDK19, potentially generating alternative kinase module assemblies (Luyties & Taatjes, 2022). (D). Model depicting dissociation of the Tail module from the Head-Middle module, forming a Tail subcomplex. (E). Two alternative conformations of the Mediator complex: the extended conformation (MED^{E}, left panel) and the bent conformation (MED^{B}, right panel), potentially associated with different MED16 isoforms (shown in red). (F). Intrinsically disordered regions (IDRs) within Mediator subunits, indicated by dotted lines and circles.

promoter escape (Nayak & Taatjes, 2022; Richter et al., 2022; Soutourina, 2018). By contrast, the kinase module acts as a dynamic regulatory unit that modulates Mediator activity and transcription output (Fig. 1B and C).

In both yeast and metazoans, the kinase module contains four subunits. In vertebrates, however, this module has acquired additional paralogous components, including MED12L, MED13L, and CDK19, corresponding to MED12, MED13 and CDK8, respectively. These paralogs retain substantial sequence similarity to their canonical counterparts (approximately 77 %, 67 %, and 57 %, respectively), yet assemble into mutually exclusive kinase modules, suggesting the existence of functionally distinct MKM variants (Luyties & Taatjes, 2022) (Fig. 1C).

Although structural biology has provided increasingly detailed views of Mediator architecture, these structures largely capture static snapshots. However, biochemical, genomic, and developmental studies indicate that Mediator is a highly dynamic complex whose composition can vary across cellular states and developmental contexts. This compositional plasticity is likely fundamental to its ability to generate context-dependent transcriptional outputs.

1.2 Dynamic assembly of the kinase module and transcriptional control

One of the best-characterized forms of Mediator flexibility is the reversible association of the CDK8 kinase module with cMED. *In vitro* biochemical studies in *Saccharomyces cerevisiae* showed that recombinant MKM binds to cMED and prevents cMED from engaging Pol II-containing PIC (Osman et al., 2021; Tsai et al., 2013). Importantly, MKM kinase activity weakens the interaction between MKM and cMED, thereby facilitating module dissociation and permitting cMED to interact with the PIC to stimulate transcription initiation and promoter escape (Osman et al., 2021; Tsai et al., 2013). Cryo-EM and cross-linking mass spectrometry further demonstrated that MKM interferes with Pol II binding to the Middle module of cMED, providing a structural basis for this antagonism (Tsai et al., 2013).

Elegant genetic experiments in yeast provided in vivo support for this model (Jeronimo & Robert, 2014; Jeronimo et al., 2016). Kin28, the yeast homolog of human cyclin-dependent kinase (CDK7), phosphorylates Ser5 of the Pol II CTD and is required for promoter escape and PIC turnover (Chen et al., 2018; Soutourina, 2018). Loss of *Kin28* function stabilizes Mediator at core promoters, revealing that Mediator complexes lacking MKM are preferentially associated with PIC in vivo (Jeronimo & Robert,

2014; Jeronimo et al., 2016; Wong et al., 2014). Together, these findings support a model in which MKM transiently and reversibly associates with cMED, thereby regulating the window for Pol II recruitment and PIC assembly, and promoter escape (Fig. 1B).

The existence of vertebrate-specific module paralogs raises an important unresolved question: do alternative kinase modules confer distinct biological functions? MED13L is a particularly intriguing candidate in this regard (Sato et al., 2004). Human genetics provides indirect support for functional divergence, as MED13L mutations are associated with congenital heart defects such as transposition of the great arteries (TGA), whereas MED13 homologs have distinct developmental roles in model organisms, including roles in *Drosophila* eye-antennal disc development (Muncke et al., 2003; Sato et al., 2004). However, direct structural and functional comparisons between canonical MKM and potential MED13L-containing MKM assemblies remain lacking (Fig. 1C).

1.3 Beyond repression: Context-dependent activating functions of CDK8

The kinase module has long been viewed primarily as a transcriptional repressor because of its mutually exclusive association with Pol II-bound Mediator (Knuesel et al., 2009; Osman et al., 2021; Tsai et al., 2013). However, this view is now clearly incomplete. A growing body of work shows that the kinase catalytic subunit CDK8/CDK19 can also positively regulate transcription in a highly context-dependent manner (Donner et al., 2010; Galbraith et al., 2013). *In vitro* studies have shown that CDK8 can phosphorylate both Ser2 and Ser5 residues of the Pol II CTD, with a preference for Ser5 (Ramanathan et al., 2001; Rickert et al., 1999), although its precise contribution to CTD phosphorylation in vivo remains unclear. Besides, CDK8 also phosphorylates a range of transcriptional effectors downstream of signaling pathways, including SREBP in lipid metabolism, SMAD proteins in TGF-β signaling, and STAT transcription factors in Wnt/β-catenin signaling contexts (Luyties & Taatjes, 2022). In addition, despite its antagonism with PIC-bound Pol II, CDK8 can modulate promoter-proximal pausing and facilitate productive elongation through functional interactions with P-TEFb (Donner et al., 2010; Galbraith et al., 2013).

Recent work further suggests that CDK8-mediated transcriptional activation depends primarily on kinase activity, whereas transcriptional repression may be an emergent property of the intact kinase module rather than of CDK8 enzymatic function alone (Chen et al., 2023). Thus, the biological output of MKM cannot be reduced to a simple repressive role; in some contexts, it also acts as a versatile signaling-responsive regulator of transcriptional dynamics.

1.4 MED26 and other subunit-specific regulatory modes

An important implication of Mediator compositional flexibility is that individual subunits can define distinct regulatory states. MED26 is a notable example. It was initially reported to exist in Mediator complexes lacking the CDK8 module (Taatjes et al., 2002), and multiple subsequent mass spectrometry studies supported the mutual exclusivity of MED26 and the MKM (Fig. 1B) (Sato et al., 2004; Takahashi et al., 2011; Zhu et al., 2024). Functionally, MED26 serves as a molecular switch between transcription initiation and elongation because it interacts with both the transcription initiation machinery and the super-elongation complex (SEC) (Takahashi et al., 2011). MED26 also occupies snRNA genes, where it recruits the Little Elongation Complex (LEC) and is essential for snRNA transcription (Takahashi et al., 2015).

Notably, MED26 appears dispensable in fully purified in vitro transcription systems but is required for transcription supported by nuclear extracts, suggesting that its function depends on additional cofactors present in the native nuclear environments (Cevher et al., 2014). More recently, genomic analyses in erythroblasts showed that MED26 is strongly associated with promoter-proximal pausing and contributes to global transcription repression during terminal erythroid differentiation (Zhu et al., 2024). Mechanistically, MED26 preferentially interacts with pausing factors compared with MED1, thereby helping maintain a paused Pol II state (Zhu et al., 2024). Intriguingly, MED26 contains a large intrinsically disordered region (IDR), spanning amino acids 88–480, and recent structural work suggests that an IDR in the kinase module subunit MED13 can block the MED26-CTD interaction, providing a possible mechanistic basis for the antagonism between MED26 and MKM (Zhao et al., 2024). Overall, current evidence supports a model in which MED26 marks a Mediator state specialized for regulating pausing and pause release, although more acute perturbation and mechanistically direct experiments will be needed to establish causality across developmental systems.

Besides MED26, several other metazoan-specific subunits further expand the regulatory complexity of the Mediator complex, including MED23, MED25, MED28, and MED30. MED28 and MED30 belong to the Head module, whereas MED23 and MED25 are Tail subunits. MED30 is critical for core Mediator stability and for cardiomyocyte transcriptional programs (Tan et al., 2021). MED23 and MED25, together with MED16 and MED24 form the "Saddle" region of the Tail, increasing both the size and complexity of metazoan Mediator relative to its yeast counterpart (Malik & Roeder, 2010). Although MED23 does not directly contact the core, deletion studies of MED23 suggest that it contributes to tail integrity, whereas MED25 appears more dispensable for structural stability but plays important roles in transcription factor recruitment (Currie et al., 2017; El Khattabi et al., 2019; Nayak & Taatjes, 2022). In erythropoiesis, MED25 expression increases during terminal differentiation and cooperates with MED1 to regulate a subset of erythroid genes, highlighting a coordinated role of these Mediator subunits in erythroid transcriptional control (Pope & Bresnick, 2013). Collectively, these metazoan-specific Mediator subunits likely provide an expanded interface for developmental and signaling inputs, thereby increasing regulatory plasticity in complex organisms.

1.5 Architectural backbones and conformational flexibility of the Tail

Within cMED, MED14 has long been regarded as the structural backbone because it spans all three core modules and contacts at least 14 subunits across a distance of more than 270 Å (Chen et al., 2021). More recent work indicates that MED17 may play a similarly central architectural role, interacting with 15 subunits and contributing not only to the integrity of the Head module but also to the overall organization of the complex (Chen et al., 2021).

The Tail module is particularly relevant for understanding Mediator plasticity. Although not essential for cell viability, Tail deletion slows proliferation and alters transcriptional responsiveness (El Khattabi et al., 2019). Historically, the Tail module has been viewed as a major interface for signal transduction pathways and transcription factor inputs (Malik & Roeder, 2010). MED16, MED23 and MED24 form a tightly associated Tail submodule that contacts multiple transcription factors, whereas MED16 serves as the principal link between the Tail and the Head-Middle-MED14 core (Chen et al., 2021; El Khattabi et al., 2019; Saleh et al., 2021). Deletion of Med16 disconnects Tail from cMED, and in yeast, Tail subunits such as Med2, Med3, and Med15 can persist as a stable subcomplex in the absence of Med16 (El Khattabi et al., 2019; Zhang et al., 2004) (Fig. 1D).

Cryo-EM studies have further revealed that the Tail is the most conformationally flexible module, adopting at least two distinct structural states termed the bent (MED^B) and extended (MED^E) conformations (Chen et al., 2021; Nayak & Taatjes, 2022). Intriguingly, human MED16 has two major isoforms that may correspond to these alternative conformations (Fig. 1E). The bent MED^B conformation was only recently resolved and may represent an additional regulatory state of the Mediator complex (El Khattabi et al., 2019; Zhao et al., 2021). Although the functional significance of these Tail states remains unknown, their existence suggests that conformational switching may provide an additional regulatory layer through which Mediator adapts to different transcriptional contexts.

1.6 Intrinsically disordered regions in Mediator and their functional implications

Despite major advances in Mediator structural biology, a substantial fraction of the complex remains unresolved in cryo-EM maps, indicating the presence of large intrinsically disordered regions (IDRs) (Richter et al., 2022). Structural studies often remove these disordered regions to improve sample homogeneity and facilitate structure determination, indirectly highlighting the contribution of these regions to Mediator complexity (Richter et al., 2022). As the biological importance of disordered proteins has become increasingly appreciated, these unresolved regions have emerged as key candidates for understanding Mediator function beyond static structural scaffolding. Notably, IDRs within Mediator subunits have been shown to promote liquid-liquid phase separation (LLPS) under physiologically relevant conditions, linking Mediator structural plasticity to the formation of transcriptional condensates (Boija et al., 2018; Nayak & Taatjes, 2022; Sabari et al., 2018).

Computational analyses indicate that many Mediator subunits contain extensive IDRs, particularly within the Middle and Tail modules (Toth-Petroczy et al., 2008). Interestingly, the overall degree of disorder in Mediator increases from yeast to humans and exceeds that of many other multiprotein assemblies of comparable size, suggesting evolutionary selection for disordered-dependent functions (Toth-Petroczy et al., 2008). In mammalian Mediator, large IDRs (>500 amino acids, a.a.) are particularly prominent in MED1 (a.a. 518–1581), MED14 (a.a. 1–50 and 969–1454), MED15 (a.a. 1–615), MED25 (a.a. 199–747), and MED26 (a.a. 88–480) (Chen et al., 2021; Richter et al., 2022; Toth-Petroczy et al., 2008) (Fig. 1F).

Among these, MED1 has been widely regarded as a representative subunit for studying the relationship between transcriptional regulation and phase separation (Boija et al., 2018; Sabari et al., 2018). IDRs in transcription factor activation domains (ADs) and in Mediator coactivators can drive condensate formation through multivalent weak interactions, thereby promoting transcriptional activation at super-enhancer associated genes (Boija et al., 2018). In particular, MED1 is enriched at super-enhancers and helps organize high-density transcriptional condensates associated with cell identity genes (Sabari et al., 2018; Zamudio et al., 2019). However, Mediator subunits are not present in equal stoichiometry within condensates, and emerging evidence suggests that distinct subunits can endow these assemblies with specialized biochemical functions.

Recent work supports this idea for MED26. In erythroblasts, MED26 and MED1 do not fully colocalize across the genome, and MED26-enriched chromatin regions are highly associated with transcriptional pausing (Zhu et al., 2024). MED26 exhibits robust phase separation capacity both in vitro and in vivo, with a lower saturation concentration than MED1, and its condensates preferentially recruit pausing factors (Zhu et al., 2024). This suggests that different Mediator subunits may nucleate biophysically distinct condensates with specialized functional outputs (Zhu et al., 2024). Similar observations have been reported for other subunits. MED25 forms liquid-like condensates in plants and is essential for phytochrome signaling (Guo et al., 2023), while MED15 forms nuclear foci whose assembly depends on both its Q-rich IDR and an adjacent hydrophobic region (Shi et al., 2021). Although MED14 is also predicted to contain substantial IDRs, experimental evidence for their functional roles remains limited. This may reflect the architectural centrality of MED14, which acts as an essential structural backbone of the Mediator complex.

Among Mediator subunits containing extensive intrinsically disordered regions, MED15 plays an important structural role by linking the Tail module to the remaining Mediator complex (Tsai et al., 2014). In contrast, MED1, MED26 and MED25 appear largely dispensable for maintaining Mediator structural integrity (Chen et al., 2021; Zhang et al., 2005). This distinction suggests that these IDR-rich subunits may function primarily as regulatory adaptors rather than structural scaffolds (Fig. 1F).

Based on these observations, we propose a conceptual model in which IDR-rich Mediator subunits act as modular "adaptors". In this view, these compositionally variable disordered subunits (MED1, MED26, MED25) provide flexible interfaces that selectively connect Mediator to different transcriptional regulators and confer specialized condensate properties, while the remaining core Mediator complex provides a stable structural platform. Much like interchangeable adaptors on a common device, these subunits could both diversify upstream inputs and encode distinct downstream outputs. This "adaptor" concept remains speculative, but it offers a helpful framework for organizing future studies on how Mediator disorder contributes to transcriptional specificity.

2. Phase separation and transcriptional condensates

2.1 Biomolecular condensates in gene regulation

Liquid-Liquid Phase Separation (LLPS) refers to the spontaneous demixing of a multicomponent system into two coexisting phases: a condensed phase enriched for selected biomolecules and a dilute phase in which those components are present at lower concentration (Boeynaems et al., 2018; Mehta & Zhang, 2022). In cells, LLPS is widely invoked as a mechanism for forming membraneless organelles, often referred to as biomolecular condensates, through multivalent interactions among proteins and nucleic acids. These assemblies often display liquid-like properties such as spherical morphology, rapid internal dynamics, and fusion upon contact (Alberti et al., 2019).

Transcriptional regulation has emerged as a paradigmatic biological context in which condensate-like organization may operate (Hnisz et al., 2017). Rather than relying solely on linear recruitment mechanisms, transcription factors, coactivators, Pol II, and general transcription factors can concentrate into spatially restricted hubs, increasing the local density of regulatory molecules and thereby facilitating gene regulation. This framework is especially relevant to super-enhancers and other regulatory loci that drive high-level expression of cell identity genes.

2.2 Super-enhancers and transcriptional condensates

Many transcription factors and PIC-associated factors, including Mediator, Pol II, and TFIID, contain IDRs that can promote condensate formation. Live-cell imaging in mouse embryonic stem cells showed that Mediator and Pol II dynamically cluster into condensate-like assemblies at super-enhancer-regulated genes (Cho et al., 2018; Sabari et al., 2018). Super-enhancers are operationally defined as clusters of closely spaced enhancer elements with exceptionally high occupancy of coactivators, lineage-defining transcription factors, or active chromatin marks such as H3K27ac (Pott & Lieb, 2015). Genes associated with super-enhancers are often critical for cell identity or oncogenic transformation (Hnisz et al., 2013; Whyte et al., 2013).

Mechanistically, transcription factors contribute both DNA-binding specificity and activation domains capable of engaging coactivator IDRs. Through these multivalent interactions, distant regulatory elements can be brought into close spatial proximity, favoring the assembly of transcriptional condensates enriched for Mediator and Pol II (Boija et al., 2018). These assemblies are therefore not merely passive accumulations of proteins, but emergent structures that spatially organize transcriptional activity. Importantly, small molecules can partition selectively into condensates based on physicochemical properties independent of canonical target affinity, suggesting a potential framework for condensate-based therapeutic strategies (Klein et al., 2020).

2.3 Dynamic transcription condensate composition during transcription initiation, pausing, and elongation

Transcription initiation begins with assembly of the pre-initiation complex (PIC) at gene promoters. General transcription factors, including TFIID, TFIIB, TFIIA, TFIIE, and TFIIH, cooperate with Pol II to establish a functional initiation complex (Sainsbury et al., 2015; Soutourina, 2018). The Mediator complex plays a central role in this process by linking enhancer-bound transcription factors with promoter-associated transcription machinery and stabilizing PIC assembly. Through these interactions, Mediator facilitates communication between enhancers and promoters and promotes transcription initiation.

Recent studies suggest that these regulatory factors do not assemble solely through linear recruitment but are frequently organized within transcriptional condensates. Many transcription factors, coactivators, and Pol II contain intrinsically disordered regions that promote multivalent interactions and enable the formation of dynamic biomolecular condensates at active regulatory loci. These assemblies concentrate transcriptional regulators in space, thereby facilitating efficient gene activation.

Importantly, the composition of transcriptional condensates changes dynamically during the transcription cycle. After initiation, Pol II typically pauses approximately 50–200 bp downstream of the promoter, a phenomenon known as promoter-proximal pausing (Adelman & Lis, 2012; Chen et al., 2018). This stage is regulated by pausing and elongation factors that control the transition to productive transcription. The recruitment and exchange of these regulators are thought to occur within transcriptional condensates, allowing rapid remodeling of the transcription machinery in response to regulatory signals (Guo et al., 2019).

A key mechanism underlying this dynamic behavior is the phosphorylation state of the Pol II CTD. The CTD is an intrinsically disordered region that undergoes extensive phosphorylation during transcription. Hypophosphorylated CTD preferentially associates with Mediator-containing initiation assemblies, whereas phosphorylation promotes dissociation from these complexes and facilitates association with elongation and RNA-processing factors. In addition to protein regulators, nascent RNA transcripts also influence condensate dynamics: low RNA levels can stabilize condensate formation, whereas higher RNA concentrations may promote their dissolution (Chen et al., 2015; Chen et al., 2017; Fischl et al., 2017; Wang et al., 2022) (Fig. 2).

Consistent with this dynamic view, several transcriptional regulators have been shown to form condensates that modulate different stages of transcription. For example, the pausing factor NELF can assemble into nuclear condensates under heat shock conditions and reinforce transcriptional pausing (Rawat et al., 2021). In erythroid cells, the Mediator subunit MED26 forms condensates that preferentially recruit pausing factors and help maintain global transcription pausing during terminal differentiation (Zhu et al., 2024). Conversely, elongation factors such as the super elongation complex (SEC) can assemble into phase-separated droplets that rapidly promote pause release and transcriptional elongation (Guo et al., 2019).

These observations support a model in which transcription proceeds through distinct but interconnected condensate states. One conceptual framework proposes two major types of transcriptional assemblies: promoter-associated condensates enriched in transcription factors, Mediator, and initiation machinery, and gene-body condensates enriched in elongation and RNA-processing factors (Cramer, 2019). RNA polymerase II is thought to shuttle between these assemblies as its CTD phosphorylation state changes during the transcription cycle (Cramer, 2019). Together, these findings suggest that transcription is coordinated through dynamic molecular assemblies rather than static protein complexes. Given the central role of Mediator in organizing transcriptional machinery and regulatory interactions, changes in Mediator composition

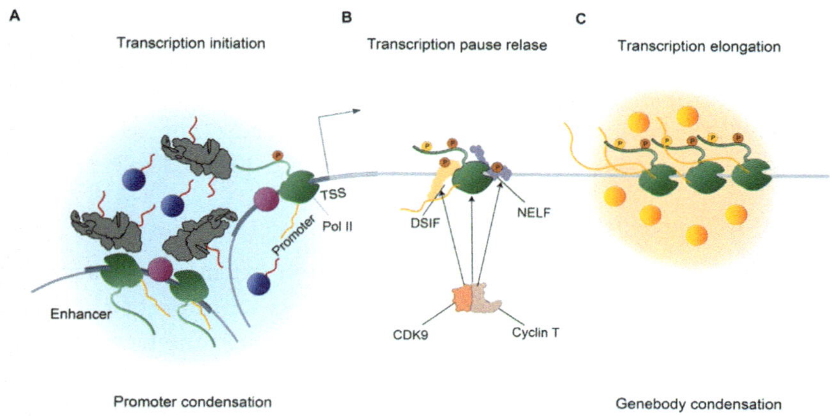

Fig. 2 Dynamic composition of transcription condensates during transcription initiation, pausing, and elongation. (A). Promoter-associated condensate (shown as circular light blue shade) composition, including transcription factors (shown as violet circles), coactivators (blue circles), the Mediator complex (grey modules), RNA polymerase II (green modules), nascent transcripts (yellow lines). During transcription initiation, IDRs within transcription factors, coactivators, Mediator, and the Pol II CTD contribute to the formation of promoter-associated condensates. (B). Mechanism of transcriptional pause release: CDK9-Cyclin T1 (shown as orange and pink modules) phosphorates pausing factors DSIF (show as yellow modules) and NELF (show as dark blue modules), leading to their dissociation from Pol II. At the same time, CDK9 also phosphorates Ser2 residues of the Pol II CTD (shown in small yellow circles), promoting the transition from paused transcription to productive elongation. (C). Condensate remodeling during transcription elongation. As nascent RNA elongates, promoter-associated condensates gradually disassemble, while elongation and RNA-processing factors (shown as dark yellow circle) are recruited, facilitating the formation of gene-body transcription condensate (shown as circular yellow shade).

may significantly influence the formation and functional properties of transcription condensates (Fig. 2). This concept is particularly relevant in developmental systems such as hematopoiesis, where lineage-specific transcription programs require precise and dynamic regulation.

3. Mediator complex and phase separation in hematopoiesis and erythropoiesis

The Mediator complex has traditionally been viewed as a general bridge linking transcription factors (TFs) to Pol II, thereby regulating transcription globally. However, recent systematic analyses of Mediator subunit depletion have revealed that individual subunits often exert highly specific effects in certain developmental contexts, suggesting that Mediator does not function solely as a uniform transcriptional scaffold (El Khattabi et al., 2019). In hematopoiesis, accumulating evidence indicates that distinct Mediator subunits contributes to lineage specification, cellular function, and hematologic malignancies (Wang et al., 2021) (Table 1).

3.1 MED1 in erythropoiesis

Among Mediator subunits, MED1 has been most extensively characterized in erythropoiesis. MED1 functions as a key coactivator of the master erythroid transcription factor GATA1 and directly interacts with the N-terminal zinc finger domain of GATA1 to promote transcriptional activation (Stumpf et al., 2006). Consistent with this role, deletion of *Med1* in mice severely impairs erythroid development (Minami et al., 2008).

Interestingly, MED25 is also highly expressed in erythroid cells and is upregulated during human erythroid maturation. MED25 cooperates with MED1 to regulate a subset of genes that are largely independent of GATA1 (Pope & Bresnick, 2013). These findings represent one of the earliest examples showing that specific Mediator subunits can interact with lineage-defining transcription factors to influence lineage development.

Evidence suggests that this regulatory relationship may be evolutionarily conserved. In *Drosophila*, MED1 interacts with GATA-family transcription factors Serpent (Srp) to regulate hematopoietic crystal cell development (Immarigeon et al., 2019). In addition, MED1 may also have GATA1-independent roles in erythropoiesis, although the underlying mechanisms remain poorly understood (Mizuta et al., 2014). Beyond erythroid cells, MED1 has also been implicated in the function of several lymphoid cell types, including regulatory T cells and invariant natural killer T (iNKT) cells (Chaudhuri et al., 2024; Yue et al., 2011).

3.2 The kinase module in hematologic malignancies and immune cell development

The CDK kinase module of Mediator represents another component with prominent functions in hematopoiesis, largely mediated by the catalytic subunits CDK8 and CDK19. Pharmacological inhibition of CDK8/CDK19 has been shown to relieve differentiation blocks in myelodysplastic syndrome (MDS) and acute myeloid leukemia (AML), promoting differentiation of leukemia stem cells (LSCs) (Ajibade et al., 2023; Pakulska et al., 2021). Although these findings have generated interest in CDK8/19 inhibitors as potential therapeutic agents, the precise mechanisms underlying these effects

Table 1 Mediator subunits associated with hematopoietic development and disorders.

Mediator module	Mediator subunit	Relevant hematopoiesis function/dysfunction	Identified target TF/cofactor	References
Middle	MED1	Erythroid lineage commitment	GATA1	Stumpf et al. (2006), Stumpf et al. (2010)
		Tumor specific differentiation of Treg	Possibly BATF, IRF4 and AP-1	Chaudhuri et al. (2024)
		iNKT differentiation	USP22	Yue et al. (2011), Zhang et al. (2020)
	MED26	Erythroid maturation	PAF1, NELF	Zhu et al. (2024)
MKM	CDK8/CDK19	AML	Undetermined	Pelish et al. (2015)
		Tem to Treg differentiation	STAT5	Akamatsu et al. (2019)
	MED12	Maintenance of HSC	Possibly RUNX1 and GATA2 etc.	Aranda-Orgilles et al. (2016)
	MED12& CCNC	T/NK cell activation	COMPASS complex, Possibly STAT and AP-1	Arce et al. (2025), Biederstädt et al. (2025)
Tail	MED23	General T cell function	MEF2	Kasper et al. (2014)

remain incompletely understood. Recent studies further indicate that CDK8/CDK19 inhibition can restore T-cell homeostasis in primary immune thrombocytopenia. Mechanistically, inhibition of CDK8/CDK19 increases regulatory T cell (Treg) frequencies and ameliorates thrombocytopenia in a STAT5-dependent manner (Wang et al., 2026). Current evidence suggests that CDK8/CDK19 inhibition may activate super-enhancers associated with tumor suppressor or lineage-determining genes (Pelish et al., 2015), and kinase-independent function of CDK8 have also been reported in leukemia (Menzl et al., 2019).

Besides hematologic malignancies, MKM activity also influences immune cell differentiation. For instance, CDK8/19 promotes the conversion of effector or memory T cells toward regulatory T cells by reducing STAT5 phosphorylation (Akamatsu et al., 2019). In contrast to CDK8/CDK19, the MKM subunit MED12 has been shown to safeguard normal hematopoiesis by maintaining hematopoietic stem cell (HSC) self-renewal (Aranda-Orgilles et al., 2016). Deletion of *Med12* results in acute bone marrow failure due to the loss of specific enhancer activity, whereas depletion of other MKM subunits does not produce a comparable phenotype. This observation highlights the functional non-equivalence of individual Mediator kinase module subunits.

Recent studies have further revealed unexpected roles for MKM components in immunotherapy. Genetic inactivation of *MED12* or *CCNC* enhances the activity of CAR-T cells by increasing core Mediator occupancy at STAT- and AP-1-regulated enhancers (Freitas et al., 2022). Similar effects have been reported in CAR-NK cells following deletion of *MED12*, *MED13*, or *CCNC* (Biederstädt et al., 2025). In T cells, MED12 has also been shown to recruit the COMPASS methyltransferase complex, suggesting that its regulatory functions extend beyond the canonical kinase activity of CDK8/19 (Arce et al., 2025). Interestingly, depletion of several core Mediator subunits, such as MED11, MED14, and MED31, produces similar transcriptional effects, suggesting that MED12 may also function as a positive regulator of transcription through mechanisms independent of the kinase module.

3.3 Other Mediator subunits in hematopoiesis

Compared with MED1 and the kinase module, fewer studies have examined the roles of other Mediator subunits in hematopoiesis. MED23 has been implicated in lymphoid cell function and HSC differentiation in mice (Chen et al., 2018; Kasper et al., 2014). In T cell, MED23 appears to regulate gene expression through interactions with the transcription factor

MEF2, whereas other studies suggest that MED23 may influence cell fate by modulating histone H2B monoubiquitination through recruitment of E3 ubiquitin ligase complexes. However, the precise mechanisms underlying these effects remain to be clarified.

MED26 is an optional Mediator subunit that is mutually exclusive with the kinase module (Chao et al., 2024). Recent works shows that MED26 promotes terminal erythroid maturation through condensate formation and enhanced promoter-proximal Pol II pausing as discussed above (Zhu et al., 2024). Global regulation of transcriptional pausing has previously been implicated in terminal erythropoiesis (Lv et al., 2023). However, how such global transcriptional changes give rise to lineage-specific developmental outcomes remains an important unresolved question.

3.4 Conceptual challenges

Together, these studies indicate that only a subset of Mediator subunits have well-defined roles in hematopoiesis, and the underlying mechanisms remain incompletely understood. A central conceptual challenge is how perturbation of a generally required transcription complex such as Mediator can lead to highly lineage-specific phenotypes.

One commonly proposed model is that individual Mediator subunits achieve specificity through interactions with lineage-defining transcription factors, as illustrated by the MED1-GATA1 interaction. However, many Mediator subunits interact with numerous transcription factors and cofactors, making it difficult to determine which interactions are functionally most important (Allen & Taatjes, 2015). Moreover, these interactions are often weak and transient, further complicating mechanistic analysis.

Another possibility is that certain Mediator subunits function semi-independently of the canonical Mediator complex or form alternative subcomplexes with distinct regulatory properties. Some biochemical studies have suggested variable stoichiometry among Mediator subunits, raising the possibility that alternative assemblies may contribute to transcription regulation (Zhang et al., 2005). However, direct experimental evidence for such mechanisms remains limited.

3.5 Phase separation in normal hematopoiesis

Beyond Mediator itself, an increasing number of proteins involved in transcriptional regulation and chromatin organization participate in phase separation during hematopoiesis. However, studies directly addressing phase separation in normal hematopoiesis remain relatively limited.

One example involves the myeloid master transcription factor CEBPA, which can drive three-dimensional genome remodeling through IDR-mediated phase separation and form condensates with MED1 to activate myeloid transcriptional programs (Christou-Kent et al., 2023). The nuclear speckle scaffold protein SON has also been shown to regulate hematopoietic stem cell maintenance and engraftment potential, in part by suppressing the expression of the inflammatory chemokine CCL5 (Luo et al., 2023).

In hematopoietic stem and progenitor cells, ASXL1, an epigenetic regulator frequently mutated in myeloid malignancies and involved in H2AK119 ubiquitination, interacts with the RNA-binding protein NONO and participates in phase separation within paraspeckles. Disruption of the ASXL1 intrinsically disordered region impairs paraspeckle assembly and is associated with defective hematopoietic stem and progenitor cell reconstitution (Yamamoto et al., 2021).

3.6 Phase separation in hematologic malignancies

Phase separation has also emerged as an important mechanism in blood cancers, particularly acute myeloid leukemia (AML). Several leukemia-associated proteins contain intrinsically disordered regions and are capable of forming biomolecular condensates that influence transcriptional and post-transcriptional regulation.

Mutations in NPM1 are among the most common genetic alterations in AML. The encoded protein nucleophosmin is an abundant nucleolar protein with multiple cellular functions, including histone chaperone activity and ribosome biogenesis (Falini et al., 2024). Leukemia-associated NPM1 mutants acquire additional nuclear export signals and become predominantly localized in the cytoplasm (Falini et al., 2024). Despite this mislocalization, NPM1 mutations drive leukemogenesis through activation of HOX gene expression programs (Brunetti et al., 2018). Although the precise mechanism remains debated, several studies suggest that aberrant condensate formation involving mutant NPM1 contributes to its oncogenic function (Datar et al., 2023; Zhu et al., 2025).

Another frequently altered gene in leukemia is NUP98, which forms fusion proteins with a variety of chromatin regulators following chromosome translocations. The fusion proteins typically retain the C-terminal FG-repeat domain of NUP98, an IDR with phase separation potential. In many cases, the fusion partners contribute DNA-binding or chromatin regulatory domains. For example, HOX family partners usually retain their DNA-binding domains, whereas other partners contain coiled-coil motifs or epigenetic reader and writer domains, such as PHD or SET domains. Similar to NPM1 mutations, NUP98 fusion proteins frequently activate *HOXA* gene expression programs (Michmerhuizen et al., 2020; SM et al., 2011).

Biophysical studies have shown that FG-repeat domains of NUP98 can undergo phase separation, and accumulating evidence indicates that NUP98 fusion proteins exploit this property to recruit transcriptional cofactors through condensate formation (Broder & Görlich, 2015; Michmerhuizen et al., 2020). For instance, the NUP98-HOXA9 fusion protein forms nuclear condensates that enhances chromatin binding, promote target gene activation, and facilitate long-range chromatin interactions (Chandra et al., 2022). Similar condensate-mediated transcriptional regulation has also been reported for other NUP98 fusion proteins, such as NUP98-KDM5A (Terlecki-Zaniewicz et al., 2021). These observations indicate that FG-repeat region of NUP98 provides a conserved phase-separation module that drives oncogenic transcriptional programs across multiple NUP98 fusion proteins (Chandra et al., 2022). Similar condensate-dependent mechanisms have also been described for other leukemia-associated fusion proteins, such as SET::NUP214, which forms phase-separated assemblies with XPO1 (CRM1) and KMT2A (MLL1) (Oka et al., 2023).

Condensate-mediated transcriptional activation is also observed in other leukemia-associated fusion proteins. In acute promyelocytic leukemia (APL), the PML-RARα fusion protein forms abnormal nuclear micro-speckles that recruit transcriptional coactivators such as BRD4, thereby activating super-enhancers and broad promoters critical for leukemogenesis (Zhang et al., 2024). In this context, disruption of normal PML nuclear bodies plays an important role. The neddylation of PML-RARα interferes with PML nuclear body nucleation, thereby altering condensate organization and contributing to leukemic transformation (Shao et al., 2022).

In addition to transcriptional regulators, RNA-binding proteins involved in RNA metabolism and translation can also form condensates that contributes to leukemia progression. The m^6A reader YTHDC1 forms enlarged condensates in AML cells within nuclear speckles and super-enhancer-associated compartments, which stabilizes specific transcripts and inhibits cellular differentiation (Cheng et al., 2021). Fibrillarin (FBL), a nucleolar RNA-binding protein involved in rRNA modification, has also been identified as a regulator of AML cell survival, likely through IDR-mediated condensate formation supporting oncogenic translation programs (Yang et al., 2024). Similarly, the cytoplasmic RNA-binding protein PABPC1 promotes translation in chronic myelogenous leukemia (CML) blast crisis by binding mRNAs with long and structured 5' UTRs and recruiting translation initiating factors through phase separation (Sun et al., 2025).

Beyond the functions of individual proteins, emerging evidence suggests that higher-order condensates themselves may serve as important regulatory hubs in leukemia. Several leukemia-associated proteins, including mutant NPM1, NUP98 fusion proteins, KMT2A, and the nuclear export factor XPO1, have been reported to co-localize in a membraneless organelle termed the coordinating body (C body) (Datar et al., 2025). This structure appears to be essential for AML cell survival and may provide a unified mechanistic framework linking multiple oncogenic pathways. Similarly, cytoplasmic P bodies, which depend on the RNA helicase DDX6 for their formation, act as reservoirs for mRNA metabolism. Disruption of P bodies impairs AML cell survival while having minimal effects on normal hematopoiesis (Bi et al., 2025; Kodali et al., 2024).

Together, these findings suggest that biomolecular condensates play a central role in the regulation of oncogenic transcriptional and RNA-processing programs in leukemia. Rather than acting solely through individual mutant proteins, leukemia-associated factors can reorganize transcriptional and RNA-processing machinery into aberrant condensates that sustain malignant gene expression programs. These findings raise the possibility that dysregulation of condensate organization may represent a broader principle in hematopoietic biology.

3.7 Phase separation in hematopoietic aging

While aberrant biomolecular condensates have been increasingly implicated in leukemia, emerging evidence suggests that dysregulation of condensate dynamics may also contribute to hematopoietic aging. The mechanisms underlying hematopoietic aging are complex and multifactorial, involving alterations in transcriptional regulation, chromatin organization, and stem cell homeostasis. Here we focus on recent studies suggesting that changes in phase separation may influence hematopoietic stem cell function, including reduced self-renewal capacity and lineage-biased differentiation. Although this area remains relatively unexplored, condensate dysregulation may represent an important and previously underappreciated aspect of hematopoietic aging.

Aberrant protein condensates have been implicated in multiple aging processes and age-related diseases. Recent studies show that the RNA-binding protein FUS is elevated in aged HSCs. Increased levels of FUS promote a phase transition that generates condensates with reduced molecular fluidity, which sequester the chromatin organizer CTCF and disrupt topologically associating domain (TAD) insulation, leading to activation of aging-associated gene expression programs (Tang et al., 2024).

In addition to such protein-specific changes, global alterations in condensate biology have also been observed in aged HSCs. Reduced production of proteins containing IDRs has been linked to decreased expression of the chaperone peptidyl-prolyl isomerase A (PPIA). Loss of PPIA leads to impaired condensate formation and defects in protein translation, and *Ppia* knockout mice recapitulate several phenotypes associated with hematopoietic aging (Maneix et al., 2024). These findings highlight the importance of phase separation in maintaining normal HSC function and suggest that systemic changes in condensate dynamics may contribute to the age-related decline in hematopoietic function.

Together, these observations raise the possibility that condensate organization represents an additional regulatory layer in hematopoietic biology. However, the specific roles of different condensate types in aging HSCs remain poorly defined and represent an important direction for future investigation.

4. Perspective

4.1 Addressing Mediator regulatory mechanisms at high temporal resolution

Most mechanistic studies of Mediator function rely on prolonged loss-of-function approaches such as CRISPR-mediated knockout (Ran et al., 2013) or shRNA-mediated knockdown (Rao et al., 2009). These perturbations typically operate on timescales of several days and are therefore well suited for developmental or phenotypic studies, but are less ideal for dissecting rapid transcriptional regulatory mechanisms.

Recently developed protein degradation systems enable much higher temporal resolution. For example, the auxin-inducible degron (AID) system and the dTAG system allow rapid degradation of target proteins within hours or even minutes (Nabet et al., 2018; Natsume et al., 2016). In the AID system, an auxin-inducible degron tag is fused to the target protein, and recognized by the *Oryza sativa* E3 ligase OsTIR1 upon auxin treatment, triggering rapid degradation (Natsume et al., 2016). Similarly, the dTAG system employs an FKBP12^{F36V} degron fused to the protein of interest, enabling ligand-induced recruitment of the cereblon (CRBN) E3 ubiquitin ligase and acute protein depletion (Nabet et al., 2018).

At the transcriptional level, conventional RNA-seq primarily measures steady-state transcript levels accumulated over time. In contrast, mechanistic studies of transcription regulation require measurements of nascent transcription (Adelman & Lis, 2012). Over the past decade, several methods, including GRO-seq, PRO-seq and TT-seq, have been developed to profile nascent transcripts by incorporating labeled nucleotides into newly synthesized RNA followed by biochemical enrichment and sequencing (Adelman & Lis, 2012; Core et al., 2008; Mahat et al., 2016; Schwalb et al., 2016).

Combining rapid protein degradation systems with nascent transcription assays will allow Mediator-dependent transcriptional events to be examined with much higher temporal precision.

4.2 Tracking endogenous transcription condensations

4.2.1 Experimental challenges in identifying phase separation

A major challenge in the field is the reliable identification of phase separation in living cells, particularly for interactions mediated by IDRs. Fluorescence recovery after bleaching (FRAP) is widely used to assess molecular mobility within condensates (Ganser & Myong, 2020). However, FRAP cannot distinguish phase separation from other forms of weak multivalent interactions such as protein-DNA binding or transient macromolecular clustering (Alberti et al., 2019; Muzzopappa et al., 2022).

Chemical perturbation using 1,6-hexanediol treatment is another commonly used approach, but it disrupts weak hydrophobic interactions broadly and is highly nonspecific, making it unsuitable as a standalone validation method (Alberti et al., 2019; Gao et al., 2022). More rigorous approaches, such as half-bleaching experiments that assess preferential molecular movement within phase boundaries, provide stronger evidence for phase-separated behavior (Muzzopappa et al., 2022).

Importantly, not all biomolecular condensates exhibit purely liquid-like behavior. Some undergo transitions into viscoelastic, gel-like, or even solid-like states with distinct molecular dynamics and network-like structures (Tanaka, 2022). Such phase transitions are particularly common in RNA-rich condensates. Consequently, commonly used assays such as FRAP may not always accurately reflect phase behavior. Additional observations, including droplet fusion or measurements using atom force microscopy (AFM), can provide complementary evidence, although these approaches are often difficult to apply in living cells.

From a biochemical perspective, conventional affinity purification approaches, such as immunoprecipitation (IP) or pulldown assays, are poorly suited for detecting interactions mediated by IDRs, which are typically weak and transient (Zhou, 2012). Moreover, these methods are highly sensitive to protein abundance and buffer conditions that may disrupt condensates, limiting their ability to capture the full interaction landscape within condensates.

In vitro reconstruction therefore remains a powerful strategy for studying condensate formation and generating phase diagrams. However, purified systems may not fully recapitulate the crowded nuclear environment (Gao et al., 2022). For example, the chromosomal passenger complex (CPC) readily undergoes phase separation under standard in vitro conditions but dissolves in cytomimetic environments (Hedtfeld et al., 2024).

To overcome these limitations, proximal labelling approaches have emerged as effective tools for mapping transient interaction networks within condensates (Qin et al., 2021). Photocatalytic proximal labelling further improves spatial and temporal resolution. For example, μMAP has been used to monitor stress granule dynamics through photocatalytic biotin labeling of G3BP1, revealing new interaction partners and mechanisms of ubiquitin-dependent stress granule disassembly (Pan et al., 2025). Genetically encoded photoactivated proximality labelling approaches have also been developed to capture RNA components within stress granules (Ren et al., 2023).

Advanced imaging approaches are providing new insights into the dynamics of transcriptional condensates, particularly in living cells. Confocal microscopy remains a commonly used method for visualizing condensates in cells; however, its spatial resolution is limited by the diffraction limit of light, and many transcriptional condensates fall below this size threshold. Beyond simply tracking proteins themselves, various environment-sensitive fluorescent probes have been developed to measure the biophysical properties of condensates, including viscosity, molecular clustering, and phase transitions (Owyong et al., 2023). Nevertheless, these probes are often insufficient for directly detecting LLPS.

More advanced imaging techniques, including single particle tracking (SPT) and fluorescence correlation spectroscopy (FCS), provide deeper insights into condensate dynamics both in vitro and in vivo (Ji et al., 2023).

FCS enables quantitative measurement of protein concentration in situ and characterization of molecular diffusion dynamics, often providing more detailed information than FRAP (Incicco et al., 2023). Its extension, fluorescence cross-correlation spectroscopy (FCCS), allows detection of interactions between proteins through correlated molecular diffusion (Bacia et al., 2006). Meanwhile, SPT can resolve the diffusion behavior of individual molecules within condensates, and theoretical studies suggest that tracking molecular dynamics across phase boundaries can reveal partition coefficients and boundary properties (Bo et al., 2021). In transcriptional regulation, SPT has also been used to directly observe transcription factor target-searching dynamics (Izeddin et al., 2014). Whether phase separation facilitates transcription factor trapping and thereby enhances transcriptional activation remains a subject of debate (Yang et al., 2024), but SPT-based approaches may provide a way to address this question.

Imaging approaches can also probe molecular interactions within condensates. Single-molecule Förster resonance energy transfer (smFRET) and fluorescence lifetime microscopy (FLIM) enable analysis of protein interactions and conformational changes in highly dynamic environments (Ha et al., 2024). In particular, FILM-FRET measurements allow detection of molecular interactions largely independent of protein concentration (Torrado et al., 2024).

These advanced imaging methods have been widely applied in vitro to investigate condensate biophysics, revealing highly dynamic molecular organization (Galvanetto et al., 2023; Wu et al., 2025). However, their application in living cells remains limited due to technical challenges in instrumentation and data analysis. For example, combined FCS, FCCS and FLIM imaging has been used to monitor stress granule assembly and track interactions between G3BP1 and FUS in living cells (Perego et al., 2023).

Although still underexplored in transcriptional systems, those approaches hold great promise for future studies. Recent live-cell super-resolution imaging has linked transcriptional bursting to the presence of transcriptional condensates (Du et al., 2024), and SPT may further reveal how transcription factors and coactivators dynamically operate within these structures (Budhathoki et al., 2025).

4.3 Design of Mediator "adaptor" proteins based on the molecular grammar of intrinsically disordered regions

The emerging framework of LLPS has provided a new conceptual basis for understanding transcriptional regulation. Beyond explaining how transcriptional condensates form, this framework also opens the possibility of engineering gene regulatory systems based on sequence features of IDRs. The Mediator complex, which contains multiple large IDRs, may represent an ideal platform for exploring such principles. In recent years, biochemical and biophysical studies have begun to decode the "molecular grammar" that governs phase separation, providing a foundation for the rational desgin of artifical phase-separating regulators (Calvario et al., 2025; De La Cruz et al., 2024; Dignon et al., 2020; Rekhi et al., 2024; Wake et al., 2025; Wang et al., 2018).

Early models suggested that aromatic residues such as tyrosine and positively charged residues such as arginine function as canonical "stickers" that drive LLPS through π-π and cation-π interactions (Wake et al., 2025; Wang et al., 2018). However, subsequent studies of IDR-containing proteins such as FUS and NCOA1 have expanded this view, indicating that phase behavior arises from a more complex and continuous spectrum of residue contributions rather than a simple binary classification (Wake et al., 2025). Polar residues such as glutamine can promote phase separation through hydrogen bonding and interactions with aromatic or charged residues, whereas glycine residues may facilitate LLPS by increasing backbone flexibility and enabling favorable contacts. Between neighboring interaction motifs (Dignon et al., 2020; Wake et al., 2025). This balance between interaction-promoting residues and sequence flexibility ultimately determines the thermodynamic landscape of condensate formation.

The Mediator-condensate axis provides a compelling biological context in which these principles may operate. For example, recent structural work suggests that the IDR of MED13 within the MKM module can compete with the IDR of MED26 for interactions with the RNA polymerase II CTD, thereby modulating transcriptional output (Zhao et al., 2024).

Sequence analyses also reveal distinct compositional biases among Mediator IDRs: MED1 and MED26 are enriched in proline and serine residues, whereas MED15 and MED25 display strong proline-glutamine enrichment. Although the functional implications of these sequence features remain incompletely understood, they likely contribute to the diverse condensate behaviors observed among different Mediator subunits.

A deeper understanding of IDR molecular grammar may ultimately enable the rational design of synthetic Mediator "adaptor" molecules capable of modulating transcriptional condensates. Such engineered adaptors could potentially tune condensate assembly, recruit specific transcriptional regulators, or shape transcriptional outputs in a programmable manner. In the context of hematopoiesis, this strategy may provide a conceptual framework of manipulating lineage-specific transcriptional programs through synthetic regulation of the Mediator-condensate axis.

5. Summary and conclusions

The Mediator complex functions as a central integrator of transcriptional regulation, bridging enhancer-bound transcription factors with RNA polymerase II at gene promoters. Through this role, Mediator coordinates signal integration and transcriptional output across diverse developmental systems. In hematopoiesis and erythropoiesis, accumulating evidence demonstrates that Mediator plays a critical role in fine-tuning transcriptional programs that govern hematopoietic stem cell maintenance, lineage commitment, and terminal differentiation. The central theme emerging from this chapter is that Mediator achieves this regulatory versatility by integrating its molecular architecture and dynamic subunit composition with the emerging mechanism of liquid-liquid phase separation, establishing a functional "Mediator-condensate axis" that organizes transcriptional regulation in hematopoiesis.

Mediator's modular structure, including the Head, Middle, Tail, and kinase modules, is inherently dynamic. The presence of large IDRs within several Mediator subunits enables flexible interactions with transcription factors, RNA polymerase II, and additional regulatory proteins. These IDR-containing subunits can act as regulatory "adaptors," allowing Mediator to assemble into transcriptional condensates at enhancers and promoters. Such condensates provide spatial organization for transcriptional machinery and are particulary important at super-enhancers (SEs), where high local concentrations of transcriptional regulators drive robust expression of cell identity genes.

The hematopoietic system provides a clear example of how specialized Mediator subunits contributes to lineage-specific transcriptional programs. The coactivator MED1 are promotes erythroid progenitor expansion by facilitating the formation and stabilization of SE-associated condensates that activate erythroid gene expression. In contrast, MED26 plays a distinct role during terminal erythroid differentiation by regulating Pol II pausing and global transcription repression necessary for cellular maturation. In addition, the Mediator kinase module, including MED12 and CDK8, acts as an important signaling interface that regulates hematopoietic stem cell maintenance and contributes to transcriptional dysregulation in hematological malignancies. Disruption of these Mediator components can therefore lead to profound defects in hematopoietic development and disease.

Looking forward, a deeper understanding of the Mediator-condensate axis will require approaches capable of resolving transcriptional regulation at high spatial and temporal resolution in live cells. In particular, future studies should focus on how IDRs of Mediator subunits dynamically control condensate assembly, partner recruitment, and transcriptional output in response to developmental and signaling cues. Deciphering these IDR-driven interaction networks will provide an important framework for understanding how transcriptional condensates are organized in normal hematopoiesis.

Finally, the recognition that LLPS-driven transcriptional condensates underlie super-enhancer activity suggests new opportunities for therapeutic intervention. Because many oncogenes in hematological malignancies (e.g., acute myeloid leukemia, AML) depend on SE-associated condensates for their high expression levels, targeting the Mediator-condensate axismay represent a promising strategy for precision medicine. Small-molecules or engineered peptides capable of selectively disrupting pathological condensates could therefore provide new avenues for the treatment of hematopoietic disorders.

Acknowledgement

We thank all members of the Lee Laboratory for helpful comments and discussions. This work was supported by the National Key Research and Development Program of China (grants 2022YFA1103300 and 2024YFA1106900 to H.Y.L.) and the National Natural Science Foundation of China (grant 82370124 to H.Y.L.). H.Y.L. also acknowledges support from the Beijing Advanced Center of RNA Biology, the Peking-Tsinghua Center for Life Sciences, and the School of Life Sciences, Peking University.

References

Adelman, K., & Lis, J. T. (2012). Promoter-proximal pausing of RNA polymerase II: Emerging roles in metazoans. *Nature Reviews. Genetics, 13*(10), 720–731. https://doi.org/10.1038/nrg3293.

Ajibade, O., Pakulska, U., Rabinovich, E., Fromowitz, A., Sahu, S., Aminov, S., ... Verma, A. (2023). Novel clinically useful inhibitor of mediator complex, RVU120, relives differentiation block in MDS/AML. *Blood, 142*, 3225. https://doi.org/10.1182/blood-2023-186227.

Akamatsu, M., Mikami, N., Ohkura, N., Kawakami, R., Kitagawa, Y., Sugimoto, A., ... Sakaguchi, S. (2019). Conversion of antigen-specific effector/memory T cells into Foxp3-expressing Treg cells by inhibition of CDK8/19. *Science Immunology, 4*(40), eaaw2707. https://doi.org/10.1126/sciimmunol.aaw2707.

Alberti, S., Gladfelter, A., & Mittag, T. (2019). Considerations and challenges in studying liquid-liquid phase separation and biomolecular condensates. *Cell, 176*(3), 419–434. https://doi.org/10.1016/j.cell.2018.12.035.

Allen, B. L., & Taatjes, D. J. (2015). The Mediator complex: A central integrator of transcription. *Nature Reviews. Molecular Cell Biology, 16*(3), 155–166. https://doi.org/10.1038/nrm3951.

Aranda-Orgilles, B., Saldaña-Meyer, R., Wang, E., Trompouki, E., Fassl, A., Lau, S., ... Aifantis, I. (2016). MED12 regulates HSC-specific enhancers independently of Mediator kinase activity to control hematopoiesis. *Cell Stem Cell, 19*(6), 784–799. https://doi.org/10.1016/j.stem.2016.08.004.

Arce, M. M., Umhoefer, J. M., Arang, N., Kasinathan, S., Freimer, J. W., Steinhart, Z., ... Marson, A. (2025). Central control of dynamic gene circuits governs T cell rest and activation. *Nature, 637*(8047), 930–939. https://doi.org/10.1038/s41586-024-08314-y.

Bacia, K., Kim, S. A., & Schwille, P. (2006). Fluorescence cross-correlation spectroscopy in living cells. *Nature Methods, 3*(2), 83–89. https://doi.org/10.1038/nmeth822.

Bi, H., Li, W., Ren, L., Zhang, H., Dong, L., Chan, A., ... Su, R. (2025). DDX6 undergoes phase separation to modulate metabolic plasticity and chemoresistance. *Nature Communications,* 260. https://doi.org/10.1038/s41467-025-66966-4.

Biederstädt, A., Basar, R., Park, J.-M., Uprety, N., Shrestha, R., Reyes Silva, F., ... Rezvani, K. (2025). Genome-wide CRISPR screens identify critical targets to enhance CAR-NK cell antitumor potency. *Cancer Cell, 43*(11), 2069–2088.e2011. https://doi.org/10.1016/j.ccell.2025.07.021.

Bo, S., Hubatsch, L., Bauermann, J., Weber, C. A., & Jülicher, F. (2021). Stochastic dynamics of single molecules across phase boundaries. *Physical Review Research, 3*(4), 043150. https://doi.org/10.1103/PhysRevResearch.3.043150.

Boeynaems, S., Alberti, S., Fawzi, N. L., Mittag, T., Polymenidou, M., Rousseau, F., ... Fuxreiter, M. (2018). Protein phase separation: A new phase in cell biology. *Trends in Cell Biology, 28*(6), 420–435. https://doi.org/10.1016/j.tcb.2018.02.004.

Boija, A., Klein, I. A., Sabari, B. R., Dall'Agnese, A., Coffey, E. L., Zamudio, A. V., ... Young, R. A. (2018). Transcription factors activate genes through the phase-separation capacity of their activation domains. *Cell, 175*(7), 1842–1855.e1816. https://doi.org/10.1016/j.cell.2018.10.042.

Broder, H., & Görlich, S. D. (2015). Nup98 FG domains from diverse species spontaneously phase-separate into particles with nuclear pore-like permselectivity. *eLife, 4.* https://doi.org/10.7554/eLife.04251.

Brunetti, L., Gundry, M. C., Sorcini, D., Guzman, A. G., Huang, Y.-H., Ramabadran, R., ... Goodell, M. A. (2018). Mutant NPM1 maintains the Leukemic State through HOX expression. *Cancer Cell, 34*(3), https://doi.org/10.1016/j.ccell.2018.08.005.

Budhathoki, A., Pandey, G., Medhanie, F. A., Galeota-Sprung, J., Aporvari, M. S., Nguyen, E., & Spille, J.-H. (2025). BPS2025 - In vivo single-particle tracking in diffraction-sized biomolecular condensates. *Biophysical Journal, 124*(3, Supplement 1), 570a. https://doi.org/10.1016/j.bpj.2024.11.2963.

Calvario, J., Antunes, D., Cipriano, R., Kalafatovic, D., Mausa, G., & Pina, A. S. (2025). Understanding biases in liquid-liquid phase separation: Investigating amino acid enrichments in phase-separating proteins toward peptide design. *Biomacromolecules, 26*(11), 7247–7264. https://doi.org/10.1021/acs.biomac.4c00224.

Cevher, M. A., Shi, Y., Li, D., Chait, B. T., Malik, S., & Roeder, R. G. (2014). Reconstitution of active human core Mediator complex reveals a critical role of the MED14 subunit. *Nature Structural & Molecular Biology, 21*(12), 1028–1034. https://doi.org/10.1038/nsmb.2914.

Chandra, B., Michmerhuizen, N. L., Shirnekhi, H. K., Tripathi, S., Pioso, B. J., Baggett, D. W., ... Kriwacki, R. W. (2022). Phase separation mediates NUP98 fusion oncoprotein leukemic transformation. *Cancer Discovery, 12*(4), 1152–1169. https://doi.org/10.1158/2159-8290.CD-21-0674.

Chao, T.-C., Chen, S.-F., Kim, H. J., Tang, H.-C., Tseng, H.-C., Xu, A., ... Tsai, K.-L. (2024). Structural basis of the human transcriptional Mediator regulated by its dissociable kinase module. *Molecular Cell, 84*(20), 3932–3949.e3910. https://doi.org/10.1016/j.molcel.2024.09.001.

Chaudhuri, S. M., Weinberg, S. E., Wang, D., Yalom, L. K., Montauti, E., Iyer, R., ... Fang, D. (2024). Mediator complex subunit 1 architects a tumorigenic Treg cell program independent of inflammation. *Cell Reports Medicine, 5*(3), 101441. https://doi.org/10.1016/j.xcrm.2024.101441.

Chen, F. X., Smith, E. R., & Shilatifard, A. (2018). Born to run: Control of transcription elongation by RNA polymerase II. *Nature Reviews. Molecular Cell Biology, 19*(7), 464–478. https://doi.org/10.1038/s41580-018-0010-5.

Chen, F. X., Woodfin, A. R., Gardini, A., Rickels, R. A., Marshall, S. A., Smith, E. R., ... Shilatifard, A. (2015). PAF1, a molecular regulator of promoter-proximal pausing by RNA polymerase II. *Cell, 162*(5), 1003–1015. https://doi.org/10.1016/j.cell.2015.07.042.

Chen, F. X., Xie, P., Collings, C. K., Cao, K., Aoi, Y., Marshall, S. A., ... Shilatifard, A. (2017). PAF1 regulation of promoter-proximal pause release via enhancer activation. *Science (New York, N. Y.), 357*(6357), 1294–1298. https://doi.org/10.1126/science.aan3269.

Chen, M., Li, J., Zhang, L., Wang, L., Cheng, C., Ji, H., ... Roninson, I. B. (2023). CDK8 and CDK19: Positive regulators of signal-induced transcription and negative regulators of Mediator complex proteins. *Nucleic Acids Research, 51*(14), 7288–7313. https://doi.org/10.1093/nar/gkad538.

Chen, X., Yin, X., Li, J., Wu, Z., Qi, Y., Wang, X., ... Xu, Y. (2021). Structures of the human Mediator and Mediator-bound preinitiation complex. *Science (New York, N. Y.), 372*(6546), https://doi.org/10.1126/science.abg0635.

Chen, X., Zhao, J., Gu, C., Cui, Y., Dai, Y., Song, G., ... Liu, X. (2018). Med23 serves as a gatekeeper of the myeloid potential of hematopoietic stem cells. *Nature Communications, 9*(1), 3746. https://doi.org/10.1038/s41467-018-06282-2.

Cheng, Y., Xie, W., Pickering, B. F., Chu, K. L., Savino, A. M., Yang, X., ... Kharas, M. G. (2021). N6-Methyladenosine on mRNA facilitates a phase-separated nuclear body that suppresses myeloid leukemic differentiation. *Cancer Cell, 39*(7), 958–972.e958. https://doi.org/10.1016/j.ccell.2021.04.017.

Cho, W. K., Spille, J. H., Hecht, M., Lee, C., Li, C., Grube, V., & Cisse, I. I. (2018). Mediator and RNA polymerase II clusters associate in transcription-dependent condensates. *Science (New York, N. Y.), 361*(6400), 412–415. https://doi.org/10.1126/science.aar4199.

Christou-Kent, M., Cuartero, S., Garcia-Cabau, C., Ruehle, J., Naderi, J., Erber, J., ... Stik, G. (2023). CEBPA phase separation links transcriptional activity and 3D chromatin hubs. *Cell Reports, 42*(8), 112897. https://doi.org/10.1016/j.celrep.2023.112897.

Core, L. J., Waterfall, J. J., & Lis, J. T. (2008). Nascent RNA sequencing reveals widespread pausing and divergent initiation at human promoters. *Science (New York, N. Y.), 322*(5909), 1845–1848. https://doi.org/10.1126/science.1162228.

Cramer, P. (2019). Organization and regulation of gene transcription. *Nature, 573*(7772), 45–54. https://doi.org/10.1038/s41586-019-1517-4.

Currie, S. L., Doane, J. J., Evans, K. S., Bhachech, N., Madison, B. J., Lau, D. K. W., ... Graves, B. J. (2017). ETV4 and AP1 transcription factors form multivalent interactions with three sites on the MED25 activator-interacting domain. *Journal of Molecular Biology, 429*(20), 2975–2995. https://doi.org/10.1016/j.jmb.2017.06.024.

Datar, G., Anand, A., Sadek, M., Potolitsyna, E., Dollinger, C., Laserna, E., ... Goodell, M. (2023). Nuclear condensates are a therapeutic vulnerability in NPM1-mutant acute myeloid leukemia. *Blood, 142*(Supplement 1), https://doi.org/10.1182/blood-2023-191107.

Datar, G. K., Khabusheva, E., Anand, A., Beale, J., Sadek, M., Chen, C.-W., ... Riback, J. A. (2025). Disparate leukemia mutations converge on nuclear phase-separated condensates. *Cell, 188*(25), 7118–7136.e7121. https://doi.org/10.1016/j.cell.2025.10.010.

De La Cruz, N., Pradhan, P., Veettil, R. T., Conti, B. A., Oppikofer, M., & Sabari, B. R. (2024). Disorder-mediated interactions target proteins to specific condensates. *Molecular Cell, 84*(18), 3497–3512.e3499. https://doi.org/10.1016/j.molcel.2024.08.017.

Dignon, G. L., Best, R. B., & Mittal, J. (2020). Biomolecular phase separation: From molecular driving forces to macroscopic properties. *Annual Review of Physical Chemistry, 71*, 53–75. https://doi.org/10.1146/annurev-physchem-071819-113553.

Donner, A. J., Ebmeier, C. C., Taatjes, D. J., & Espinosa, J. M. (2010). CDK8 is a positive regulator of transcriptional elongation within the serum response network. *Nature Structural & Molecular Biology, 17*(2), 194–201. https://doi.org/10.1038/nsmb.1752.

Du, M., Stitzinger, S. H., Spille, J.-H., Cho, W.-K., Lee, C., Hijaz, M., ... Cissé, I. I. (2024). Direct observation of a condensate effect on super-enhancer controlled gene bursting. *Cell, 187*(2), 331–344.e317. https://doi.org/10.1016/j.cell.2023.12.005.

El Khattabi, L., Zhao, H., Kalchschmidt, J., Young, N., Jung, S., Van Blerkom, P., ... Casellas, R. (2019). A pliable mediator acts as a functional rather than an architectural bridge between promoters and enhancers. *Cell, 178*(5), 1145–1158.e1120. https://doi.org/10.1016/j.cell.2019.07.011.

Falini, B., Sorcini, D., Perriello, V. M., Sportoletti, P., Falini, B., Sorcini, D., ... Sportoletti, P. (2024). Functions of the native NPM1 protein and its leukemic mutant. *Leukemia: Official Journal of the Leukemia Society of America, Leukemia Research Fund, U. K, 39*(2), 276–290. https://doi.org/10.1038/s41375-024-02476-4.

Fischl, H., Howe, F. S., Furger, A., & Mellor, J. (2017). Paf1 has distinct roles in transcription elongation and differential transcript fate. *Molecular Cell, 65*(4), 685–698.e688. https://doi.org/10.1016/j.molcel.2017.01.006.

Freitas, K. A., Belk, J. A., Sotillo, E., Quinn, P. J., Ramello, M. C., Malipatlolla, M., ... Mackall, C. L. (2022). Enhanced T cell effector activity by targeting the Mediator kinase module. *Science (New York, N. Y.), 378*(6620), eabn5647. https://doi.org/10.1126/science.abn5647.

Galbraith, M. D., Allen, M. A., Bensard, C. L., Wang, X., Schwinn, M. K., Qin, B., ... Espinosa, J. M. (2013). HIF1A employs CDK8-mediator to stimulate RNAPII elongation in response to hypoxia. *Cell, 153*(6), 1327–1339. https://doi.org/10.1016/j.cell.2013.04.048.

Galvanetto, N., Ivanović, M. T., Chowdhury, A., Sottini, A., Nüesch, M. F., Nettels, D., ... Schuler, B. (2023). Extreme dynamics in a biomolecular condensate. *Nature, 619*(7971), 876–883. https://doi.org/10.1038/s41586-023-06329-5.

Ganser, L. R., & Myong, S. (2020). Methods to study phase-separated condensates and the underlying molecular interactions. *Trends in Biochemical Sciences, 45*(11), 1004–1005. https://doi.org/10.1016/j.tibs.2020.05.011.

Gao, Y., Li, X., Li, P., & Lin, Y. (2022). A brief guideline for studies of phase-separated biomolecular condensates. *Nature Chemical Biology, 18*(12), 1307–1318. https://doi.org/10.1038/s41589-022-01204-2.

Guo, Q., Jing, Y., Gao, Y., Liu, Y., Fang, X., & Lin, R. (2023). The PIF1/PIF3-MED25-HDA19 transcriptional repression complex regulates phytochrome signaling in Arabidopsis. *The New Phytologist, 240*(3), 1097–1115. https://doi.org/10.1111/nph.19205.

Guo, Y. E., Manteiga, J. C., Henninger, J. E., Sabari, B. R., Dall'Agnese, A., Hannett, N. M., ... Young, R. A. (2019). Pol II phosphorylation regulates a switch between transcriptional and splicing condensates. *Nature, 572*(7770), 543. https://doi.org/10.1038/s41586-019-1464-0.

Ha, T., Fei, J., Schmid, S., Lee, N. K., Gonzalez, R. L., Paul, S., & Yeou, S. (2024). Fluorescence resonance energy transfer at the single-molecule level. *Nature Reviews Methods Primers, 4*(1), 21. https://doi.org/10.1038/s43586-024-00298-3.

Hedtfeld, M., Dammers, A., Koerner, C., & Musacchio, A. (2024). A validation strategy to assess the role of phase separation as a determinant of macromolecular localization. *Molecular Cell, 84*(9), 1783–1801.e1787. https://doi.org/10.1016/j.molcel.2024.03.022.

Hnisz, D., Abraham, B. J., Lee, T. I., Lau, A., Saint-Andre, V., Sigova, A. A., ... Young, R. A. (2013). Super-enhancers in the control of cell identity and disease. *Cell, 155*(4), 934–947. https://doi.org/10.1016/j.cell.2013.09.053.

Hnisz, D., Shrinivas, K., Young, R. A., Chakraborty, A. K., & Sharp, P. A. (2017). A phase separation model for transcriptional control. *Cell, 169*(1), 13–23. https://doi.org/10.1016/j.cell.2017.02.007.

Immarigeon, C., Bernat-Fabre, S., Augé, B., Faucher, C., Gobert, V., Haenlin, M., ... Boube, M. (2019). Drosophila mediator subunit Med1 is required for GATA-dependent developmental processes: Divergent binding interfaces for conserved coactivator functions. *Molecular and Cellular Biology, 39*(7), e00477–00418. https://doi.org/10.1128/MCB.00477-18.

Incicco, J. J., Roy, D., Stuchell-Brereton, M. D., & Soranno, A. (2023). Fluorescence correlation spectroscopy and phase separation. In H.-X. Zhou, J.-H. Spille, & P. R. Banerjee (Eds.). *Phase-separated biomolecular condensates: Methods and protocols* (pp. 161–198). Springer US. https://doi.org/10.1007/978-1-0716-2663-4_8.

Izeddin, I., Récamier, V., Bosanac, L., Cissé, I. I., Boudarene, L., Dugast-Darzacq, C., ... Darzacq, X. (2014). Single-molecule tracking in live cells reveals distinct target-search strategies of transcription factors in the nucleus. *eLife, 3*, e02230. https://doi.org/10.7554/eLife.02230.

Jeronimo, C., Langelier, M. F., Bataille, A. R., Pascal, J. M., Pugh, B. F., & Robert, F. (2016). Tail and kinase modules differently regulate core mediator recruitment and function in vivo. *Molecular Cell, 64*(3), 455–466. https://doi.org/10.1016/j.molcel.2016.09.002.

Jeronimo, C., & Robert, F. (2014). Kin28 regulates the transient association of Mediator with core promoters. *Nature Structural & Molecular Biology, 21*(5), 449–455. https://doi.org/10.1038/nsmb.2810.

Ji, J., Wang, W., & Chen, C. (2023). Single-molecule techniques to visualize and to characterize liquid-liquid phase separation and phase transition. *Acta Biochimica et Biophysica Sinica, 55*(7), 1023–1033. https://doi.org/10.3724/abbs.2023028.

Kang, Y. K., Guermah, M., Yuan, C. X., & Roeder, R. G. (2002). The TRAP/Mediator coactivator complex interacts directly with estrogen receptors alpha and beta through the TRAP220 subunit and directly enhances estrogen receptor function in vitro. *Proceedings of the National Academy of Sciences of the United States of America, 99*(5), 2642–2647. https://doi.org/10.1073/pnas.261715899.

Kasper, L. H., Fukuyama, T., & Brindle, P. K. (2014). T-Cells null for the MED23 subunit of mediator express decreased levels of KLF2 and inefficiently populate the peripheral lymphoid organs. *PLoS One, 9*(7), e102076. https://doi.org/10.1371/journal.pone.0102076.

Klein, I. A., Boija, A., Afeyan, L. K., Hawken, S. W., Fan, M., Dall'Agnese, A., ... Young, R. A. (2020). Partitioning of cancer therapeutics in nuclear condensates. *Science (New York, N. Y.), 368*(6497), 1386–1392. https://doi.org/10.1126/science.aaz4427.

Knuesel, M. T., Meyer, K. D., Bernecky, C., & Taatjes, D. J. (2009). The human CDK8 subcomplex is a molecular switch that controls Mediator coactivator function. *Genes & Development, 23*(4), 439–451. https://doi.org/10.1101/gad.1767009.

Kodali, S., Proietti, L., Valcarcel, G., López-Rubio, A. V., Pessina, P., Eder, T., ... Di Stefano, B. (2024). RNA sequestration in P-bodies sustains myeloid leukaemia. *Nature Cell Biology, 26*(10), 1745–1758. https://doi.org/10.1038/s41556-024-01489-6.

Luo, H., Cortés-López, M., Tam, C. L., Xiao, M., Wakiro, I., Chu, K. L., ... Kharas, M. G. (2023). SON is an essential m6A target for hematopoietic stem cell fate. *Cell Stem Cell, 30*(12), 1658–1673.e1610. https://doi.org/10.1016/j.stem.2023.11.006.

Luyties, O., & Taatjes, D. J. (2022). The Mediator kinase module: An interface between cell signaling and transcription. *Trends in Biochemical Sciences, 47*(4), 314–327. https://doi.org/10.1016/j.tibs.2022.01.002.

Lv, X., Murphy, K., Murphy, Z., Getman, M., Rahman, N., Nakamura, Y., ... Steiner, L. A. (2023). HEXIM1 is an essential transcription regulator during human erythropoiesis. *Blood, 142*(25), 2198–2215. https://doi.org/10.1182/blood.2022019495.

Mahat, D. B., Kwak, H., Booth, G. T., Jonkers, I. H., Danko, C. G., Patel, R. K., ... Lis, J. T. (2016). Base-pair-resolution genome-wide mapping of active RNA polymerases using precision nuclear run-on (PRO-seq). *Nature Protocols, 11*(8), 1455–1476. https://doi.org/10.1038/nprot.2016.086.

Malik, S., Guermah, M., Yuan, C. X., Wu, W., Yamamura, S., & Roeder, R. G. (2004). Structural and functional organization of TRAP220, the TRAP/mediator subunit that is targeted by nuclear receptors. *Molecular and Cellular Biology, 24*(18), 8244–8254. https://doi.org/10.1128/MCB.24.18.8244-8254.2004.

Malik, S., & Roeder, R. G. (2010). The metazoan Mediator co-activator complex as an integrative hub for transcriptional regulation. *Nature Reviews. Genetics, 11*(11), 761–772. https://doi.org/10.1038/nrg2901.

Maneix, L., Iakova, P., Lee, C. G., Moree, S. E., Lu, X., Datar, G. K., ... Catic, A. (2024). Cyclophilin a supports translation of intrinsically disordered proteins and affects haematopoietic stem cell ageing. *Nature Cell Biology, 26*(4), 593–603. https://doi.org/10.1038/s41556-024-01387-x.

Mehta, S., & Zhang, J. (2022). Liquid-liquid phase separation drives cellular function and dysfunction in cancer. *Nature Reviews. Cancer, 22*(4), 239–252. https://doi.org/10.1038/s41568-022-00444-7.

Menzl, I., Zhang, T., Berger-Becvar, A., Grausenburger, R., Heller, G., Prchal-Murphy, M., ... Sexl, V. (2019). A kinase-independent role for CDK8 in BCR-ABL1+ leukemia. *Nature Communications, 10*(1), 4741. https://doi.org/10.1038/s41467-019-12656-x.

Michmerhuizen, N. L., Klco, J. M., & Mullighan, C. G. (2020). Mechanistic insights and potential therapeutic approaches for NUP98-rearranged hematologic malignancies. *Blood, 136*(20), 2275–2289. https://doi.org/10.1182/blood.2020007093.

Minami, T., Horie, O., Inoue, K., Toji, S., Urahama, N., & Ito, M. (2008). GATA-1-specific coactivator MED1/TRAP220, transcriptional mediator subunit, mediates erythroid differentiation in K562 erythroleukemia cells. *Blood, 112*(11), 2259. https://doi.org/10.1182/blood.V112.11.2259.2259.

Mizuta, S., Minami, T., Fujita, H., Kaminaga, C., Matsui, K., Ishino, R., ... Ito, M. (2014). CCAR1/CoCoA pair-mediated recruitment of the Mediator defines a novel pathway for GATA1 function. *Genes to Cells, 19*(1), 28–51. https://doi.org/10.1111/gtc.12104.

Muncke, N., Jung, C., Rudiger, H., Ulmer, H., Roeth, R., Hubert, A., ... Rappold, G. (2003). Missense mutations and gene interruption in PROSIT240, a novel TRAP240-like gene, in patients with congenital heart defect (transposition of the great arteries). *Circulation, 108*(23), 2843–2850. https://doi.org/10.1161/01.CIR.0000103684.77636.CD.

Muzzopappa, F., Hummert, J., Anfossi, M., Tashev, S. A., Herten, D.-P., & Erdel, F. (2022). Detecting and quantifying liquid–liquid phase separation in living cells by model-free calibrated half-bleaching. *Nature Communications, 13*(1), 7787. https://doi.org/10.1038/s41467-022-35430-y.

Nabet, B., Roberts, J. M., Buckley, D. L., Paulk, J., Dastjerdi, S., Yang, A., ... Bradner, J. E. (2018). The dTAG system for immediate and target-specific protein degradation. *Nature Chemical Biology, 14*(5), 431–441. https://doi.org/10.1038/s41589-018-0021-8.

Natsume, T., Kiyomitsu, T., Saga, Y., & Kanemaki, M. T. (2016). Rapid protein depletion in human cells by auxin-inducible degron tagging with short homology donors. *Cell Reports, 15*(1), 210–218. https://doi.org/10.1016/j.celrep.2016.03.001.

Nayak, S., & Taatjes, D. J. (2022). SnapShot: Mediator complex structure. *Cell, 185*(18), 3458.e3451. https://doi.org/10.1016/j.cell.2022.07.011.

Oka, M., Otani, M., Miyamoto, Y., Oshima, R., Adachi, J., Tomonaga, T., ... Yoneda, Y. (2023). Phase-separated nuclear bodies of nucleoporin fusions promote condensation of MLL1/CRM1 and rearrangement of 3D genome structure. *Cell Reports, 42*(8), https://doi.org/10.1016/j.celrep.2023.112884.

Osman, S., Mohammad, E., Lidschreiber, M., Stuetzer, A., Bazso, F. L., Maier, K. C., ... Cramer, P. (2021). The Cdk8 kinase module regulates interaction of the mediator complex with RNA polymerase II. *The Journal of Biological Chemistry, 296*, 100734. https://doi.org/10.1016/j.jbc.2021.100734.

Owyong, T. C., Zhao, J., & Hong, Y. (2023). Small molecule fluorescent probes for the study of protein phase separation. *Current Opinion in Chemical Biology, 76*, 102354. https://doi.org/10.1016/j.cbpa.2023.102354.

Pakulska, U., Obacz, M., Goller, K., Combik, M., Keska, K., Zaucha, J., ... Rzymski, T. (2021). RVU120 (SEL120) CDK8/19 Inhibitor - a drug candidate for the treatment of MDS can induce erythroid differentiation. *Blood, 138*, 1518. https://doi.org/10.1182/blood-2021-150662.

Pan, C., Knutson, S. D., Huth, S. W., & MacMillan, D. W. C. (2025). μMap proximity labeling in living cells reveals stress granule disassembly mechanisms. *Nature Chemical Biology, 21*(4), 490–500. https://doi.org/10.1038/s41589-024-01721-2.

Pelish, H. E., Liau, B. B., Nitulescu, I. I., Tangpeerachaikul, A., Poss, Z. C., Da Silva, D. H., ... Shair, M. D. (2015). Mediator kinase inhibition further activates super-enhancer-associated genes in AML. *Nature, 526*(7572), 273–276. https://doi.org/10.1038/nature14904.

Perego, E., Zappone, S., Castagnetti, F., Mariani, D., Vitiello, E., Rupert, J., ... Vicidomini, G. (2023). Single-photon microscopy to study biomolecular condensates. *Nature Communications, 14*(1), 8224. https://doi.org/10.1038/s41467-023-43969-7.

Pope, N. J., & Bresnick, E. H. (2013). Establishment of a cell-type-specific genetic network by the mediator complex component Med1. *Molecular and Cellular Biology, 33*(10), 1938–1955. https://doi.org/10.1128/MCB.00141-13.

Pott, S., & Lieb, J. D. (2015). What are super-enhancers? *Nature Genetics, 47*(1), 8–12. https://doi.org/10.1038/ng.3167.

Qin, W., Cho, K. F., Cavanagh, P. E., & Ting, A. Y. (2021). Deciphering molecular interactions by proximity labeling. *Nature Methods, 18*(2), 133–143. https://doi.org/10.1038/s41592-020-01010-5.

Ramanathan, Y., Rajpara, S. M., Reza, S. M., Lees, E., Shuman, S., Mathews, M. B., & Pe'ery, T. (2001). Three RNA polymerase II carboxyl-terminal domain kinases display distinct substrate preferences. *The Journal of Biological Chemistry, 276*(14), 10913–10920. https://doi.org/10.1074/jbc.M010975200.

Ran, F. A., Hsu, P. D., Wright, J., Agarwala, V., Scott, D. A., & Zhang, F. (2013). Genome engineering using the CRISPR-Cas9 system. *Nature Protocols, 8*(11), 2281–2308. https://doi.org/10.1038/nprot.2013.143.

Rao, D. D., Vorhies, J. S., Senzer, N., & Nemunaitis, J. (2009). siRNA vs. shRNA: Similarities and differences. *Advanced Drug Delivery Reviews, 61*(9), 746–759. https://doi.org/10.1016/j.addr.2009.04.004.

Rawat, P., Boehning, M., Hummel, B., Aprile-Garcia, F., Pandit, A. S., Eisenhardt, N., ... Sawarkar, R. (2021). Stress-induced nuclear condensation of NELF drives transcriptional downregulation. *Molecular Cell, 81*(5), 1013–1026.e1011. https://doi.org/10.1016/j.molcel.2021.01.016.

Rekhi, S., Garcia, C. G., Barai, M., Rizuan, A., Schuster, B. S., Kiick, K. L., & Mittal, J. (2024). Expanding the molecular language of protein liquid-liquid phase separation. *Nature Chemistry, 16*(7), 1113–1124. https://doi.org/10.1038/s41557-024-01489-x.

Ren, Z., Tang, W., Peng, L., & Zou, P. (2023). Profiling stress-triggered RNA condensation with photocatalytic proximity labeling. *Nature Communications, 14*(1), 7390. https://doi.org/10.1038/s41467-023-43194-2.

Richter, W. F., Nayak, S., Iwasa, J., & Taatjes, D. J. (2022). The Mediator complex as a master regulator of transcription by RNA polymerase II. *Nature Reviews. Molecular Cell Biology, 23*(11), 732–749. https://doi.org/10.1038/s41580-022-00498-3.

Rickert, P., Corden, J. L., & Lees, E. (1999). Cyclin C/CDK8 and cyclin H/CDK7/p36 are biochemically distinct CTD kinases. *Oncogene, 18*(4), 1093–1102. https://doi.org/10.1038/sj.onc.1202399.

Sabari, B. R., Dall'Agnese, A., Boija, A., Klein, I. A., Coffey, E. L., Shrinivas, K., ... Young, R. A. (2018). Coactivator condensation at super-enhancers links phase separation and gene control. *Science (New York, N. Y.), 361*(6400), https://doi.org/10.1126/science.aar3958.

Sainsbury, S., Bernecky, C., & Cramer, P. (2015). Structural basis of transcription initiation by RNA polymerase II. *Nature Reviews. Molecular Cell Biology, 16*(3), 129–143. https://doi.org/10.1038/nrm3952.

Saleh, M. M., Jeronimo, C., Robert, F., & Zentner, G. E. (2021). Connection of core and tail Mediator modules restrains transcription from TFIID-dependent promoters. *PLoS Genetics, 17*(8), e1009529. https://doi.org/10.1371/journal.pgen.1009529.

Sato, S., Tomomori-Sato, C., Parmely, T. J., Florens, L., Zybailov, B., Swanson, S. K., ... Conaway, R. C. (2004). A set of consensus mammalian mediator subunits identified by multidimensional protein identification technology. *Molecular Cell, 14*(5), 685–691. https://doi.org/10.1016/j.molcel.2004.05.006.

Schwalb, B., Michel, M., Zacher, B., Fruhauf, K., Demel, C., Tresch, A., ... Cramer, P. (2016). TT-seq maps the human transient transcriptome. *Science (New York, N. Y.), 352*(6290), 1225–1228. https://doi.org/10.1126/science.aad9841.

Shao, X., Chen, Y., Xu, A., Xiang, D., Wang, W., Du, W., ... Ying, M. (2022). Deneddylation of PML/RARα reconstructs functional PML nuclear bodies via orchestrating phase separation to eradicate APL. *Cell Death & Differentiation, 29*(8), https://doi.org/10.1038/s41418-022-00955-8.

Shi, Y., Chen, J., Zeng, W. J., Li, M., Zhao, W., Zhang, X. D., & Yao, J. (2021). Formation of nuclear condensates by the Mediator complex subunit Med15 in mammalian cells. *BMC Biology, 19*(1), 245. https://doi.org/10.1186/s12915-021-01178-y.

SM, G., CI, S., & PD, A. (2011). NUP98 gene fusions and hematopoietic malignancies: common themes and new biologic insights. *Blood, 118*(24), 6247–6257. https://doi.org/10.1182/blood-2011-07-328880.

Soutourina, J. (2018). Transcription regulation by the Mediator complex. *Nature Reviews. Molecular Cell Biology, 19*(4), 262–274. https://doi.org/10.1038/nrm.2017.115.

Stumpf, M., Waskow, C., Krötschel, M., van Essen, D., Rodriguez, P., Zhang, X., ... Borggrefe, T. (2006). The mediator complex functions as a coactivator for GATA-1 in erythropoiesis via subunit Med1/TRAP220. *Proceedings of the National Academy of Sciences, 103*(49), 18504–18509. https://doi.org/10.1073/pnas.0604494103.

Stumpf, M., Yue, X., Schmitz, S., Luche, H., Reddy, J. K., & Borggrefe, T. (2010). Specific erythroid-lineage defect in mice conditionally deficient for Mediator subunit Med1. *Proceedings of the National Academy of Sciences, 107*(50), 21541–21546. https://doi.org/10.1073/pnas.1005794107.

Sun, C., Xu, X., Chen, Z., Zhou, F., Wang, W., Chen, J., ... Ma, Y. (2025). Selective translational control by PABPC1 phase separation regulates blast crisis and therapy resistance in chronic myeloid leukaemia. *Nature Cell Biology, 27*(4), 683–695. https://doi.org/10.1038/s41556-024-01607-4.

Taatjes, D. J., Naar, A. M., Andel, F., 3rd, Nogales, E., & Tjian, R. (2002). Structure, function, and activator-induced conformations of the CRSP coactivator. *Science (New York, N. Y.), 295*(5557), 1058–1062. https://doi.org/10.1126/science.1065249.

Takahashi, H., Parmely, T. J., Sato, S., Tomomori-Sato, C., Banks, C. A., Kong, S. E., ... Conaway, J. W. (2011). Human mediator subunit MED26 functions as a docking site for transcription elongation factors. *Cell, 146*(1), 92–104. https://doi.org/10.1016/j.cell.2011.06.005.

Takahashi, H., Takigawa, I., Watanabe, M., Anwar, D., Shibata, M., Tomomori-Sato, C., ... Hatakeyama, S. (2015). MED26 regulates the transcription of snRNA genes through the recruitment of little elongation complex. *Nature Communications, 6*, 5941. https://doi.org/10.1038/ncomms6941.

Tan, C., Zhu, S., Chen, Z., Liu, C., Li, Y. E., Zhu, M., ... Fang, X. (2021). Mediator complex proximal Tail subunit MED30 is critical for Mediator core stability and cardiomyocyte transcriptional network. *PLoS Genetics, 17*(9), e1009785. https://doi.org/10.1371/journal.pgen.1009785.

Tanaka, H. (2022). Viscoelastic phase separation in biological cells. *Communications Physics, 5*(1), 167. https://doi.org/10.1038/s42005-022-00947-7.

Tang, B., Wang, X., He, H., Chen, R., Qiao, G., Yang, Y., ... Wang, J. (2024). Aging-disturbed FUS phase transition impairs hematopoietic stem cells by altering chromatin structure. *Blood, 143*(2), 124–138. https://doi.org/10.1182/blood.2023020539.

Terlecki-Zaniewicz, S., Humer, T., Eder, T., Schmoellerl, J., Heyes, E., Manhart, G., ... Grebien, F. (2021). Biomolecular condensation of NUP98 fusion proteins drives leukemogenic gene expression. *Nature Structural & Molecular Biology, 28*(2), 190–201. https://doi.org/10.1038/s41594-020-00550-w.

Torrado, B., Pannunzio, B., Malacrida, L., & Digman, M. A. (2024). Fluorescence lifetime imaging microscopy. *Nature Reviews Methods Primers, 4*(1), 80. https://doi.org/10.1038/s43586-024-00358-8.

Toth-Petroczy, A., Oldfield, C. J., Simon, I., Takagi, Y., Dunker, A. K., Uversky, V. N., & Fuxreiter, M. (2008). Malleable machines in transcription regulation: The mediator complex. *PLoS Computational Biology, 4*(12), e1000243. https://doi.org/10.1371/journal.pcbi.1000243.

Tsai, K. L., Sato, S., Tomomori-Sato, C., Conaway, R. C., Conaway, J. W., & Asturias, F. J. (2013). A conserved Mediator-CDK8 kinase module association regulates Mediator-RNA polymerase II interaction. *Nature Structural & Molecular Biology, 20*(5), 611–619. https://doi.org/10.1038/nsmb.2549.

Tsai, K. L., Tomomori-Sato, C., Sato, S., Conaway, R. C., Conaway, J. W., & Asturias, F. J. (2014). Subunit architecture and functional modular rearrangements of the transcriptional mediator complex. *Cell, 158*(2), 463. https://doi.org/10.1016/j.cell.2014.06.036.

Wake, N., Weng, S. L., Zheng, T., Wang, S. H., Kirilenko, V., Mittal, J., & Fawzi, N. L. (2025). Expanding the molecular grammar of polar residues and arginine in FUS phase separation. *Nature Chemical Biology, 21*(7), 1076–1088. https://doi.org/10.1038/s41589-024-01828-6.

Wang, J., Choi, J. M., Holehouse, A. S., Lee, H. O., Zhang, X., Jahnel, M., ... Hyman, A. A. (2018). A molecular grammar governing the driving forces for phase separation of Prion-like RNA binding proteins. *Cell, 174*(3), 688–699.e616. https://doi.org/10.1016/j.cell.2018.06.006.

Wang, Y. M., Zhou, H., Leng, S., Ma, J. J., Li, H., Li, G., ... Liu, X. G. (2026). CDK8/CDK19 inhibition restores T-cell homeostasis in primary immune thrombocytopenia. *Blood*. https://doi.org/10.1182/blood.2025031332.

Wang, Z., Song, A., Xu, H., Hu, S., Tao, B., Peng, L., ... Chen, F. X. (2022). Coordinated regulation of RNA polymerase II pausing and elongation progression by PAF1. *Science Advances, 8*(13), eabm5504. https://doi.org/10.1126/sciadv.abm5504.

Wang, Z., Wang, P., Li, Y., Peng, H., Zhu, Y., Mohandas, N., & Liu, J. (2021). Interplay between cofactors and transcription factors in hematopoiesis and hematological malignancies. *Signal Transduction and Targeted Therapy, 6*(1), 24. https://doi.org/10.1038/s41392-020-00422-1.

Whyte, W. A., Orlando, D. A., Hnisz, D., Abraham, B. J., Lin, C. Y., Kagey, M. H., ... Young, R. A. (2013). Master transcription factors and mediator establish super-enhancers at key cell identity genes. *Cell, 153*(2), 307–319. https://doi.org/10.1016/j.cell.2013.03.035.

Wong, K. H., Jin, Y., & Struhl, K. (2014). TFIIH phosphorylation of the Pol II CTD stimulates mediator dissociation from the preinitiation complex and promoter escape. *Molecular Cell, 54*(4), 601–612. https://doi.org/10.1016/j.molcel.2014.03.024.

Wu, T., King, M. R., Qiu, Y., Farag, M., Pappu, R. V., & Lew, M. D. (2025). Single-fluorogen imaging reveals distinct environmental and structural features of biomolecular condensates. *Nature Physics, 21*(5), 778–786. https://doi.org/10.1038/s41567-025-02827-7.

Yamamoto, K., Goyama, S., Asada, S., Fujino, T., Yonezawa, T., Sato, N., ... Kitamura, T. (2021). A histone modifier, ASXL1, interacts with NONO and is involved in paraspeckle formation in hematopoietic cells. *Cell Reports, 36*(8), 109576. https://doi.org/10.1016/j.celrep.2021.109576.

Yang, J., Chung, C.-I., Koach, J., Liu, H., Navalkar, A., He, H., ... Shu, X. (2024). MYC phase separation selectively modulates the transcriptome. *Nature Structural & Molecular Biology, 31*(10), 1567–1579. https://doi.org/10.1038/s41594-024-01322-6.

Yang, L., Zhang, Z., Jiang, P., Kong, D., Yu, Z., Shi, D., ... Qian, P. (2024). Phase separation-competent FBL promotes early pre-rRNA processing and translation in acute myeloid leukaemia. *Nature Cell Biology, 26*(6), 946–961. https://doi.org/10.1038/s41556-024-01420-z.

Yue, X., Izcue, A., & Borggrefe, T. (2011). Essential role of Mediator subunit Med1 in invariant natural killer T-cell development. *Proceedings of the National Academy of Sciences, 108*(41), 17105–17110. https://doi.org/10.1073/pnas.1109095108.

Zamudio, A. V., Dall'Agnese, A., Henninger, J. E., Manteiga, J. C., Afeyan, L. K., Hannett, N. M., ... Young, R. A. (2019). Mediator condensates localize signaling factors to key cell identity genes. *Molecular Cell, 76*(5), 753–766.e756. https://doi.org/10.1016/j.molcel.2019.08.016.

Zhang, F., Sumibcay, L., Hinnebusch, A. G., & Swanson, M. J. (2004). A triad of subunits from the Gal11/tail domain of Srb mediator is an in vivo target of transcriptional activator Gcn4p. *Molecular and Cellular Biology, 24*(15), 6871–6886. https://doi.org/10.1128/MCB.24.15.6871-6886.2004.

Zhang, X., Krutchinsky, A., Fukuda, A., Chen, W., Yamamura, S., Chait, B. T., & Roeder, R. G. (2005). MED1/TRAP220 exists predominantly in a TRAP/ Mediator subpopulation enriched in RNA polymerase II and is required for ER-mediated transcription. *Molecular Cell, 19*(1), 89–100. https://doi.org/10.1016/j.molcel.2005.05.015.

Zhang, Y., Lou, J., Liu, Y., Jin, P., Tan, Y., Song, H., ... Wang, K. (2024). Phase separation of PML/RARα and BRD4 coassembled microspeckles governs transcriptional dysregulation in acute promyelocytic leukemia. *Proceedings of the National Academy of Sciences, 121*(34), e2406519121. https://doi.org/10.1073/pnas.2406519121.

Zhang, Y., Wang, Y., Gao, B., Sun, Y., Cao, L., Genardi, S. M., ... Fang, D. (2020). USP22 controls iNKT immunity through MED1 suppression of histone H2A monoubiquitination. *Journal of Experimental Medicine, 217*(5), e20182218. https://doi.org/10.1084/jem.20182218.

Zhao, H., Li, J., Xiang, Y., Malik, S., Vartak, S. V., Veronezi, G. M. B., ... Asturias, F. J. (2024). An IDR-dependent mechanism for nuclear receptor control of Mediator interaction with RNA polymerase II. *Molecular Cell, 84*(14), 2648–2664.e2610. https://doi.org/10.1016/j.molcel.2024.06.006.

Zhao, H., Young, N., Kalchschmidt, J., Lieberman, J., El Khattabi, L., Casellas, R., & Asturias, F. J. (2021). Structure of mammalian Mediator complex reveals Tail module architecture and interaction with a conserved core. *Nature Communications, 12*(1), 1355. https://doi.org/10.1038/s41467-021-21601-w.

Zhou, H.-X. (2012). Intrinsic disorder: Signaling via highly specific but short-lived association. *Trends in Biochemical Sciences, 37*(2), 43–48. https://doi.org/10.1016/j.tibs.2011.11.002.

Zhu, S., Zhang, X., Li, N., Zhao, X., Li, M., Xie, S., ... Lee, H.-Y. (2024). MED26-enriched condensates drive erythropoiesis through modulating transcription pausing. *eLife*. https://doi.org/10.7554/elife.102023.1.

Zhu, Y., Xu, W., Wang, R., Hu, L., Huang, X., Zhu, H., & Li, P. (2025). Novel strategy against AML by rescuing loss of phase separation of pathogenic NPM1c. *bioRxiv*. https://doi.org/10.1101/2025.04.08.646961.

CHAPTER FOUR

Cell fate decision and lineage specification mechanisms in hematopoiesis

In-Won Lee[1], Yesenia Earth[1], and Yoon-A Kang*
Division of Hematology, Department of Medicine, School of Medicine, Washington University in St. Louis, St. Louis, MO, United States
*Corresponding author. e-mail address: yoonakang@wustl.edu

Contents

Abstract

Red blood cells constitute over 99 % of all blood cells and 2–3 million erythrocytes are generated every second in human body. The process of generating red blood cells, erythropoiesis, is tightly regulated at the multiple levels of the hematopoietic hierarchy to fulfil this intensive demand for oxygen carrying cells. Recent studies uncovered that lineage specification begins at the multipotent stages such as

[1] These authors equally contribute to this work.

Current Topics in Developmental Biology, Volume 170
ISSN 0070-2153, https://doi.org/10.1016/bs.ctdb.2026.06.001

hematopoietic stem cell (HSC) and multipotent progenitor (MPP) populations. In this review, we discuss lineage-biased HSCs and MPPs, the erythroid linage bias in HSCs and MPPs, how erythroid fate decisions are made in common myeloid progenitors (CMPs) and megakaryocyte-erythroid progenitors (MEPs), and what alters erythroid lineage commitment in aging and disease conditions.

1. Introduction

Blood production is controlled by the differentiation rates of hematopoietic stem cells (HSCs) and downstream multipotent progenitors (MPPs) (Seita & Weissman, 2010). HSCs give rise to MPPs, which then differentiate into either myeloid- or lymphoid committed progenitors that produce all type of mature blood cells (Fig. 1) (Seita & Weissman, 2010). In the traditional hematopoietic tree, lineage specification was thought to be a gradual stepwise loss of lineage potency and gain of certain lineage commitment (Orkin & Zon, 2008; Seita & Weissman, 2010). However, recent advances of single cell technologies and experimental tools revealed that hematopoietic populations are better represented as a continuum instead of discrete populations, and there are bypass mechanisms to fulfill emergency hematopoiesis (Buenrostro et al., 2018; Haas et al., 2015; Laurenti & Gottgens, 2018; Paul et al., 2015; Velten et al., 2017; Wilson et al., 2015; Yamamoto et al., 2013). Indeed, many studies have identified myeloid- or megakaryocyte (Meg)-biased HSCs based on transplantation experiments and lineage tracing analyses (Challen et al., 2010; Haas et al., 2015; Pei et al., 2020; Sanjuan-Pla et al., 2013; Yamamoto et al., 2013), suggesting lineage priming and specification starts from multipotent cells such as HSCs and MPPs. In this review, we discuss how initial erythroid lineage specification occurs at the adult HSC and MPP levels, how this priming/biases pass onto lineage commitment, and how these mechanisms operate under aging and disease contexts.

2. HSCs and lineage biases

Hematopoietic stem cells are most well characterized adult stem cells, which have provided a paradigm for stem cell research due to the easy access and refined surface markers (Orkin & Zon, 2008). In the 1960s, early studies demonstrated that a single clonogenic cell can self-renew and make lymphoid cells as well as myeloid-erythroid progeny using spleen colony-forming unit

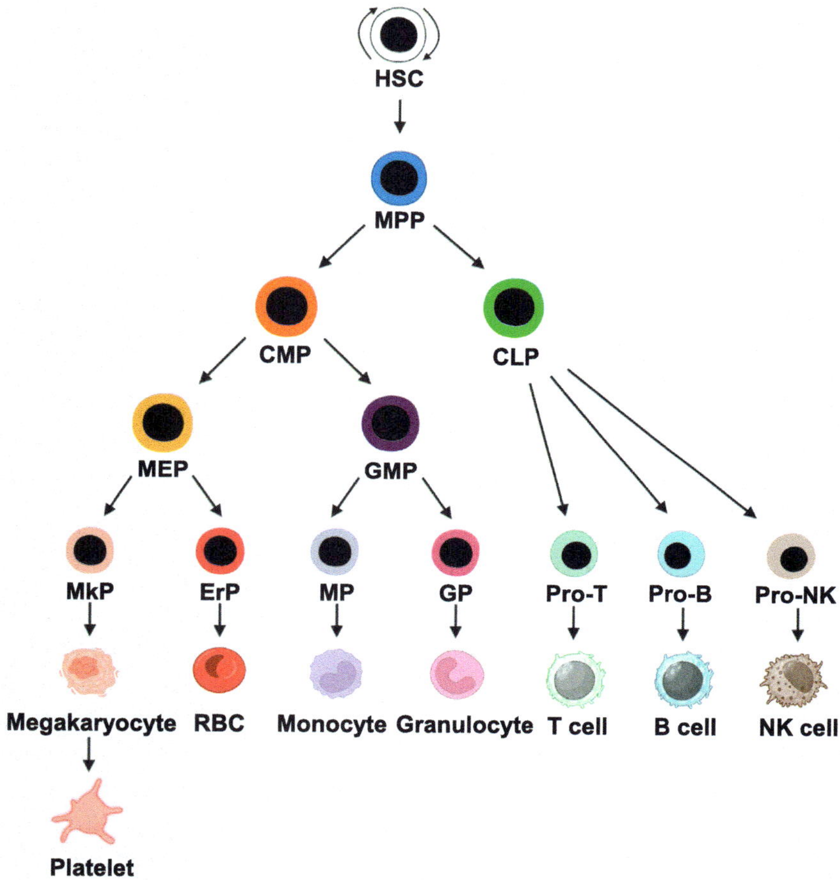

Fig. 1 Classical model of hematopoiesis. In the traditional tree-like model, hematopoiesis is depicted as a stepwise process progressing through discrete intermediate stages. HSCs lose self-renewal capacity and differentiate into MPPs, which bifurcate into CMPs and CLPs, where myeloid versus lymphoid fate decisions occur. Further lineage specification at the CMP and CLP stages gives rise to MEPs and GMPs, as well as Pro-T, Pro-B, and Pro-NK progenitors, respectively. Subsequent binary fate decisions at the MEP and GMP stages generate MkPs or ErPs, and MPs or GPs, respectively. HSC, hematopoietic stem cell; MPP, multipotent progenitor; CMP, common myeloid progenitor; CLP, common lymphoid progenitor; MEP, megakaryocyte-erythroid progenitor; GMP, granulocyte-macrophage progenitor; MkP, megakaryocyte progenitor; ErP, erythroid progenitor; MP, monocyte progenitor; GP, granulocyte progenitor; RBC, red blood cell; Pro-T, progenitor T cells; Pro-B, progenitor B cells; Pro-NK, progenitor natural killer cells; NK cell, natural killer cell.

(CFU-S) assays and cytological studies, providing evidence for the concept of blood stem cells (Becker et al., 1963; Siminovitch et al., 1963; Till & Mc, 1961; Wu et al., 1967; Wu et al., 1968). The initial effort to prospectively isolated HSCs by size and density evolved to using modern technologies like fluorescence-activated cell sorting (FACS) and monoclonal antibodies in the 1980s (Iscove, 1990; Iscove et al., 1972; Muller-Sieburg et al., 1986; Spangrude et al., 1988). In the 1990s, further marker refinement enabled distinction between lifelong self-renewing long-term HSCs (LT-HSCs) and transient short-term HSCs (ST-HSCs), and single cell transplantation experiments truly demonstrated that a single blood stem cell can generate the whole blood lineage (Morrison et al., 1997; Morrison & Weissman, 1994; Osawa et al., 1996). The current consensus of immunophenotypic LT-HSCs in mouse is $Lin^{-}Sca\text{-}1^{+}c\text{-}Kit^{+}Flk2^{-}CD48^{-}CD150^{+}CD34^{-}$ bone marrow cells and $Lin^{-}CD34^{+}CD38^{-}CD90^{+}CD45RA^{-} \pm CD49f^{+}$ cells in humans (Anjos-Afonso & Bonnet, 2023; Olson et al., 2020).

2.1 Lineage-biased HSCs in mice

Early lineage tracing work using random inactivation of polymorphic X chromosome glucose-6-phosphate-dehydrogenase (G6PD) alleles in heterozygous female Safari cats provided the first quantitative model for HSC behavior in native hematopoiesis, which suggested stochastic hematopoiesis in vivo (Abkowitz et al., 1996). Subsequent studies have demonstrated the heterogeneity of HSC pools and the presence of lineage bias in HSCs. Müller-Sieburg et al. showed that HSCs are intrinsically different in their self-renewal capacity and lineage output by studying single clones and, in particular, they characterized myeloid-biased HSCs, confirming lineage-biased HSCs (Muller-Sieburg et al., 2004; Muller-Sieburg et al., 2002). Single-cell transplantation assays followed by serial transplantation further validated that LT-HSCs are heterogeneous population with distinct stable clones, indicating that the lineage bias is a solid feature of HSCs and their lineage output is predetermined and stably inherited (Dykstra et al., 2007). Differential capacity for hoechst dye efflux could distinguish myeloid-biased HSCs and lymphoid-biased HSCs and this lineage bias was stable upon aging or serial transplantation (Challen et al., 2010). Nakauchi group reported that $CD150^{high}$ LT-HSCs are a myeloid-biased stable component of HSC pool using clonal analysis via single-cell transplantation and paired daughter cell assays, which was confirmed by an independent group (Beerman et al., 2010; Morita et al., 2010; Yamamoto et al., 2013). Yamamoto et al. found myeloid-restricted progenitors with long-term repopulating activity, which are committed to megakaryocytes, megakaryocyte-erythrocyte, or common myeloid cells within HSC pool (Yamamoto et al., 2013). $CD41^+$ HSCs are myeloid-biased, exhibiting myeloerythroid and megakaryocyte gene priming in contrast to $CD41^-$ HSCs, which produced more lymphoid-biased progenies (Gekas & Graf, 2013). CD49b also distinguishes myeloid-biased HSCs ($CD49b^-$ HSCs) versus lymphoid-biased HSCs ($CD49b^+$ HSCs) (Somuncular et al., 2022). Studies using more advanced single cell lineage tracing techniques confirm the myeloid-biased HSCs. Pei et al. showed that HSCs can be categorized into three groups: multilineage HSCs, myeloid-erythroid restricted HSCs, and differentiation-inactive HSCs (which give rise to another HSCs through symmetric division) using Polylox barcodes, which enable fate mapping and capturing of transcriptome from each cell (Pei et al., 2017; Pei et al., 2020). Megakaryocyte-biased HSCs are also evident, and they can be marked by the expression of von Willebrand Factor (vWF) in mice (Carrelha et al., 2024; Carrelha et al., 2018; Haas et al., 2015; Rodriguez-Fraticelli et al., 2018; Sanjuan-Pla et al., 2013).

2.2 Lineage-biased HSCs in humans

In humans, single-cell resolution HSC study integrating flow cytometric, transcriptomic and functional data indicated that 'continuum of low-primed undifferentiated hematopoietic stem and progenitor cells' (CLOUD-HSPCs) give rise to unipotent cells, confirming the continuum model of hematopoiesis at the transcriptome level found in a mice system (Velten et al., 2017). Combined single-cell RNA-sequencing data with chromatin accessibility landscape also demonstrated that HSC differentiation is a continuous process rather than a step-by-step procedure (Buenrostro et al., 2018) (Fig. 2). Clonal assays together with statistical computation found that lineage bias occurs at the HSC level and initial dendritic cell bias is marked by IRF8 expression, which is enforced by the FLT3L during cell division along the dendritic cell specification (Lee et al., 2017). Single-cell level gene expression profiling using Fluidigm platform showed that adult bone marrow HSCs express key megakaryocyte lineage factors, indicating the presence of megakaryocyte-biased HSCs (Notta et al., 2016). In vitro single cell differentiation culture coupled with index-sorting and single-cell RNA-sequencing revealed that myeloid/lymphoid lineage restriction already occurs at the HSC level (Belluschi et al., 2018). Most recent studies using naturally occurring mitochondrial DNA mutations or somatic mutations also demonstrated the lineage biases in HSC during native hematopoiesis (Weng et al., 2024; Yoshizato et al., 2025). HSC clonal output was negatively correlated with

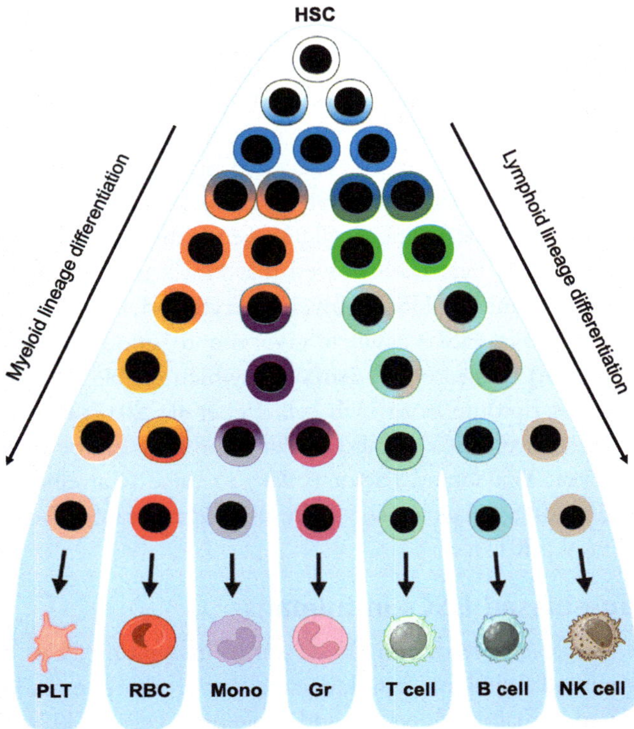

Fig. 2 Revised continuum model of hematopoiesis. In the revised continuum model of hematopoiesis, differentiation is depicted as a continuous and dynamic process rather than a strictly hierarchical, stepwise progression. HSCs gradually transition through a spectrum of lineage-primed states without passing through rigidly defined intermediate progenitors. Lineage bias emerges early within the HSC and MPP compartments, with cells exhibiting transcriptional priming toward specific lineages prior to overt commitment. Rather than discrete bifurcation points, lineage fate decisions reflect progressive restriction along interconnected developmental trajectories, ultimately giving rise to mature blood cell lineages. HSC, hematopoietic stem cell; PLT, platelet; RBC, red blood cell; Mono, monocyte; Gr, granulocyte; NK cell, natural killer cell.

lymphoid lineage bias and positively correlated with erythroid and myeloid lineages, echoing previous finding in mouse system (Muller-Sieburg et al., 2004; Muller-Sieburg et al., 2002; Weng et al., 2024; Yamamoto et al., 2013). Somatic single-nucleotide variants analysis across a robust number of clones from healthy aged individuals uncovered that HSCs are composed of multi-lineage clones, myeloid-lineage clones, and myeloid plus B cell lineage without T cell clones in human native hematopoiesis (Yoshizato et al., 2025).

Of note, authors explained that megakaryocyte clone was not detected in their study unlike mice work and previous human studies probably due to the very low number of megakaryocytes (<0.1 %) in the mononuclear bone marrow cells used for their study.

2.3 Erythroid-biased HSCs

Array of studies using either traditional clonal assays/single-cell transplantation or advanced single cell clonal analysis in native mouse and human hematopoiesis consistently showed the presence of myeloid, lymphoid, or megakaryocyte-biased HSCs. Although megakaryocyte lineage restricted self-renewing cells in the phenotypic HSC compartment are well characterized in molecular and cellular levels, it remains elusive whether erythroid lineage restricted clones are present in HSC pools (Carrelha et al., 2024; Carrelha et al., 2018; Haas et al., 2015; Notta et al., 2016; Rodriguez-Fraticelli et al., 2018; Sanjuan-Pla et al., 2013). Most studies showed that erythroid lineage bias accompanied by other myeloid lineages such as megakaryocyte-erythrocyte bias, myeloid-erythroid bias, or common myeloid bias (Gekas & Graf, 2013; Pei et al., 2017; Pei et al., 2020; Yamamoto et al., 2013). Of note, quantitative analyses of mature cell production in absolute number using in vivo clonal assays uncovered robust erythroid potential of HSCs and MPPs, dominating the production (Boyer et al., 2019). Given that erythroid bias is always accompanied by other myeloid lineage bias, this study suggests that erythroid production may be a default setting and other cues that imprint non-erythroid lineage bias may make HSCs deviate from erythroid production.

2.4 Establishment, maintenance, and execution of lineage bias

There are studies showing that lineage-restricted HSCs exhibit lineage plasticity; platelet-restricted production in primary transplantation followed by additional myeloid and lymphoid lineage differentiation in vitro culture and in secondary transplantation (Carrelha et al., 2018) or during serial transplantations of aged-specific latent-HSCs (Yamamoto et al., 2018). Bias switching, where progenies produce different lineages from their ancestor is occasionally observed (Benz et al., 2012; Lee et al., 2017). However, the lineage bias of HSCs is generally intrinsically stable and heritable in both mice and humans (Challen et al., 2010; Dykstra et al., 2007; Lee et al., 2017; Pei et al., 2020; Spencer Chapman et al., 2024; Weng et al., 2024; Yamamoto et al., 2013; Yoshizato et al., 2025). This stable lineage biases are likely mediated by epigenetic mechanisms in particular, genome-wide DNA methylation or chromatin accessibility variation (Li et al., 2023; Weng et al., 2024; Yu et al., 2017). The fact that CD49b$^-$ myeloid-biased and CD49b$^+$ lymphoid-biased HSCs are transcriptionally similar but have distinct chromatin accessibility also indicates the involvement of epigenetic mechanisms in establishing lineage bias of HSCs (Somuncular et al., 2022).

At the cellular level, myeloid-biased HSCs express lower levels of IL-7 receptor, and consequently, their progeny fail to respond to IL-7 in vitro, indicating that myeloid-biased HSCs exhibit their myeloid-bias by not responding to the key lymphopoiesis cytokine IL-7 (Muller-Sieburg et al., 2004). In line with this finding, common lymphoid progenitors generated from lymphoid-deficient-HSC express decreased level of key lymphoid genes such as *Pax5, Il7r, Tcf3,* and *Ikaros* (Benz et al., 2012). Furthermore, TGF-β1, cytokine secreted by platelets promotes myeloid-biased HSC proliferation and myeloid differentiation, likely providing self-enforcing positive feedback (Challen et al., 2010). In support of this idea, CD41 knockout mice exhibited multilineage HSC-derived hematopoietic defects that were long-term transplantable and partially mediated by reduced exposure of HSCs to important platelet-released cytokines such as TGF-β1 (Gekas & Graf, 2013). Similarly, thrombopoietin (TPO), the key megakaryocyte/platelet cytokine is critical for platelet-biased HSC maintenance (Sanjuan-Pla et al., 2013). Lymphoid-biased HSC maintenance appears to be mediated by mitofusin 2 (Mfn2), which tethers mitochondria to endoplasmic reticulum thereby regulating intracellular calcium levels and transcription factor activity (Luchsinger et al., 2016). Altogether, multiple studies indicate that lineage bias is maintained by epigenetic mechanisms and reinforced at the cellular level by interaction with external cues.

How initial lineage biases in HSCs are established remains unclear. Most widely accepted notion is stochastic models. Somatic mutations accumulate with cell divisions during early embryogenesis as producing complex genetic mosaics (Ju et al., 2017; Uchimura et al., 2022). It is likely that HSCs further acquire somatic mutations as they divide obtaining unique sets of genetic and epigenetic alterations, which could confer lineage biases by instructing lineage-specific gene expression programs. In addition to this intrinsic factors, external environmental cues can contribute to lineage bias in HSCs as well. The spatial localization of HSCs within the bone marrow defines a unique microenvironment, called the niche, which regulates HSC lineage bias through diverse cell-extrinsic signals (e.g., cytokines and cell–cell interactions) determined by neighboring cell types (Comazzetto et al., 2021; Gao et al., 2018; Morrison & Scadden, 2014; Pinho & Frenette, 2019). Indeed, cytokines can actively instruct lineage choices as well as promoting cell survival and proliferation (Iwasaki-Arai et al., 2003; Kondo et al., 2000; Mossadegh-Keller et al., 2013; Rieger et al., 2009). Numerous studies indicate the niche effect on HSC lineage biases. Pre-transplantation conditioning was reported to induce HSC lineage bias (Lu et al., 2019). Sister cell experiments revealed that cells with very similar transcriptional profiles can pre-committed to different lineages (Weinreb et al., 2020). Proportion of lymphoid-deficient HSCs increased in the fetal bone marrow compared to fetal liver (Benz et al., 2012). In zebrafish, miR-128 regulates Wnt and Notch signaling in the aorta-gonad-mesonephros (AGM). Lowering Wnt signaling resulted in erythroid-biased HSPCs and increasing Notch signaling lead to lymphoid-biased HSPCs with long-term consequence, suggesting that adult HSPC lineage biases arise in the AGM endothelium (Ghersi et al., 2023). Asymmetric distribution of fate determinants has been suggested to cause asymmetric HSC divisions (Beckmann et al., 2007; Brummendorf et al., 1998; Wu et al., 2007). However, there are no direct evidence yet demonstrating that asymmetric distribution of fate determinants causes lineage bias in HSCs.

3. MPPs and lineage biases

MPP populations are immediately downstream of HSCs in the hematopoietic hierarchy, which can generate all hematopoietic lineage cells but lacking lifelong self-renewing capacity. MPPs have been identified based on their distinct surface markers in mouse bone marrow (Christensen & Weissman, 2001; Morrison et al., 1997; Morrison & Weissman, 1994) and human cord blood as well as bone marrow (Majeti et al., 2007). Due to the lack of long-term reconstitution ability in the transplantation setting, MPP populations have been overlooked for a while. However, lineage tracing analyses using barcoding systems revealed that MPPs contribute to blood production for a long period time in native hematopoiesis, highlighting their functional significance in blood production (Rodriguez-Fraticelli et al., 2018; Sun et al., 2014). This finding together with the identification and characterization of lineage-biased MPP populations fuel the research underlying lineage specification mechanisms at the top of the hematopoietic hierarchy (Cabezas-Wallscheid et al., 2014; Pietras et al., 2015; Rodriguez-Fraticelli et al., 2018).

3.1 Lineage-biased MPPs in mice

Currently, six MPP subsets have been identified, MPP1, MPP2, MPP3, MPP4, MPP5, and MPP6 (Cabezas-Wallscheid et al., 2014; Oguro et al., 2013; Pietras et al., 2015; Sommerkamp et al., 2021; Wilson et al., 2008). MPP1 have short-term self-renewal capacity with multipotency (Cabezas-Wallscheid et al., 2014; Wilson et al., 2008). They are also known as ST-HSC and can generate MPP2, MPP3 and MPP4 (Cabezas-Wallscheid et al., 2014; Pietras et al., 2015). MPP2 preferentially produce megakaryocytes and MPP3 are biased toward granulocyte/macrophage output (Cabezas-Wallscheid et al., 2014; Pietras et al., 2015). Within MPP3, $Fc\gamma R^{+}$ MPP3 with high endoplasmic reticulum volume ($Fc\gamma R^{+}/ER^{high}$ MPP3) are a transitional population toward granulocyte-macrophage progenitor (GMP) commitment, representing the myeloid bias of MPP3 (Kang et al., 2023). $Fc\gamma R^{-}$ MPP3 with low endoplasmic reticulum volume ($Fc\gamma R^{-}/ER^{low}$ MPP3) are a true multipotent subset with erythroid bias (unpublished data). MPP4 are the most abundant and lymphoid-biased population in steady-state but under regenerative/inflammatory/disease conditions, they are molecularly reprogrammed toward myeloid output (Kang et al., 2020; Pietras et al., 2015; Reynaud et al., 2011). MPP5 and MPP6 are most recently discovered functionally intermediate populations between HSCs and lineage-biased MPP2–4 subsets (Sommerkamp et al., 2021). MPP5 produce myeloid cells rapidly for a short period time and contribute to long-term lymphoid lineage reconstitution in a transplantation setting (Sommerkamp et al., 2021). The functional differences between MPP1/ST-HSC and MPP6 are unclear (Sommerkamp et al., 2021). Immunophenotypes and functional characteristics of each MPP subset are summarized in Table 1 and revised hematopoiesis model with lineage-biased HSCs and MPPs is depicted in Fig. 3. A brief overview of current MPP subsets is also reviewed in (Challen et al., 2021).

MPP2–4 subsets are independently produced from HSCs (Pietras et al., 2015), indicating that the underlying mechanisms will shed light on lineage specification process from lineage-biased HSC pool. However, how these MPP subsets are generated from HSCs are poorly understood. Increased ratio of Wnt over Notch1 signaling in HSCs promoted MPP3 production (Kang et al., 2020). Blocking Notch2 signaling by in vivo administration of anti-Notch2 antibody reduced MPP4 population, but it is unclear whether the effect is through HSCs or MPP4 itself (Kang et al., 2020). Given that HSC lineage bias is largely heritable, it is plausible that

Table 1 Immunophenotype, lineage output and functionally overlapped populations from previous literatures of murine multipotent progenitor (MPP) subsets.

Population	Immunophenotype	Output	Overlap
MPP 1[a]	Lin^- $Sca\text{-}1^+$ $c\text{-}Kit^+$ $CD34^+$ $CD48^-$ $CD150^+$ $CD135^-$	Multilineage	MPP (Akala et al., 2008) ITRC (Benveniste et al., 2010) $CD150^-$ $CD34^-$ KSL (Morita et al., 2010) MPP-1, MPP-2 & MPP-3 (Oguro et al., 2013) Fraction III (Yamamoto et al., 2013) ST-HSC (Pietras et al. 2015)
MPP 2[a,b]	Lin^- $Sca\text{-}1^+$ $c\text{-}Kit^+$ $CD34^+$ $CD48^+$ $CD150^+$ $CD135^-$	Meg-biased	$GATA\text{-}1^+$ MPP (Arinobu et al. 2007) HPC-2 (Oguro et al. 2013)
MPP 3[a,b]	Lin^- $Sca\text{-}1^+$ $c\text{-}Kit^+$ $CD34^+$ $CD48^+$ $CD150^-$ $CD135^-$	GM-dominant	$PU.1^+$ MPP (Arinobu et al. 2007) HPC-1 (Oguro et al. 2013)
MPP 4[a,b]	Lin^- $Sca\text{-}1^+$ $c\text{-}Kit^+$ $CD34^+$ $CD48^+$ $CD150^-$ $CD135^+$	Lymphoid-biased	LMPP (Adolfsson et al., 2005) MPP (Akala et al., 2008) HPC-1 (Oguro et al. 2013)
MPP 5[c]	Lin^- $Sca\text{-}1^+$ $c\text{-}Kit^+$ $CD34^+$ $CD48^-$ $CD150^-$ $CD135^-$	Myeloid and lymphoid	MPP1 (Cabezas-Wallscheid et al., 2014) ST-HSC (Pietras et al., 2015)

(continued)

Table 1 Immunophenotype, lineage output and functionally overlapped populations from previous literatures of murine multipotent progenitor (MPP) subsets. (*cont'd*)

Population	Immunophenotype	Output	Overlap
MPP 6[c]	Lin^- $Sca\text{-}1^+$ $c\text{-}Kit^+$ $CD34^-$ $CD48^-$ $CD150^-$ $CD135^-$	Multilineage	MPP1 (Cabezas-Wallscheid et al., 2014) ST-HSC (Pietras et al., 2015)

Meg, megakaryocyte; GM, granulocyte/macrophage.
[a]Cabezas-Wallscheid, N., Klimmeck, D., Hansson, J., Lipka, D. B., Reyes, A., Wang, Q., ... Trumpp, A. (2014). Identification of regulatory networks in hscs and their immediate progeny via integrated proteome, transcriptome, and DNA methylome analysis. *Cell Stem Cell, 15*(4), 507–522.
[b]Pietras, E. M., Reynaud, D., Kang, Y. A., Carlin, D., Calero-Nieto, F. J., Leavitt, A. D., ... Passegue, E. (2015). Functionally distinct subsets of lineage-biased multipotent progenitors control blood production in normal and regenerative conditions. *Cell Stem Cell, 17*(1), 35–46.
[c]Sommerkamp, P., Romero-Mulero, M. C., Narr, A., Ladel, L., Hustin, L., Schonberger, K., ... Cabezas-Wallscheid, N. (2021). Mouse multipotent progenitor 5 cells are located at the interphase between hematopoietic stem and progenitor cells. *Blood, 137*(23), 3218–3224.

myeloid-biased HSCs preferentially generate myeloid-biased MPP3, whereas megakaryocyte-biased HSCs give rise to megakaryocyte-biased MPP2. This raises important questions. What are the specific roles of Wnt and Notch1 signaling in directing the generation of MPP3 from HSCs? Can Wnt or Notch1 signaling redirect lymphoid-biased HSCs toward the production of MPP3? Moreover, do distinct signaling pathways selectively promote the expansion or specification of particular lineage-biased HSC subsets? What molecular mechanisms underlie these processes? Addressing these questions will be essential to elucidate the mechanisms governing the lineage specification from lineage-biased HSCs to lineage-biased MPP subpopulations.

3.2 Lineage-biased MPPs in humans

In humans, MPP is immunophenotypically defined as $Lin^-CD34^+CD38^-CD45RA^-CD90^-CD49f^-$ (Notta et al., 2011). Lineage-biased MPP subpopulations are less characterized compared to mice. Lymphoid-primed multipotential progenitors (LMPPs) were first identified as $Lin^-CD34^+CD38^-CD90^-CD45RA^+CD10^-$, which displayed lineage potential of lymphoid and myeloid (GM), but not megakaryocyte and erythroid (Goardon et al., 2011). Subsequently, MPP was further sub-fractionated into MPP F1, MPP F2, and MPP F3 using the markers already linked to megakaryocyte and erythroid lineage specification, CD110 and CD71 respectively (Notta et al., 2016). Single

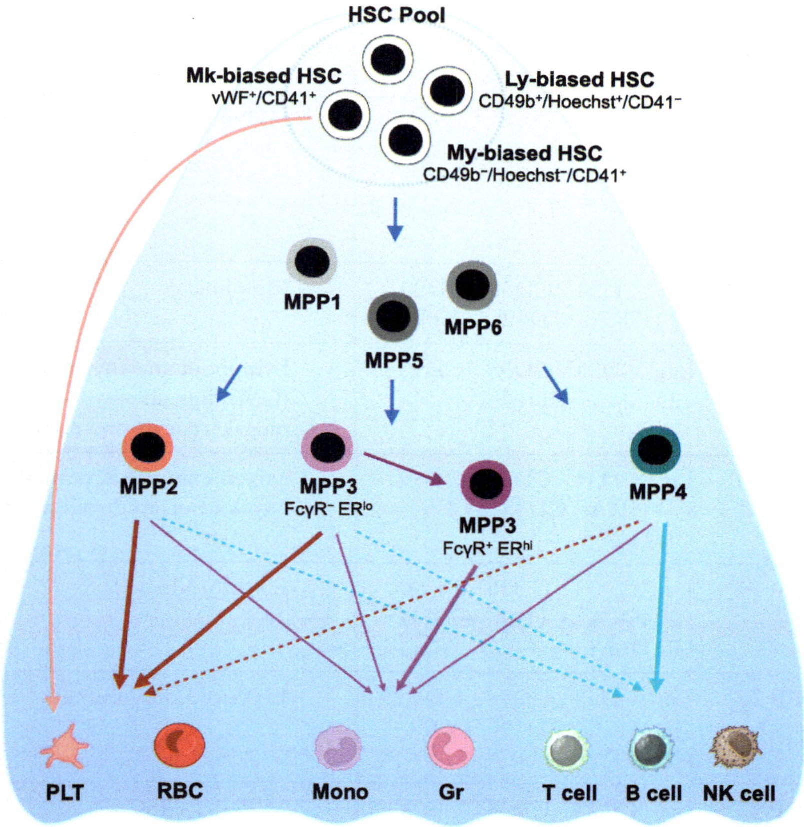

Fig. 3 Hematopoiesis model incorporated with lineage-biased HSCs and MPP populations. The HSC pool consists of Mk, My, and Ly-biased HSC states. Within the HSC subsets, Mk-biased HSCs (vWF$^+$/CD41$^+$) have direct platelet output. HSCs differentiate through MPP1, 5, and 6 into MPP2, 3, and 4, which have enhanced lineage-biases. MPP2 mainly shows Mk-biased output as maintaining other myeloid lineage production. MPP3 is functionally subdivided based on FcγR expression and ER volume. FcγR$^-$ MPP3 with low ER volume (FcγR$^-$ ER^{lo} MPP3) has multilineage output, whereas FcγR$^+$ MPP3 with high ER volume (FcγR$^+$ER^{hi} MPP3) exhibits predominantly granulocyte and macrophage output. MPP4 is a lymphoid-primed MPP with prominent lymphoid output in steady state while reprogrammed toward myeloid output under stress and disease conditions. HSC, hematopoietic stem cell; Mk, megakaryocyte; My, myeloid; Ly, lymphoid; vWF, von Willebrand factor; MPP, multipotent progenitor; ER, endoplasmic reticulum; PLT, platelet; RBC, red blood cell; Mono, monocyte; Gr, granulocyte; NK cell, natural killer cell.

Table 2 Immunophenotype and lineage output of human hematopoietic stem cell (HSC) and multipotent progenitor (MPP) subpopulations.

Population	Immunophenotype	Output
HSC	Lin^{-} $CD34^{+}$ $CD38^{-}$ $CD90^{+}$ $CD45RA^{-}$ (Majeti et al. 2007) Lin^{-} $CD34^{+}$ $CD38^{-}$ $CD90^{+}$ $CD45RA^{-}$ $CD49f^{+}$ (Notta et al., 2011)	Multilineage
MPP[a]	Lin^{-} $CD34^{+}$ $CD38^{-}$ $CD90^{-}$ $CD45RA^{-}$ $CD49f^{-}$	Multilineage
LMPP[b]	Lin^{-} $CD34^{+}$ $CD38^{-}$ $CD90^{-}$ $CD45RA^{+}$ $CD10^{-}$	Lymphoid and myeloid (GM), but no megakaryocyte and erythroid
MPP F1[c]	Lin^{-} $CD34^{+}$ $CD38^{-}$ $CD90^{-}$ $CD45RA^{-}$ $CD49f^{-}$ $CD71^{-}$ $CD110^{-}$	Mixed colonies as dominant, Myeloid or erythroid only colonies
MPP F2[c]	Lin^{-} $CD34^{+}$ $CD38^{-}$ $CD90^{-}$ $CD45RA^{-}$ $CD49f^{-}$ $CD71^{+}$ $CD110^{-}$	Erythroid-biased
MPP F3[c]	Lin^{-} $CD34^{+}$ $CD38^{-}$ $CD90^{-}$ $CD45RA^{-}$ $CD49f^{-}$ $CD71^{+}$ $CD110^{+}$	Erythroid only output
MPP 1[d]	Lin^{-} $CD34^{+}$ $CD38^{dim/-}$ $CD90^{-}$ $CD45RA^{-}$ $CD2^{-}$ $CLL1^{-}$ $CD69^{+}$	MPP/ST-HSC
MPP 2[d]	Lin^{-} $CD34^{+}$ $CD38^{dim}$ $CD90^{-}$ $CD45RA^{-}$ $CD2^{-}$ $CLL1^{-}$ $CD69^{-}$	Erythroid-biased
MPP 3[d]	Lin^{-} $CD34^{+}$ $CD38^{dim/-}$ $CD90^{-}$ $CD45RA^{-}$ $CD2^{-}$ $CLL1^{+}$ $CD123^{-}$ $CD117^{+}$	Myeloid-biased
LMPP 1[d]	Lin^{-} $CD34^{+}$ $CD38^{dim/-}$ $CD90^{-}$ $CD45RA^{+}$ $CD2^{-}$ $CLL1^{-}$ $CD69^{+/dim}$	Strong lymphoid potential
LMPP 2[d]	Lin^{-} $CD34^{+}$ $CD38^{dim/-}$ $CD90^{-}$ $CD45RA^{+}$ $CD2^{+}$	Lymphoid and myeloid potential

LMPP 3[d]	Lin^- $CD34^+$ $CD38^{dim/-}$ $CD90^-$ $CD45RA^+$ $CD2^-$ $CLL1^+$ $CD69^-$	Mostly GMP

HSC, hematopoietic stem cell; MPP, multipotent progenitor; LMPP, lymphoid-primed multipotent progenitor; GM, granulocyte/macrophage; MPP F, multipotent progenitor fraction; ST-HSC, short-term hematopoietic stem cell; GMP, granulocyte-macrophage progenitor.

[a]Notta, F., Doulatov, S., Laurenti, E., Poeppl, A., Jurisica, I., & Dick, J. E. (2011). Isolation of single human hematopoietic stem cells capable of long-term multilineage engraftment. *Science, 333*(6039), 218–221.

[b]Goardon, N., Marchi, E., Atzberger, A., Quek, L., Schuh, A., Soneji, S., ... Vyas, P. (2011). Coexistence of LMPP-like and GMP-like leukemia stem cells in acute myeloid leukemia. *Cancer Cell, 19*(1), 138–152.

[c]Notta, F., Zandi, S., Takayama, N., Dobson, S., Gan, O. I., Wilson, G., ... Dick, J. E. (2016). Distinct routes of lineage development reshape the human blood hierarchy across ontogeny. *Science, 351*(6269), aab2116.

[d]Ediriwickrema, A., Nakauchi, Y., Fan, A. C., Kohnke, T., Hu, X., Luca, B. A., ... Majeti, R. (2025). A single-cell framework identifies functionally and molecularly distinct multipotent progenitors in adult human hematopoiesis. *Cell Rep, 44*(9), 116236.

cell in vitro assay supporting myeloid, erythroid, or megakaryocyte lineage differentiation showed that MPP F1 ($CD71^-CD110^-$ MPP) had mixed colonies as a dominant lineage output with myeloid only and erythroid only output; MPP F2 ($CD71^+CD110^-$ MPP) had mostly erythroid only colonies with megakaryocyte-erythroid colonies, exhibiting erythroid lineage bias; MPP F3 ($CD71^+CD110^+$ MPP) had erythroid only output, indicating erythroid lineage commitment (Notta et al., 2016). Recent study unbiasedly found markers to define lineage-biased MPP subsets via concurrent single-cell RNA-sequencing and surface marker expression analysis using antibody-derived tags (ADTs; scADT-sequencing) (Ediriwickrema et al., 2025). Ediriwickrema et al. characterized $CLL1^-CD69^+$ MPP as truly multipotent (MPP1/ST-HSC), $CLL1^-CD69^-$ MPP as erythroid-biased (MPP2), and $CLL1^+$ MPP as myeloid-biased (MPP3). They also divided LMPP into three subpopulations. $CD2^-CLL1^-$ LMPP (LMPP1) are the most lymphoid-biased, $CD2^+$ LMPP (LMPP2) have both lymphoid and myeloid potential, and $CD2^-CLL1^+$ LMPP (LMPP3) are mostly GMPs. Immunophenotypes and functional characteristics of each MPP subset are summarized in Table 2. The mechanisms underlying the generation of these lineage-biased MPP subsets and LMPP subpopulations from HSCs remain unknown. Therefore, further studies are warranted to define the molecular and cellular mechanisms driving their specification.

4. Lineage fate decision in common myeloid progenitor (CMP)

4.1 Functional controversy of immunophenotypically defined CMP

Historically, the first lineage bifurcation (myeloid versus lymphoid) was thought to happen downstream of MPP where CMP and common lymphoid progenitor (CLP) emerge (Fig. 1). Akashi et al. divided myeloid progenitor (MP; $Lin^-Sca\text{-}1^-c\text{-}Kit^+$) population into three group, CMP ($Lin^-Sca\text{-}1^-c\text{-}Kit^+IL\text{-}7R\alpha^-CD34^+Fc\gamma R^{low}$), GMP ($Lin^-Sca\text{-}1^-c\text{-}Kit^+IL\text{-}7R\alpha^-CD34^+Fc\gamma R^{high}$), and megakaryocyte-erythroid progenitor (MEP; $Lin^-Sca\text{-}1^-c\text{-}Kit^+IL\text{-}7R\alpha^-CD34^-Fc\gamma R^{low}$) (Akashi et al., 2000). In vitro single cell assays and in vivo transplantations showed that CMP gives rise to GMP and MEP without lymphoid lineage, demonstrating multipotent myeloid differentiation capacity without long-term self-renewal (Akashi et al., 2000). Although the CMP model has been very useful to conceptualize cell fate decision and lineage specification mechanisms, further studies revealed that CMP compartment is heterogenous and cells with true mixed myeloid potential are rare (Paul et al., 2015; Pronk et al., 2007). Pronk et al. showed that CMP can be subfractionated into pre-granulocyte/macrophage (Pre-GM; $Lin^-Sca\text{-}1^-c\text{-}Kit^+CD150^-Endoglin^-CD41^-$) and pre-megakaryocyte/erythroid (Pre-MegE; $Lin^-Sca\text{-}1^-c\text{-}Kit^+$ $CD150^+Endoglin^-CD41^-$) progenitors, which were restricted to GMP colony formation and erythroid, megakaryocytic, or megakaryocyte/erythroid colonies, respectively (Pronk et al., 2007). Multiplex single-cell RT-PCR analysis of lineage-associated transcripts using Pre-MegE, Pre-GM, megakaryocyte progenitor (MkP), and pre-colony forming unit-erythroid (CFU-E) further validated that cells with multiple lineage markers were rare, supporting early transcriptional priming and arguing against a broadly multipotent CMP state (Pronk et al., 2007). Single-cell RNA-sequencing analysis of 2730 MPs found that individual myeloid progenitor cells clustered into lineage-specific differentiation programs, verifying that lineage bias/commitment already occurs at the upstream of MP stage (Paul et al., 2015). In vivo tracing of lentiviral barcoded CMPs also revealed that most individual CMPs had either erythroid only or myeloid only output after transplantation, confirming that "common" myeloid progenitors are a transcriptionally and functionally heterogeneous population that is largely already committed to specific lineages (Perie et al., 2015).

4.2 Lineage fate decision mechanisms in functionally defined CMP

Although initially immunophenotypically defined CMP population appears to be a mixture of already committed cells, it does not necessarily mean that there are no common myeloid progenitor population. Single cell studies repeatedly showed the presence of cells with multi-myeloid lineage potentials, which are true CMP residing at the HSC and MPP levels. The question is whether the previously established mechanisms as myeloid versus erythroid lineage decision at the CMP level also work at the HSCs and MPPs. PU.1 and C/EBPα are known to drive GM lineage in uncommitted progenitor populations (Laiosa et al., 2006). Increased PU.1 level in HSCs by IL-1β treatment drove myeloid differentiation (Chavez et al., 2021; Pietras et al., 2016). PU.1-deficeint MPP3 failed to differentiate into myeloid lineage during in vitro IL-1 stimulation (Chavez et al., 2023). Retroviral transduction of C/EBPα into human CD34^{+} cells promoted granulocytic differentiation while inhibiting erythrocyte differentiation (Cammenga et al., 2003). Similarly, HSCs transduced with GATA1, the master regulator of erythroid differentiation, only produced megakaryocyte/erythroid colonies in vitro and Ter119^{+} erythroid cells in vivo, clearly indicating that GATA1 directs HSCs into megakaryocyte/erythroid lineage commitment (Iwasaki et al., 2003). Since GATA1 and PU.1 suppress expression of each other while promoting their own expression and they are mutually exclusively expressed during erythroid versus myeloid differentiation, the antagonism between GATA1 and PU.1 has been considered to be the determinant of erythroid versus myeloid lineage fate decision (Wolff & Humeniuk, 2013). Live imaging of individual HSC in vitro differentiation using fluorescent tagged GATA1 and PU.1 reporter mice showed that PU.1 was expressed in HSCs but GATA1 was not detected (Hoppe et al., 2016). However, the detection sensitivity of fluorescent tagged GATA1 and PU.1 was challenged by another study, which performed the quantification of endogenous lineage specific transcription factors using mass cytometry time of flight (CyTOF) in a single cell level (Palii et al., 2019), implying the potential limitation of live imaging of fluorescent tagged proteins when their expression level is low. *Gata1* expression in HSCs was repeatedly shown by qRT-PCR analysis, microarray or single-cell RNA-sequencing although low level (Arinobu et al., 2007; Gekas & Graf, 2013; Nestorowa et al., 2016; Pei et al., 2017; Pei et al., 2020; Wu et al., 2020; Yan et al., 2021). Altogether, these data indicate that it is plausible that GATA1 and PU.1 cross-antagonism works at the

HSC level. However, further studies are required to demonstrate this cross-antagonism and identify additional regulators governing initial erythroid versus myeloid fate decisions in HSCs.

It has been reported that GATA1$^+$ MPP (Lin$^-$Sca-1$^+$c-Kit$^+$CD34$^+$GATA1$^+$) had myeloerythroid potential and PU.1$^+$ MPP (Lin$^-$Sca-1$^+$c-Kit$^+$CD34$^+$PU.1$^+$) exhibited myelolymphoid potential without megakaryocyte and erythroid lineage, suggesting that GATA1 and PU.1 level bifurcate myeloerythroid and myelolymphoid lineage from HSCs (Arinobu et al., 2007). GATA1$^+$ MPP aligns with current MPP2 and FcγR$^-$ MPP3, whereas PU.1$^+$ MPP corresponds to FcγR$^+$ MPP3 (Kang et al., 2023; Pietras et al., 2015). Although GATA1$^+$ versus PU.1$^+$ MPP distinction provided the evidence that GATA1 and PU.1 activation primes HSCs toward certain lineage potentials, further dissection of PU.1 level within GATA1$^+$ MPP and their respective lineage potential will warrant the antagonistic function of GATA1 and PU.1 in erythroid versus myeloid lineage output in HSC and MPP levels. GATA1-FOG1 complex drives megakaryocyte and erythroid gene expression programs while repressing alternative fate regulators by recruiting a nucleosome remodeling and deacetylase (NuRD) complex to ensure megakaryocyte and erythroid lineage development (Dore & Crispino, 2011). Whether this GATA1-FOG1-NuRD regulatory axis functions at MPPs remains to be validated. BACH2 promotes erythroid commitment while suppressing myeloid fate through repression of C/EBPβ in the LSK population (Kato et al., 2018). Elucidating whether GATA1 and BACH2 functionally cooperate to promote erythropoiesis at the HSC and MPP levels will be important for the comprehensive understanding of early erythroid lineage commitment mechanisms (Fig. 4).

Extrinsic factors such as cytokines in the bone marrow niche also contribute to fate decision between erythroid and myeloid lineage. Although the function of erythropoietin (EPO) in erythropoiesis is well established, whether EPO signaling is instructive or permissive has been on debate. Growing evidence show that EPO has instructive roles in HSCs and MPPs. High systemic level of EPO transcriptionally reprograms HSCs and MPPs toward erythroid lineage in vivo, resulting in increased erythroid output at the expense of myeloid lineage (Grover et al., 2014). Single-cell RNA sequencing analysis of HSPCs upon ex vivo EPO treatment showed that EPO increased erythroid gene expression in MPP1 and MPP2. Accordingly, transplantation of barcoded EPO-exposed MPP2 verified the enrichment of myeloid-erythroid-biased clones (Eisele et al., 2022). Thrombopoietin

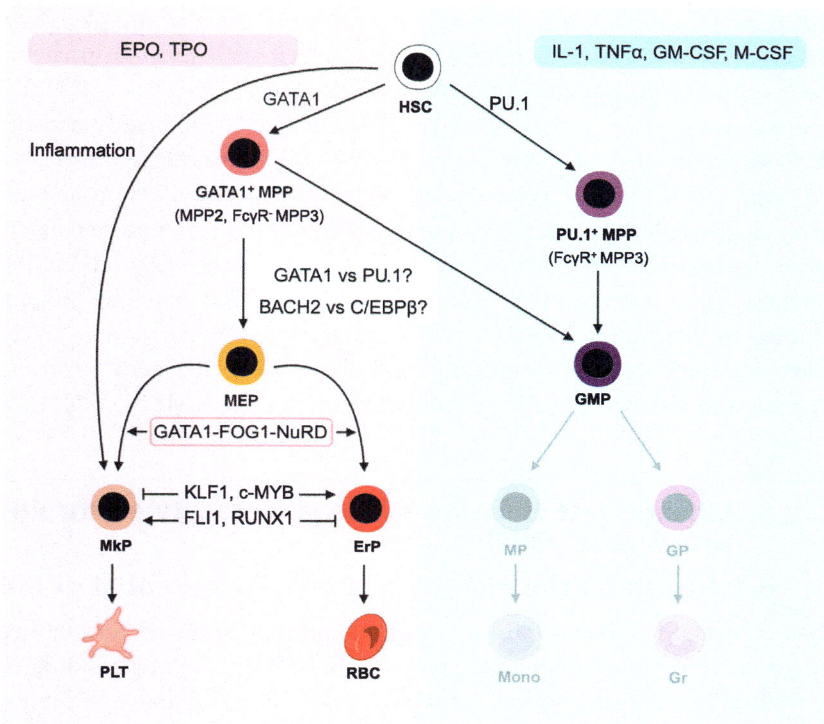

Fig. 4 Erythroid lineage commitment regulated by transcription factors and cytokine signaling. Schematic diagram highlighting key cytokines and transcription factors that govern lineage fate decisions leading to RBC production from HSCs. Erythropoietin (EPO) signaling, which is indispensable for erythropoiesis, can bias HSCs and MPP toward the erythroid lineage. Thrombopoietin (TPO), the essential cytokine for HSC maintenance and megakaryocyte development, has synergistic effects with EPO in enhancing erythropoiesis. In contrast, inflammatory/myeloid cytokines such as IL-1, TNFα, GM-CSF, and M-CSF drive HSCs towards the myeloid lineage. HSCs have a direct differentiation pathway toward MkP under inflammatory conditions. MPP with high GATA1 expression (GATA1$^+$ MPP), functionally corresponding to MPP2 and FcγR$^-$ MPP3 subsets, generates either MEP or GMP. MPP with high PU.1 expression (PU.1$^+$ MPP3), functionally overlapping with FcγR$^+$ MPP3, differentiates into GMP. The transcriptional mechanisms regulating MPP differentiation into MEP versus GMP remain incompletely defined. The classical cross-antagonisms between GATA1 and PU.1 and/or between BACH2 and C/EBPβ likely function at the MPP level. GATA1 promotes megakaryocyte-erythroid differentiation by antagonizing PU.1 function that fosters myeloid differentiation. Similarly, BACH2 promotes erythropoiesis by repressing C/EBPβ, which increases myelopoiesis. GATA1-FOG1 complex drives megakaryocyte and erythroid gene expression programs while suppressing alternative fate regulators by recruiting a nucleosome remodeling and deacetylase (NuRD) complex. C-MYB and KLF1 reinforce erythroid differentiation by

(Continued)

(TPO), the essential cytokine for HSC maintenance and megakaryocyte development was shown to have synergistic effects with EPO, in particular, under stress/disease condition (Kaushansky et al., 1995; Zou et al., 2022). In contrast, macrophage colony stimulating factor (M-CSF) directly activated PU.1-driven myeloid differentiation program in HSCs both in vitro and in vivo and these effects were independent of survival or proliferation (Mossadegh-Keller et al., 2013). Upon inflammation, granulocyte-macrophage colony stimulating factor (GM-CSF)-receptor was increased on HSCs and HSC gene expression profile was skewed toward myeloid program, resulting in their differentiation toward GMP (Regan-Komito et al., 2020). Taken together, cytokine signaling drives lineage specific gene expression program and instructs lineage specification in HSCs and MPPs (Fig. 4).

5. Lineage fate decision in megakaryocyte-erythroid progenitor (MEP)

5.1 Revisiting the identity and lineage potential of MEP

MEP is regarded as the final lineage decision point for erythroid commitment. However, current immunophenotypically defined MEP are mostly composed of erythroid or megakaryocyte committed progenitors leaving true bipotent MEP rare in this population (Xavier-Ferrucio & Krause, 2018). There are also multiple human MEP gating strategies, indicating the heterogeneity of this population (Kwon et al., 2021). Most recent gating strategy devised by Krause group for human MEP improved the purity of true bipotent MEP (Sanada et al., 2016). Roughly half of MEP isolated with this gating strategy ($CD34^{+}Lin^{-}CD135^{-}CD45Ra^{-}CD38^{mid}CD110^{+}CD36^{-}CD41^{-}$) gave rise to megakaryocyte/erythroid (Mk/E) colonies, 25 % of MEP produced erythroid colonies, and 20 % of MEP formed megakaryocyte colonies in their dual CFU-Mk/E megacult assays (Sanada et al., 2016). Of note, the same MEP gave rise to 90 % of erythroid colonies in methylcellulose assays (Sanada et al., 2016), indicating that the assay method could affect the lineage output

Fig. 4—Cont'd repressing megakaryocytic transcriptional networks. FLI1 and RUNX1 antagonistically act on KLF1 thereby promoting megakaryopoiesis and suppressing erythroid differentiation. HSC, hematopoietic stem cell; MPP, multipotent progenitor; MEP, megakaryocyte-erythroid progenitor; GMP, granulocyte-macrophage progenitor; MkP, megakaryocyte progenitor; ErP, erythroid progenitor; MP, monocyte progenitor; GP, granulocyte progenitor; PLT, platelet; RBC, red blood cell; Mono, monocyte; Gr, granulocyte.

and what we observed in vitro assays may not reflect the real lineage potential of these cells in vivo. How many megakaryocytes are generated through MEP or directly from HSCs is unclear. Lineage tracing study in native hematopoiesis suggested that direct production from HSCs is the major route (Rodriguez-Fraticelli et al., 2018), but it could be attributed to technical aspects such as sequencing depth for a population with low frequency or transposon tagging efficiency for a transient population, etc. Nonetheless, many studies demonstrated the existence of functionally bipotent MEP (Akashi et al., 2000; Pronk et al., 2007; Sanada et al., 2016; Yamamoto et al., 2013).

5.2 Megakaryocyte versus erythroid fate determinants in MEP

Cross-antagonism between KLF1 and FLI1 is a well-known fate determinant in MEP, where KLF1 promotes erythropoiesis at the expense of megakaryocytic differentiation by inhibiting FLI1-driven megakaryocyte gene expression (Frontelo et al., 2007; Starck et al., 2003; Tallack & Perkins, 2010). Single-cell CyTOF data confirmed the co-expression of KLF1 and FLI1 in human cord blood derived erythroid-biased MEP (E-MEP) cells, further supporting this model (Palii et al., 2019). Single-cell RNA-sequencing study combined with fate assays and computational analysis also predicted that KLF1 and FLI1 as key regulators of erythroid versus megakaryocyte fate decision (Tusi et al., 2018). Whether this mechanism also works at the upstream of MEP needs further investigation because KLF1 was known to be absent in HSCs and MPPs while FLI1 is expressed in the hematopoietic stem and progenitor cells (Li et al., 2015; Siatecka & Bieker, 2011). However, single-cell RNA-sequencing data may provide evidence for KLF1 and FLI1 co-expression in the subset of HSCs and MPPs.

C-MYB promotes erythroid differentiation while suppressing megakaryopoiesis via KLF1 and LMO2 expression and miR-486–3p/MAF axis (Bianchi et al., 2015; Bianchi et al., 2010). This regulatory function of c-MYB was verified in adult human MEP (Sanada et al., 2016). Given c-MYB is highly expressed in HSPC populations and decreased toward terminal differentiation of mature cells (Greig et al., 2008), c-MYB may exert the same regulatory function at the HSC and MPP levels. Indeed, multiple studies imply the role of c-MYB in fate decision upstream of MEP (Bianchi et al., 2015; Bianchi et al., 2010; Emambokus et al., 2003).

RUNX1 cooperates with FLI1, GATA1, and FOG1 to promotes megakaryocyte differentiation while repressing KLF1-dependent erythroid gene expression, thereby locking in megakaryocyte fate (Elagib et al., 2003; Huang et al., 2009; Kuvardina et al., 2015; Kwon et al., 2024).

Given the essential role of RUNX1 in megakaryocyte-biased HSC differentiation into megakaryocytes (Wang et al., 2023), it will be interesting to investigate whether RUNX1 also functions in the megakaryocyte priming of HSCs.

Interestingly, cell cycle speed also regulates erythroid versus megakaryocyte fate decision as MEP with higher cell cycle rate prefers erythroid over megakaryocyte lineage (Lu et al., 2018). Previous study using fetal liver progenitor cells showed that PU.1 prolongs cell cycle length, leading to the accumulation of PU.1 itself and thereby resulting in myelopoiesis over B lymphopoiesis (Kueh et al., 2013). This study indicates that different cell cycle rate could cause accumulation or depletion of lineage-specific transcription factors, depending on their synthesis rates and half-lives. Accordingly, cell cycle duration may also influence lineage bias in HSCs and MPPs.

6. Erythropoiesis in aging and myeloid malignancy

6.1 Erythropoiesis in aging

Although myeloid-biased HSCs increase with age in both mice and humans, it is not pan-myeloid expansion and erythroid production rather decreases upon aging (Beerman et al., 2010; Gekas & Graf, 2013; Grover et al., 2016; Kyriacou & Shibeeb, 2025; Pang et al., 2011; Yamamoto et al., 2018). It has been suggested that aged HSCs have reduced responsiveness to EPO based on the elevated EPO level in elderly (Groarke & Young, 2019). However, more direct causes of reduced erythropoiesis in elderly are likely attributed to increased inflammation in aged bone marrow. Chronic low-grade inflammation called "inflammaging" is caused by increased pro-myeloid/inflammatory cytokines with age (Ferrucci & Fabbri, 2018). In inflammatory conditions, EPO signaling was reduced and negatively correlated with the circulating levels of IL-1 and IL-6 (Okonko et al., 2013). IL-6 suppressed EPO dependent erythroid maturation in TF-1 cell line (McCranor et al., 2014). IFNγ inhibits erythropoiesis via upregulating IRF-1-PU.1 axis in erythroid precursors (Libregts et al., 2011). TNFα and TNF-related apoptosis-inducing ligand (TRAIL) prohibit erythroid lineage differentiation from adult peripheral blood $CD34^+$ hematopoietic progenitor cells (Xiao et al., 2002; Zamai et al., 2000). This "anemia of inflammation" also could be caused by upstream lineage skewing at the HSC and MPP levels given the function of these cytokines in driving myeloid differentiation from HSCs and MPPs (Pietras, 2017; Pietras et al., 2016; Reynaud et al., 2011; Yamashita & Passegue, 2019). Indeed, blocking IL-1 signaling in aged mice restored erythroid related defects (Mitchell et al., 2023).

6.2 Erythropoiesis in myeloid malignancy

Myeloid malignancies are often accompanied by anemia (Weiss et al., 2019). Although this "anemia of chronic disease" is mediated by many factors including inflammatory signals and genetic mutations directly causing survival and maturation defects of erythroid progenitors, the skewed lineage bias at the HSC and MPP level also contributes to anemia in myeloid cancer. Myelodysplastic syndromes (MDS) are clonal HSC disorders characterized by accumulation of blasts in the bone marrow and cytopenias including anemia (Tefferi & Vardiman, 2009). Single-cell RNA-sequencing analysis of HSPCs from MDS patients versus healthy donors revealed the lower expression level of *GATA1* and *KLF1* in MDS HSPCs, indicating reduced erythroid transcriptional priming (Zeng et al., 2024). ATAC-sequencing analysis of MDS patient $CD34^+CD38^-$ cells showed that chromatin accessibility decreased at GATA target sites but increased at CEBP target sites compared to normal controls, supporting skewed lineage bias at the transcriptional regulatory level in MDS HSPCs (Oshima et al., 2025). Moreover, single-cell RNA-sequencing showed the expansion of granulo-monocytic-biased HSCs in MDS patients, indicating a shift toward myelopoiesis at the HSC level (Rodriguez-Sevilla et al., 2024). The reduced *BACH2* expression in MDS could, in part, contribute to anemia and myeloid skewing in MDS patients (Kato et al., 2018). Together, these data demonstrate that shift of lineage bias at the HSC and MPP level contributes to anemia in myeloid malignancy.

7. Concluding remarks

It is now clear that lineage specification begins at the level of multipotent HSCs and MPPs. Lineage-biased progenitors subsequently reinforce their lineage fidelity to generate specific blood cells. Erythroid bias is detectable in HSCs, often in conjunction with other myeloid programs such as megakaryocyte or granulocyte/macrophage (GM) lineages. In

MPPs, both megakaryocyte-biased MPP2 and GM-biased MPP3 retain robust erythroid potential. These observations suggest that megakaryocyte versus erythroid and GM versus erythroid fate specification needs to happen at the HSC and MPP stages.

Previous studies investigating megakaryocyte versus erythroid or GM versus erythroid fate decisions in cell lines or primary CMP and MEP populations have provided important mechanistic insights, particularly because many of the implicated transcription factors are also expressed in HSCs and MPPs. However, these inferred mechanisms require direct validation in HSCs and MPPs, as these stem and progenitor cells possess distinct epigenetic landscapes, metabolic states, and cell-cycle dynamics compared with downstream populations, all of which can influence transcription factor function. In addition, cell type-specific differences in the composition of transcriptional activator and repressor complexes may further modulate lineage outcomes.

Single-cell omics studies have generated an enormous amount of descriptive information, including gene regulatory networks that potentially govern lineage fate decisions. The next critical step is to translate these datasets into mechanistic understanding through functional experimentation. Continued advances in single-cell technologies will accelerate this transition. A deeper understanding of lineage specification mechanisms in HSCs and MPPs will provide opportunities to develop targeted therapeutic strategies aimed at correcting lineage imbalances and improving hematopoietic disorders at the apex of the hematopoietic hierarchy.

Acknowledgements

None.

References

Abkowitz, J. L., Catlin, S. N., & Guttorp, P. (1996). Evidence that hematopoiesis may be a stochastic process in vivo. *Nature Medicine, 2*, 190–197.

Adolfsson, J., Mansson, R., Buza-Vidas, N., Hultquist, A., Liuba, K., Jensen, C. T., ... Jacobsen, S. E. (2005). Identification of Flt3+ lympho-myeloid stem cells lacking erythro-megakaryocytic potential a revised road map for adult blood lineage commitment. *Cell, 121*(2), 295–306.

Akala, O. O., Park, I.-K., Qian, D., Pihalja, M., Becker, M. W., & Clarke, M. F. (2008). Long-term haematopoietic reconstitution by Trp53-/-p16Ink4a-/-p19Arf-/- multipotent progenitors. *Nature, 453*(7192), 228–232.

Akashi, K., Traver, D., Miyamoto, T., & Weissman, I. L. (2000). A clonogenic common myeloid progenitor that gives rise to all myeloid lineages. *Nature, 404*, 193–197.

Anjos-Afonso, F., & Bonnet, D. (2023). Human CD34+ hematopoietic stem cell hierarchy: How far are we with its delineation at the most primitive level? *Blood, 142*, 509–518.

Arinobu, Y., Mizuno, S., Chong, Y., Shigematsu, H., Iino, T., Iwasaki, H., ... Akashi, K. (2007). Reciprocal activation of GATA-1 and PU.1 marks initial specification of hematopoietic stem cells into myeloerythroid and myelolymphoid lineages. *Cell Stem Cell, 1*, 416–427.

Becker, A. J., Mc, C. E., & Till, J. E. (1963). Cytological demonstration of the clonal nature of spleen colonies derived from transplanted mouse marrow cells. *Nature, 197*, 452–454.

Beckmann, J., Scheitza, S., Wernet, P., Fischer, J. C., & Giebel, B. (2007). Asymmetric cell division within the human hematopoietic stem and progenitor cell compartment: Identification of asymmetrically segregating proteins. *Blood, 109*, 5494–5501.

Beerman, I., Bhattacharya, D., Zandi, S., Sigvardsson, M., Weissman, I. L., Bryder, D., & Rossi, D. J. (2010). Functionally distinct hematopoietic stem cells modulate hematopoietic lineage potential during aging by a mechanism of clonal expansion. *Proceedings of the National Academy of Sciences of the United States of America, 107*, 5465–5470.

Belluschi, S., Calderbank, E. F., Ciaurro, V., Pijuan-Sala, B., Santoro, A., Mende, N., ... Laurenti, E. (2018). Myelo-lymphoid lineage restriction occurs in the human haematopoietic stem cell compartment before lymphoid-primed multipotent progenitors. *Nature Communications, 9*, 4100.

Benveniste, P., Frelin, C., Janmohamed, S., Barbara, M., Herrington, R., Hyam, D., & Iscove, N. N. (2010). Intermediate-term hematopoietic stem cells with extended but time-limited reconstitution potential. *Cell Stem Cell, 6*(1), 48–58.

Benz, C., Copley, M. R., Kent, D. G., Wohrer, S., Cortes, A., Aghaeepour, N., ... Eaves, C. J. (2012). Hematopoietic stem cell subtypes expand differentially during development and display distinct lymphopoietic programs. *Cell Stem Cell, 10*, 273–283.

Bianchi, E., Bulgarelli, J., Ruberti, S., Rontauroli, S., Sacchi, G., Norfo, R., ... Manfredini, R. (2015). MYB controls erythroid versus megakaryocyte lineage fate decision through the miR-486-3p-mediated downregulation of MAF. *Cell Death and Differentiation, 22*, 1906–1921.

Bianchi, E., Zini, R., Salati, S., Tenedini, E., Norfo, R., Tagliafico, E., ... Ferrari, S. (2010). c-Myb supports erythropoiesis through the transactivation of KLF1 and LMO2 expression. *Blood, 116*, e99–e110.

Boyer, S. W., Rajendiran, S., Beaudin, A. E., Smith-Berdan, S., Muthuswamy, P. K., Perez-Cunningham, J., ... Forsberg, E. C. (2019). Clonal and quantitative in vivo assessment of hematopoietic stem cell differentiation reveals strong erythroid potential of multipotent cells. *Stem Cell Reports, 12*, 801–815.

Brummendorf, T. H., Dragowska, W., Zijlmans, J., Thornbury, G., & Lansdorp, P. M. (1998). Asymmetric cell divisions sustain long-term hematopoiesis from single-sorted human fetal liver cells. *The Journal of Experimental Medicine, 188*, 1117–1124.

Buenrostro, J. D., Corces, M. R., Lareau, C. A., Wu, B., Schep, A. N., Aryee, M. J., ... Greenleaf, W. J. (2018). Integrated single-cell analysis maps the continuous regulatory landscape of human hematopoietic differentiation. *Cell, 173*, 1535–1548.e1516.

Cabezas-Wallscheid, N., Klimmeck, D., Hansson, J., Lipka, D. B., Reyes, A., Wang, Q., ... Trumpp, A. (2014). Identification of regulatory networks in HSCs and their immediate progeny via integrated proteome, transcriptome, and DNA methylome analysis. *Cell Stem Cell, 15*, 507–522.

Cammenga, J., Mulloy, J. C., Berguido, F. J., MacGrogan, D., Viale, A., & Nimer, S. D. (2003). Induction of C/EBPalpha activity alters gene expression and differentiation of human CD34+ cells. *Blood, 101*, 2206–2214.

Carrelha, J., Mazzi, S., Winroth, A., Hagemann-Jensen, M., Ziegenhain, C., Hogstrand, K., ... Jacobsen, S. E. W. (2024). Alternative platelet differentiation pathways initiated by nonhierarchically related hematopoietic stem cells. *Nature Immunology, 25*, 1007–1019.

Carrelha, J., Meng, Y., Kettyle, L. M., Luis, T. C., Norfo, R., Alcolea, V., ... Jacobsen, S. E. W. (2018). Hierarchically related lineage-restricted fates of multipotent haematopoietic stem cells. *Nature, 554*, 106–111.

Challen, G. A., Boles, N. C., Chambers, S. M., & Goodell, M. A. (2010). Distinct hematopoietic stem cell subtypes are differentially regulated by TGF-beta1. *Cell Stem Cell, 6*, 265–278.

Challen, G. A., Pietras, E. M., Wallscheid, N. C., & Signer, R. A. J. (2021). Simplified murine multipotent progenitor isolation scheme: Establishing a consensus approach for multipotent progenitor identification. *Experimental Hematology, 104*, 55–63.

Chavez, J. S., Rabe, J. L., Loeffler, D., Higa, K. C., Hernandez, G., Mills, T. S., ... Pietras, E. M. (2021). PU.1 enforces quiescence and limits hematopoietic stem cell expansion during inflammatory stress. *The Journal of Experimental Medicine, 218*.

Chavez, J. S., Rabe, J. L., Nino, K. E., Wells, H. H., Gessner, R. L., Mills, T. S., ... Pietras, E. M. (2023). PU.1 is required to restrain myelopoiesis during chronic inflammatory stress. *Frontiers in Cell and Developmental Biology, 11*, 1204160.

Christensen, J. L., & Weissman, I. L. (2001). Flk-2 is a marker in hematopoietic stem cell differentiation: A simple method to isolate long-term stem cells. *Proceedings of the National Academy of Sciences of the United States of America, 98*, 14541–14546.

Comazzetto, S., Shen, B., & Morrison, S. J. (2021). Niches that regulate stem cells and hematopoiesis in adult bone marrow. *Developmental Cell, 56*, 1848–1860.

Dore, L. C., & Crispino, J. D. (2011). Transcription factor networks in erythroid cell and megakaryocyte development. *Blood, 118*, 231–239.

Dykstra, B., Kent, D., Bowie, M., McCaffrey, L., Hamilton, M., Lyons, K., ... Eaves, C. (2007). Long-term propagation of distinct hematopoietic differentiation programs in vivo. *Cell Stem Cell, 1*, 218–229.

Ediriwickrema, A., Nakauchi, Y., Fan, A. C., Kohnke, T., Hu, X., Luca, B. A., ... Majeti, R. (2025). A single-cell framework identifies functionally and molecularly distinct multipotent progenitors in adult human hematopoiesis. *Cell Reports, 44*, 116236.

Eisele, A. S., Cosgrove, J., Magniez, A., Tubeuf, E., Tenreira Bento, S., Conrad, C., ... Perie, L. (2022). Erythropoietin directly remodels the clonal composition of murine hematopoietic multipotent progenitor cells. *Elife, 11*.

Elagib, K. E., Racke, F. K., Mogass, M., Khetawat, R., Delehanty, L. L., & Goldfarb, A. N. (2003). RUNX1 and GATA-1 coexpression and cooperation in megakaryocytic differentiation. *Blood, 101*, 4333–4341.

Emambokus, N., Vegiopoulos, A., Harman, B., Jenkinson, E., Anderson, G., & Frampton, J. (2003). Progression through key stages of haemopoiesis is dependent on distinct threshold levels of c-Myb. *The EMBO Journal, 22*, 4478–4488.

Ferrucci, L., & Fabbri, E. (2018). Inflammageing: Chronic inflammation in ageing, cardiovascular disease, and frailty. *Nature Reviews Cardiology, 15*, 505–522.

Frontelo, P., Manwani, D., Galdass, M., Karsunky, H., Lohmann, F., Gallagher, P. G., & Bieker, J. J. (2007). Novel role for EKLF in megakaryocyte lineage commitment. *Blood, 110*, 3871–3880.

Gao, X., Xu, C., Asada, N., & Frenette, P. S. (2018). The hematopoietic stem cell niche: From embryo to adult. *Development (Cambridge, England), 145*.

Gekas, C., & Graf, T. (2013). CD41 expression marks myeloid-biased adult hematopoietic stem cells and increases with age. *Blood, 121*, 4463–4472.

Ghersi, J. J., Baldissera, G., Hintzen, J., Luff, S. A., Cheng, S., Xia, I. F., ... Nicoli, S. (2023). Haematopoietic stem and progenitor cell heterogeneity is inherited from the embryonic endothelium. *Nature Cell Biology, 25*, 1135–1145.

Goardon, N., Marchi, E., Atzberger, A., Quek, L., Schuh, A., Soneji, S., ... Vyas, P. (2011). Coexistence of LMPP-like and GMP-like leukemia stem cells in acute myeloid leukemia. *Cancer Cell, 19*, 138–152.

Greig, K. T., Carotta, S., & Nutt, S. L. (2008). Critical roles for c-Myb in hematopoietic progenitor cells. *Seminars in Immunology, 20*, 247–256.

Groarke, E. M., & Young, N. S. (2019). Aging and hematopoiesis. *Clinics in Geriatric Medicine, 35*, 285–293.

Grover, A., Mancini, E., Moore, S., Mead, A. J., Atkinson, D., Rasmussen, K. D., ... Nerlov, C. (2014). Erythropoietin guides multipotent hematopoietic progenitor cells toward an erythroid fate. *The Journal of Experimental Medicine, 211*, 181–188.

Grover, A., Sanjuan-Pla, A., Thongjuea, S., Carrelha, J., Giustacchini, A., Gambardella, A., ... Nerlov, C. (2016). Single-cell RNA sequencing reveals molecular and functional platelet bias of aged haematopoietic stem cells. *Nature Communications, 7*, 11075.

Haas, S., Hansson, J., Klimmeck, D., Loeffler, D., Velten, L., Uckelmann, H., ... Essers, M. A. (2015). Inflammation-induced emergency megakaryopoiesis driven by hematopoietic stem cell-like megakaryocyte progenitors. *Cell Stem Cell, 17*, 422–434.

Hoppe, P. S., Schwarzfischer, M., Loeffler, D., Kokkaliaris, K. D., Hilsenbeck, O., Moritz, N., ... Schroeder, T. (2016). Early myeloid lineage choice is not initiated by random PU.1 to GATA1 protein ratios. *Nature, 535*, 299–302.

Huang, H., Yu, M., Akie, T. E., Moran, T. B., Woo, A. J., Tu, N., ... Cantor, A. B. (2009). Differentiation-dependent interactions between RUNX-1 and FLI-1 during megakaryocyte development. *Molecular and Cellular Biology, 29*, 4103–4115.

Iscove, N. (1990). Haematopoiesis. Searching for stem cells. *Nature, 347*, 126–127.

Iscove, N. N., Messner, H., Till, J. E., & McCulloch, E. A. (1972). Human marrow cells forming colonies in culture: Analysis by velocity sedimentation and suspension culture. *Series Haematologica, 5*, 37–49.

Iwasaki, H., Mizuno, S., Wells, R. A., Cantor, A. B., Watanabe, S., & Akashi, K. (2003). GATA-1 converts lymphoid and myelomonocytic progenitors into the megakaryocyte/erythrocyte lineages. *Immunity, 19*, 451–462.

Iwasaki-Arai, J., Iwasaki, H., Miyamoto, T., Watanabe, S., & Akashi, K. (2003). Enforced granulocyte/macrophage colony-stimulating factor signals do not support lymphopoiesis, but instruct lymphoid to myelomonocytic lineage conversion. *The Journal of Experimental Medicine, 197*, 1311–1322.

Ju, Y. S., Martincorena, I., Gerstung, M., Petljak, M., Alexandrov, L. B., Rahbari, R., ... Stratton, M. R. (2017). Somatic mutations reveal asymmetric cellular dynamics in the early human embryo. *Nature, 543*, 714–718.

Kang, Y. A., Paik, H., Zhang, S. Y., Chen, J. J., Olson, O. C., Mitchell, C. A., ... Passegue, E. (2023). Secretory MPP3 reinforce myeloid differentiation trajectory and amplify myeloid cell production. *The Journal of Experimental Medicine, 220*.

Kang, Y. A., Pietras, E. M., & Passegue, E. (2020). Deregulated Notch and Wnt signaling activates early-stage myeloid regeneration pathways in leukemia. *The Journal of Experimental Medicine, 217*.

Kato, H., Itoh-Nakadai, A., Matsumoto, M., Ishii, Y., Watanabe-Matsui, M., Ikeda, M., ... Igarashi, K. (2018). Infection perturbs Bach2- and Bach1-dependent erythroid lineage 'choice' to cause anemia. *Nature Immunology, 19*, 1059–1070.

Kaushansky, K., Broudy, V. C., Grossmann, A., Humes, J., Lin, N., Ren, H. P., ... Sprugel, K. H. (1995). Thrombopoietin expands erythroid progenitors, increases red cell production, and enhances erythroid recovery after myelosuppressive therapy. *The Journal of Clinical Investigation, 96*, 1683–1687.

Kondo, M., Scherer, D. C., Miyamoto, T., King, A. G., Akashi, K., Sugamura, K., & Weissman, I. L. (2000). Cell-fate conversion of lymphoid-committed progenitors by instructive actions of cytokines. *Nature, 407*, 383–386.

Kueh, H. Y., Champhekar, A., Nutt, S. L., Elowitz, M. B., & Rothenberg, E. V. (2013). Positive feedback between PU.1 and the cell cycle controls myeloid differentiation. *Science (New York, N. Y.), 341*, 670–673.

Kuvardina, O. N., Herglotz, J., Kolodziej, S., Kohrs, N., Herkt, S., Wojcik, B., ... Lausen, J. (2015). RUNX1 represses the erythroid gene expression program during megakaryocytic differentiation. *Blood, 125*, 3570–3579.

Kwon, N., Lu, Y. C., Thompson, E. N., Mancuso, R. I., Wang, L., Zhang, P. X., & Krause, D. S. (2024). CDK9 phosphorylates RUNX1 to promote megakaryocytic fate in megakaryocytic-erythroid progenitors. *Blood, 144*, 1800–1812.

Kwon, N., Thompson, E. N., Mayday, M. Y., Scanlon, V., Lu, Y. C., & Krause, D. S. (2021). Current understanding of human megakaryocytic-erythroid progenitors and their fate determinants. *Current Opinion in Hematology, 28*, 28–35.

Kyriacou, R. P., & Shibeeb, S. (2025). Red blood cells and human aging: Exploring their biomarker potential. *Diagnostics (Basel), 15*.

Laiosa, C. V., Stadtfeld, M., & Graf, T. (2006). Determinants of lymphoid-myeloid lineage diversification. *Annual Review of Immunology, 24*, 705–738.

Laurenti, E., & Gottgens, B. (2018). From haematopoietic stem cells to complex differentiation landscapes. *Nature, 553*, 418–426.

Lee, J., Zhou, Y. J., Ma, W., Zhang, W., Aljoufi, A., Luh, T., ... Liu, K. (2017). Lineage specification of human dendritic cells is marked by IRF8 expression in hematopoietic stem cells and multipotent progenitors. *Nature Immunology, 18*, 877–888.

Li, L., Bowling, S., McGeary, S. E., Yu, Q., Lemke, B., Alcedo, K., ... Camargo, F. D. (2023). A mouse model with high clonal barcode diversity for joint lineage, transcriptomic, and epigenomic profiling in single cells. *Cell, 186*, 5183–5199.e5122.

Li, Y., Luo, H., Liu, T., Zacksenhaus, E., & Ben-David, Y. (2015). The ets transcription factor Fli-1 in development, cancer and disease. *Oncogene, 34*, 2022–2031.

Libregts, S. F., Gutierrez, L., de Bruin, A. M., Wensveen, F. M., Papadopoulos, P., van Ijcken, W., ... Nolte, M. A. (2011). Chronic IFN-gamma production in mice induces anemia by reducing erythrocyte life span and inhibiting erythropoiesis through an IRF-1/PU.1 axis. *Blood, 118*, 2578–2588.

Lu, R., Czechowicz, A., Seita, J., Jiang, D., & Weissman, I. L. (2019). Clonal-level lineage commitment pathways of hematopoietic stem cells in vivo. *Proceedings of the National Academy of Sciences of the United States of America, 116*, 1447–1456.

Lu, Y. C., Sanada, C., Xavier-Ferrucio, J., Wang, L., Zhang, P. X., Grimes, H. L., ... Krause, D. S. (2018). The molecular signature of megakaryocyte-erythroid progenitors reveals a role for the cell cycle in fate specification. *Cell Reports, 25*, 2083–2093.e2084.

Luchsinger, L. L., de Almeida, M. J., Corrigan, D. J., Mumau, M., & Snoeck, H. W. (2016). Mitofusin 2 maintains haematopoietic stem cells with extensive lymphoid potential. *Nature, 529*, 528–531.

Majeti, R., Park, C. Y., & Weissman, I. L. (2007). Identification of a hierarchy of multipotent hematopoietic progenitors in human cord blood. *Cell Stem Cell, 1*, 635–645.

McCranor, B. J., Kim, M. J., Cruz, N. M., Xue, Q. L., Berger, A. E., Walston, J. D., ... Roy, C. N. (2014). Interleukin-6 directly impairs the erythroid development of human TF-1 erythroleukemic cells. *Blood Cells, Molecules & Diseases, 52*, 126–133.

Mitchell, C. A., Verovskaya, E. V., Calero-Nieto, F. J., Olson, O. C., Swann, J. W., Wang, X., ... Passegue, E. (2023). Stromal niche inflammation mediated by IL-1 signalling is a targetable driver of haematopoietic ageing. *Nature Cell Biology, 25*, 30–41.

Morita, Y., Ema, H., & Nakauchi, H. (2010). Heterogeneity and hierarchy within the most primitive hematopoietic stem cell compartment. *The Journal of Experimental Medicine, 207*, 1173–1182.

Morrison, S. J., & Scadden, D. T. (2014). The bone marrow niche for haematopoietic stem cells. *Nature, 505*, 327–334.

Morrison, S. J., Wandycz, A. M., Hemmati, H. D., Wright, D. E., & Weissman, I. L. (1997). Identification of a lineage of multipotent hematopoietic progenitors. *Development (Cambridge, England), 124*, 1929–1939.

Morrison, S. J., & Weissman, I. L. (1994). The long-term repopulating subset of hematopoietic stem cells is deterministic and isolatable by phenotype. *Immunity, 1*, 661–673.
Mossadegh-Keller, N., Sarrazin, S., Kandalla, P. K., Espinosa, L., Stanley, E. R., Nutt, S. L., ... Sieweke, M. H. (2013). M-CSF instructs myeloid lineage fate in single haematopoietic stem cells. *Nature, 497*, 239–243.
Muller-Sieburg, C. E., Cho, R. H., Karlsson, L., Huang, J. F., & Sieburg, H. B. (2004). Myeloid-biased hematopoietic stem cells have extensive self-renewal capacity but generate diminished lymphoid progeny with impaired IL-7 responsiveness. *Blood, 103*, 4111–4118.
Muller-Sieburg, C. E., Cho, R. H., Thoman, M., Adkins, B., & Sieburg, H. B. (2002). Deterministic regulation of hematopoietic stem cell self-renewal and differentiation. *Blood, 100*, 1302–1309.
Muller-Sieburg, C. E., Whitlock, C. A., & Weissman, I. L. (1986). Isolation of two early B lymphocyte progenitors from mouse marrow: A committed pre-pre-B. *Cell, 44*, 653–662.
Nestorowa, S., Hamey, F. K., Pijuan Sala, B., Diamanti, E., Shepherd, M., Laurenti, E., ... Gottgens, B. (2016). A single-cell resolution map of mouse hematopoietic stem and progenitor cell differentiation. *Blood, 128*, e20–e31.
Notta, F., Doulatov, S., Laurenti, E., Poeppl, A., Jurisica, I., & Dick, J. E. (2011). Isolation of single human hematopoietic stem cells capable of long-term multilineage engraftment. *Science (New York, N. Y.), 333*, 218–221.
Notta, F., Zandi, S., Takayama, N., Dobson, S., Gan, O. I., Wilson, G., ... Dick, J. E. (2016). Distinct routes of lineage development reshape the human blood hierarchy across ontogeny. *Science (New York, N. Y.), 351*, aab2116.
Oguro, H., Ding, L., & Morrison, S. J. (2013). SLAM family markers resolve functionally distinct subpopulations of hematopoietic stem cells and multipotent progenitors. *Cell Stem Cell, 13*, 102–116.
Okonko, D. O., Marley, S. B., Anker, S. D., Poole-Wilson, P. A., & Gordon, M. Y. (2013). Erythropoietin resistance contributes to anaemia in chronic heart failure and relates to aberrant JAK-STAT signal transduction. *International Journal of Cardiology, 164*, 359–364.
Olson, O. C., Kang, Y. A., & Passegue, E. (2020). Normal hematopoiesis is a balancing act of self-renewal and regeneration. *Cold Spring Harbor Perspectives in Medicine, 10*.
Orkin, S. H., & Zon, L. I. (2008). Hematopoiesis: An evolving paradigm for stem cell biology. *Cell, 132*, 631–644.
Osawa, M., Hanada, K., Hamada, H., & Nakauchi, H. (1996). Long-term lymphohematopoietic reconstitution by a single CD34-low/negative hematopoietic stem cell. *Science (New York, N. Y.), 273*, 242–245.
Oshima, M., Takayama, N., Nakajima-Takagi, Y., Shinoda, D., Itokawa, N., Kurosawa, S., ... Iwama, A. (2025). Chromatin accessibility in stem cells unveils progressive transcriptional alterations in myelodysplastic syndrome. *Nature Communications, 16*, 10726.
Palii, C. G., Cheng, Q., Gillespie, M. A., Shannon, P., Mazurczyk, M., Napolitani, G., ... Brand, M. (2019). Single-cell proteomics reveal that quantitative changes in co-expressed lineage-specific transcription factors determine cell fate. *Cell Stem Cell, 24*, 812–820.e815.
Pang, W. W., Price, E. A., Sahoo, D., Beerman, I., Maloney, W. J., Rossi, D. J., ... Weissman, I. L. (2011). Human bone marrow hematopoietic stem cells are increased in frequency and myeloid-biased with age. *Proceedings of the National Academy of Sciences of the United States of America, 108*, 20012–20017.
Paul, F., Arkin, Y., Giladi, A., Jaitin, D. A., Kenigsberg, E., Keren-Shaul, H., ... Amit, I. (2015). Transcriptional heterogeneity and lineage commitment in myeloid progenitors. *Cell, 163*, 1663–1677.

Pei, W., Feyerabend, T. B., Rossler, J., Wang, X., Postrach, D., Busch, K., ... Rodewald, H. R. (2017). Polylox barcoding reveals haematopoietic stem cell fates realized in vivo. *Nature, 548*, 456–460.

Pei, W., Shang, F., Wang, X., Fanti, A. K., Greco, A., Busch, K., ... Rodewald, H. R. (2020). Resolving fates and single-cell transcriptomes of hematopoietic stem cell clones by PolyloxExpress barcoding. *Cell Stem Cell, 27*, 383–395.e388.

Perie, L., Duffy, K. R., Kok, L., de Boer, R. J., & Schumacher, T. N. (2015). The branching point in erythro-myeloid differentiation. *Cell, 163*, 1655–1662.

Pietras, E. M. (2017). Inflammation: A key regulator of hematopoietic stem cell fate in health and disease. *Blood, 130*, 1693–1698.

Pietras, E. M., Mirantes-Barbeito, C., Fong, S., Loeffler, D., Kovtonyuk, L. V., Zhang, S., ... Passegue, E. (2016). Chronic interleukin-1 exposure drives haematopoietic stem cells towards precocious myeloid differentiation at the expense of self-renewal. *Nature Cell Biology, 18*, 607–618.

Pietras, E. M., Reynaud, D., Kang, Y. A., Carlin, D., Calero-Nieto, F. J., Leavitt, A. D., ... Passegue, E. (2015). Functionally distinct subsets of lineage-biased multipotent progenitors control blood production in normal and regenerative conditions. *Cell Stem Cell, 17*, 35–46.

Pinho, S., & Frenette, P. S. (2019). Haematopoietic stem cell activity and interactions with the niche. *Nature Reviews. Molecular Cell Biology, 20*, 303–320.

Pronk, C. J., Rossi, D. J., Mansson, R., Attema, J. L., Norddahl, G. L., Chan, C. K., ... Bryder, D. (2007). Elucidation of the phenotypic, functional, and molecular topography of a myeloerythroid progenitor cell hierarchy. *Cell Stem Cell, 1*, 428–442.

Regan-Komito, D., Swann, J. W., Demetriou, P., Cohen, E. S., Horwood, N. J., Sansom, S. N., & Griseri, T. (2020). GM-CSF drives dysregulated hematopoietic stem cell activity and pathogenic extramedullary myelopoiesis in experimental spondyloarthritis. *Nature Communications, 11*, 155.

Reynaud, D., Pietras, E., Barry-Holson, K., Mir, A., Binnewies, M., Jeanne, M., ... Passegue, E. (2011). IL-6 controls leukemic multipotent progenitor cell fate and contributes to chronic myelogenous leukemia development. *Cancer Cell, 20*, 661–673.

Rieger, M. A., Hoppe, P. S., Smejkal, B. M., Eitelhuber, A. C., & Schroeder, T. (2009). Hematopoietic cytokines can instruct lineage choice. *Science (New York, N. Y.), 325*, 217–218.

Rodriguez-Fraticelli, A. E., Wolock, S. L., Weinreb, C. S., Panero, R., Patel, S. H., Jankovic, M., ... Camargo, F. D. (2018). Clonal analysis of lineage fate in native haematopoiesis. *Nature, 553*, 212–216.

Rodriguez-Sevilla, J. J., Ganan-Gomez, I., Ma, F., Chien, K., Del Rey, M., Loghavi, S., ... Colla, S. (2024). Hematopoietic stem cells with granulo-monocytic differentiation state overcome venetoclax sensitivity in patients with myelodysplastic syndromes. *Nature Communications, 15*, 2428.

Sanada, C., Xavier-Ferrucio, J., Lu, Y. C., Min, E., Zhang, P. X., Zou, S., ... Krause, D. S. (2016). Adult human megakaryocyte-erythroid progenitors are in the CD34+CD38mid fraction. *Blood, 128*, 923–933.

Sanjuan-Pla, A., Macaulay, I. C., Jensen, C. T., Woll, P. S., Luis, T. C., Mead, A., ... Jacobsen, S. E. (2013). Platelet-biased stem cells reside at the apex of the haematopoietic stem-cell hierarchy. *Nature, 502*, 232–236.

Seita, J., & Weissman, I. L. (2010). Hematopoietic stem cell: Self-renewal versus differentiation. *Wiley Interdisciplinary Reviews: Systems Biology and Medicine, 2*, 640–653.

Siatecka, M., & Bieker, J. J. (2011). The multifunctional role of EKLF/KLF1 during erythropoiesis. *Blood, 118*, 2044–2054.

Siminovitch, L., McCulloch, E. A., & Till, J. E. (1963). The distribution of colony-forming cells among spleen colonies. *Journal of Cellular and Comparative Physiology, 62*, 327–336.

Sommerkamp, P., Romero-Mulero, M. C., Narr, A., Ladel, L., Hustin, L., Schonberger, K., ... Cabezas-Wallscheid, N. (2021). Mouse multipotent progenitor 5 cells are located at the interphase between hematopoietic stem and progenitor cells. *Blood, 137*, 3218–3224.
Somuncular, E., Hauenstein, J., Khalkar, P., Johansson, A. S., Dumral, O., Frengen, N. S., ... Luc, S. (2022). CD49b identifies functionally and epigenetically distinct subsets of lineage-biased hematopoietic stem cells. *Stem Cell Reports, 17*, 1546–1560.
Spangrude, G. J., Heimfeld, S., & Weissman, I. L. (1988). Purification and characterization of mouse hematopoietic stem cells. *Science (New York, N. Y.), 241*, 58–62.
Spencer Chapman, M., Wilk, C. M., Boettcher, S., Mitchell, E., Dawson, K., Williams, N., ... Campbell, P. J. (2024). Clonal dynamics after allogeneic haematopoietic cell transplantation. *Nature, 635*, 926–934.
Starck, J., Cohet, N., Gonnet, C., Sarrazin, S., Doubeikovskaia, Z., Doubeikovski, A., ... Morle, F. (2003). Functional cross-antagonism between transcription factors FLI-1 and EKLF. *Molecular and Cellular Biology, 23*, 1390–1402.
Sun, J., Ramos, A., Chapman, B., Johnnidis, J. B., Le, L., Ho, Y. J., ... Camargo, F. D. (2014). Clonal dynamics of native haematopoiesis. *Nature, 514*, 322–327.
Tallack, M. R., & Perkins, A. C. (2010). Megakaryocyte-erythroid lineage promiscuity in EKLF null mouse blood. *Haematologica, 95*, 144–147.
Tefferi, A., & Vardiman, J. W. (2009). Myelodysplastic syndromes. *The New England Journal of Medicine, 361*, 1872–1885.
Till, J. E., & Mc, C. E. (1961). A direct measurement of the radiation sensitivity of normal mouse bone marrow cells. *Radiation Research, 14*, 213–222.
Tusi, B. K., Wolock, S. L., Weinreb, C., Hwang, Y., Hidalgo, D., Zilionis, R., ... Socolovsky, M. (2018). Population snapshots predict early haematopoietic and erythroid hierarchies. *Nature, 555*, 54–60.
Uchimura, A., Matsumoto, H., Satoh, Y., Minakuchi, Y., Wakayama, S., Wakayama, T., ... Yagi, T. (2022). Early embryonic mutations reveal dynamics of somatic and germ cell lineages in mice. *Genome Research, 32*, 945–955.
Velten, L., Haas, S. F., Raffel, S., Blaszkiewicz, S., Islam, S., Hennig, B. P., ... Steinmetz, L. M. (2017). Human haematopoietic stem cell lineage commitment is a continuous process. *Nature Cell Biology, 19*, 271–281.
Wang, C., Tu, Z., Cai, X., Wang, W., Davis, A. K., Nattamai, K., ... Zheng, Y. (2023). A critical role of RUNX1 in governing megakaryocyte-primed hematopoietic stem cell differentiation. *Blood Advances, 7*, 2590–2605.
Weinreb, C., Rodriguez-Fraticelli, A., Camargo, F. D., & Klein, A. M. (2020). Lineage tracing on transcriptional landscapes links state to fate during differentiation. *Science (New York, N. Y.), 367*.
Weiss, G., Ganz, T., & Goodnough, L. T. (2019). Anemia of inflammation. *Blood, 133*, 40–50.
Weng, C., Yu, F., Yang, D., Poeschla, M., Liggett, L. A., Jones, M. G., ... Sankaran, V. G. (2024). Deciphering cell states and genealogies of human haematopoiesis. *Nature, 627*, 389–398.
Wilson, A., Laurenti, E., Oser, G., van der Wath, R. C., Blanco-Bose, W., Jaworski, M., ... Trumpp, A. (2008). Hematopoietic stem cells reversibly switch from dormancy to self-renewal during homeostasis and repair. *Cell, 135*, 1118–1129.
Wilson, N. K., Kent, D. G., Buettner, F., Shehata, M., Macaulay, I. C., Calero-Nieto, F. J., ... Gottgens, B. (2015). Combined single-cell functional and gene expression analysis resolves heterogeneity within stem cell populations. *Cell Stem Cell, 16*, 712–724.
Wolff, L., & Humeniuk, R. (2013). Concise review: Erythroid versus myeloid lineage commitment: Regulating the master regulators. *Stem Cells, 31*, 1237–1244.

Wu, A. M., Till, J. E., Siminovitch, L., & McCulloch, E. A. (1967). A cytological study of the capacity for differentiation of normal hemopoietic colony-forming cells. *Journal of Cellular Physiology, 69*, 177–184.

Wu, A. M., Till, J. E., Siminovitch, L., & McCulloch, E. A. (1968). Cytological evidence for a relationship between normal hemotopoietic colony-forming cells and cells of the lymphoid system. *The Journal of Experimental Medicine, 127*, 455–464.

Wu, M., Kwon, H. Y., Rattis, F., Blum, J., Zhao, C., Ashkenazi, R., ... Reya, T. (2007). Imaging hematopoietic precursor division in real time. *Cell Stem Cell, 1*, 541–554.

Wu, Z., Gao, S., Diamond, C., Kajigaya, S., Chen, J., Shi, R., ... Young, N. S. (2020). Sequencing of RNA in single cells reveals a distinct transcriptome signature of hematopoiesis in GATA2 deficiency. *Blood Advances, 4*, 2656–2670.

Xavier-Ferrucio, J., & Krause, D. S. (2018). Concise review: Bipotent megakaryocytic-erythroid progenitors: Concepts and controversies. *Stem Cells, 36*, 1138–1145.

Xiao, W., Koizumi, K., Nishio, M., Endo, T., Osawa, M., Fujimoto, K., ... Sawada, K. (2002). Tumor necrosis factor-alpha inhibits generation of glycophorin A+ cells by CD34+ cells. *Experimental Hematology, 30*, 1238–1247.

Yamamoto, R., Morita, Y., Ooehara, J., Hamanaka, S., Onodera, M., Rudolph, K. L., ... Nakauchi, H. (2013). Clonal analysis unveils self-renewing lineage-restricted progenitors generated directly from hematopoietic stem cells. *Cell, 154*, 1112–1126.

Yamamoto, R., Wilkinson, A. C., Ooehara, J., Lan, X., Lai, C. Y., Nakauchi, Y., ... Nakauchi, H. (2018). Large-scale clonal analysis resolves aging of the mouse hematopoietic stem cell compartment. *Cell Stem Cell, 22*, 600–607.e604.

Yamashita, M., & Passegue, E. (2019). TNF-alpha coordinates hematopoietic stem cell survival and myeloid regeneration. *Cell Stem Cell, 25*, 357–372.e357.

Yan, B., Yang, J., Kim, M. Y., Luo, H., Cesari, N., Yang, T., ... Qiu, Y. (2021). HDAC1 is required for GATA-1 transcription activity, global chromatin occupancy and hematopoiesis. *Nucleic Acids Research, 49*, 9783–9798.

Yoshizato, T., Nilsson, C., Grasso, F., Hogstrand, K., Mazzi, S., Winroth, A., ... Jacobsen, S. E. W. (2025). Stable clonal contribution of lineage-restricted stem cells to human hematopoiesis. *Nature Genetics, 57*, 3088–3100.

Yu, V. W. C., Yusuf, R. Z., Oki, T., Wu, J., Saez, B., Wang, X., ... Scadden, D. T. (2017). Epigenetic memory underlies cell-autonomous heterogeneous behavior of hematopoietic stem cells. *Cell, 168*, 944–945.

Zamai, L., Secchiero, P., Pierpaoli, S., Bassini, A., Papa, S., Alnemri, E. S., ... Zauli, G. (2000). TNF-related apoptosis-inducing ligand (TRAIL) as a negative regulator of normal human erythropoiesis. *Blood, 95*, 3716–3724.

Zeng, X., Wang, Y., Dai, M., Li, W., Huang, Q., Qin, L., ... Qi, L. (2024). Single-cell transcriptomics dissects the transcriptome alterations of hematopoietic stem cells in myelodysplastic neoplasms. *Journal of Translational Medicine, 22*, 359.

Zou, H., Xu, P., Wong, R. S. M., & Yan, X. (2022). A novel combination therapy of erythropoietin and thrombopoietin to treat erythropoietin-resistance anemia. *Pharmaceutical Research, 39*, 1249–1265.

CHAPTER FIVE

Hijacking the host: Red blood cell remodeling and extracellular vesicle release during *Babesia* infection

Divya Beri, Yi-Tian Dai, Pratap Vydyam, Marilis Rodriguez, and Cheryl A. Lobo*
Laboratory of Blood-Borne Parasites, Lindsley F. Kimball Research Institute, New York Blood Center, Rye, NY, United States
*Corresponding author. e-mail address: clobo@nybc.org

Contents

Current Topics in Developmental Biology, Volume 170
ISSN 0070-2153, https://doi.org/10.1016/bs.ctdb.2026.06.003

Abstract

The red blood cells (RBC) of various species serve as the host cells for the asexual stage of *Babesia* causing hemolysis and anemia, hallmark symptoms of parasitic infections that target RBCs. After *Babesia* invasion, RBCs are progressively modified by the parasite to suit their requirements of rapid proliferation and egress, while avoiding host immune consequences. This involves a complete morphological, biochemical and rheological overhaul of the RBC whereby the parasite confers on the host RBC, a new set of properties. *Babesia* infected RBCs exhibit increased membrane rigidity, reduced deformability, increased osmotic resistance, high vesiculation rates and in some species increased adhesiveness to the vascular endothelium. In this chapter we describe our current understanding of these changes in the *Babesia*-infected RBC that contribute to the modified RBC architecture. Further work must be directed to elucidate the structural and functional relationships of the pathogen-RBC networks at play in the infected cell, as these interactions may present targets for the development of novel therapeutics.

1. Introduction

1.1 Human Babesiosis-the pathogens

Babesia spp. are tick-transmitted apicomplexan erythrocytic protozoan parasites that cause disease in diverse vertebrate hosts including cattle, horses, dogs and humans. Although Victor Babes first identified *Babesia* in 1888 as the cause of the fatal red water fever in Eastern European cattle, human babesiosis was not described until 1957 in a Croatian herdsman, realizing the zoonotic potential of the hemoparasite. (Skrabalo & Deanovic, 1957). More human infections followed over the next few decades, and at least four taxonomically classified *Babesia* species (*B. divergens, B. duncani, B. microti*, and *B. venatorum*) have now been confirmed as zoonotic pathogens, with some others that have not yet been identified to species (Hunfeld et al., 2008; Kumar et al., 2021; Ord & Lobo, 2015).

Despite exhibiting a complex life cycle spanning dual hosts, a tick vector and a mammalian host, *Babesia* has successfully established itself as the second most common blood-borne parasites of mammals after trypanosomes. Most human babesiosis cases are in the United States with *Babesia microti* causing most infections (Yang et al., 2021). In January 2011, the Centers for Disease Control and Prevention (CDC) made babesiosis a nationally notifiable condition (Gray & Herwaldt, 2019). Since then, in addition to high incidence, drug-resistant *Babesia* has been reported (Holbrook et al., 2023; Krause et al., 2024; Marcos & Wormser, 2023; Rogers et al., 2023) While humans are dead-end hosts for *Babesia*, human-to-human transmission of *Babesia* spp. can occur from blood transfusion (TTB) (Leiby, 2006) and the first case of TTB was reported in 1980 in Boston (Jacoby et al., 1980). Today, TTB still poses a threat despite the availability of a blood screening test, because of the lack of a universal mandate to test blood collections (Drews et al., 2023b; Gerber et al., 1994).

1.2 Babesia parasite biology

Erythrocytic invasion is the central pathogenic step in the life cycle of *Babesia* (Fig. 1). When *Babesia* spp. sporozoites are first injected from a tick bite into

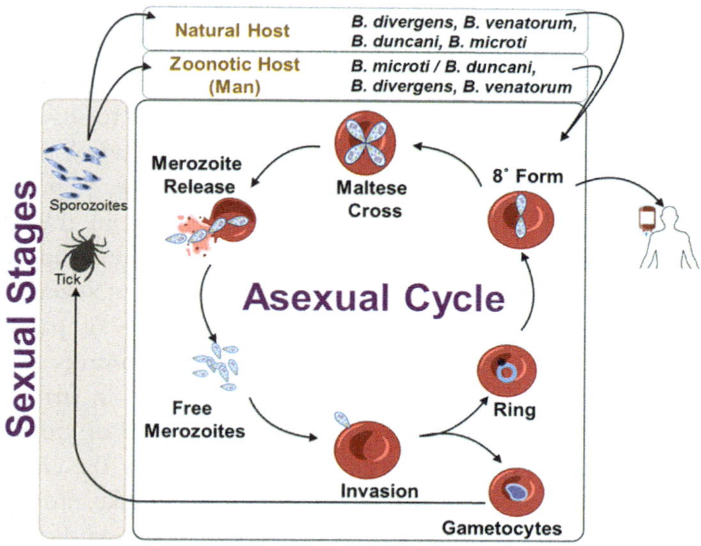

Fig. 1 Life cycle of *Babesia* parasites, emphasizing the pathological asexual stages of *Babesia*: The *Babesia* life cycle alternates between vertebrate hosts (including humans, cattle, mice, and deer) and the definitive tick vector. During a blood meal, an infected tick introduces sporozoites into the host (natural and zoonotic), where they invade RBCs and undergo asexual replication with significant population expansion, which is responsible for symptomatic disease. Although humans are considered dead-end hosts, transmission can occur through blood transfusion.

the human host, they target the host RBCs immediately, unlike malaria parasites which are required to undergo an exo-erythrocytic phase in hepatic cells. Further, infected RBCs remain circulating in peripheral blood stream, including regularly passing through the hosts' spleen, and it has been shown that most human species do not sequester to the fine capillaries of the bone marrow or organs. It is the parasite's ability to first recognize and then invade host RBCs that is central to human babesiosis and *Babesia* invades RBCs using multiple complex interactions between parasite proteins and the host cell surface, which are not fully elucidated yet (Cursino-Santos et al., 2014b; Gonzalez et al., 2019; Gonzalez et al., 2025; Li et al., 2019; Lobo et al., 2012; Lobo, 2005; Sun et al., 2011). Once inside the RBC, the parasite begins a cycle of maturation and growth. The early stages of the cycle are morphologically indistinguishable from *Plasmodium* spp., with both appearing as ring-like parasites. Replication occurs by binary fission, where one ring forms divides into two, often referred to as "figure eight" form (Fig. 1). Fission may occur again, giving ride to the tetrad form know as a "Maltese Cross" (Homer et al., 2000; Krampitz, 1979). Both these morphological forms are unique to *Babesia* spp. and are the basis of definitive diagnosis by microscopy. Once the parasites have concluded division, the resulting merozoites egress from the RBCs and seek new RBCs to invade, perpetuating the intracellular cycle of infection and resulting pathology (Fig. 1).

In vitro, it has been shown that the *Babesia divergens* population can expand inside the RBCs or egress at multiple points in the life cycle (Cursino-Santos et al., 2016). Using synchronized parasite populations, it has been shown that there is a sequential progression of seven morphological forms of *B. divergens* in culture. The dynamics of parasite proliferation and differentiation are maintained through controls that ensure that the infected RBC sub-populations are maintained in strict ratios to enable rapid movement between new invasion events or further intra-RBC development and replication cycles, as dictated by the environment of the parasite (Cursino-Santos et al., 2016). Thus, unlike most parasites, *Babesia* exhibits plasticity in its life cycle enabling it to swiftly respond to environmental conditions like host RBCs and nutritional availability (Cursino-Santos et al., 2016; Lobo et al., 2019).

As *Babesia,* are intraerythrocytic protozoan parasites, most symptoms are associated with hemolysis due to the cyclical replication of *Babesia* in the RBCs, causing acute and chronic hemolytic anemia, with added complications due to the release of toxins and waste products into the bloodstream. Further damage can be caused by cytokine storms as the host's immune system responds to infection. In addition, infected hosts become sick and die because the parasites dramatically alter the structure and function of the RBCs in which they reside and replicate. *Babesia* infected RBCs can cause microvascular obstruction, partly due to parasitic molecules on the RBC surface (Aikawa et al., 1985; Fathi et al., 2024; Hakimi et al., 2020) exacerbating hemolysis. Parasitemias do not always relate directly to the degree of anemia, suggesting that erythrocyte destruction is not only due to lysis of infected cells or their removal by splenic and liver macrophages, but also due to lysis of bystander cells which might be a significant contributing factor to the process (Krause et al., 2007). This in turn can lead to anemia, jaundice, weakness, and depending on parasite burden, organ damage, neurological complications, and death have also been reported.

Babesia replication within RBCs is accompanied by striking modifications to the morphological, biochemical and biophysical architecture of the infected host RBC. Successful survival of the parasite within a terminally differentiated host cell presents multiple challenges. *Babesia* must maintain the mechanical integrity of the host RBC under the severe hemodynamic stress of the circulation, obtain nutrition from outside the host RBC, avoid the host immune system and escape splenic destruction. Mature RBCs lack de novo protein and lipid synthesis apparatus as well as protein trafficking pathways, necessitating *Babesia* to exploit its own biochemical machinery. Modification of the normal structure and function of RBCs by *Babesia*, particularly the RBC's rheological, mechanical and adhesive properties, appears to be vital for the parasite; however, these modifications also underlie the pathogenesis of the severe and frequently fatal complications of babesiosis (Jajosky & Stowell, 2025). The accumulation of infected RBCs in the microvasculature, perturbation of blood flow or partial or complete obstruction of blood vessels by rheologically abnormal infected RBCs has severe and often fatal clinical complications. In this chapter, we will highlight what is known about how *Babesia* spp. modify the RBC to allow for successful invasion, proliferation and egress in the host RBC. As *Plasmodium* and *Babesia* share a common host cell, at times we also draw upon the more extensive information available for *Plasmodium*, which has been studied more comprehensively. Specifically, we focus on *Babesia*-driven cytoskeletal modification of the RBC, describing the parasite exportome driven remodeling of the RBC, changes in RBC vesiculation and will discuss the impact these parasitic processes have on the host RBC cell, including changes in deformability, osmotic fragility and loss of cell surface area leading to loss of cellular viability and hemolysis.

2. Host red blood cell remodeling: A critical step for parasite survival

2.1 Introduction to red blood cell remodeling: What is known

The non-nucleated erythrocyte is unique among cells in the human body as the cell membrane is its sole and only structural component which accounts for all its diverse antigenic, transport, and mechanical functionalities (Mohandas & Gallagher, 2008). An important and physiologically distinguishing feature of the RBCs is its ability to undergo large deformations during its continuous passage through the narrow capillaries of the microvasculature. Several elegant studies have documented that splenic sequestration of abnormal red cells with reduced deformability accounts for the decreased life span and resulting hemolytic anemia in several red cell disorders and parasitic diseases such as malaria. On the cellular level, it is now widely accepted that the biological functions of cells are related to their mechanical properties (Lai et al., 2015). For RBCs, cytoskeleton affords mechanical stability to the cell, and these have been extensively researched over the last few decades (Discher et al., 1994; Hale et al., 2011; Lenormand et al., 2001).

The RBC membrane exhibits unique material behavior. It is highly elastic, rapidly responds to applied fluid stresses in the order of milliseconds and is stronger than steel in terms of structural resistance. While a normal red cell can deform with linear extensions of up to approximately 250 %, a 3 % to 4 % increase in surface area results in hemolysis (Safeukui et al., 2018), as has been shown in previous studies as well as our recent study of *Babesia*-infected wildtype and sickle red cells (Beri et al., 2025b). Hence an important feature of induced red cell deformations is that they involve no significant change in membrane surface area. These unusual membrane material properties are the result of an evolution-driven "engineering" process which is composed of a plasma membrane envelope composed of cholesterol and phospholipids which is anchored to a 2-dimensional elastic network of skeletal proteins through tethering sites on cytoplasmic domains of transmembrane proteins embedded in the lipid bilayer (Hansen et al., 1997; Mohandas & Gallagher, 2008). Therefore, along with the lipid bilayer (expertly reviewed in (Mohandas & Gallagher, 2008)), membrane proteins and skeletal proteins play a key role in maintaining RBC plasma membrane stability and afford these unique properties to RBCs as enlisted above.

2.2 Parasite exportome and its role in remodeling the host cell

The proteins exported by the intra-RBC parasite to the RBC cytosoo, membrane or exported out of the host cell are collectively referred to as the "exportome" in this Review. While decades of work in P. *falciparum* and related malaria parasites have unraveled the amazing world of parasite proteomes, much less is known about *Babesia* exportomes and most current studies are extrapolations from what is known from the malaria parasite. Several of these exported proteins function to remodel the host cell. These remodeling proteins extensively modify the RBC membrane and cytoskeleton and help in formation of malaria parasite-induced novel organelles such as 'Maurer's Cleft (MC), tubulovesicular network (TVN) and parasitophorous vacuole membrane (PVM). After invasion into the erythrocyte, the malaria parasite is found within a parasitophorous-vacuole (PV) enclosed by a PVM – a feature that is absent in *Babesia spp* – where the PVM is dissolved rapidly after invasion. *Plasmodium* proteins are found in the PV, the PVM, the host-cell cytoplasm and the erythrocyte membrane. The screening of the parasite proteome/genome for proteins that fulfils the discriminative criteria associated with remodeling proteins identified a unique PEXEL/VTS motif that targets a repertoire of ~300–400 exported proteins in *Plasmodium falciparum* – counting towards ~8–10 % of all proteins in the parasite (Chaubey et al., 2014; Siau et al., 2016). In *P. falciparum* each stage of malaria infection, specific parasite protein orchestrates exact and complex changes in the host cell to achieve efficient infection. Much of our knowledge on *Babesia* exportome is from *B. bovis* – the leading parasite causing babesiosis in cattle and detailed in Section 2.5. Like malaria, studies on exportome in human babesiosis and the remodeling of host RBCs can provide an efficient window into developing targeted therapeutics, and thus, constitutes an important area of future research. Details on *Plasmodium* and *Babesia* exportome and comparisons between the two Apicomplexans are expanded in Sections 2.4, 2.5 and 2.6 below.

2.3 Components of core RBC cytoskeleton

2.3.1 RBC membrane proteins

More than 50 transmembrane proteins of various abundance ranging from a few hundred to a million copies per red cell have been well characterized, several of which serve as blood antigens. The membrane proteins are the major workhorses of the human RBCs and exhibit wide functional heterogeneity, serving as proteins supporting the mechanical stability of RBCs, as transport proteins, as adhesion proteins involved in cellular communication and as signaling receptors.

Membrane proteins with transport function include band 3 (anion transporter), aquaporin 1 (water transporter), Glut1 (glucose and L-dehydroascorbic acid transporter), Kidd antigen protein (urea transporter), RhAG (gas transporter, probably of carbon dioxide), Na+-K+-ATPase, Ca++ ATPase, Na+-K+−2Cl− cotransporter, Na+-Cl− cotransporter, Na+-K+ cotransporter, K+-Cl− cotransporter, and Gardos Channel. Membrane proteins with adhesive function include ICAM-4, which interacts with integrins and Lu, the laminin-binding protein. Of direct relevance to structural integrity of the membrane and therefore this Review are 2 macromolecular complexes of membrane proteins, one ankyrin-based, and the other protein 4.1R-based. The intricate interactions of these membrane proteins have been laid out in previous Reviews (Mohandas & Gallagher, 2008). Briefly, band 3 and RhAG link the lipid bilayer to the membrane skeleton through the interaction of their cytoplasmic domains with ankyrin, and glycophorin C, XK, Rh factor protein, and Duffy through their interaction with protein 4.1R. These membrane protein linkages with skeletal proteins have been implicated to play a role in regulating cohesion between lipid bilayer and membrane skeleton and thus enable the RBCs to maintain its membrane surface area by preventing overt membrane vesiculation. Central to the stability and functional regulation of the RBC membrane is band 3, its most abundant protein at approximately 1.2 × 106 copies per cell (Satchwell & Toye, 2021). Band 3 has been shown to have direct contribution to gas exchange through the electroneutral exchange of chloride ions with the bicarbonate product (HCO3-) of carbon dioxide metabolism, serve as the major site of membrane cytoskeletal connectivity and a hub for hemoglobin and glycolytic enzyme binding. Compelling evidence from a study showed that band 3 and its interactions act as the RBC molecular switch to sense the oxygenated/ deoxygenated state, imparting an appropriate metabolic response as required by the cell (Lewis et al., 2009). A recent report published in 2021 reported the unequivocal importance of Band3 in maintaining glycolytic flux in RBCs due to its interaction with GAPDH, a key glycolytic enzymes (Issaian et al., 2021). The study also demonstrated that loss of band 3 or mutation of the N-terminal GAPDH binding site caused a failure of pentose phosphate pathway activation, with concurrent defects in glutathione recycling and evidence of increasing oxidation products – thereby highlighting the critical role played by Band 3 in RBC metabolism.

2.3.2 Cytoskeletal proteins

The principal protein constituents of the 2-dimensional spectrin-based membrane skeletal network are α- and β-spectrin, actin, protein 4.1R, adducin, dematin, tropomyosin, and tropomodulin. A unique structural feature of the long filamentous spectrin is its large number of triple-helical repeats of 106 amino acids, 20 in α-spectrin and 16 in β-spectrin. First discovered in erythrocytes, these triple-helical bundles define a spectrin super family of proteins that includes dystrophin, actinin, and utrophin (Speicher & Marchesi, 1984). α- and β-spectrin form an antiparallel heterodimer through strong lateral interaction between repeats 19 and 20 near the C-terminus of α-spectrin with repeats 1 and 2 near the N-terminus of β-spectrin. The 36 triple-helical repeats of spectrin are structurally heterogeneous in terms of their thermal stability (An et al., 2006). Spectrin tetramer, the major structural component of the 2-dimensional skeletal network, is formed by the lateral interaction of a solitary helix at the N-terminus of the α-chain from 1 dimer with 2 helices at the C-terminus of the β-chain from the other to create a stable triple-helical repeat (DeSilva et al., 1992; Lecomte et al., 1999). The other end of the 100 nm-long spectrin dimer forms a junctional complex with F-actin and protein 4.1R (Ungewickell et al., 1979). While actin interacts weakly with N-terminus of β-spectrin; the interaction is greatly enhanced by protein 4.1R (Ohanian et al., 1984). The spectrin dimer-dimer interaction and the spectrin-actin-protein 4.1R junctional complex are key regulators of membrane mechanical stability and play a critical role in preventing deformation-induced membrane fragmentation as the cell encounters high fluid shear stresses in circulation (Mohandas & Gallagher, 2008). Therefore, together these membrane components (lipid bilayer, membrane proteins and skeletal proteins) carry out the wide diverse functional roles of the RBCs which includes the cytoskeletal meshwork responsible for allowing for deformability during microcirculation and providing mechanical stability to the RBCs.

2.4 Plasmodium exportome and its role in host RBC remodeling

Interaction of *Plasmodium*-specific proteins with these key cytoskeletal elements at different stages of infection have been investigated thoroughly. A family of erythrocyte Duffy binding-like (DBL) proteins, located within the micronemes, plays a crucial role in the binding of merozoites to the host cell. In P. *falciparum*, six homologous DBL genes have been found, including erythrocyte binding antigen-175 (EBA-175), EBA-181 (JESEBL), EBA-140

(BAEBL), EBA-165 (PEBL), MAEBL and erythrocyte binding ligand-like 1 protein. Recent research work utilizing P. *falciparum* phage display libraries and use of recombinant proteins have identified a protein-protein interaction between erythrocyte membrane protein 4.1 (4.1R) and EBA-181 (Lanzillotti & Coetzer, 2004; Lanzillotti & Coetzer, 2006; Lauterbach et al., 2003). Another report highlighted the interaction between ring parasite–infected erythrocyte surface antigen (RESA) and spectrin and that this interaction stabilizes the spectrin tetramer relative to the dimer state. Further, it was shown that RESA-induced stabilization of the tetramer is accompanied by a large elevation in mechanical stability of the membrane, by reduced susceptibility to heat-induced vesiculation, and increased resistance to further invasion of the same RBC by a second malaria parasite (Pei et al., 2007a). Another key malaria exported protein knob-associated histidine-rich protein (KAHRP) was shown to bind to ankyrin at its band 3-binding site (Magowan et al., 2000) as well as to spectrin (Pei et al., 2005). As ankyrin R binds to repeats 14–15 of β-spectrin which is also close to the spectrin tetramerization site this arrangement suggests that it is possible for a single molecule of KAHRP to engage in stabilizing two reactions at the membrane skeleton, one with ankyrin R and one with spectrin. Another study showed that PfEMP3 (an exported *Plasmodium* protein) and the spectrin chain of RBCs – this interaction was shown to be close to the spectrin αβ-heterodimer that forms the junctions of the membrane skeletal network by ternary interaction with actin and protein 4.1R. The group showed that this interaction destabilizes the junction and thereby the RBC membrane. Their findings imply that disruption of the skeleton network by malaria parasite exported proteins that are expressed at late stages of malaria development (schizonts) could potentially provide a mechanism for merozoite egress from erythrocytes (Pei et al., 2007b). There are a total of 400 proteins exported by the malaria parasite and these studies have shown only a fraction of interactions that occur at different life stages. Several studies have illustrated the relationship between mechanical properties of the membranes and the behavior of the cells. While earlier studies showed that the adhesiveness, and hence the mobility of RBCs, is strongly affected by the stiffness of the plasma membrane (Efremov & Cao, 2011), recent experiments have now revealed that the stiffening of RBCs after malaria parasites invasion correlates with the structural transformation in the cytoskeleton of the infected RBCs (Glenister et al., 2002; Nash et al., 1989; Paulitschke & Nash, 1993; Xu et al., 2013). Understanding these interaction dynamics and identification of more such interactions and roles in host-parasite interactions is an ongoing research subject. These studies are invaluable to dissect host-parasite interactions at the molecular level as well as appreciate the mechanical, molecular and physical changes imposed by the parasite on its host RBCs.

Band 3 has also been shown to be important for the malarial parasites *Plasmodium falciparum* and *Plasmodium vivax* (Alam et al., 2016; Kesely et al., 2016). The malaria parasite activates the process of band 3 phosphorylation for its own benefit; the resultant destabilization of the membrane allows for egress of its merozoites and has been implicated in the ability of *P. falciparum* to form knobs and adhere to the microvasculature (Pantaleo et al., 2017), Using synchronized *P. falciparum* culture models, the authors showed a progressive increase in band 3 phosphorylation from the ring to the trophozoite stage until the schizont stage. This is accompanied by an increase in band 3–containing microparticle formation and an increase in hemolysis. By adding a Syk tyrosine kinase inhibitor (TKI) that interferes with phosphorylation of band 3, the authors demonstrated that they could inhibit these post-translational modifications on RBC band 3 and thereby decrease hemolysis, microparticle formation, and suppress merozoite egress from the schizont stage leading to marked decreases in overall parasitemia as the trapped merozoites cannot egress and ultimately die. Therefore, band 3. Given the critical role of band 3, there is a potential for other uses of Syk inhibition in hematology. Many red cell membrane disorders involve band 3, including hereditary spherocytosis and glucose-6-phosphate dehydrogenase favism to Streptococcus-mediated hemolysis (Arese et al., 2012; Hassoun et al., 1997; Higashi et al., 2016; Perrotta et al., 2008), as well as pathologies of sickle cell anemia, and therefore, there might be clinical benefit from therapies that decrease band 3 clustering and allow for membrane stabilization, decreased microparticle formation, and decreased free hemoglobin (Hierso et al., 2017).

2.5 *Babesia* exportome and its role in host RBC remodeling

Most of the *Babesia* exportome studies have been conducted on *B. bovis*, the parasite causing babesiosis in cattle and linked to huge economic and livestock loss in the last few decades and expanded in Section 2.5. Currently, there are no reports on the interaction of parasite proteins with the RBC cytoskeleton over different stages in its lifecycle. However, recent work from our group has showed that upon infection, like the malaria parasite, *B. divergens* imposes biomechanical changes on the RBCs and leads to lowered deformability and alteration in response to osmolar stress. These changes are exacerbated in genetic defects which affect the RBC membrane such as sickle cell disease (Beri et al., 2025b; Jajosky & Stowell, 2025). Based on our current understanding of *Babesia*'s remodeling of host RBCs and extrapolations from what is known from *Plasmodium falciparum*, we suggest the model shown in Fig. 2: the left side, the key cytoskeletal proteins are shown in uninfected RBCs. On the right, possible changes on these key proteins and action of parasite exported proteases are shown during invasion, growth and egress. We believe that the interaction of proteases with host cytoskeletal elements and changes in post translational modifications, especially phosphorylation of Band3, modifies the cytoskeletal framework of the host cell during intracellular growth of the parasite, and eventual dismantling of the network to facilitate hemolysis and egress of mature merozoites. Further studies on the parasite exportome and its interactions with RBC cytoskeleton can serve as an important resource for treatment of hematological consequences of babesiosis and future informed chemotherapeutic development against babesiosis.

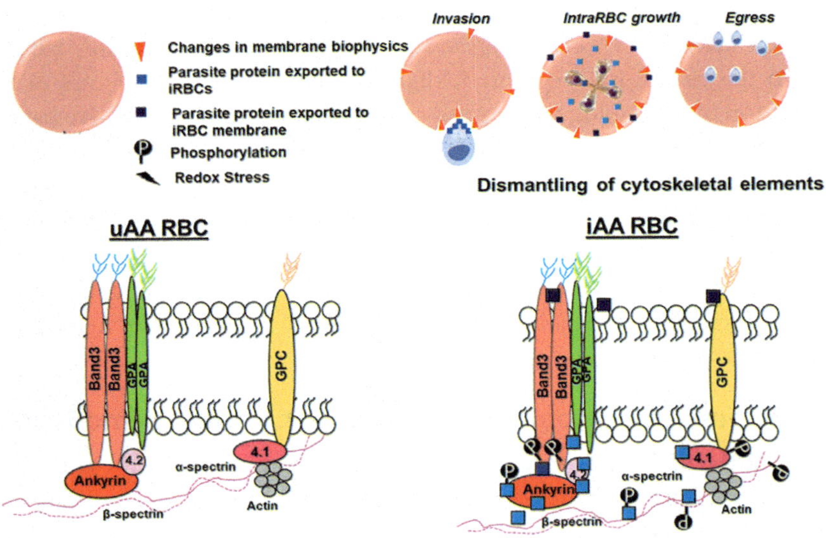

Fig. 2 Proposed hypothesis/model for cytoskeleton rearrangements upon *Babesia* invasion, growth and egress: Uninfected RBCs (left) showing normal RBC cytoskeletal arrangement. Babesia-infected RBCs have 3 developmental stages: invasion, growth, and egress all of which induce cytoskeleton remodeling due to interaction of parasite proteins with RBC membrane proteins, leading to changes in phosphorylation and cleavage of RBC proteins and dismantling of the RBC cytoskeleton.

2.6 Host cell remodeling with a focus on *Babesia bovis* and comparisons to *Plasmodium falciparum*

In this section, we describe what is known of the exportome in *B. bovis*, the most widely studied babesiosis causing parasite which causes disease in cattle. Clinically and economically, *B. bovis* is an important parasite, as it is known for using cytoadhesion of host onto endothelial cells to facilitate immune evasion and other pathological consequences of the disease, – similar to pathological manifestations in *P. falciparum*. The lethality of these two parasites is fundamentally rooted in their ability to remodel the host RBC (Hakimi et al., 2022; Sojka et al., 2022). The core triad of invasion, replication, and virulence in these apicomplexan parasites is directly associated to parasite-mediated modifications in the host erythrocyte cytoskeleton and membrane architecture (Cardoso et al., 2016; Frenal & Soldati-Favre, 2009; Warncke & Beck, 2019). Both genera share notable similarities in morphology, metabolism, and intracellular development, particularly during the asexual intraerythrocytic stages, where extensive host–parasite interactions are vital to establish nutrient acquisition systems, evade immune clearance, and ensure mechanical flexibility of infected RBCs (Djokic et al., 2021; Paul et al., 2016; Si et al., 2023). This molecular level orchestration is mediated by a subset of multigene family located at telomeric regions in *Plasmodium* (var) and *Babesia* (ves), encode a major variant surface antigens (PfEMP1 and VESA1, respectively) that mediate antigenic variation and cytoadhesion. These gene families represent parallel evolutionary solutions to maintaining persistent infection in their vertebrate hosts (Hviid et al., 2024).

Following invasion, both parasites remodel by exporting a large repertoire of proteins to the RBC cytoplasm and surface, transforming a metabolically inactive cell into a dynamic environment supportive of parasite growth (Hakimi et al., 2022; Hutchings et al., 2007; Lobo et al., 2012; Warncke & Beck, 2019). In *P. falciparum*, protein export occurs through the parasitophorous vacuole membrane (PVM) using the *Plasmodium* translocon of exported proteins (PTEX) complex. Key surface antigens such as PfEMP1, RIFIN, and STEVOR are anchored via the knob-associated histidine-rich protein (KAHRP), forming electron-dense protrusions knobs that facilitate cytoadhering to vascular endothelium and mediate immune evasion (Bullen et al., 2022; Cutts et al., 2017; Elsworth et al., 2016; Jonsdottir et al., 2021; Miyazaki et al., 2021). In contrast, *B. bovis* remodels host RBCs through the secretion of variant erythrocyte surface antigen 1 (VESA1) via spherical bodies rather than a PVM-bound translocon. These parasites generate ridge-like surface modifications analogous to knobs, supporting adhesion of infected erythrocytes to endothelial cells (Allred et al., 2000; Fathi et al., 2024; Hakimi et al., 2022; Han et al., 2021). This chapter explores the remodeling mechanisms associated with the *Babesia bovis* VESA family, compares them with the *Plasmodium falciparum* PfEMP1 system, at molecular level, and highlights associated gaps and functional parallels in both parasite lineages. Comparatively, despite the extensive genomic studies on closely related *Babesia* parasites (*B. duncani B. microti, and B. MO1*), is well reported, the functional dissection of the of the virulence factors are less reported and their exportomes remain largely uncharted (Baltaitis Iu et al., 1986; Garg et al., 2014; Gonzalez et al., 2019; Puri et al., 2021; Singh et al., 2024). We also detailed on the plausible functional equivalents of the exportome in these parasite species.

Babesia bovis is the most severe among bovine babesiosis-causing parasites, adding complexity to the disease with cerebral babesiosis in cattle, which has a higher fatality rate due to parasite sequestration in brain endothelial cells (Bock et al., 2004; Gohil et al., 2010). Because terminally differentiated RBCs lack de novo translational machinery and endogenous protein trafficking pathways, all modifications in RBCs post-infection are exclusively orchestrated by the parasite. These parasite-driven changes confer critical advantages to *B.* bovis for immune evasion, primarily by avoiding antibody-mediated immune responses or direct splenic clearance of iRBCs. Systematic modification of iRBC surface architecture is achieved by expressing a repertoire of genome-encoded surface ligands, notably variant surface proteins, exported to the iRBCs. Morphological alterations, such as surface "ridges," consist of diverse parasite-secreted proteins in the cytosol and cytoskeleton, serving as cytoadhesion sites to endothelial cells (Fathi et al., 2024; Hutchings et al., 2007). Parasites enact invasive, adaptive, and altering processes through their exportome. These ridges differ from knob structures in *Plasmodium* iRBCs (further discussed) (Pelle et al., 2015). Unlike *Plasmodium, B. bovis* modifies iRBCs via a vesicle-driven, rapid, and less complex pathway, lacking a persistent PVM or Maurer's clefts. More recently, Ueti MW et al. (2021) expanded the genome to include 326 additional functional genes (Brayton et al., 2007; Gohil et al., 2010; Jones et al., 1997). The exact mechanisms for many exportome proteins remain unclear, yet their roles in nutrient exchange, cytoskeletal alteration, and endothelial attachment have been demonstrated. A major exportome family, variant erythrocyte surface antigen 1 (VESA1) proteins, are heterodimers encoded by multicopy ves1α and ves1β gene families, first identified as mediators of antigenic variation in *B. bovis* (Allred et al., 2000). Latest genomic assemblies report 133 *ves1* genes: 81 ves1α, 48 ves1β, and 4 unclassified (Brayton et al., 2007; Hakimi et al., 2025; Ueti et al., 2021; Wang et al., 2012). Ves1α and ves1β are divergently transcribed from a bidirectional promoter at a locus of active transcription (LAT). Antigenic variation in *B. bovis* operates through epigenetic regulation and recombination, generating *ves1* gene mosaicism (Al-Khedery & Allred, 2006; Hakimi et al., 2022). Structurally, VESA1 proteins possess a single transmembrane domain, a large, cysteine- and lysine-rich extracellular domain, and a short cytoplasmic tail, functionally resembling *Plasmodium* PfEMP1 (Allred et al., 2000; Hakimi et al., 2022). VESA1 clustering on membrane ridges mediates cytoadhesion and contributes to cerebral babesiosis pathology (Fathi et al., 2024). Adjacent to the ves1 cluster, *B. bovis* contains the *smorf* multigene family, regulatory analogs to *Plasmodium* RIFIN/STEVOR, with 44 members, making it the second largest in the *B. bovis* genome. *Smorf* genes (300 bp–1 kb), arranged head-to-head or tail-to-head within ves clusters, were unexpectedly discovered during genome sequencing, underlining their potential in parasite adaptation (Allred, 2019; Brayton et al., 2007; Ferreri et al., 2012). Four spherical body proteins (SBP1–SBP4) in the exportome are localized to "spherical bodies," encoded by multi-copy genes and implicated in ridge formation on iRBCs (Fathi et al., 2024). Their precise molecular function remains unresolved. Studies show overexpressed truncated SBP2 (SBP2t11) reduces parasite invasion of brain endothelial cells (Fathi et al., 2024; Gallego-Lopez et al., 2019), while A. Fathi et al. demonstrated that SBP3 interacts with host Band 3 and that knockdown of SBP3 impairs cytoadherence and ridge formation (Fathi et al., 2024).

Recent studies demonstrate that *Babesia* parasites export a diverse repertoire of proteins and extracellular vesicles (EVs) that extensively remodel host RBCs and contribute to disease pathogenesis and detailed under Section 3.3. Despite these advances, substantial knowledge gaps persist in defining the full *Babesia* exportome and its underlying trafficking machinery, particularly in zoonotic species such as *B. divergens, B. duncani, B. microti, and B. MO1.* Genomic analyses reveal that the *B. duncani* genome encodes 747 multigene family members, including 397 species-specific genes (BdUMGFs). Members of the largest family, BdUMGF1, are predominantly subtelomeric and transcriptionally repressed, suggesting a role in antigenic variation and immune evasion analogous to virulence gene families in *Plasmodium* (Singh et al., 2023). Proteomic studies have identified two secreted antigens, BdV234 and BdV38, which are released into the host erythrocyte environment via vesicle-mediated export and represent promising biomarkers for active *B. duncani* infection (Chand et al., 2024). In *B. MO1*, distinct classes of ves genes have evolved, with several among the most highly expressed multigene families, indicating a likely role in host RBC modification and pathogenicity (Singh et al., 2024). While VESA1 is well characterized in *B. bovis* as a mediator of cytoadhesion and immune evasion, its expanded and divergent counterparts in B. MO1 may fulfill analogous functions. Additionally, expanded families of major facilitator superfamily (MFS) transporters in *B. MO1 and B. divergens* but not *B. bovis* suggest species-specific adaptations for nutrient uptake (Singh et al., 2024). Integrating genomic, transcriptomic, and proteomic datasets will be essential to construct a comprehensive exportome map and to identify conserved targets for therapeutic and vaccine development.

3. Extracellular vesicles (EV): Mediators of cell-to-cell communication in parasitic systems

3.1 Introduction to EVs

Extracellular vesicles (EVs) represent a heterogeneous group of membrane-bound vesicles derived from both endosomal and plasma membrane origins that overlap in size, content, and function. (Raposo & Stoorvogel, 2013) EVs are categorized depending on their size or their release mechanism. If formed by the plasma membrane's outward budding, they are called ectosomes (Sun et al., 2021). In a second release mechanism the endosomal membrane buds inward to form microsomes or multivesicular bodies (MVBs). The outer microsome membrane then fuses with the plasma membrane to release exosomes (Abels & Breakefield, 2016). Vesicle size is also used to discriminate EVs. While shedding micro vesicle's diameter ranges from 200 to 1000 nm, exosomes diameters span 30–150 nm (Abels & Breakefield, 2016) (Fig. 3). Thus, the size of these vesicles can range from 30 to 1000 nm. In parasitic infections, this heterogeneity complicates precise classification but also underlies the diverse biological roles EVs fulfill—ranging from immune modulation to inter-parasite communication. Importantly, EVs carry molecular fingerprints reflective of their source, making them valuable noninvasive biomarkers of infection. (Kumar et al., 2024). Advances in "EV-omics"—integrated proteomic, lipidomic, exposomic, and transcriptomic profiling—have revealed parasite-specific EV signatures with diagnostic and prognostic potential in malaria,

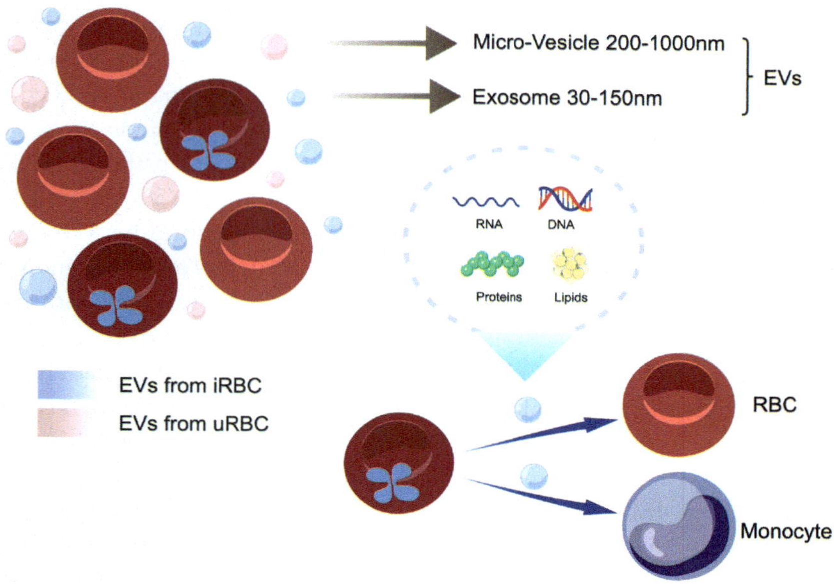

Fig. 3 Extracellular vesicles (EV) release and communication between uninfected and infected RBCs: uninfected RBCs (light red) and *Babesia*-infected iRBCs (dark red) both release EVs, including exosomes and microvesicles. EV cargo includes DNA, RNA, proteins and lipids and these have been known to serve as a communication tool to target host RBCs as well as immune cells.

leishmaniasis, trypanosomiasis, and amebiasis (Aggio et al., 2025; Babatunde et al., 2020; Costa et al., 2025). Moreover, EV secretion and uptake appear to engage conserved pathways such as tetraspanins, heat-shock proteins, and ESCRT components, while also exploiting parasite-specific adaptations that shape tropism and host-cell targeting (Moyano et al., 2019; Salas et al., 2021). These insights, together with growing data on EV-mediated antigen delivery and immune priming, highlight the emerging potential of parasite-derived EVs as vaccine platforms and therapeutic targets (Bhardwaj et al., 2019).

3.2 Role of EVs in parasitic protozoa

In the last few years, there has been many reports on the release of EVs from major human parasitic pathogens including *Plasmodium*, Leishmania, Giardia, Trypanosoma, Schistosoma, and Fasciola species (Marti & Johnson, 2016; Moyano et al., 2019). These parasites employ a dual-host life cycle, necessitating quick adaptation to changing host environments. EVs produced and secreted by parasite infected cells can fuse with both uninfected and infected host cells to deliver cargo that modulates the recipient cells

(Fig. 3). EVs are a key mechanism that these parasites use to persist in the human host by not only regulating host immune responses but by provide sensing mechanisms within the parasite population (Cipriano & Hajduk, 2018). EVs are central to parasite survival, host immune modulation, and intercellular communication across diverse protozoan pathogens. EVs including micro-vesicles (200–1000 nm) and exosomes (50–150 nm) are increasingly recognized as key mediators of communication between protozoan parasites and their hosts (Sharma et al., 2023). In *Leishmania*-induced cutaneous leishmaniasis, parasitic EVs (*Leish*EXOs), carry GP63 virulence factor that modulate macrophage cytokine production and antigen presentation, thereby promoting parasite persistence and inflammatory pathology (da Silva Lira Filho et al., 2021). *Entamoeba histolytica*, the causative agent of amebiasis, releases EVs that activate innate immune cells, modulate reactive oxygen species (ROS) production, delay neutrophil responses, and deliver proteins such as GalNA-c lectin and calreticulin that influence host cell death and tissue invasion (Diaz-Godinez et al., 2022). In trypanosomatids, *Trypanosoma brucei* EVs transfer proteins and RNAs to host immune cells, inducing differentiation of both M1/M2 macrophages and regulatory $CD4^+/CD8^+$ T cells (Dias-Guerreiro et al., 2021), while *Trypanosoma cruzi* trypomastigote EVs alter host gene expression, upregulating ubiquitin-related pathways (SUMO1/2, Ube2C), downregulating Rho-GTPases, and reducing apoptosis (Cornet-Gomez et al., 2023). *Plasmodium falciparum*-derived EVs (*Pf*EVs) carry DNA cargo supporting genetic exchange and their uptake is dictated by host membrane mechanics, with stiffer T cell membranes favoring direct fusion and more deformable monocytes internalizing *Pf*EVs via endocytosis, highlighting the interplay between host biophysics and parasite signaling (Alfandari et al., 2025; Kozela et al., 2025).

3.3 EVs in *Babesia* sp: Current knowledge and information gaps

The last decade has seen a large volume of research on EVs in parasitic protozoa as elucidated above, specifically Apicomplexan parasites reviewed in (Haldar & Bhattacharjee, 2025). However, study of EVs in *Babesia* spp is in its infancy and there are only a handful of studies performed in EVs in human babesiosis. EVs derived from *Babesia*-infected erythrocytes are markedly more abundant than those from uninfected cells and contain parasite-derived proteins, lipids, and nucleic acids, supporting roles in intercellular communication, immune modulation, and dissemination of virulence factors beyond the parasite. In 2019, Thekkiniath et al. demonstrated a model for vesicular-mediated export of parasite antigens in *B. microti* in which parasite antigens are packaged into EVs and released into the extracellular environment, facilitating cell-to-cell communication in a manner reminiscent of malaria parasites (Thekkiniath et al., 2019). In 2022, our group published the first study in EVs in *Babesia* spp (Beri et al., 2022). The study characterized EVs from cultures of *B. divergens* and plasma of mice infected with *B. microti*. Quantitative analysis of these vesicles revealed a correlation with parasite infection loads with infected cells releasing about 2–4 folds higher EVs than uninfected RBCs. Speaking to the functional relevance of EVs in *Babesia*, we showed that labeled EVs are taken up by other uninfected RBCs in culture as well as monocytes. Importantly, the study showed a clean methodology based on Image flow cytometry to differentiate between host-derived and parasite-derived EVs by quantitatively estimating EVs with that *Babesia*-specific signature markers (Bd37 in *B. divergens* and Bm2 in *B. microti*). Our results further confirmed that EVs released by this system are immunogenic to both infected human and mouse sera.

Recent study has shown the uptake of EVs released by *B. microti* ex vivo vulture by macrophages resulting in increase in NF-κB activity and pro-inflammatory cytokines (Hagos et al., 2025). Results from this study reinstated previous results and provided direct insights into the mechanisms of intercellular communication between *Babesia* and macrophages, which are critical to the induction of the innate immune response against the parasite in the mammalian host. Another interesting recent study has demonstrated the presence of BdP50, a putative *B. divergens* glycosylphosphatidylinositol-anchored protein in EVs secreted by the parasite (Hagos et al., 2025). BdP50 binds to RBCs and possibly mediates interactions of free merozoites and Bd-derived EVs with the host cell as evidenced by 88 % loss of invasion capacity of free merozoites in RBCs in the presence of anti-BdP50 antibodies. The study elegantly showed that Bd-derived EVs, in addition to promoting parasite growth, displays BdP50 that can mimic the merozoite surface to likely attenuate the blocking effect of host-generated antibodies, thereby ensuring the parasite survival during subsequent rounds of invasion and growth within the host system. Thereby, this opens up new and targeted avenues for development of chemotherapeutics to block parasite invasion – thereby demonstrating the value of studying EVs beyond merely tools as studying cell-to-cell communication.

EVs carry several biomolecules including proteins, lipids, and nucleic acids that are the key players in the interactions between parasite and hosts

(Chen et al., 2025). (Fig. 2) Very few studies have been done to characterize the EVs in *Babesia*. Our study provided a detailed microRNA (miRNA) analysis. Two different platforms were used to analyze miRNA – next-generation sequencing (NGS) and microarray analysis to provide a overcome the inherent drawbacks of each system. demonstrating that several miRNA are upregulated in Bd-derived EVs, pointing to a potential role in influencing disease outcome. NGS analysis provided a total of 805 known miRNAs and 169 novel miRNAs, which have not previously been mapped, while the classical microarray platform identified 234 miRNA, of which 56 were shared between the two methods. Our analysis revealed interesting parallels with malaria – for example, both our platforms detected an upregulation of miR-4454, which has been associated with severe thrombocytopenia in P. *vivax*–infected human plasma (Santos et al., 2021). We also observed a 3.5-fold increase in miR-4497 in iRBC-derived EVs which has previously associated with splenic sequestration malaria (Gupta et al., 2021); however, their role in *Babesia* infection is not understood. Thus, our results have provided a strong rationale to further study the role of miRNAs present in the EVs derived from *Babesia*-infected RBCs to attain a better understanding of disease pathogenesis, mechanism of parasite survival and intercellular communication (Beri et al., 2022). Another recent study (Gonzalez et al., 2025) has provided the first proteome of Bd-infected RBCs' derived EVs, of which 305 were derived from the parasite. Among the human proteins present in *Bd*-derived EVs, many proteins were involved in metabolic, cell differentiation and metal ion binding processes or were identified as cytosol components. Interestingly, the most abundant functional related human proteins from uRBC-derived EVs were different from those present in *Bd*-derived EVs – thereby suggesting that there are significant parasite-induced alterations in EVs produced by the host cell. *Babesia divergens* protein cargo included proteins related to biological process, molecular function and cellular components – and require further studies to elucidate the roles of these proteins in cellular communication, parasite survival and propagation in the host.

Our recent study on *Babesia*-infected wildtype host cells (AA) and RBCs from sickle cell disease (SCD) patients (SS) has shown that the number of EVs is higher in uninfected SS cells as compared to uninfected AA cells; and there is a further dramatic increase in infected AA and further in infected SS cells. We corroborated these increase in EV production in uninfected and infected SS as compared to corresponding to

AA cells in terms of loss of projected surface area; and showed that infected SS cells lose as much as 8 % of their surface area upon infection. Loss of surface area and decrease in the overall surface area/volume ratio in RBCs is correlated to cellular deformability (Namvar et al., 2021), and thereby we correlated this loss of surface area to our observation of drastically reduced deformability of uninfected and infected SS cells.

In conclusion, EVs in parasitic systems can play roles ranging from cell-to cell communication, transfer of biomolecules, protection against immune system as well as cellular biomechanics. The studies on EVs from *Babesia* have set a promising rationale for further detailed EV studies for better understanding of host-parasite interactions in babesiosis.

3.4 Pathways of EV biogenesis in uninfected and infected RBCs

EV biogenesis in parasite infected RBCs exhibits greater heterogeneity than the canonical endosomal model implies. In nucleated cells, exosomes are typically generated through ESCRT-dependent budding in multi-vesicular bodies, while micro-vesicles arise from outward blebbing of the plasma membrane—mechanisms that are well defined. (Catalano & O'Driscoll, 2020). In contrast, anucleate RBSs depend on alternative routes of vesiculation driven by oxidative stress, calcium influx, cytoskeletal protease activation, and lipid domain reorganization (Thangaraju et al., 2020). For instance, malaria EVs are released via membrane budding associated with cytoskeletal remodeling, independent of the ESCRT system, and the equilibrium between membrane and endocytic uptake is determined by the degree of endocytic engagement (Dey et al., 2024). Likewise, other protozoan parasites such as *Toxoplasma gondii* and *Giardia lamblia* employ parasite-specific secretory routes to produce EVs, and encystation-specific vesicles (ESVs) enriched in effector molecules and virulence factors (Cruz-Bustos et al., 2025; Wampfler et al., 2014). Emerging evidence further suggests that the ESCRT members remain involved in some protozoan EV biogenesis despite apparent deviations from the classical pathway. For example, ESCRT I-III in *Trichomonas vaginalis* (Salas et al., 2021) and *Giardia intestinalis* (Moyano et al., 2019) and hypothesized that parasites may "kidnap" host ESCRT machinery for their own EV formation (Cruz Camacho et al., 2023). This points that in parasite infected cells, EV generation represents a mosaic of canonical endosomal budding, membrane-shed ectosome release, and parasite-specific budding events.

4. Red cell deformability – physiological importance in blood borne parasitic infections

4.1 Introduction to RBC deformability

During their average 120-days lifespan in circulation, erythrocytes can maintain their shape, and stability in response to external stresses encountered during multiple passages through the smallest capillaries and the sinusoids of the spleen in the human body. This is largely attributed to the attribute that healthy RBCs can deform readily in response to shear stress in the circulation facilitating their efficient passage through capillaries. RBCs also export vasoactive mediators in response to shear stress and other physiological and pathological stimuli. Deoxygenation of RBC hemoglobin leads to the export of vasodilator and antiadhesive S-nitroso thiols (SNOs) and adenosine triphosphate (ATP) in parallel with oxygen transport in the respiratory cycle. Together, these mediated responses to shear stress and oxygen offloading promote the efficient flow of blood cells and in turn optimize oxygen delivery (McMahon, 2019).

RBC deformability is governed by three major factors: (i) surface area to volume ratio (SA:V), (ii) the internal viscosity of the RBCs, and (iii) the rheological properties of the membrane (Renoux et al., 2019). A high SA: V ratio indicates that the cell has more surface area relative to its internal volume, which allows it to bend, fold, and stretch more extensively without generating excessive internal pressure during cytosolic fluid displacement and as it passes through thin capillaries and splenic slits (Renoux et al., 2019), (Renoux et al., 2019); while a low SA:V ratio results in a more rigid cell due to reduced membrane area or increased internal volume, both of which promote a more rapid and pronounced buildup of internal pressure. This pressure increase can ultimately be responsible for membrane rupture and hemolysis (Yazdanbakhsh et al., 2025). The second factor – internal viscosity of RBC – refers to the thickness and resistance to flow of the intracellular fluid. This property in the RBCs is uniquely and primarily dependent on the concentration of hemoglobin as it lacks other organelles, their deformability is especially sensitive to changes in hemoglobin concentration, which can increase due to factors such as dehydration and high-altitude adaptation (Clark, 1989). Increase in concentration of hemoglobin can potentially increase cytoplasmic viscosity, limiting the ability of the internal contents to rearrange during deformation and thereby reducing overall cell flexibility and deformability. Finally, the membrane of RBCs is critical in determining the extent of deformability of the cell by providing the cell with its elastic and stretchable properties, as has been outlined in detailed under Section 1.2. The lipid bilayer, membrane proteins and cytoskeletal network together ensure that RBCs can undergo repeated deformation and to return to its original biconcave shape after mechanical stress throughout their lifecycle in circulation.

4.2 Measuring deformability of RBCs: A historical lookback at available methodology

Cell assessment techniques used to evaluate deformability generally fall into two main categories: single-cell and bulk flow methods, each encompassing a range of techniques with distinct advantages and limitations. And have been expertly reviewed by (Yazdanbakhsh et al., 2025). Briefly, these techniques have been described below, with an emphasis on LORRCA-based ektacytometry analysis.

Micropipette and filterability methods that allowed early single-cell study of RBC rheology in the 1960s continue to yield valuable insights into RBC deformability (Canham & Burton, 1968). Atomic Force Microscopy (AFM) is another method amenable for single-cell deformability measurements. This is a highly sensitive method. This method utilizes a scanning probe mounted on a flexible cantilever. The probe tip contacts the RBC surface and applies a controlled, incremental force to induce deformation. As the cell deforms, the cantilever bends, resulting in a shift in its position. A laser beam, reflected off the back of the cantilever onto a photodetector, changes angle in response to this deflection. By tracking the laser deflection, the surface indentation of the RBC can be quantified, enabling the calculation of its mechanical properties. Alternatively, optical tweezer techniques have been now widely used due to their use in single-cell investigation (Pesen et al., 2025; Zhu et al., 2020). Optical force is applied to opposite ends of a visualized cell, and one of the laser beams is moved inward, deforming the cell. The degree of physical displacement of the cell in the perpendicular axis is determined visually. In a variation of the optical tweezer technique, both beams are directed at the cell, well within its boundaries, trapping the cell in place. Then one beam is slowly moved in a direction away from the other, stretching the cell. This technique is extremely sensitive due to its ability to apply precise force on the RBCs; however, the low throughput nature of this technique often limits its utilization for large number of samples.

In more recent times, ektacytometry has been utilized to measure deformability which uses the optical (laser) diffraction pattern to determine the change in length and width of a population of RBCs subjected to varying shear stresses (Rabai et al., 2014). The first automated ektacytometery or viscodiffractometry equipment was developed by Dr. Bessis, Dr. Narla and Dr. Feo, along with colleagues in France in the 1970s (Bessis et al., 1980; Bessis et al., 1982; Mohandas et al., 1980). By subjecting red cells to varying shear stress in suspending media of different osmolalities, one could identify separate contributions of membrane viscoelastic properties, internal viscosity, and surface area-to-volume ratio to overall cellular deformability. The influence of drugs on red cell deformability as well as red cell disorders such as sickle cell anemia, thalassemia and G6PD-defieicncy could also be investigated. Analysis of the diffraction pattern of red cells that are aligned in the ektacytometer at minimal shear stress enables improved measurement of red cell diameter. Modern computerized image analysis, which also saw an upsurge in the 1990s, further improved the reliability of this measurement. Therefore, the first developers of the automated ektacytometer realized its huge importance in providing a sensitive assessment of the overall functional integrity of living red cells or subpopulations of red cells. Ektacytometry has been employed for decades now and has proved to be an unvaluable tool for studies on RBC-related diseases, RBC storage and stability in blood banks (Chasis et al., 1989). The first generation ektacytometer produced by Technicon relied on self-made consumables and a more flexible, but less user-friendly interface (Clark et al., 1983). Laser-assisted Optical Rotational Cell Analyzer (LORRCA) is a common equipment used in scientific research in recent times to measure RBC rheology. Ektacytometry provides a convenient measure of red cell deformability in response to alterations in shear stress (measured in pascals (Pa)) or suspending medium osmolality. Pertinent parameters of red cell deformability include the maximum elongation index (EI Max), a measure of the maximum deformability of a red cell in response to increasing shear stress, and shear stress ½ ($SS_{½}$), the shear stress required to achieve half maximal deformability. Osmotic gradient ektacytometry has several informative parameters. These include the elongation index minimum (EI_{min}), a measure of surface-to-volume ratio and the osmolality at which it occurs (O_{min}), which is a measure of osmotic fragility. EI_{max} and the osmolality at which it occurs (O (EI_{max})) provide information on membrane flexibility and cell surface area. Half maximal elongation in the hypertonic arm of the osmotic gradient is represented by EI_{hyper}. EI_{hyper} and the osmolality at which it occurs, $O_{hyper,}$ provide information about the intracellular viscosity of the red cell which is determined by hemoglobin concentration. commonly measured elongation index (EI) reports the change in length relative to the width as the RBC is stretched progressively by (typically incremental) shear stress and the EI_{max} value corresponding to maximal shear stress (30 Pa in most systems including the LORRCA, as detailed above) achievable by the RBCs. Healthy human RBCs usually have a deformability close to 0.8 while, it decreases in pathologies or other RBC-related diseases (Beri et al., 2025b).

Ektacytometry has been widely used to assess RBC deformability in inheritable RBC disorders such as hereditary spherocytosis (HS) and hereditary elliptocytosis (HE), structural hemoglobinopathies such as sickle cell anemia, lifestyle-related diseases such as diabetes mellitus (Ebenuwa et al., 2025), as well as pathogenic diseases such as malaria and more recently babesiosis (Beri et al., 2025b; Da Costa et al., 2016; Dondorp et al., 1999).

4.3 Splenic clearance of RBCs: Stiff vs Rigid

The spleen is in the left superior abdomen and is the largest lymphatic organ in the human body. It plays a key role in the human immune system by protecting the body from pathogenic microorganisms reaching the bloodstream, through innate phagocytosis or adaptive responses operated by lymphocytes and antibodies (Li et al., 2018). The spleen also serves as a filter that can remove RBCs from circulation and therefore serves as one of the most important physiological removal system for old, damaged or pathogenic RBCs.

RBCs can be recognized as altered when changes in their shape, size, or surface are detected, or their deformability is impaired because of such changes. The most well-studied molecular alterations in senescent RBCs are externalization of phosphatidyl-serine (PS) on the outer leaflet of the phospholipid bilayer of the RBC membrane or fixation of antibody on surface antigens, as occurs in transfusion mismatch. Sensing of these alterations by immune cells, predominantly the macrophages, occurs with a strong predominance in the spleen and liver (Jandl et al., 1956). By contrast, recognition of altered size, shape, and deformability is considered a spleen-specific function. Surface area loss and reduced deformability are hallmarks of aging RBCs (Waugh et al., 1992). A classic example of a genetic RBC disease causing an increased splenic retention has been reported in HS disease which is a genetic disorder resulting in dysfunctional membrane proteins that play a role in transforming the shape of the RBC from a normal discocyte to a sphere. It occurs due to defects in critical cytoskeletal proteins such as band 3, ankyrin and spectrin connecting the RBC membrane. It leads to increased vesiculation, which in turn, causes a gradual reduction in cell surface area by as much as 20 % (and therefore reduction in RBC deformability) compared with that of a healthy RBC and eventually causes significantly elevated RBC retention rates in the spleen because of their increased sphericity and eventual hemolysis Increased in vitro mechanical retention of RBCs collected in splenectomized subjects further supports this inference (Prendki et al., 2012). In parasitic diseases such as malaria and babesiosis, splenomegaly has been reported (Henry et al., 2022; Yi et al., 2018). In malaria, several groups have reported entrapment of parasitized RBCs (which have lowered deformability) in the spleen and suggested the role of this action in providing natural resistance against malaria, especially in endemic zones such as regions of Africa (Henry et al., 2022).

4.4 Effect of blood-borne parasites on RBC deformability

Blood borne parasite pose a major challenge to RBC's deformability and overall cellular health. Hemoparasites such as *Plasmodium* and *Babesia* spp are known to not only export hundreds of proteins to the host cell (including the membrane) but also increase in apparent RBC viscosity and biophysical characteristics (Fedosov et al., 2011). While there is a large body of data on how the malaria parasite decreases RBC deformability and affects splenic entrapment of uninfected as well as RBCs and its pathogenic effect on vaso-occlusion and hemolytic damage, very little is known about changes of RBC upon *Babesia* infection. In malaria, for each of the major human parasite species, P. *falciparum*, P. *vivax* and *P. knowlesi* – the effect of parasite growth on RBC deformability has been described: (Barber et al., 2018; Depond et al., 2019; Fedosov et al., 2011; Guizouarn & Barshtein, 2020; Rathnam et al., 2025; Safeukui et al., 2018). Several pathologies associated with malaria, specifically and importantly cerebral malaria and properties of cytoadherence have been to some extent attributed to modifications imparted to host RBCs by their RBCs, including mechanical and biophysical changes.

A recent study from our lab now has shed light on changes in deformability upon *Babesia divergens* infection (Beri et al., 2025b). We used LORRCA to define changes in human RBCs upon *Babesia* infection. Our ektacytometry analysis demonstrated that wild type infected AA cells had a significant reduction of deformability under both shear stress and osmolarity gradients. Our results showed a direct correlation between parasite load and deformability – therefore, higher the parasite loads, higher was the reduction in EI_{max} which correlated to deformability. Thus, our results imply that upon infection, *Babesia* modifies the host RBCs such that there is a biomechanical shift in the host cells towards increased rigidity. As explained above, rigid

RBCs cannot pass through thin splenic slits, and therefore, get entrapped in the spleen. This observation can explain why mice infected with *Babesia microti* have splenomegaly, as has been shown by us previously (Yi et al., 2018). Several questions remain: which parasite stage affords maximum damage on RBCs, the extent of damage to the host and bystander cells (uninfected cells in a milieu of infected cells) as well as measurement of rheological parameters at a single cell level. It also remains to be seen if the splenic entrapped RBCs are infected, and if the splenic macrophages recognize them or is the spleen the hiding place for *Babesia*-infected RBCs. Thus, rheological changes afforded on the host RBC upon *Babesia* infection can explain some of the important pathological outcomes of the disease, and further research in this area will allow us to understand pathology of this parasite.

In our current study, we used Image stream cytometric technology to quantify changes in projected surface area upon infection. Our results indicated a decrease in projected surface area. We also employed a tunable resistive pulse sensor to measure release of extra-cellular vesicles (EV) from these host RBCs, before and after infection, we showed that there is a significant increased rate of EV production – these observations proved that reduction in projected surface area due to infection and thus, a decrease in Surface area/volume ratio upon infection – which as explained above is directly linked to reduction in deformability. Additionally, detailed analysis of the metabolomic changes in the *Babesia*-infected RBC has shown a significant hijacking of RBC membrane cholesterol and phosphor lipids from the RBC membrane into the membranes of the expanding parasite populations. This provides a potential biomechanical and biomolecular insight into understanding the pathogenic effects of infection of membranes via changes in host cell deformability (Beri et al., 2023b; Beri et al., 2025a). These recent studies in babesiosis demonstrate the critical need to study parasitic infections in a multipronged way, with an equal emphasis on host-parasite interactions and the multiple ways the parasite hijacks the host to serve its own needs and evade the immune system.

5. Hemoglobinopathies and *Babesia*

5.1 Challenges to parasite proliferation growth in hostile RBC environments

Patients that are particularly susceptible to hemolysis are those suffering from hemoglobinopathies such as sickle cell anemia and thalassemia. These individuals require frequent blood transfusions and Transfusion-transmitted babesiosis remains a problem in the US because of the lack of universal donor testing (Bloch et al., 2023; Drews et al., 2023a; Lobo et al., 2013; Lobo et al., 2020). SCD patients are particularly vulnerable to TTB given their high frequency of transfusion, requiring blood procurement from states where *Babesia* screening is not mandatory (Beri et al., 2021; Costa et al., 2023; Karkoska et al., 2018).

Studies on *Babesia* spp. in SCD suggest that these hemoglobinopathies impede intraerythrocytic growth of *Babesia* parasites (Beri et al., 2023a; Cursino-Santos et al., 2019) It is believed that multiple mechanisms of resistance against parasite proliferation in sickle cells may operate to confer protection, including enhanced oxidative stress, increased membrane damage, and a shorter red cell life span because of parasite modifications enhancing sickling of infected RBCs, making them more prone to splenic removal, heightened RBC bystander effect leading to massive hemolysis and therefore unfavorable conditions for the parasite to grow and proliferate (Beri et al., 2021; Beri et al., 2023a). Using a combination of image flow cytometry and machine learning algorithms, characteristics of host cells in SCD preferred by *Babesia* were studied. In particular it was shown that oxygenation and fetal hemoglobin played a key role in determining the sickling rate of RBCs which in turn determined the choice of host RBC that *Babesia* exploited during infection (Beri et al., 2023a). Parasites unequivocally proliferated in cells with fetal hemoglobin and multiplication rates matched those in wild type RBCs when hypoxic conditions were replaced by normoxia or hyperoxia (Beri et al., 2023a). In addition to host RBC parameters, it was found that *Babesia* can alter its constituent sub-population structure remaining as 1 N form in hostile RBCs like the sickle RBC in hypoxia. Thus, the proliferation dynamics are controlled by both the environment as well as the parasites replication machinery, factors that can allow persistence of infection in various hosts (Cursino-Santos et al., 2014a; Cursino-Santos et al., 2016; Cursino-Santos et al., 2019).

5.2 Impact of parasite infection on stressed RBCs

The impaired rheology of SCD is a well-known hallmark of disease pathophysiology. Host SCD RBCs that are already structurally compromised (Leal et al., 2018) are even more susceptible to parasite-induced RBC membrane modifications, compared to wild type RBCs. Recent work has shown that *Babesia* infection of SCD RBCs significantly decreased RBC deformability under shear, osmolality and oxygen stressors (Beri et al., 2025b).

These occurred as a function of both parasite load and oxygenation. Deformability regulates rheology, RBC lifespan and oxygen transport and thus this exacerbation of RBC rheological properties from both primary (SCD) and secondary (infection) causes can lead to premature RBC lysis or splenic retention and premature removal from the circulation.

Mechanisms leading to this heavily altered rheological profile in *Babesia*-infected SCD RBCs were defined through study of vesiculation rates in these host cells. It has been shown known that *Babesia* infected RBCS release up to 10 times the number of EVs as uninfected RBCs (Beri et al., 2022). When comparing the output of EVs from uninfected and infected SCD RBCs, the number of EVs is higher in uninfected SS cells as compared to uninfected AA cells (Fibach, 2021); however, there is a further dramatic increase in SS cells upon *Babesia* infection (Beri et al., 2025b). Probing a potential cause of this heightened EV release, it was demonstrated that this extra vesiculation led to an 8 % of their surface area upon infection. Loss of surface area and decrease in the overall surface area/volume ratio in RBCs is correlated to cellular deformability (Namvar et al., 2021), and thus this loss of RBC surface area may lead to the drastically reduced deformability observed in infected SCD RBCs. The increased vesiculation rates can in part be explained by the fragility of SCD RBC membranes which are in constant contact with both the sickle hemoglobin and its toxic by-products. Upon infection, parasite remodeling of the RBC cytoskeleton imposes a further strain on the SCD membrane potentially increasing the release of EVs and contributing to loss of cellular deformability (Beri et al., 2025b).

6. Conclusion and future directions

The patho-biology of babesiosis is a direct consequence of the cyclical replication of parasites in host red cells. The parasite's ability to first recognize and then invade host RBCs is central to the disease process, and thus the steps in the erythrocytic asexual cycle provide vulnerable points for chemotherapeutic or immunotherapeutic interventions. However, there is still scarce information available on how *Babesia* proteins interface with the RBC cytoskeleton to provide a framework that allows hemodynamic stability, nutrient import and protection from the host immune system. Such information is fundamental to learning more about the biology of *Babesia* spp. and understanding the cellular and molecular mechanisms by which they cause infection and disease. Identification and elucidation of the role of *Babesia* proteins that are intimately involved in parasite virulence together with mapping the RBC proteins they interact with can help open novel strategies to fight this important zoonotic disease. Additionally, given the transfusion threat that *Babesia* still presents to vulnerable populations like sickle cell disease patients, it is imperative to study how parasite infection in these hosts differs from disease in wildtype hosts. Special attention must be paid to the role of the RBC membrane in SCD that leads to high sickling rates, hyper-vesiculation and premature host cell lysis of not only infected but also bystander red cells in this infected host population. While initial studies have indicated that EVs play multifaceted roles in *Babesia* pathogenesis and host-parasite interactions and immune responses, we need more research into their mechanisms of action. Studying the precise molecular mechanisms that regulate EV release from infected RBCs as well as their molecular cargo in detail may hasten their use in biomarker discovery and vaccine development.

Acknowledgements

None.

Financial support and sponsorship

This work was supported by the National Institutes of Health (NIH), National Heart, Lung, and Blood Institute (Grant P01 HL149626 [C.A.L.]) and the NIH, National Institute of Allergy and Infectious Diseases (Grants R61AI187095 [C.A.L.] and 5K99AI182422–02 [D.B.]).

Conflicts of interest

There are no conflicts of interest.

References

Abels, E. R., & Breakefield, X. O. (2016). Introduction to extracellular vesicles: biogenesis, RNA cargo selection, content, release, and uptake. *Cellular and Molecular Neurobiology, 36*(3), 301–312. https://doi.org/10.1007/s10571-016-0366-z.

Aggio, J. B., Vedam, V. V., Nisimura, L. M., da Silva, R. V., Lovo-Martins, M. I., Borges, B. S., ... Wowk, P. F. (2025). Trypanosomatid extracellular vesicles as potential immunogens for Chagas disease. *International Journal of Molecular Sciences, 26*(4), https://doi.org/10.3390/ijms26041544.

Aikawa, M., Rabbege, J., Uni, S., Ristic, M., & Miller, L. H. (1985). Structural alteration of the membrane of erythrocytes infected with *Babesia bovis*. *The American Journal of Tropical Medicine and Hygiene, 34*(1), 45–49. https://doi.org/10.4269/ajtmh.1985.34.45.

Al-Khedery, B., & Allred, D. R. (2006). Antigenic variation in *Babesia bovis* occurs through segmental gene conversion of the ves multigene family, within a bidirectional locus of active transcription. *Molecular Microbiology, 59*(2), 402–414. https://doi.org/10.1111/j.1365-2958.2005.04993.x.

Alam, M. S., Rathore, S., Tyagi, R. K., & Sharma, Y. D. (2016). Host-parasite interaction: multiple sites in the Plasmodium vivax tryptophan-rich antigen PvTRAg38 interact with the erythrocyte receptor band 3. *FEBS Letters, 590*(2), 232–241. https://doi.org/10.1002/1873-3468.12053.

Alfandari, D., Rosenhek-Goldian, I., Kozela, E., Nevo, R., Senprun, M. B., Moisieiev, A., ... Regev-Rudzki, N. (2025). Host immune cell membrane deformability governs the uptake route of malaria-derived extracellular vesicles. *ACS Nano, 19*(10), 9760–9778. https://doi.org/10.1021/acsnano.4c07503.

Allred, D. R. (2019). Variable and variant protein multigene families in *Babesia bovis* persistence. *Pathogens, 8*(2), https://doi.org/10.3390/pathogens8020076.

Allred, D. R., Carlton, J. M., Satcher, R. L., Long, J. A., Brown, W. C., Patterson, P. E., ... Stroup, S. E. (2000). The ves multigene family of *B. bovis* encodes components of rapid antigenic variation at the infected erythrocyte surface. *Molecular Cell, 5*(1), 153–162. https://doi.org/10.1016/s1097-2765(00)80411-6.

An, X., Guo, X., Zhang, X., Baines, A. J., Debnath, G., Moyo, D., ... Mohandas, N. (2006). Conformational stabilities of the structural repeats of erythroid spectrin and their functional implications. *The Journal of Biological Chemistry, 281*(15), 10527–10532. https://doi.org/10.1074/jbc.M513725200.

Arese, P., Gallo, V., Pantaleo, A., & Turrini, F. (2012). Life and death of glucose-6-phosphate dehydrogenase (G6PD) deficient erythrocytes - role of redox stress and band 3 modifications. *Transfusion Medicine and Hemotherapy, 39*(5), 328–334. https://doi.org/10.1159/000343123.

Babatunde, K. A., Yesodha Subramanian, B., Ahouidi, A. D., Martinez Murillo, P., Walch, M., & Mantel, P. Y. (2020). Role of extracellular vesicles in cellular cross talk in malaria. *Frontiers in Immunology, 11*, 22. https://doi.org/10.3389/fimmu.2020.00022.

Baltaitis Iu, V., Korsunovskii, A. I., Sheveliuk, S. B., Rupitsev, A. A., & Matiashin, M. I. (1986). [Quantitative evaluation of the endoscopic data in nonspecific ulcerative colitis]. *Vrachebnoe Delo,* (2), 16–18. Retrieved from . https://www.ncbi.nlm.nih.gov/pubmed/3705561.

Barber, B. E., Russell, B., Grigg, M. J., Zhang, R., William, T., Amir, A., ... Yeo, T. W. (2018). Reduced red blood cell deformability in *Plasmodium knowlesi* malaria. *Blood Advances, 2*(4), 433–443. https://doi.org/10.1182/bloodadvances.2017013730.

Beri, D., Rodriguez, M., & Lobo, C. A. (2025a). Shared host, distinct invaders: metabolomic footprints of *Plasmodium* and *Babesia* in host red cells. *Current Opinion in Hematology, 32*(6), 350–356. https://doi.org/10.1097/MOH.0000000000000891.

Beri, D., Rodriguez, M., Singh, M., Liu, Y., Rasquinha, G., An, X., ... Lobo, C. A. (2022). Identification and characterization of extracellular vesicles from red cells infected with *Babesia divergens* and *Babesia microti*. *Frontiers in Cellular and Infection Microbiology, 12*, 962944. https://doi.org/10.3389/fcimb.2022.962944.

Beri, D., Rodriguez, M., Singh, M., McLaughlin, D., Liu, Y., Zhong, H., ... Lobo, C. A. (2025b). Babesiosis and sickle red blood cells: loss of deformability, altered osmotic fragility, and hypervesiculation. *Blood, 145*(19), 2202–2213. https://doi.org/10.1182/blood.2024027602.

Beri, D., Singh, M., Rodriguez, M., Barbu-Stevanovic, M., Rasquinha, G., Mendelson, A., ... Lobo, C. A. (2023a). Elucidating parasite and host-cell factors enabling *Babesia* infection in sickle red cells under hypoxic/hyperoxic conditions. *Blood Advances,* 7(4), 649–663. https://doi.org/10.1182/bloodadvances.2022008159.

Beri, D., Singh, M., Rodriguez, M., Goyal, N., Rasquinha, G., Liu, Y., ... Lobo, C. A. (2023b). Global metabolomic profiling of host red blood cells infected with *Babesia divergens* reveals novel antiparasitic target pathways. *Microbiology Spectrum, 11*(2), e0468822. https://doi.org/10.1128/spectrum.04688-22.

Beri, D., Singh, M., Rodriguez, M., Yazdanbakhsh, K., & Lobo, C. A. (2021). Sickle cell anemia and *Babesia* infection. *Pathogens, 10*(11), https://doi.org/10.3390/pathogens10111435.

Bessis, M., Feo, C., & Jones, E. (1982). Quantitation of red cell deformability during progressive deoxygenation and oxygenation in sickling disorders (the use of an automated Ektacytometer). *Blood Cells, 8*(1), 17–28. Retrieved from . https://www.ncbi.nlm.nih.gov/pubmed/7115975.

Bessis, M., Mohandas, N., & Feo, C. (1980). Automated ektacytometry: a new method of measuring red cell deformability and red cell indices. *Blood Cells, 6*(3), 315–327. Retrieved from . https://www.ncbi.nlm.nih.gov/pubmed/7397390.

Bhardwaj, N., Yadav, R., Sampath Kumar, V., & Khan, S. A. (2019). Melanotic neuroectodermal tumour of infancy presenting as a lytic lesion in femur: a rare tumour at a rare site with an unusual behaviour. *BMJ Case Reports, 12*(10), https://doi.org/10.1136/bcr-2019-231959.

Bloch, E. M., Busch, M. P., Corash, L. M., Dodd, R., Hailu, B., Kleinman, S., ... Katz, L. (2023). Leveraging donor populations to study the epidemiology and pathogenesis of transfusion-transmitted and emerging infectious diseases. *Transfusion Medicine Reviews, 37*(4), 150769. https://doi.org/10.1016/j.tmrv.2023.150769.

Bock, R., Jackson, L., de Vos, A., & Jorgensen, W. (2004). Babesiosis of cattle. *Parasitology, 129*, S247–S269. https://doi.org/10.1017/s0031182004005190.

Brayton, K. A., Lau, A. O., Herndon, D. R., Hannick, L., Kappmeyer, L. S., Berens, S. J., ... Nene, V. M. (2007). Genome sequence of *Babesia bovis* and comparative analysis of apicomplexan hemoprotozoa. *PLoS Pathogens, 3*(10), 1401–1413. https://doi.org/10.1371/journal.ppat.0030148.

Bullen, H. E., Sanders, P. R., Dans, M. G., Jonsdottir, T. K., Riglar, D. T., Looker, O., ... Crabb, B. S. (2022). The *Plasmodium falciparum* parasitophorous vacuole protein P113 interacts with the parasite protein export machinery and maintains normal vacuole architecture. *Molecular Microbiology, 117*(5), 1245–1262. https://doi.org/10.1111/mmi.14904.

Canham, P. B., & Burton, A. C. (1968). Distribution of size and shape in populations of normal human red cells. *Circulation Research, 22*(3), 405–422. https://doi.org/10.1161/01.res.22.3.405.

Cardoso, R., Soares, H., Hemphill, A., & Leitao, A. (2016). Apicomplexans pulling the strings: manipulation of the host cell cytoskeleton dynamics. *Parasitology, 143*(8), 957–970. https://doi.org/10.1017/S0031182016000524.

Catalano, M., & O'Driscoll, L. (2020). Inhibiting extracellular vesicles formation and release: a review of EV inhibitors. *Journal of Extracellular Vesicles, 9*(1), 1703244. https://doi.org/10.1080/20013078.2019.1703244.

Chand, M., Vydyam, P., Pal, A. C., Thekkiniath, J., Darif, D., Li, Z., ... Ben Mamoun, C. (2024). A set of diagnostic tests for detection of active *Babesia duncani* infection. *International Journal of Infectious Diseases: IJID: Official Publication of the International Society for Infectious Diseases, 147*, 107178. https://doi.org/10.1016/j.ijid.2024.107178.

Chasis, J. A., Prenant, M., Leung, A., & Mohandas, N. (1989). Membrane assembly and remodeling during reticulocyte maturation. *Blood, 74*(3), 1112–1120. Retrieved from . https://www.ncbi.nlm.nih.gov/pubmed/2752157.

Chaubey, S., Grover, M., & Tatu, U. (2014). Endoplasmic reticulum stress triggers gametocytogenesis in the malaria parasite. *The Journal of Biological Chemistry, 289*(24), 16662–16674. https://doi.org/10.1074/jbc.M114.551549.

Chen, Y., Douanne, N., Wu, T., Kaur, I., Tsering, T., Erzingatzian, A., ... Burnier, J. V. (2025). Translating insights from extracellular vesicles to biomimetic synthetic vesicles for biomedical applications. *Sci Adv, 11*(9), eads5249. https://doi.org/10.1126/sciadv.ads5249.

Cipriano, M. J., & Hajduk, S. L. (2018). Drivers of persistent infection: pathogen-induced extracellular vesicles. *Essays in Biochemistry, 62*(2), 135–147. https://doi.org/10.1042/EBC20170083.

Clark, M. R. (1989). Mean corpuscular hemoglobin concentration and cell deformability. *Annals of the New York Academy of Sciences, 565*, 284–294. https://doi.org/10.1111/j.1749-6632.1989.tb24176.x.

Clark, M. R., Mohandas, N., & Shohet, S. B. (1983). Osmotic gradient ektacytometry: comprehensive characterization of red cell volume and surface maintenance. *Blood, 61*(5), 899–910. Retrieved from . https://www.ncbi.nlm.nih.gov/pubmed/6831052.

Cornet-Gomez, A., Retana Moreira, L., Kronenberger, T., & Osuna, A. (2023). Extracellular vesicles of trypomastigotes of *Trypanosoma cruzi* induce changes in ubiquitin-related processes, cell-signaling pathways and apoptosis. *Sci Rep, 13*(1), 7618. https://doi.org/10.1038/s41598-023-34820-6.

Costa, V., Mercure-Corriveau, N., Gourneau, J., Tobian, A. A. R., Jones, J. M., Lauriello, A., ... Bloch, E. M. (2023). Transfusion-transmitted babesiosis in a patient with sickle cell disease undergoing chronic red cell exchange. *Transfusion, 63*(3), 652–655. https://doi.org/10.1111/trf.17244.

Costa, V. F. A., Chometon, T. Q., Gomes de Castro, K. K., Ponte, M. S. G., Pimentel, M. I. F., Lyra, M. R., ... Bertho, A. L. (2025). Extracellular Vesicle Release from Immune Cells in Cutaneous Leishmaniasis: Modulation by Leishmania (V.) braziliensis and Reversal by Antimonial Therapy. *Pathogens, 14*(8), https://doi.org/10.3390/pathogens14080771.

Cruz-Bustos, T., Feix, A. S., Hummel, K., Schlosser, S., Razzazi-Fazeli, E., & Joachim, A. (2025). The proteomic landscape of *Toxoplasma gondii* extracellular vesicles across diverse host cell types. *Front Cell Infect Microbiol, 15*, 1565684. https://doi.org/10.3389/fcimb.2025.1565684.

Cruz Camacho, A., Alfandari, D., Kozela, E., & Regev-Rudzki, N. (2023). Biogenesis of extracellular vesicles in protozoan parasites: The ESCRT complex in the trafficking fast lane? *PLoS Pathogens, 19*(2), e1011140. https://doi.org/10.1371/journal.ppat.1011140.

Cursino-Santos, J. R., Alhassan, A., Singh, M., & Lobo, C. A. (2014a). *Babesia*: impact of cold storage on the survival and the viability of parasites in blood bags. *Transfusion, 54*(3), 585–591. https://doi.org/10.1111/trf.12357.

Cursino-Santos, J. R., Halverson, G., Rodriguez, M., Narla, M., & Lobo, C. A. (2014b). Identification of binding domains on red blood cell glycophorins for *Babesia divergens*. *Transfusion, 54*(4), 982–989. https://doi.org/10.1111/trf.12388.

Cursino-Santos, J. R., Singh, M., Pham, P., Rodriguez, M., & Lobo, C. A. (2016). *Babesia divergens* builds a complex population structure composed of specific ratios of infected cells to ensure a prompt response to changing environmental conditions. *Cellular Microbiology, 18*(6), 859–874. https://doi.org/10.1111/cmi.12555.

Cursino-Santos, J. R., Singh, M., Senaldi, E., Manwani, D., Yazdanbakhsh, K., & Lobo, C. A. (2019). Altered parasite life-cycle processes characterize Babesia divergens infection in human sickle cell anemia. *Haematologica, 104*(11), 2189–2199. https://doi.org/10.3324/haematol.2018.214304.

Cutts, E. E., Laasch, N., Reiter, D. M., Trenker, R., Slater, L. M., Stansfeld, P. J., & Vakonakis, I. (2017). Structural analysis of P. falciparum KAHRP and PfEMP1 complexes with host erythrocyte spectrin suggests a model for cytoadherent knob protrusions. *PLoS Pathogens, 13*(8), e1006552. https://doi.org/10.1371/journal.ppat.1006552.

Da Costa, L., Suner, L., Galimand, J., Bonnel, A., Pascreau, T., Couque, N., ... French Society of, H. (2016). Diagnostic tool for red blood cell membrane disorders: Assessment of a new generation ektacytometer. *Blood Cells, Molecules & Diseases, 56*(1), 9–22. https://doi.org/10.1016/j.bcmd.2015.09.001.

da Silva Lira Filho, A., Fajardo, E. F., Chang, K. P., Clement, P., & Olivier, M. (2021). Leishmania exosomes/extracellular vesicles containing GP63 are essential for enhanced cutaneous leishmaniasis development upon co-inoculation of Leishmania amazonensis and its exosomes. *Front Cell Infect Microbiol, 11*, 709258. https://doi.org/10.3389/fcimb.2021.709258.

Depond, M., Henry, B., Buffet, P., & Ndour, P. A. (2019). Methods to investigate the deformability of RBC during malaria. *Front Physiol, 10*, 1613. https://doi.org/10.3389/fphys.2019.01613.

DeSilva, T. M., Peng, K. C., Speicher, K. D., & Speicher, D. W. (1992). Analysis of human red cell spectrin tetramer (head-to-head) assembly using complementary univalent peptides. *Biochemistry, 31*(44), 10872–10878. https://doi.org/10.1021/bi00159a030.

Dey, S., Mohapatra, S., Khokhar, M., Hassan, S., & Pandey, R. K. (2024). Extracellular vesicles in malaria: shedding light on pathogenic depths. *ACS Infect Dis, 10*(3), 827–844. https://doi.org/10.1021/acsinfecdis.3c00649.

Dias-Guerreiro, T., Palma-Marques, J., Mourata-Goncalves, P., Alexandre-Pires, G., Valerio-Bolas, A., Gabriel, A., ... Santos-Gomes, G. (2021). African trypanosomiasis: extracellular vesicles shed by *Trypanosoma brucei brucei* manipulate host mononuclear cells. *Biomedicines, 9*(8), https://doi.org/10.3390/biomedicines9081056.

Diaz-Godinez, C., Rios-Valencia, D. G., Garcia-Aguirre, S., Martinez-Calvillo, S., & Carrero, J. C. (2022). Immunomodulatory effect of extracellular vesicles from *Entamoeba histolytica* trophozoites: Regulation of NETs and respiratory burst during confrontation with human neutrophils. *Front Cell Infect Microbiol, 12*, 1018314. https://doi.org/10.3389/fcimb.2022.1018314.

Discher, D. E., Mohandas, N., & Evans, E. A. (1994). Molecular maps of red cell deformation: hidden elasticity and in situ connectivity. *Science (New York, N. Y.), 266*(5187), 1032–1035. https://doi.org/10.1126/science.7973655.

Djokic, V., Rocha, S. C., & Parveen, N. (2021). Lessons learned for pathogenesis, immunology, and disease of erythrocytic parasites: *Plasmodium* and *Babesia*. *Front Cell Infect Microbiol, 11*, 685239. https://doi.org/10.3389/fcimb.2021.685239.

Dondorp, A. M., Angus, B. J., Chotivanich, K., Silamut, K., Ruangveerayuth, R., Hardeman, M. R., ... White, N. J. (1999). Red blood cell deformability as a predictor of anemia in severe falciparum malaria. *The American Journal of Tropical Medicine and Hygiene, 60*(5), 733–737. https://doi.org/10.4269/ajtmh.1999.60.733.

Drews, S. J., Kjemtrup, A. M., Krause, P. J., Lambert, G., Leiby, D. A., Lewin, A., ... Bloch, E. M. (2023a). Transfusion-transmitted *Babesia* spp.: a changing landscape of epidemiology, regulation, and risk mitigation. *Journal of Clinical Microbiology, 61*(10), e0126822. https://doi.org/10.1128/jcm.01268-22.

Drews, S. J., Wendel, S., Leiby, D. A., Tonnetti, L., Ushiro-Lumb, I., O'Brien, S. F., ... on behalf of the International Society of Blood Transfusion Working Party Parasite Subgroup. (2023). Climate change and parasitic risk to the blood supply. *Transfusion, 63*(3), 638–645. https://doi.org/10.1111/trf.17234.

Ebenuwa, I., Violet, P. C., Teng, S., Greene, T., Munyan, N., Berman, R., ... Levine, M. (2025). Altered RBC aggregability in diabetes: a threshold for pathophysiological structure-function RBC changes. *Cardiovascular Diabetology – Endocrinology Reports, 11*(1), 44. https://doi.org/10.1186/s40842-025-00256-2.

Efremov, A., & Cao, J. (2011). Bistability of cell adhesion in shear flow. *Biophysical Journal, 101*(5), 1032–1040. https://doi.org/10.1016/j.bpj.2011.07.026.

Elsworth, B., Sanders, P. R., Nebl, T., Batinovic, S., Kalanon, M., Nie, C. Q., ... Gilson, P. R. (2016). Proteomic analysis reveals novel proteins associated with the *Plasmodium* protein exporter PTEX and a loss of complex stability upon truncation of the core PTEX component, PTEX150. *Cellular Microbiology, 18*(11), 1551–1569. https://doi.org/10.1111/cmi.12596.

Fathi, A., Hakimi, H., Sakaguchi, M., Yamagishi, J., Kawazu, S. I., & Asada, M. (2024). Critical role of Babesia bovis spherical body protein 3 in ridge formation on infected red blood cells. *PLoS Pathogens, 20*(11), e1012294. https://doi.org/10.1371/journal.ppat.1012294.

Fedosov, D. A., Caswell, B., Suresh, S., & Karniadakis, G. E. (2011). Quantifying the biophysical characteristics of *Plasmodium-falciparum*-parasitized red blood cells in microcirculation. *Proceedings of the National Academy of Sciences of the United States of America, 108*(1), 35–39. https://doi.org/10.1073/pnas.1009492108.

Ferreri, L. M., Brayton, K. A., Sondgeroth, K. S., Lau, A. O., Suarez, C. E., & McElwain, T. F. (2012). Expression and strain variation of the novel "small open reading frame" (smorf) multigene family in *Babesia bovis*. *International Journal for Parasitology, 42*(2), 131–138. https://doi.org/10.1016/j.ijpara.2011.10.004.

Fibach, E. (2021). The redox balance and membrane shedding in RBC production, maturation, and senescence. *Frontiers in Physiology, 12*, 604738. https://doi.org/10.3389/fphys.2021.604738.

Frenal, K., & Soldati-Favre, D. (2009). Role of the parasite and host cytoskeleton in apicomplexa parasitism. *Cell Host & Microbe, 5*(6), 602–611. https://doi.org/10.1016/j.chom.2009.05.013.

Gallego-Lopez, G. M., Lau, A. O. T., O'Connor, R. M., Ueti, M. W., Cooke, B. M., Laughery, J. M., ... Suarez, C. E. (2019). Up-regulated expression of spherical body protein 2 truncated copy 11 in Babesia bovis is associated with reduced cytoadhesion to vascular endothelial cells. *International Journal for Parasitology, 49*(2), 127–137. https://doi.org/10.1016/j.ijpara.2018.05.015.

Garg, A., Stein, A., Zhao, W., Dwivedi, A., Frutos, R., Cornillot, E., & Ben Mamoun, C. (2014). Sequence and annotation of the apicoplast genome of the human pathogen *Babesia microti*. *PLoS One, 9*(10), e107939. https://doi.org/10.1371/journal.pone.0107939.

Gerber, M. A., Shapiro, E. D., Krause, P. J., Cable, R. G., Badon, S. J., & Ryan, R. W. (1994). The risk of acquiring Lyme disease or babesiosis from a blood transfusion. *The Journal of Infectious Diseases, 170*(1), 231–234. https://doi.org/10.1093/infdis/170.1.231.

Glenister, F. K., Coppel, R. L., Cowman, A. F., Mohandas, N., & Cooke, B. M. (2002). Contribution of parasite proteins to altered mechanical properties of malaria-infected red blood cells. *Blood, 99*(3), 1060. https://doi.org/10.1182/blood.v99.3.1060.

Gohil, S., Kats, L. M., Sturm, A., & Cooke, B. M. (2010). Recent insights into alteration of red blood cells by *Babesia bovis*: moovin' forward. *Trends in Parasitology, 26*(12), 591–599. https://doi.org/10.1016/j.pt.2010.06.012.

Gonzalez, L. M., Estrada, K., Grande, R., Jimenez-Jacinto, V., Vega-Alvarado, L., Sevilla, E., ... Montero, E. (2019). Comparative and functional genomics of the protozoan parasite *Babesia divergens* highlighting the invasion and egress processes. *PLoS Neglected Tropical Diseases, 13*(8), e0007680. https://doi.org/10.1371/journal.pntd.0007680.

Gonzalez, L. M., Revuelta, B., Gil, A., Terron, M. C., Sachse, M. C., Sotillo, J., ... Montero, E. (2025). Elucidating the functional role of the novel BdP50 protein and extracellular vesicles in the human erythrocyte infection by *Babesia divergens*. *PLoS Neglected Tropical Diseases, 19*(8), e0013401. https://doi.org/10.1371/journal.pntd.0013401.

Gray, E. B., & Herwaldt, B. L. (2019). Babesiosis Surveillance - United States, 2011-2015. *MMWR. Surveillance Summaries: Morbidity and Mortality Weekly Report. Surveillance Summaries / CDC, 68*(6), 1–11. https://doi.org/10.15585/mmwr.ss6806a1.

Guizouarn, H., & Barshtein, G. (2020). Editorial: red blood cell vascular adhesion and deformability. *Frontiers in Physiology, 11*, 657. https://doi.org/10.3389/fphys.2020.00657.

Gupta, H., Rubio, M., Sitoe, A., Varo, R., Cistero, P., Madrid, L., ... Mayor, A. (2021). Plasma microRNA profiling of *Plasmodium falciparum* biomass and association with severity of malaria disease. *Emerging Infectious Diseases, 27*(2), 430–442. https://doi.org/10.3201/eid2702.191795.

Hagos, B., Brasov, I., Branscome, H., Rashid, S., Bradford, R., Leonelli, J., ... Molestina, R. E. (2025). Activation of macrophages by extracellular vesicles derived from *Babesia*-infected red blood cells. *Infection and Immunity, 93*(5), e0033324. https://doi.org/10.1128/iai.00333-24.

Hakimi, H., Templeton, T. J., Sakaguchi, M., Yamagishi, J., Miyazaki, S., Yahata, K., ... Asada, M. (2020). Novel *Babesia bovis* exported proteins that modify properties of infected red blood cells. *PLoS Pathogens, 16*(10), e1008917. https://doi.org/10.1371/journal.ppat.1008917.

Hakimi, H., Yamagishi, J., Kawazu, S. I., & Asada, M. (2022). Advances in understanding red blood cell modifications by *Babesia*. *PLoS Pathogens, 18*(9), e1010770. https://doi.org/10.1371/journal.ppat.1010770.

Hakimi, H., Yamagishi, J., Sakaguchi, M., Fathi, A., Hakimi, J. S., Verocai, G. G., ... Asada, M. (2025). ves1alpha genes expression is the major determinant of *Babesia bovis*-infected erythrocytes cytoadhesion to endothelial cells. *PLoS Pathogens, 21*(4), e1012583. https://doi.org/10.1371/journal.ppat.1012583.

Haldar, K., & Bhattacharjee, S. (2025). Vesicular mechanisms of drug resistance in apicomplexan parasites. *Microbiology and Molecular Biology Reviews: MMBR, 89*(1), e0001024. https://doi.org/10.1128/mmbr.00010-24.

Hale, J. P., Winlove, C. P., & Petrov, P. G. (2011). Effect of hydroperoxides on red blood cell membrane mechanical properties. *Biophysical Journal, 101*(8), 1921–1929. https://doi.org/10.1016/j.bpj.2011.08.053.

Han, Z., Nie, Z., Shu, X., Zheng, Y., Luo, W., Zhang, H., ... Zhao, J. (2021). Identification of a novel variant erythrocyte surface antigen-1 (VESA1) in Babesia orientalis. *Parasitology Research, 120*(8), 2863–2872. https://doi.org/10.1007/s00436-021-07194-9.

Hansen, J. C., Skalak, R., Chien, S., & Hoger, A. (1997). Influence of network topology on the elasticity of the red blood cell membrane skeleton. *Biophysical Journal, 72*(5), 2369–2381. https://doi.org/10.1016/S0006-3495(97)78882-9.

Hassoun, H., Vassiliadis, J. N., Murray, J., Njolstad, P. R., Rogus, J. J., Ballas, S. K., ... Palek, J. (1997). Characterization of the underlying molecular defect in hereditary spherocytosis associated with spectrin deficiency. *Blood, 90*(1), 398–406. Retrieved from . https://www.ncbi.nlm.nih.gov/pubmed/9207476.

Henry, B., Volle, G., Akpovi, H., Gineau, L., Roussel, C., Ndour, P. A., ... Buffet, P. (2022). Splenic clearance of rigid erythrocytes as an inherited mechanism for splenomegaly and natural resistance to malaria. *EBioMedicine, 82*, 104167. https://doi.org/10.1016/j.ebiom.2022.104167.

Hierso, R., Lemonne, N., Villaescusa, R., Lalanne-Mistrih, M. L., Charlot, K., Etienne-Julan, M., ... Hardy-Dessources, M. D. (2017). Exacerbation of oxidative stress during sickle vaso-occlusive crisis is associated with decreased anti-band 3 autoantibodies rate and increased red blood cell-derived microparticle level: a prospective study. *British Journal of Haematology, 176*(5), 805–813. https://doi.org/10.1111/bjh.14476.

Higashi, D. L., Biais, N., Donahue, D. L., Mayfield, J. A., Tessier, C. R., Rodriguez, K., ... Lee, S. W. (2016). Activation of band 3 mediates group A Streptococcus streptolysin S-based beta-haemolysis. *Nature Microbiology, 1*, 15004. https://doi.org/10.1038/nmicrobiol.2015.4.

Holbrook, N. R., Klontz, E. H., Adams, G. C., Schnittman, S. R., Issa, N. C., Bond, S. A., ... Lemieux, J. E. (2023). *Babesia microti* variant with multiple resistance mutations detected in an immunocompromised patient receiving atovaquone prophylaxis. *Open Forum Infectious Diseases, 10*(3), ofad097. https://doi.org/10.1093/ofid/ofad097.

Homer, M. J., Aguilar-Delfin, I., Telford, S. R., 3rd, Krause, P. J., & Persing, D. H. (2000). Babesiosis. *Clinical Microbiology Reviews, 13*(3), 451–469. https://doi.org/10.1128/CMR.13.3.451.

Hunfeld, K. P., Hildebrandt, A., & Gray, J. S. (2008). Babesiosis: recent insights into an ancient disease. *International Journal for Parasitology, 38*(11), 1219–1237. https://doi.org/10.1016/j.ijpara.2008.03.001.

Hutchings, C. L., Li, A., Fernandez, K. M., Fletcher, T., Jackson, L. A., Molloy, J. B., ... Cooke, B. M. (2007). New insights into the altered adhesive and mechanical properties of red blood cells parasitized by *Babesia bovis*. *Molecular Microbiology, 65*(4), 1092–1105. https://doi.org/10.1111/j.1365-2958.2007.05850.x.

Hviid, L., Jensen, A. R., & Deitsch, K. W. (2024). PfEMP1 and var genes - Still of key importance in *Plasmodium falciparum* malaria pathogenesis and immunity. *Advances in Parasitology, 125*, 53–103. https://doi.org/10.1016/bs.apar.2024.02.001.

Issaian, A., Hay, A., Dzieciatkowska, M., Roberti, D., Perrotta, S., Darula, Z., ... D'Alessandro, A. (2021). The interactome of the N-terminus of band 3 regulates red blood cell metabolism and storage quality. *Haematologica, 106*(11), 2971–2985. https://doi.org/10.3324/haematol.2020.278252.

Jacoby, G. A., Hunt, J. V., Kosinski, K. S., Demirjian, Z. N., Huggins, C., Etkind, P., ... Spielman, A. (1980). Treatment of transfusion-transmitted babesiosis by exchange transfusion. *The New England Journal of Medicine, 303*(19), 1098–1100. https://doi.org/10.1056/NEJM198011063031906.

Jajosky, R. P., & Stowell, S. R. (2025). *Babesia* not only is invasive but creates RBC inflexibility. *Blood, 145*(19), 2114–2115. https://doi.org/10.1182/blood.2025028446.

Jandl, J. H., Greenberg, M. S., Yonemoto, R. H., & Castle, W. B. (1956). Clinical determination of the sites of red cell sequestration in hemolytic anemias. *The Journal of Clinical Investigation, 35*(8), 842–867. https://doi.org/10.1172/JCI103338.

Jones, S. H., Lew, A. E., Jorgensen, W. K., & Barker, S. C. (1997). *Babesia bovis*: genome size, number of chromosomes and telomeric probe hybridisation. *International Journal for Parasitology, 27*(12), 1569–1573. https://doi.org/10.1016/s0020-7519(97)00099-4.

Jonsdottir, T. K., Gabriela, M., Crabb, B. S., & Gilson, P. R. (2021). Defining the essential exportome of the malaria parasite. *Trends in Parasitology, 37*(7), 664–675. https://doi.org/10.1016/j.pt.2021.04.009.

Karkoska, K., Louie, J., Appiah-Kubi, A. O., Wolfe, L., Rubin, L., Rajan, S., & Aygun, B. (2018). Transfusion-transmitted babesiosis leading to severe hemolysis in two patients with sickle cell anemia. *Pediatric Blood & Cancer, 65*(1), https://doi.org/10.1002/pbc.26734.

Kesely, K. R., Pantaleo, A., Turrini, F. M., Olupot-Olupot, P., & Low, P. S. (2016). Inhibition of an erythrocyte tyrosine kinase with imatinib prevents *Plasmodium falciparum* egress and terminates parasitemia. *PLoS One, 11*(10), e0164895. https://doi.org/10.1371/journal.pone.0164895.

Kozela, E., Petrovich-Kopitman, E., Berger, Y., Camacho, A. C., Shoham, Y., Morandi, M. I., ... Regev-Rudzki, N. (2025). Spectral flow cytometry for detecting DNA cargo in malaria parasite-derived extracellular vesicles. *The Journal of Biological Chemistry, 301*(5), 108481. https://doi.org/10.1016/j.jbc.2025.108481.

Krampitz, H. E. (1979). *Babesia microti*: morphology, distribution and host relationship in Germany. *Zentralblatt fur Bakteriologie, Mikrobiologie, und Hygiene, 244*(2-3), 411–415. Retrieved from . https://www.ncbi.nlm.nih.gov/pubmed/388948.

Krause, P. J., Daily, J., Telford, S. R., Vannier, E., Lantos, P., & Spielman, A. (2007). Shared features in the pathobiology of babesiosis and malaria. *Trends in Parasitology, 23*(12), 605–610. https://doi.org/10.1016/j.pt.2007.09.005.

Krause, P. J., Rogers, R., Shah, M. K., Kang, H., Parsonnet, J., Kodama, R., & Vannier, E. (2024). Tafenoquine for Relapsing Babesiosis: A Case Series. *Clinical Infectious Diseases: an Official Publication of the Infectious Diseases Society of America, 79*(1), 130–137. https://doi.org/10.1093/cid/ciae238.

Kumar, A., O'Bryan, J., & Krause, P. J. (2021). The global emergence of human babesiosis. *Pathogens, 10*(11), https://doi.org/10.3390/pathogens10111144.

Kumar, M. A., Baba, S. K., Sadida, H. Q., Marzooqi, S. A., Jerobin, J., Altemani, F. H., ... Bhat, A. A. (2024). Extracellular vesicles as tools and targets in therapy for diseases. *Signal Transduction and Targeted Therapy, 9*(1), 27. https://doi.org/10.1038/s41392-024-01735-1.

Lai, L., Xu, X., Lim, C. T., & Cao, J. (2015). Stiffening of red blood cells induced by cytoskeleton disorders: a joint theory-experiment study. *Biophysical Journal, 109*(11), 2287–2294. https://doi.org/10.1016/j.bpj.2015.10.036.

Lanzillotti, R., & Coetzer, T. L. (2004). Myosin-like sequences in the malaria parasite *Plasmodium falciparum* bind human erythrocyte membrane protein 4.1. *Haematologica, 89*(10), 1168–1171. Retrieved from . https://www.ncbi.nlm.nih.gov/pubmed/15477199.

Lanzillotti, R., & Coetzer, T. L. (2006). The 10 kDa domain of human erythrocyte protein 4.1 binds the *Plasmodium falciparum* EBA-181 protein. *Malaria Journal, 5*, 100. https://doi.org/10.1186/1475-2875-5-100.

Lauterbach, S. B., Lanzillotti, R., & Coetzer, T. L. (2003). Construction and use of *Plasmodium falciparum* phage display libraries to identify host parasite interactions. *Malaria Journal, 2*(1), 47. https://doi.org/10.1186/1475-2875-2-47.

Leal, J. K. F., Adjobo-Hermans, M. J. W., & Bosman, G. (2018). Red blood cell homeostasis: mechanisms and effects of microvesicle generation in health and disease. *Frontiers in Physiology, 9*, 703. https://doi.org/10.3389/fphys.2018.00703.

Lecomte, M. C., Nicolas, G., Dhermy, D., Pinder, J. C., & Gratzer, W. B. (1999). Properties of normal and mutant polypeptide fragments from the dimer self-association sites of human red cell spectrin. *European Biophysics Journal: EBJ, 28*(3), 208–215. https://doi.org/10.1007/s002490050201.

Leiby, D. A. (2006). Babesiosis and blood transfusion: flying under the radar. *Vox Sanguinis, 90*(3), 157–165. https://doi.org/10.1111/j.1423-0410.2006.00740.x.

Lenormand, G., Henon, S., Richert, A., Simeon, J., & Gallet, F. (2001). Direct measurement of the area expansion and shear moduli of the human red blood cell membrane skeleton. *Biophysical Journal, 81*(1), 43–56. https://doi.org/10.1016/S0006-3495(01)75678-0.

Lewis, I. A., Campanella, M. E., Markley, J. L., & Low, P. S. (2009). Role of band 3 in regulating metabolic flux of red blood cells. *Proceedings of the National Academy of Sciences of the United States of America, 106*(44), 18515–18520. https://doi.org/10.1073/pnas.0905999106.

Li, H., Lu, L., Li, X., Buffet, P. A., Dao, M., Karniadakis, G. E., & Suresh, S. (2018). Mechanics of diseased red blood cells in human spleen and consequences for hereditary blood disorders. *Proceedings of the National Academy of Sciences of the United States of America, 115*(38), 9574–9579. https://doi.org/10.1073/pnas.1806501115.

Li, M., Ao, Y., Guo, J., Nie, Z., Liu, Q., Yu, L., ... Zhao, J. (2019). Surface antigen 1 is a crucial secreted protein that mediates *Babesia microti* invasion into host cells. *Frontiers in Microbiology, 10*, 3046. https://doi.org/10.3389/fmicb.2019.03046.

Lobo, C. A. (2005). *Babesia divergens* and *Plasmodium falciparum* use common receptors, glycophorins A and B, to invade the human red blood cell. *Infection and Immunity, 73*(1), 649–651. https://doi.org/10.1128/IAI.73.1.649-651.2005.

Lobo, C. A., Cursino-Santos, J. R., Alhassan, A., & Rodrigues, M. (2013). *Babesia*: an emerging infectious threat in transfusion medicine. *PLoS Pathogens, 9*(7), e1003387. https://doi.org/10.1371/journal.ppat.1003387.

Lobo, C. A., Cursino-Santos, J. R., Singh, M., & Rodriguez, M. (2019). *Babesia divergens*: a drive to survive. *Pathogens, 8*(3), https://doi.org/10.3390/pathogens8030095.

Lobo, C. A., Rodriguez, M., & Cursino-Santos, J. R. (2012). *Babesia* and red cell invasion. *Current Opinion in Hematology, 19*(3), 170–175. https://doi.org/10.1097/MOH.0b013e328352245a.

Lobo, C. A., Singh, M., & Rodriguez, M. (2020). Human babesiosis: recent advances and future challenges. *Current Opinion in Hematology, 27*(6), 399–405. https://doi.org/10.1097/MOH.0000000000000606.

Magowan, C., Nunomura, W., Waller, K. L., Yeung, J., Liang, J., Van Dort, H., ... Mohandas, N. (2000). *Plasmodium falciparum* histidine-rich protein 1 associates with the band 3 binding domain of ankyrin in the infected red cell membrane. *Biochimica Biophysica Acta, 1502*(3), 461–470. https://doi.org/10.1016/s0925-4439(00)00069-7.

Marcos, L. A., & Wormser, G. P. (2023). Relapsing Babesiosis With Molecular Evidence of Resistance to Certain Antimicrobials Commonly Used to Treat *Babesia microti* Infections. *Open Forum Infectious Diseases, 10*(8), ofad391. https://doi.org/10.1093/ofid/ofad391.

Marti, M., & Johnson, P. J. (2016). Emerging roles for extracellular vesicles in parasitic infections. *Current Opinion in Microbiology, 32*, 66–70. https://doi.org/10.1016/j.mib.2016.04.008.

McMahon, T. J. (2019). Red blood cell deformability, vasoactive mediators, and adhesion. *Frontiers in Physiology, 10*, 1417. https://doi.org/10.3389/fphys.2019.01417.

Miyazaki, S., Chitama, B. A., Kagaya, W., Lucky, A. B., Zhu, X., Yahata, K., ... Kaneko, O. (2021). Plasmodium falciparum SURFIN(4.1) forms an intermediate complex with PTEX components and Pf113 during export to the red blood cell. *Parasitology International, 83*, 102358. https://doi.org/10.1016/j.parint.2021.102358.

Mohandas, N., Clark, M. R., Jacobs, M. S., Groner, W., & Shohet, S. B. (1980). Ektacytometric analysis of factors regulating red cell deformability. *Blood Cells, 6*(3), 329–334. Retrieved from . https://www.ncbi.nlm.nih.gov/pubmed/6156730.

Mohandas, N., & Gallagher, P. G. (2008). Red cell membrane: past, present, and future. *Blood, 112*(10), 3939–3948. https://doi.org/10.1182/blood-2008-07-161166.

Moyano, S., Musso, J., Feliziani, C., Zamponi, N., Frontera, L. S., Ropolo, A. S., ... Touz, M. (2019). Exosome biogenesis in the protozoa parasite *Giardia lamblia*: a model of reduced interorganellar crosstalk. *Cells, 8*(12), https://doi.org/10.3390/cells8121600.

Namvar, A., Blanch, A. J., Dixon, M. W., Carmo, O. M. S., Liu, B., Tiash, S., ... Tilley, L. (2021). Surface area-to-volume ratio, not cellular viscoelasticity, is the major determinant of red blood cell traversal through small channels. *Cellular Microbiology, 23*(1), e13270. https://doi.org/10.1111/cmi.13270.

Nash, G. B., O'Brien, E., Gordon-Smith, E. C., & Dormandy, J. A. (1989). Abnormalities in the mechanical properties of red blood cells caused by *Plasmodium falciparum. Blood, 74*(2), 855–861. Retrieved from . https://www.ncbi.nlm.nih.gov/pubmed/2665857.

Ohanian, V., Wolfe, L. C., John, K. M., Pinder, J. C., Lux, S. E., & Gratzer, W. B. (1984). Analysis of the ternary interaction of the red cell membrane skeletal proteins spectrin, actin, and 4.1. *Biochemistry, 23*(19), 4416–4420. https://doi.org/10.1021/bi00314a027.

Ord, R. L., & Lobo, C. A. (2015). Human babesiosis: pathogens, prevalence, diagnosis and treatment. *Current Clinical Microbiology Reports, 2*(4), 173–181. https://doi.org/10.1007/s40588-015-0025-z.

Pantaleo, A., Kesely, K. R., Pau, M. C., Tsamesidis, I., Schwarzer, E., Skorokhod, O. A., ... Turrini, F. M. (2017). Syk inhibitors interfere with erythrocyte membrane modification during P falciparum growth and suppress parasite egress. *Blood, 130*(8), 1031–1040. https://doi.org/10.1182/blood-2016-11-748053.

Paul, A. S., Moreira, C. K., Elsworth, B., Allred, D. R., & Duraisingh, M. T. (2016). Extensive shared chemosensitivity between malaria and babesiosis blood-stage parasites. *Antimicrobial Agents and Chemotherapy, 60*(8), 5059–5063. https://doi.org/10.1128/AAC.00928-16.

Paulitschke, M., & Nash, G. B. (1993). Membrane rigidity of red blood cells parasitized by different strains of Plasmodium falciparum. *The Journal of Laboratory and Clinical Medicine, 122*(5), 581–589. Retrieved from . https://www.ncbi.nlm.nih.gov/pubmed/8228577.

Pei, X., An, X., Guo, X., Tarnawski, M., Coppel, R., & Mohandas, N. (2005). Structural and functional studies of interaction between *Plasmodium falciparum* knob-associated histidine-rich protein (KAHRP) and erythrocyte spectrin. *The Journal of Biological Chemistry, 280*(35), 31166–31171. https://doi.org/10.1074/jbc.M505298200.

Pei, X., Guo, X., Coppel, R., Bhattacharjee, S., Haldar, K., Gratzer, W., ... An, X. (2007a). The ring-infected erythrocyte surface antigen (RESA) of Plasmodium falciparum stabilizes spectrin tetramers and suppresses further invasion. *Blood, 110*(3), 1036–1042. https://doi.org/10.1182/blood-2007-02-076919.

Pei, X., Guo, X., Coppel, R., Mohandas, N., & An, X. (2007b). *Plasmodium falciparum* erythrocyte membrane protein 3 (PfEMP3) destabilizes erythrocyte membrane skeleton. *The Journal of Biological Chemistry, 282*(37), 26754–26758. https://doi.org/10.1074/jbc.M70161220.

Pelle, K. G., Jiang, R. H., Mantel, P. Y., Xiao, Y. P., Hjelmqvist, D., Gallego-Lopez, G. M., ... Marti, M. (2015). Shared elements of host-targeting pathways among apicomplexan parasites of differing lifestyles. *Cellular Microbiology, 17*(11), 1618–1639. https://doi.org/10.1111/cmi.12460.

Perrotta, S., Gallagher, P. G., & Mohandas, N. (2008). Hereditary spherocytosis. *Lancet, 372*(9647), 1411–1426. https://doi.org/10.1016/S0140-6736(08)61588-3.

Pesen, T., Akgun, B., & Unlu, M. B. (2025). Measuring the effect of repetitive stretching on the deformability of human red blood cells using optical tweezers. *Scientific Reports, 15*(1), 9060. https://doi.org/10.1038/s41598-025-93288-8.

Prendki, V., Ndour, P. A., Jais, X., Ciceron, L., Settegrana, C., & Buffet, P. (2012). Reduced deformability of circulating erythrocytes: a marker of hyposplenism. *American Journal of Hematology, 87*(10), E81–E82. https://doi.org/10.1002/ajh.23297.

Puri, A., Bajpai, S., Meredith, S., Aravind, L., Krause, P. J., & Kumar, S. (2021). *Babesia microti*: pathogen genomics, genetic variability, immunodominant antigens, and pathogenesis. *Frontiers in Microbiology, 12*, 697669. https://doi.org/10.3389/fmicb.2021.697669.

Rabai, M., Detterich, J. A., Wenby, R. B., Hernandez, T. M., Toth, K., Meiselman, H. J., & Wood, J. C. (2014). Deformability analysis of sickle blood using ektacytometry. *Biorheology, 51*(2-3), 159–170. https://doi.org/10.3233/BIR-140660.

Raposo, G., & Stoorvogel, W. (2013). Extracellular vesicles: exosomes, microvesicles, and friends. *The Journal of Cell Biology, 200*(4), 373–383. https://doi.org/10.1083/jcb.201211138.

Rathnam, J. T. T., Grigg, M. J., Dondorp, A. M., William, T., Rajasekhar, M., Rajahram, G., ... Anstey, N. M. (2025). Reduced red blood cell deformability in vivax malaria. *The Journal of Infectious Diseases, 231*(3), e566–e569. https://doi.org/10.1093/infdis/jiae490.

Renoux, C., Faivre, M., Bessaa, A., Da Costa, L., Joly, P., Gauthier, A., & Connes, P. (2019). Impact of surface-area-to-volume ratio, internal viscosity and membrane viscoelasticity on red blood cell deformability measured in isotonic condition. *Scientific Reports, 9*(1), 6771. https://doi.org/10.1038/s41598-019-43200-y.

Rogers, R., Krause, P. J., Norris, A. M., Ting, M. H., Nagami, E. H., Cilley, B., & Vannier, E. (2023). Broad antimicrobial resistance in a case of relapsing babesiosis successfully treated with tafenoquine. *Clinical Infectious Diseases: an Official Publication of the Infectious Diseases Society of America, 76*(4), 741–744. https://doi.org/10.1093/cid/ciac473.

Safeukui, I., Buffet, P. A., Deplaine, G., Perrot, S., Brousse, V., Sauvanet, A., ... Mohandas, N. (2018). Sensing of red blood cells with decreased membrane deformability by the human spleen. *Blood Advances, 2*(20), 2581–2587. https://doi.org/10.1182/bloodadvances.2018024562.

Salas, N., Coceres, V. M., Melo, T. D. S., Pereira-Neves, A., Maguire, V. G., Rodriguez, T. M., ... de Miguel, N. (2021). VPS32, a member of the ESCRT complex, modulates adherence to host cells in the parasite Trichomonas vaginalis by affecting biogenesis and cargo sorting of released extracellular vesicles. *Cellular and Molecular Life Sciences: CMLS, 79*(1), 11. https://doi.org/10.1007/s00018-021-04083-3.

Santos, M. L. S., Coimbra, R. S., Sousa, T. N., Guimaraes, L. F. F., Gomes, M. S., Amaral, L. R., ... Carvalho, L. H. (2021). The interface between inflammatory mediators and microRNAs in *Plasmodium vivax* severe thrombocytopenia. *Frontiers in Cellular and Infection Microbiology, 11*, 631333. https://doi.org/10.3389/fcimb.2021.631333.

Satchwell, T. J., & Toye, A. M. (2021). Band 3, an essential red blood cell hub of activity. *Haematologica, 106*(11), 2792–2793. https://doi.org/10.3324/haematol.2021.278643.

Sharma, M., Lozano-Amado, D., Chowdhury, D., & Singh, U. (2023). Extracellular vesicles and their impact on the biology of protozoan parasites. *Tropical Medicine and Infectious Disease, 8*(9), https://doi.org/10.3390/tropicalmed8090448.

Si, W., Fang, C., Liu, C., Yin, M., Xu, W., Li, Y., ... Sun, J. (2023). Why is *Babesia* not killed by artemisinin like *Plasmodium*. *Parasites & Vectors, 16*(1), 193. https://doi.org/10.1186/s13071-023-05783-4.

Siau, A., Huang, X., Weng, M., Sze, S. K., & Preiser, P. R. (2016). Proteome mapping of *Plasmodium*: identification of the *P. yoelii* remodellome. *Scientific Reports, 6*, 31055. https://doi.org/10.1038/srep31055.

Singh, P., Lonardi, S., Liang, Q., Vydyam, P., Khabirova, E., Fang, T., ... Mamoun, C. B. (2023). *Babesia duncani* multi-omics identifies virulence factors and drug targets. *Nature Microbiology, 8*(5), 845–859. https://doi.org/10.1038/s41564-023-01360-8.

Singh, P., Vydyam, P., Fang, T., Estrada, K., Gonzalez, L. M., Grande, R., ... Mamoun, C. B. (2024). Insights into the evolution, virulence and speciation of *Babesia* MO1 and *Babesia divergens* through multiomics analyses. *Emerging Microbes & Infections, 13*(1), 2386136. https://doi.org/10.1080/22221751.2024.2386136.

Skrabalo, Z., & Deanovic, Z. (1957). Piroplasmosis in man; report of a case. *Documenta de Medicina Geographica et Tropica, 9*(1), 11–16. Retrieved from . https://www.ncbi.nlm.nih.gov/pubmed/13427667.

Sojka, D., Jalovecka, M., & Perner, J. (2022). *Babesia, Theileria, Plasmodium* and Hemoglobin. *Microorganisms, 10*(8), https://doi.org/10.3390/microorganisms10081651.

Speicher, D. W., & Marchesi, V. T. (1984). Erythrocyte spectrin is comprised of many homologous triple helical segments. *Nature, 311*(5982), 177–180. https://doi.org/10.1038/311177a0.

Sun, M., Xue, X., Li, L., Xu, D., Li, S., Li, S. C., & Su, Q. (2021). Ectosome biogenesis and release processes observed by using live-cell dynamic imaging in mammalian glial cells. *Quantitative Imaging in Medicine and Surgery, 11*(11), 4604–4616. https://doi.org/10.21037/qims-20-1015.

Sun, Y., Moreau, E., Chauvin, A., & Malandrin, L. (2011). The invasion process of bovine erythrocyte by *Babesia divergens*: knowledge from an in vitro assay. *Veterinary Research, 42*(1), 62. https://doi.org/10.1186/1297-9716-42-62.

Thangaraju, K., Neerukonda, S. N., Katneni, U., & Buehler, P. W. (2020). Extracellular vesicles from red blood cells and their evolving roles in health, coagulopathy and therapy. *International Journal of Molecular Sciences, 22*(1), https://doi.org/10.3390/ijms22010153.

Thekkiniath, J., Kilian, N., Lawres, L., Gewirtz, M. A., Graham, M. M., Liu, X., ... Ben Mamoun, C. (2019). Evidence for vesicle-mediated antigen export by the human pathogen *Babesia microti*. *Life Science Alliance, 2*(3), https://doi.org/10.26508/lsa.201900382.

Ueti, M. W., Johnson, W. C., Kappmeyer, L. S., Herndon, D. R., Mousel, M. R., Reif, K. E., ... Brayton, K. A. (2021). Comparative analysis of gene expression between Babesia bovis blood stages and kinetes allowed by improved genome annotation. *International Journal for Parasitology, 51*(2-3), 123–136. https://doi.org/10.1016/j.ijpara.2020.08.006.

Ungewickell, E., Bennett, P. M., Calvert, R., Ohanian, V., & Gratzer, W. B. (1979). In vitro formation of a complex between cytoskeletal proteins of the human erythrocyte. *Nature, 280*(5725), 811–814. https://doi.org/10.1038/280811a0.

Wampfler, P. B., Tosevski, V., Nanni, P., Spycher, C., & Hehl, A. B. (2014). Proteomics of secretory and endocytic organelles in Giardia lamblia. *PLoS One, 9*(4), e94089. https://doi.org/10.1371/journal.pone.0094089.

Wang, X., Xiao, Y. P., Bouchut, A., Al-Khedery, B., Wang, H., & Allred, D. R. (2012). Characterization of the unusual bidirectional ves promoters driving VESA1 expression and associated with antigenic variation in *Babesia bovis*. *Eukaryotic Cell, 11*(3), 260–269. https://doi.org/10.1128/EC.05318-11.

Warncke, J. D., & Beck, H. P. (2019). Host cytoskeleton remodeling through- out the blood stages of Plasmodium falciparum. *Microbiology and Molecular Biology Reviews: MMBR, 83*(4), https://doi.org/10.1128/MMBR.00013-19.

Waugh, R. E., Narla, M., Jackson, C. W., Mueller, T. J., Suzuki, T., & Dale, G. L. (1992). Rheologic properties of senescent erythrocytes: Loss of surface area and volume with red blood cell age. *Blood, 79*(5), 1351–1358. Retrieved from . https://www.ncbi.nlm.nih.gov/pubmed/1536958.

Xu, X., Efremov, A. K., Li, A., Lai, L., Dao, M., Lim, C. T., & Cao, J. (2013). Probing the cytoadherence of malaria infected red blood cells under flow. *PLoS One, 8*(5), e64763. https://doi.org/10.1371/journal.pone.0064763.

Yang, Y., Christie, J., Koster, L., Du, A., & Yao, C. (2021). Emerging human Babesiosis with "ground zero" in North America. *Microorganisms, 9*(2), https://doi.org/10.3390/microorganisms9020440.

Yazdanbakhsh, M., Phan, C., Patel, P., Simpson, E., Wingnean, K., & Acker, J. P. (2025). RBC deformability and clinical relevance in transfusion recipients. *Transfusion and Apheresis Science: Official Journal of the World Apheresis Association: Official Journal of the European Society for Haemapheresis, 64*(6), 104278. https://doi.org/10.1016/j.transci.2025.104278.

Yi, W., Bao, W., Rodriguez, M., Liu, Y., Singh, M., Ramlall, V., ... Yazdanbakhsh, K. (2018). Robust adaptive immune response against *Babesia microti* infection marked by low parasitemia in a murine model of sickle cell disease. *Blood Advances, 2*(23), 3462–3478. https://doi.org/10.1182/bloodadvances.2018026468.

Zhu, R., Avsievich, T., Popov, A., & Meglinski, I. (2020). Optical tweezers in studies of red blood cells. *Cells, 9*(3), https://doi.org/10.3390/cells9030545.

CHAPTER SIX

How advances in chromosome conformation capture (3C) methods are reshaping our understanding of gene regulation in hematopoiesis

Hangpeng Li[a,1], Nicholas Denny[a,1], and James Davies[b,*]
[a]MRC Molecular Haematology Unit, MRC Weatherall Institute of Molecular Medicine, Radcliffe Department of Medicine, University of Oxford, Oxford, United Kingdom
[b]Oxford National Institute of Health Research Biomedical Research Centre, University of Oxford, Oxford, United Kingdom
*Corresponding author. e-mail address: james.davies@imm.ox.ac.uk

Contents

[1] Equal contribution

Current Topics in Developmental Biology, Volume 170
ISSN 0070-2153, https://doi.org/10.1016/bs.ctdb.2026.06.007

Abstract

The three-dimensional organization of the DNA within the nucleus plays a key role in regulating gene expression. Over the past two decades, advances in chromosome conformation capture (3C) technologies, in tandem with other methods, have shown that the genome forms a complex structure at multiple scales. Early studies identified large-scale structures such as chromosome territories, compartments and topologically associating domains (TADs). As the resolution of 3C techniques has improved, it has become possible to identify contacts between regulatory elements in detail and more recently, it has become possible to define intricate structures within *cis*-regulatory elements. In this chapter, we review the development of 3C-based methodologies and discuss the strengths and limitations of the different approaches. We examine how these technologies have refined our understanding of genome organization and gene regulation. Recent high-resolution studies reveal that chromatin architecture extends beyond classical domain structures to include nanoscale organization. Integration of 3C data with super-resolution imaging and molecular dynamics simulations supports a model in which genome folding is governed by the biophysical properties of chromatin.

1. Introduction

The emergence of complex multicellular organisms required the evolution of mechanisms that precisely control the where, when and to what level genes are expressed across different cellular contexts. Within chromatin, the genome is intricately organized by hundreds of millions of protein molecules to form highly dynamic structures that control gene expression. In higher eukaryotes, many genes are regulated by multi-protein complexes bound to small (200–500 bp) *cis*-regulatory elements (enhancers, promoters and insulators), which may be separated by a million DNA base pairs in the linear genome (Oudelaar & Higgs, 2021).

Over the past 20 years, our understanding of the relationship between genome structure and function has greatly increased, mainly driven by the ever-increasing resolution of new techniques in super-resolution imaging and chromatin conformation capture (3C). These approaches have revealed ever more detailed levels of chromatin organization from chromosomal territories to chromatin compartments, topologically associating domains (TADs), sub-TADs and fine-scale contacts between *cis*-regulatory elements (Oudelaar & Higgs, 2021; Nora et al., 2012; Dixon et al., 2012). Recently, it has become possible to use 3C-based approaches to identify that complex structures occur within *cis*-regulatory elements (Li et al., 2025).

Interpretation of the vast non-coding genome, which makes up over 95 % of the sequence and contains over 90 % of the variants linked to common diseases (Oudelaar & Higgs, 2021) as well as mutations causing inherited diseases (Nora et al., 2012) and malignancy (Dixon et al., 2012), remains extremely challenging. These sequences are largely unused in clinical genetics (Li et al., 2025) and the non-coding genome remains a vast untapped resource for identifying new pathways and therapeutic targets for human disease (Mukherjee et al., 1988).

Despite decades of work and major advances in genome sequencing, how chromatin-based regulatory structures function on the molecular level at individual genes remains poorly understood, and this has huge implications for human health. In this review, we will focus on how developments in chromosome conformation capture assays have informed our understanding of gene regulation, with a particular focus on hematopoiesis.

2. Overview of 3C technologies

2.1 Classical 3C studies

3C assays are based on fragmenting DNA and then using DNA ligases to join sequences in close physical proximity (Fig. 1). This reorders the linear DNA sequence such that it reflects its three-dimensional structure, which can be read out using a variety of methodologies. The proximity ligation method was developed in the late 1980s and used to detect DNA–DNA interactions in plasmids (Mukherjee et al., 1988). It was subsequently employed to demonstrate gene regulatory interactions at the rat prolactin gene using viral minichromosomes (Cullen et al., 1993).

In 2002 Dekker et al., published a seminal paper describing the use of 3C to detect the physical structure of eukaryotic chromosomes in yeast (Dekker et al., 2002). Studies in mammalian cells quickly followed (Tolhuis et al., 2002), providing the first evidence of enhancer–promoter contact at the

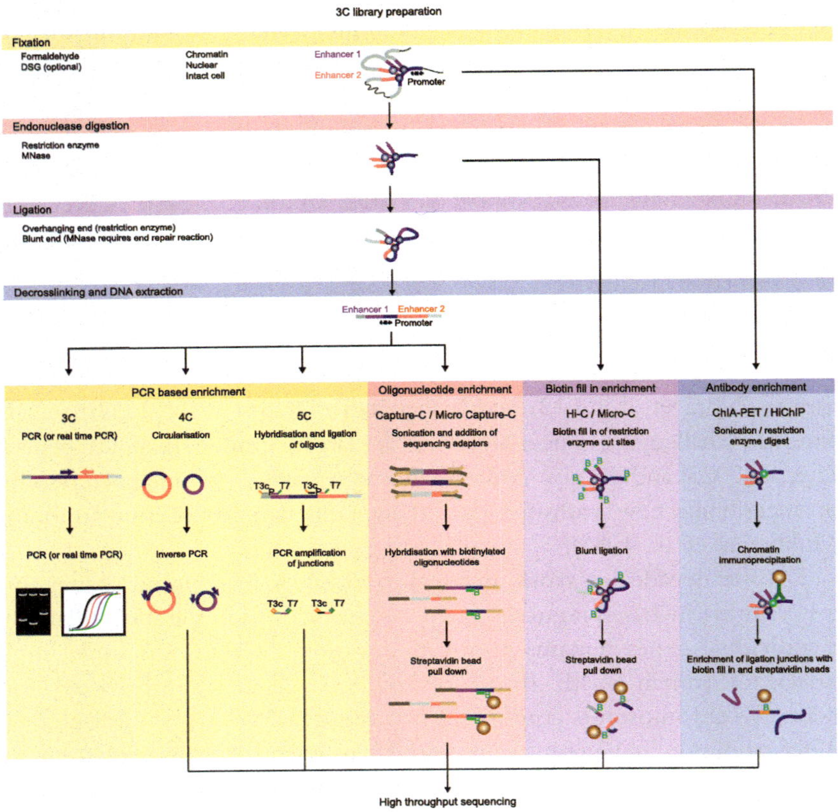

Fig. 1 A comparison of 3C methodologies. 3C assays are based on the proximity ligation principle in which formaldehyde fixed chromatin is digested using an endonuclease and ligated with DNA ligase to create DNA concatemers such that the order of fragments reflects the 3D structure of chromatin at time of fixation. The resulting DNA contains an extremely complex mixture of ligation junctions which can be used to reconstruct the proximity of sequences at the time of fixation. Simple sequencing of the resulting DNA is an inefficient method of determining structure, particularly for large genomes. Multiple different methods have been developed to enrich ligation junction and reconstruct specific regions.

murine beta-globin locus. This study used the six-cutter restriction enzyme *Hind*III to digest chromatin, which results in fragments with a mean size of 4096 bp. Initially, ligation junctions were detected in 3C libraries using polymerase chain reaction (PCR) followed by gel electrophoresis. Ligation junctions between the enhancers at the beta-globin locus and the restriction fragment containing the promoter were detected, providing evidence that enhancers physically interact with the promoter to induce transcription.

Subsequently, 3C was combined with quantitative PCR, which improved the sensitivity and quantification of the assay and was used to analyze many other loci, including the alpha-globin locus (Vernimmen et al., 2009, 2007; Zhou et al., 2006), the T-helper type 2 cytokine locus (Spilianakis & Flavell, 2004) and the immunoglobulin light chain locus (Liu & Garrard, 2005).

These early 3C methods were technically challenging because, even with 6-cutter restriction enzymes, ligation junctions are rare, stochastic events. Even for strong enhancer–promoter interactions, physical proximity only occurs in a subset of cells at any given time and within any individual cell, it is likely that every restriction fragment can ligate to multiple different partners. Consequently, any specific ligation junction represents less than 1 % of all ligation products, the majority of which are non-informative or local interactions. Therefore, detecting true ligations is difficult and often approaches the sensitivity limits of the assay. The other major limitation was that the low throughput meant that it was difficult to detect interactions with regulatory elements that had not been defined by other techniques.

2.2 Single viewpoint methods (4C)

The next major advance in the field was the development of 4C (Zhao et al., 2006; Simonis et al., 2006). This uses 3C library preparation, often with a four-cutter restriction enzyme, followed by the generation of circular DNA fragments. Inverse PCR from a 'viewpoint' is then used to amplify any interacting partners (Fig. 1). Initially, partners that interacted with a viewpoint were identified using microarray, but this was superseded by next-generation sequencing. Since all sequences interacting with a given viewpoint can be captured simultaneously, 4C circumvents the problem of rare individual ligation junctions by aggregating signal across many different ligation products. This increases the sensitivity of the assay and enables higher resolution analysis of interactions. Moreover, by capturing interactions between a single viewpoint and the rest of the genome in parallel, 4C substantially increases throughput relative to one-to-one 3C assays. Several optimized variants of 4C have been developed, differing in how circular DNA fragments are generated. Some approaches use ligation conditions to directly produce small circularized fragments that are suitable for amplification (Liu & Garrard, 2005; Zhao et al., 2006) whereas other approaches employ a second restriction enzyme digest and ligation with a different restriction enzyme (Simonis et al., 2006). Following this, inverse PCR is used to amplify fragments in the library interacting with a particular viewpoint sequence.

For many years, 4C had the highest sensitivity and resolution of all the 3C library based techniques and many important discoveries were made using 4C. It has been widely used to describe enhancer–promoter interactions and show how these change during differentiation and development (Andrey et al., 2013). Very long-range interactions between active genes have been demonstrated in *cis* using 4C (Andrey et al., 2013; de Wit et al., 2013). 4C has also been used to link potential regulatory SNPs with genes (Pasquali et al., 2014). It can be used to determine structural changes such as chromosomal rearrangements (Simonis et al., 2009), although this is probably more easily done using conventional karyotype analysis and FISH. It has also been used to demonstrate novel disease mechanisms; for example, 4C was used to show that chromosomal translocations can result in a distal enhancer becoming juxtaposed to an oncogene, leading to malignancy (Groschel et al., 2014).

However, the method has several limitations. First, with the standard protocol, it is impossible to distinguish PCR duplicates from unique ligation events because identical reads may arise either from repeated amplification of a single ligation product or from multiple cells generating the same interaction. Secondly, the 4C interaction profile depends on how well the interacting fragment can be PCR-amplified. Fragments that are shorter or have favorable sequence composition are preferentially amplified, whereas larger fragments or those with high GC content are underrepresented. As a result, read counts do not directly reflect interaction frequency, making quantitative interpretation and assessment of data quality difficult. Despite computational attempts to correct for these biases, 4C data remain prone to technical artefact.

2.3 Many vs many method 5C

Although the use of two-step PCR allows multiplexing of multiple 4C libraries in a single sequencing run, generally inverse PCR is performed with one viewpoint at a time, which limits the throughput of 4C experiments. To address this, Chromosome Conformation Capture Carbon Copy (5C) was developed as a massively multiplexed, PCR-based readout of 3C ligation junctions within selected genomic regions (Dostie et al., 2006). 5C uses a standard 3C library plus pooled forward and reverse primers tiled at restriction fragment ends. Primer pairs that flank a ligation junction are ligated and PCR-amplified, then sequenced to produce dense, locus-scale contact maps. This technique was used to determine interaction profiles at the pilot regions of the ENCODE project including the globin loci (Sanyal et al., 2012; Bau et al., 2011). In addition, analysis of the X-chromosome by 5C provided some of the first evidence for the existence of topologically associating domains (TADs) (Nora et al., 2012). Although massively multiplexed 5C can be performed, it requires significant resources for somewhat limited resolution and scale, for example, the ENCODE project had to use 5826 oligos to cover 1% of the genome. As a result, 5C has been largely superseded by hybridization capture–based strategies and higher-resolution genome-wide methods, which offer a broader scope with comparable or improved sensitivity.

2.4 All vs all methods Hi-C

"All-vs-all" chromatin conformation assays aim to measure interaction frequencies between every genomic locus and every other locus genome-wide, producing a comprehensive view of 3D genome organization. These data are usually presented as a contact matrix, in which each pixel represents the frequency of observing proximity between two genomic bins.

The most widely used all-vs-all 3C method is Hi-C (Lieberman-Aiden et al., 2009), which uses a modified method for 3C library preparation. Here a biotinylated nucleotide fill in is performed after the restriction enzyme digestion and following this blunt end ligation is then performed. After DNA extraction and sonication of the material, a streptavidin bead pull-down is performed to concentrate ligation junctions, which can then be analyzed by high-throughput sequencing (Fig. 1).

Since it was initially reported (Lieberman-Aiden et al., 2009), Hi-C experiments have produced data at improved resolution by using increased sequencing depth. Initially, 40 kb bins were used (Lieberman-Aiden et al., 2009), but this was subsequently improved to 5–10 kb (Jin et al., 2013) and even down to a 1 kb bin size (Rao et al., 2014).

The difficulty in generating high-resolution, genome-wide 3C data, compared with assays such as ChIP-seq or ATAC-seq, which produce a single value per genomic position, is that 3C requires a genome-wide interaction profile to be generated for each position. This means that changing the bin size to increase resolution requires an approximately quadratic (n^2) increase in data. As a result, producing "all-vs-all" data at 1 kb resolution requires a gargantuan effort; with the analysis of several billion read pairs and to increase this to the levels attainable with targeted approaches would require trillions of reads. To circumvent these inherent limitations of both the single viewpoint and "all vs all" approaches, a number of approaches have been developed to generate targeted enrichment of data.

2.5 Oligonucleotide capture based approaches

These approaches have been used extensively for 3C analysis and they allow thousands of targets to be enriched simultaneously. This can either be used to target specific sites in the genome, commonly gene promoters or *cis*-regulatory elements or they can be tiled sequentially across regions of interest.

Two main methods have been developed. Our group has pioneered the development of the Capture-C approach (Davies et al., 2016; Hughes et al., 2014), in which a 3C library is made and the material is directly captured with targeted oligonucleotide capture. Alternatively, the promoter capture Hi-C method (Mifsud et al., 2015; Javierre et al., 2016) includes a Hi-C like library preparation (biotin fill in and blunt end ligation in the 3C library preparation, followed by a biotin pull down) followed by targeted oligonucleotide capture. Although conceptually the pcHi-C approach might provide better enrichment of ligation junctions, this is offset by the additional complexity of the protocol, the potential inefficiencies from the blunt end ligation and the loss of library complexity caused by the two different enrichment steps. The highly efficient digestion and overhanging end ligation, combined with the targeted oligonucleotide enrichment at the restriction fragment cut site in the Capture-C approach, generate libraries with a similar proportion of ligation junctions per sequenced read but greater library complexity and depth at individual viewpoints (Davies et al., 2017).

These approaches are highly flexible in terms of design, allowing interrogation of single to up to tens of thousands of targets. However, increases in the number of capture oligonucleotides result in lower quality data from each individual viewpoint. This is in part because increases in the number of oligonucleotides cause a drop in concentration of each oligonucleotide in the hybridization step. The sensitivity of these assays starts to fall when more than ~100 oligonucleotides are included in the capture reactions.

For approaches where oligonucleotides are tiled across entire loci to determine the whole structure of the locus rather than simply the contact frequency from an individual restriction fragment, this is offset by the ability to merge data from adjacent restriction fragments. This counteracts the losses in data from the use of large numbers of oligonucleotides and potentially allows high quality data to be generated from small numbers of cells. However, this can be problematic for interpreting genome scale datasets, which can fail to detect regulatory interactions (Schoenfelder et al., 2015).

2.6 Chromatin immunoprecipitation based methods (ChIA-PET, HiChIP Plac-seq)

In these techniques chromatin immunoprecipitation (ChIP) is used to enrich the contact frequency data from sites of interest. Several different variants have been developed, including ChIA-PET (Fullwood et al., 2009; Kieffer-Kwon et al., 2013), HiChIP (Mumbach et al., 2016) and PLAC-seq (Fang et al., 2016). ChIA-PET studies were initially performed using ChIP for the oestrogen-receptor-alpha (Fullwood et al., 2009) and PolII (Kieffer-Kwon et al., 2013).

Although these techniques have become popular because they allow enrichment of protein associated chromatin interactions without predefining regions of interest, they have major limitations. First, compared to other approaches, chromatin immunoprecipitation is an inherently inefficient method and typically recovers significantly less than 1 % of the starting material, depending on the targeted protein. This markedly reduces the number of unique ligation products that can be obtained from any individual site, meaning that additional sequencing depth often results in PCR duplicates rather than informative contacts. The second problem, inherent in all of these methods, is that contacts are only determined from sites bound by the protein of interest. As a result, the interaction profiles are inherently dependent on protein binding, making it difficult to compare the contacts between conditions in which the distribution of the immunoprecipitated protein of interest changes. Furthermore, it is likely that ChIP enrichment introduces bias towards detecting interactions from sites that are strongly bound by the protein of interest. This leads to preferential detection of interactions between regions bound by the targeted protein, rather than regions that lack such binding. This is particularly problematic because many commonly used targets for these types of experiments (such as CTCF, RNA PolII, H3K27ac) frequently interact with one another in vivo, making it challenging to disentangle true chromatin contact frequencies from those disproportionately enriched due to ChIP-related bias. In contrast, targeted enrichment for specific genomic sequences has significant advantages because this generates higher-quality data without the risk of systematic biases.

2.7 Single cell 3C approaches

While these studies have discovered general principles of chromosome folding, they provide little to no information about cell-to-cell variability. This is mainly because chromatin architecture has proven more difficult to adapt to single-cell analysis than RNA-seq or ATAC-seq. RNA-seq was the first of these techniques to be successfully scaled to single cells in part because each cell contains hundreds of thousands of RNA molecules, providing multiple independent observations per expressed gene for analysis. ATAC-seq presents a greater challenge, as each of the tens of thousands of accessible chromatin regions yields only a limited number of potential transposase insertion events per cell. However, both approaches measure one-dimensional features at individual genomic loci, where each measurement reflects the state of a single region independently of others.

In contrast, chromatin conformation is intrinsically more complex. Structural information exists throughout the entire genome, rather than being confined to a defined subset of loci. This poses a fundamental limitation for single-cell applications. Unlike RNA-seq or ATAC-seq, where measurements scale linearly with the number of genomic loci, 3 C methods must interrogate pairwise interactions, resulting in a quadratically scaling search space. At the same time, each restriction fragment can participate in only a single ligation event per allele in any given cell, meaning that only one interaction is observed per locus. Consequently, each cell provides only a sparse and highly incomplete sampling of the vast space of possible chromatin interactions. As a result, even with complete sequencing of an individual cell's genome, only a sparse, incomplete representation of chromatin contacts can be recovered (Hughes & Davies, 2024).

Despite these difficulties, various single-cell Hi-C technologies, such as single-cell Hi-C and high-throughput implementations sci-Hi-C, have been developed to uncover the heterogeneity of genome architectures in the context of cell cycle, early development, and complex tissues. While these approaches substantially increase the number of cells that can be profiled, they do not overcome the fundamental sparsity of interaction data within individual cells.

A further challenge arises from the diploid nature of the genome, which complicates 3D reconstruction of autosomes because contacts from maternal and paternal alleles are typically indistinguishable. Early studies addressed this by focusing on haploid systems, such as the male X chromosome (Nagano et al., 2013) or haploid mESC cells (Stevens et al., 2017). Building on this, Dip-C enabled 3D genome structures to be resolved in an allele-specific manner from single human cells by combining a transposon-based whole-genome amplification method and a computational method to impute haplotypes from sparse data (Tan et al., 2018). More recently, single-cell Micro-C variants (scMicro-C) have improved the resolution of detected contacts, but the data remains sparse at a single-cell level (Wu et al., 2025).

In parallel, multi-omic extensions have begun to link 3D contacts to additional molecular layers. For example, sn-m3C-seq/scMethyl-HiC can simultaneously capture the chromosome conformation and DNA methylome in a cell (Li et al., 2019), while newer approaches have been developed that allow simultaneous RNA-seq (Liu et al., 2023; Zhou et al., 2024), although their resolution is typically limited to 50–100 kb (Zhou et al., 2024). Given the sparsity and noise of single-cell Hi-C datasets, specialized computational methods have been developed to address these challenges. These algorithms aim to denoise data and reconstruct 3D structure from sparse single-cell maps, and include techniques such as preprocessing/normalization, missing-data imputation, latent embeddings, clustering, and pseudo time inference (Galitsyna & Gelfand, 2021; Gong et al., 2023).

3. Recent developments – sub restriction fragment resolution methodologies

One limitation of many 3C-derived assays is that practical resolution is often inadequate below ~500 bp, because restriction enzyme fragment sizes, library complexity and read depth all constrain how precisely contacts can be localized. This becomes especially consequential when the biological questions shift from domain and loop level organization to the internal architecture of *cis*-regulatory elements, where regulatory information is encoded at the scale of nucleosomes and individual transcription factor binding sites (typically ~6–20 bp) (Lieberman-Aiden et al., 2009; Rao et al., 2014; Furey, 2012).

A major step toward this scale came from the development of assays that replace restriction enzyme digestion with MNase to fragment chromatin to mono- or di-nucleosomes, enabling contact maps with nucleosome-scale granularity and improved dynamic range for short-range interactions compared with conventional Hi-C (Rao et al., 2014; Hsieh et al., 2015, 2020; Krietenstein & Rando, 2020). However, "all-vs-all" mapping across mammalian genomes remains sequencing intensive: as bin size decreases, the number of possible locus pairs grows rapidly, and achieving sufficient counts per bin for the potential improvement in resolution over restriction enzymes demands billions of unique contacts to be determined for each sample. This has driven a complementary class of approaches that retain high-resolution 3C library construction but concentrate sequencing on subsets of loci using hybridization capture (Davies et al., 2016; Mifsud et al., 2015; Oudelaar et al., 2020).

In this context, Region Capture Micro-C (RCMC) combines MNase based 3C with densely tiled capture across regions of interest, yielding deep, nucleosome resolution contact maps within targeted regions (Goel et al., 2023). However, like most conventional 3C approaches (Rao et al., 2014; Hsieh et al., 2020), RCMC typically sequences libraries using 50 bp paired-end reads to reduce sequencing costs and increase depth. In addition, Micro-C and RCMC sequence the dinucleosomal band (~ 400 bp in size), which means that the precise ligation junction position is inferred rather than directly observed. This inference limits resolution to around 200 bp and precludes mapping subnucleosome scale contacts such as those within *cis*-regulatory elements.

Micro Capture-C (MCC) overcomes this limitation and can visualize reproducible contacts of specified genomic regions at base-pair resolution (Hua et al., 2021). By combining MNase-based fragmentation with careful preservation of cellular integrity and full sequencing of ligation junctions, MCC localizes contacts with sub-nucleosomal precision. With base-pair mapping of ligation junctions, MCC can reveal footprint-like signatures of DNA-binding proteins, similar to DNase I hypersensitivity-based footprinting. Building on this, Micro Capture-C ultra (MCCu) extends targeted mapping to generate region focused, Hi-C like maps at single base-pair resolution (Li et al., 2025). By producing multidimensional 3C data at this scale, MCCu enables chromatin architecture to be resolved with unprecedented detail. In particular, it can visualize nanoscale domains of nucleosomes and complex structural features within nucleosome-depleted regions at active regulatory elements, down to the level of individual transcription factor binding sites. Crucially, this degree of resolution allows direct integration with molecular dynamics simulations and microscopy data, providing a mechanistic framework to address the long-standing question of what underlies enhancer–promoter interactions.

4. Technical considerations

Interpretation of 3C data relies on the assumption that the number of ligation products detected is proportional to the underlying contact probabilities between genomic loci. This is dependent on a number of factors including potential biases caused by different methods of fixation and chromatin preparation.

4.1 Crosslinking

Traditionally, cells are initially fixed using formaldehyde crosslinking. It is possible to manufacture 3C libraries without this fixation step by embedding the cells in agarose plugs to preserve nuclear shape and structure. This results in similar Hi-C profiles to those generated from formaldehyde fixed material, although there is greater background noise (Rao et al., 2014).

Early 3C studies tended to use formaldehyde for crosslinking, but more recently there has been increasing use of additional crosslinking agents. This reduces noise, and for experiments with supra-kilobase resolution, is unlikely to result in alterations in resolution. For ultra-high resolution MNase-based 3C experiments, the protocols are currently discordant on this point. Micro-C based protocols use double crosslinking with formaldehyde and DSG (disuccinimidyl glutarate). In other MNase-based implementations, such as MCC and MCCu, formaldehyde-only crosslinking is retained to minimize the risk that longer crosslinking distances introduced by additional reagents could potentially blur localized contact signals.

4.2 Chromatin preparation

In classical 3C studies, researchers used preparations of chromatin to undertake experiments. This has been largely superseded because it was realized that the restriction digest and ligation could both be performed on intact nuclei, so called in situ protocols, which result in better signal through reduction of trans contacts. However, some recent protocols have returned to using extraction of chromatin and stabilization on concavalin beads.

In techniques using restriction enzymes to cut chromatin, the restriction enzyme digest is the step that often requires optimization to ensure that sufficient levels of digestion are achieved. This generally requires treatment of the nucleus with detergents to provide access for restriction enzymes to cut their recognition sites. This can be circumvented with MNase based protocols, where the digest can be performed on intact cells, which have been permeabilized with digitonin. This is possible because MNase is a smaller protein than restriction enzymes (MNase 16 kDa vs *Dpn*II 34 kDa / *Hin*dIII 35 kDa), and it does not need to access specific nucleosome or protein-bound sequences for the assay to work.

4.3 Endonuclease

The choice of endonuclease determines the maximum resolution of the data because interactions can only be detected between fragments defined by nuclease cut sites. Consequently, resolution is limited by the size and distribution of these fragments. 3C assays have largely been performed with restriction enzymes, which cleave specific DNA sequences, allowing controlled digestion of chromatin over prolonged periods of incubation without the risk of over digestion. In addition, they often generate compatible overhanging ends that facilitate efficient ligation of interacting fragments.

In contrast, non-specific nucleases such as DNaseI and MNase cleave DNA in a less predictable way and require careful titration to obtain adequate digestion without losing the majority of DNA from the sample. In addition, these enzymes require end repair prior to the ligation reaction (DNase I produces a mixture of 3' and 5' overhangs and MNase typically leaves a 3'-phosphate with a 5'-OH overhang (Cuatrecasas et al., 1967)), which can reduce ligation efficiency.

The classical 3C studies used six-cutter restriction enzymes, which produce fragments with a mean size of 4096 bp. 4C and Capture-C generally tend to use four-cutter restriction enzymes, which generate data that has 16 times the resolution (in a random DNA sequence, a four-cutter will cut with a mean fragment size of 256 bp). Both four (Rao et al., 2014; Sexton et al., 2012) and six (Lieberman-Aiden et al., 2009) cutter enzymes have been used for generating Hi-C libraries, although here the resolution is often limited by the sequencing depth rather than the choice of restriction enzyme. Importantly, restriction enzyme digestion produces fragments that follow an exponential size distribution, with many short fragments and a long tail of large fragments. This leads to uneven genomic representation, with some regions being covered by many small fragments and others by a few larger ones, resulting in substantial variation in the number of detectable ligation events and the effective resolution across the genome. This makes analysis of these data challenging because the number of observations within any region of the genome is highly variable.

To address these limitations, alternative fragmentation strategies have been explored. For example, DNase I-based approaches were developed to achieve more uniform fragmentation; however, these methods can introduce bias towards accessible chromatin regions and require careful optimization to ensure complete digestion without excessive DNA loss.

More commonly, micrococcal nuclease (MNase) has been used to fragment chromatin in a manner that is less dependent on DNA sequence and instead reflects nucleosome organization. MNase preferentially cleaves linker DNA between nucleosomes, generating relatively uniform fragments at approximately nucleosome resolution. This enables substantially higher-resolution mapping of chromatin interactions and reduces the variability in fragment sizes associated with restriction enzyme digestion.

However, MNase-based approaches also introduce their own biases, as digestion efficiency depends on chromatin accessibility and nucleosome positioning. In addition, careful titration is required to avoid over-digestion, and the resulting DNA ends still require processing prior to ligation. Despite these challenges, MNase-based methods such as Micro-C and MCC have become widely used due to their ability to provide high-resolution and more uniformly distributed contact maps.

4.4 Sequencing

To maximize sequencing depth for the same expenditure, most next generation sequencing based 3 C assays have used short sequencing reads (50 bp paired end). Whilst this is efficient, it means that the ligation junction between reads is generally not sequenced for most reported reads and interactions are inferred from the mapping position of the read pairs.

An alternative and potentially more robust approach involves sonicating libraries to ~200 bp and sequencing with 300 bp sequencing reads (150 bp paired end). This allows the entire read to be reconstructed using the region of central overlap and allows the ligation junction site to be sequenced directly. Whilst this provides additional confidence and modestly improves data quality, it probably does not make a marked difference to data quality for restriction enzyme based approaches where ligation junctions occur at predefined cut sites. However, it is particularly beneficial for MNase-based methods because cut sites are not constrained by specific sequences. Consequently, combining these approaches enables precise identification of the ligation junctions, allowing interaction resolution to be defined at base pair precision. In contrast, when ligation junctions are inferred from short read sequencing, resolution is constrained by the fragment size, which is typically around 200 bp.

5. Genome architecture as described by 3C techniques

5.1 Chromosome territories

Chromosome territories refer to the non-random spatial segregation of individual chromosomes into distinct volumes within the interphase nucleus (Fig. 2). Originally established by microscopy-based studies, this concept was later synthesized into a broader framework linking nuclear organization to gene regulation (Cremer & Cremer, 2001, 2010). In "all-vs-all" proximity ligation maps, such as Hi-C, chromosome territories are readily detectable because *cis* contacts, which happen within the same chromosome, are strongly enriched over trans contacts, which happen between chromosomes. This yields a characteristic block structure (Fig. 2) (Lieberman-Aiden et al., 2009). Because each chromosome exists as a maternal and paternal homolog, its territories can be allele-resolved. Haplotype phased approaches that assign Hi-C contacts to parental homologs show that many long-range interactions occur in *cis* on a single homolog, reflecting allele-specific regulatory landscapes, especially at loci with strong *cis*-regulatory or parent-of-origin constraints (Selvaraj et al., 2013; Leung et al., 2015).

5.2 Laminar associated domains

Within chromosome territories, genome organization is further shaped by interactions with nuclear substructures. One such example is lamina-associated domains (LADs), which are large genomic regions that preferentially contact the nuclear lamina at the nuclear periphery. Genome-wide mapping originally defined LADs as hundreds to thousands of sharply bounded domains typically spanning ~0.1–10 Mb. These regions are typically gene poor and transcriptionally inactive, consistent with a repressive nuclear subcompartment (Guelen et al., 2008).

Functionally, LADs are often linked to heterochromatin associated features, such as late replication and repressive chromatin signatures. They are thought to contribute to stabilizing a peripheral inactive environment for subsets of the genome. Importantly, lamina association is not uniform across cells. Single-cell mapping revealed a core set of LADs with high cell-to-cell consistency, interspersed with more variable LADs, implying that peripheral genome organization contains both stable scaffolding and dynamic, probabilistic components (van Steensel & Belmont, 2017; Poleshko et al., 2017). LAD organization is also developmentally plastic. During lineage commitment, large sets of genomic

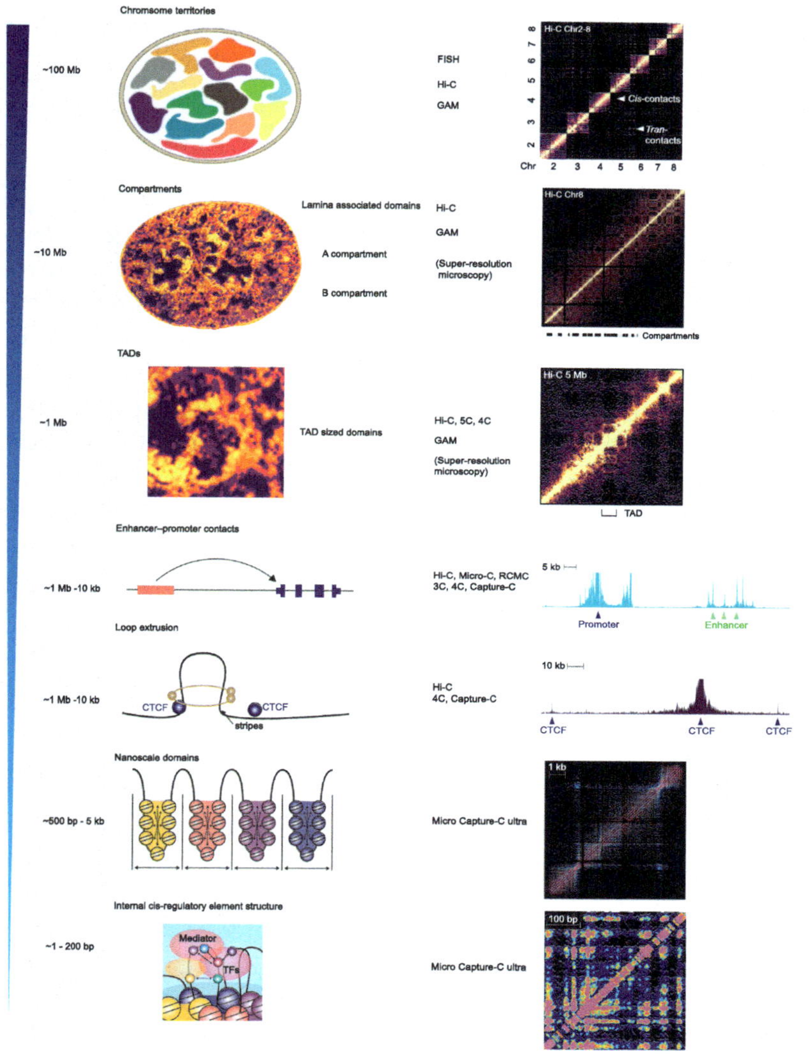

Fig. 2 Overview of different scales of genome organization. Different scales of genome organization and methods by which these structures can be detected. At a large scale, chromosome territories can be detected with microscopy and genome architecture mapping (GAM) (Beagrie et al., 2017) as well as with Hi-C. Different genome compartments, such as A/B compartments and lamina-associated domains, can be identified with 3C approaches and correlated with chromatin compaction as defined by microscopy. On a smaller scale, TADs can be defined in 3C datasets, which probably correlate with the sponge-like chromatin structures visualized by super-resolution microscopy. Multiple 3C techniques have been used to interrogate enhancer–promoter contacts and loop extrusion. More recently, Micro Capture-C has been used to define smaller-scale structures, including nanoscale domains and fine-scale structures within enhancers and promoters.

regions can gain or lose lamina association, coinciding with coordinated changes in gene activity and chromatin state (Takayama et al., 2021; Chen et al., 2019).

5.3 A and B compartments

A/B compartments were first inferred from early Hi-C maps as a plaid-like pattern of long-range contact preferences, reflecting genome-wide segregation of two interaction environments (Fig. 2) (Lieberman-Aiden et al., 2009). "A" compartments broadly correspond to open, gene-rich, transcriptionally active chromatin, whereas "B" compartments correlate with more compact, gene-poor, transcriptionally inactive chromatin. With increased depth, compartments resolve into subcompartments with different chromatin states and preferred partners, yet the A/B partition remains a dominant organizing principle (Rao et al., 2014). Compartmentalization is dynamic across cell states. During stem cell differentiation, a substantial fraction of the genome switches between A and B, and these switches track coordinated changes in gene expression and chromatin state. When combined with haplotype-resolved analyses, these maps also reveal widespread allelic biases in expression and chromatin that can be coupled to allele-specific 3D organization (Dixon et al., 2015; Greenwald et al., 2019).

5.4 Topologically associating domains (TADs) and sub-TADs

In 2012, studies reported hierarchical genome organization (Sexton et al., 2012) into 'topological domains' (Dixon et al., 2012) or 'topologically associating domains' (Nora et al., 2012) using orthogonal 3C methodologies. These studies showed that the genome segregates into genomic loci that exhibit high frequencies of internal interactions. Using Hi-C, Dixon et al. showed that the position of domain boundaries is highly conserved across different cell types and between species. Furthermore, Nora et al. showed that these boundaries are essential spacers and that disruption of boundaries leads to ectopic contact between isolated elements.

It was quickly realized that the boundaries between TADs were often enriched for CTCF bound elements. However, this is not universally the case, as many TADs are formed by other mechanisms.

Imaging studies can now partially ascribe a physical correlate to TADs. Super-resolution microscopy has shown that the nucleus forms an irregularly shaped, spongelike chromatin compartment and a complex interchromatin network of channels that transport RNA and macromolecules to the nuclear pores. Chromatin forms domains around 200–300 nm in diameter, which correspond to topologically associated domain (TAD)-sized segments (Tolhuis et al., 2002; Vernimmen et al., 2009, 2007). However, it is unlikely that chromatin domains completely correlate with TAD because TADs are formed by different mechanisms at different loci.

At the time of the discovery of TADs, techniques such as 4C had shown that, at active genes, contact between regulatory elements occurred on a much smaller scale. Subsequent work demonstrated that TADs contained smaller sub-TADs, which were detectable with techniques such as 5C (Phillips-Cremins et al., 2013). Thus, ascribing a function or physical correlate to TADs has been challenging. Overall, these observations suggest a hierarchical approach to genome organization; TADs provide a broad structural framework, whereas more localized dynamic sub-TAD interactions regulate gene expression. Thus, ascribing a physical or functional correlate to TADs has been challenging.

5.5 Enhancer–Promoter contacts

It has long been proposed that distal enhancers require spatial proximity with the promoter they control to allow the proteins at the *cis*-regulatory elements to interact with each other to control transcription. From the first discovery of enhancers, it was shown that enhancers from one gene could activate different core promoters, providing that the enhancer was placed in the same plasmid as the promoter (Banerji et al., 1981, 1983). Further evidence that physical proximity was important for regulating gene expression came from the discovery of insulator sequences, which disrupt enhancer–promoter communication (Chung et al., 1993).

Subsequently, seminal 3C studies provided direct evidence for physical contact between enhancers and promoters at the beta-globin locus (Palstra et al., 2003). This led to the idea of enhancer–promoter loop formation, and researchers identified a number of proteins that potentially mediate or stabilize these interactions, including the cohesin complex (Kagey et al., 2010), mediator complex components and transcriptional co-factors such as LDB1 (Deng et al., 2012).

As the resolution of 3C data has improved, it has become evident that the contacts between regulatory elements are highly localized. Recent micrococcal nuclease-based techniques show that the physical contacts coincide precisely with the nucleosome-depleted regions (Li et al., 2025; Goel et al., 2023; Hua et al., 2021). Perturbation studies have demonstrated that deletion of individual transcription factor binding sites can specifically reduce enhancer–promoter interactions at the motif, suggesting a direct role for transcription factors in mediating enhancer–promoter contacts (Hua et al., 2021). However, rapid depletion of the pioneer transcription factor SOX2 using a degron system, showed that more significant changes in long-range contacts occur when there are concomitant changes in chromatin accessibility, as measured by ATAC-seq (Li et al., 2025). Through integration of high-resolution 3C with molecular dynamics simulations it has been proposed that nucleosome depletion itself is a key determinant of enhancer–promoter contacts (Li et al., 2025). These findings suggest that enhancer–promoter contacts emerge from a combination of sequence-specific transcription factor binding and alteration of the biophysical properties of chromatin caused by removal of nucleosomes.

5.6 CTCF and cohesin-mediated loop extrusion

The zinc finger protein CTCF has long been recognized as a chromatin insulator (Chung et al., 1993). It is known to be enriched at TAD boundaries (Dixon et al., 2012; Rao et al., 2014) and to demarcate regulatory neighborhoods (Nora et al., 2017).

The central mechanism by which CTCF orchestrates genome architecture is thought to be cohesin-mediated loop extrusion (Fig. 2) (Rao et al., 2017; Fudenberg et al., 2016; Sanborn et al., 2015). In this model, cohesin loads onto chromatin and progressively enlarges a loop until it is halted by boundary elements, most prominently CTCF bound in a convergent motif orientation (Rao et al., 2014; Vietri Rudan et al., 2015). This orientation dependence creates a predictable asymmetry in insulation and loop formation across genomes and across evolution, consistent with directionally constrained interactions between CTCF sites. Perturbation experiments that invert or reconfigure CTCF sites further support the idea that motif orientation helps determine which loops can be stabilized (Guo et al., 2015).

Perturbing cohesin and its regulators further reinforces the kinetic logic of loop extrusion. Acute depletion of cohesin collapses most CTCF-anchored loops, whereas loss of WAPL, a key cohesin release factor, increases cohesin residence time on chromatin and shifts genome folding toward longer and more extensive loops (Rao et al., 2017; Haarhuis et al., 2017).

More recently, it has been proposed that loading of cohesin by NIPBL occurs at active regulatory elements (Hua et al., 2021). This potentially provides a mechanism by which chromatin structure is controlled by gene expression, with cohesin-mediated loop extrusion leading to insulation of active genes from surrounding genes and simultaneously increasing enhancer–promoter contacts.

Insulation is important because it prevents inappropriate gene activation. Deleting the boundaries of a TAD or a key CTCF binding sites within it, can reduce insulation and lead to fusion of adjacent domains, thereby allowing enhancers to access genes that were previously insulated (Rao et al., 2014; Nora et al., 2017). These findings have direct implications for human disease, because structural variants that delete, duplicate, or rearrange boundary elements can rewire enhancer–promoter communication and cause developmental phenotypes or oncogene activation.

Interestingly, loop extrusion alone does not fully account for fine-scale enhancer–promoter interactions or transcriptional output. Perturbation studies have consistently shown that acute cohesin degradation erases most large-scale CTCF-anchored loops and domains, yet results in relatively subtle changes in transcription and enhancer-promoter contacts (Schwarzer et al., 2017). These observations suggest that while cohesin-mediated loop extrusion establishes a mesoscale framework for genome folding, it is not the main determinant of regulatory interactions.

5.7 Nanoscale domains

Recently, nanoscale domains have been reported, which appear similar to TAD-like structures but on a ~ 100–1000 fold smaller scale, measuring upwards of ~500 bp (Li et al., 2025). These fine-scale structures are formed by clusters of nucleosomes, which are partitioned by *cis*-regulatory elements and CTCF binding sites. Chemically specific, coarse-grained molecular dynamics simulations of chromatin support the hypothesis that the formation of these structures is orchestrated by the physicochemical properties of DNA and nucleosomes. These show that nucleosome removal alters the biophysical properties of chromatin and leads to partitioning of the adjacent nucleosomes from one another. Furthermore, neutralization of the positive charges on the lysine residues of the histone proteins within the nucleosome alters the electrostatic forces and this likely also contributes to nanoscale domain formation.

5.8 Structures within *cis*-regulatory elements

Recent studies have also delineated specific structures within cis-regulatory elements <200 bp, resolving contacts between individual transcription factor motifs within enhancers and promoters (Li et al., 2025; Hua et al., 2021).

These findings suggest that *cis*-regulatory elements form intricate three-dimensional structures. This directly links interactions among proteins to the fine-scale structures within regulatory elements. This potentially allows 3C methods to be used to directly interrogate the effects of sequence variation. At promoters, fine-scale architecture is tightly coupled to the transcriptional machinery. Consistent with this, acute perturbation of key coactivators such as Mediator significantly alters nanoscale contacts around the transcription start site (TSS), in parallel with transcriptional changes.

6. Uses of 3C approaches

6.1 Interpretation of non-coding variation and disease

The primary driver of the high cost of drug discovery and development is failure, with only about 10 % of clinical programs ultimately achieving regulatory approval. This low success rate largely reflects inadequate efficacy, underscoring our limited understanding of disease mechanisms. Human genetics offers a unique opportunity to improve this situation, as hundreds of disease-associated loci now directly implicate genes and pathways involved in pathogenesis (Claussnitzer et al., 2020). Previous studies have shown that drug mechanisms with genetic support have approximately a 2.6-fold higher probability of success compared with those lacking such evidence (Minikel et al., 2024; Nelson et al., 2015).

Genome-wide association studies (GWAS) examine hundreds of thousands of genetic variants across large populations to identify those that are statistically associated with specific traits or diseases (Uffelmann & Posthuma, 2021). This approach has produced a substantial number of robust associations across a wide range of phenotypes, and to further the biological interpretation of these loci, next-generation sequencing methods are employed to capture the full spectrum of genetic variation (Belkadi et al., 2015; Ng et al., 2009; Saunders et al., 2012).

Understanding the full potential of human genetics requires moving beyond association to understanding causation, specifically, how risk variants contribute to disease. Animal models can facilitate this translation, particularly for coding variants in evolutionarily conserved genes (Cader et al., 2016; Murthy et al., 2014). However, approximately 90 % of the variants identified by GWAS fall in non-coding regions, which are often less conserved and functionally more complex (Claussnitzer et al., 2020). For instance, variants in enhancer regions are scattered around the genes they regulate, and can be located more than a million base pairs upstream or downstream from the target genes (Krivega & Dean, 2012; Chin et al., 2024). It has been found that 77 % of blood trait GWAS loci contain at least one fine-mapped variant overlapping an enhancer region, thus, a high-precision enhancer–promoter interaction map is needed for gene prioritization (Morris et al., 2023; Qi et al., 2024).

6.2 Linking GWAS variants to target genes

Mapping spatial contacts between chromatin regions has the potential to identify the target genes controlled by candidate cis-regulatory elements, which are increasingly cataloged across tissues and cell types as a framework for interpreting noncoding variation (Consortium et al., 2022). At present, 3C approaches remain the most direct experimental methods for mapping physical connectivity between *cis*-regulatory elements, providing a practical route from regulatory DNA to candidate target genes.

A classic early demonstration came from the 8q24 "gene desert" region, which harbors multiple independent associations with cancer (Pomerantz et al., 2009). The colorectal cancer risk variant rs6983267 lies within an enhancer that forms long-range 3C-detectable contacts with the *MYC* locus, supporting a model in which a noncoding GWAS signal acts through distal enhancer control of an oncogene rather than the nearest annotated gene. Similarly, the obesity-associated interval within *FTO* was long assumed to implicate *FTO* itself. However, 4C-based chromatin profiling revealed that the associated region physically contacts the *IRX3* promoter across a broad, megabase-scale regulatory landscape, shifting attention toward a distal developmental regulator as a likely effector gene at the locus (Smemo et al., 2014).

Recent advancement of 3C chemistry and capture-based designs have improved both resolution and sensitivity, enabling contact patterns to be resolved at near base-pair granularity within selected regions, approaching the scale of individual transcription factor binding sites (Li et al., 2025; Hua et al., 2021). In parallel, this highlights the potential of a more precise analysis of chromatin contact changes between haplotypes by coupling with allele specific mapping and read assignment.

7. Working toward a unified model

3C approaches have great utility for defining the relative contact frequencies between specific sequences of DNA within a population of cells. However, they do not provide information on the physical location of sequences within the nucleus or the temporal dynamics of changes. Furthermore, since 3C signals are relative to DNA sequence they do not provide real-world measurement of distances because the density of DNA is highly variable within the nucleus. It is critical, therefore, to integrate 3C approaches with other modalities to build models of chromatin architecture (Fig. 3).

7.1 Integration with super-resolution microscopy

Super-resolution microscopy studies have shown that chromatin forms irregularly shaped structures around 200–300 nm in size, which assemble to form a complex sponge-like structure on a nuclear scale (Wu et al., 2025; Li et al., 2019; Liu et al., 2023). Between these chromatin-dense structures resides an interchromatin compartment, which forms a complex network of channels that transport RNA and macromolecules to and from the nuclear pores (Wu et al., 2025; Li et al., 2019; Liu et al., 2023). The interchromatin compartment is highly enriched for RNA but has low DNA signal and conversely, the chromatin compartment is enriched for DNA but only contains low levels of RNA. The chromatin compartment itself contains regions with different levels of chromatin compaction, and it likely forms a highly dynamic continuum between very dense regions of compacted nucleosomes and DNA and regions with little or no DNA. These differing regions have been shown to correlate with different functional properties {Miron, 2020 #52}.

RNA polymerase II localizes to the interchromatin space and the adjacent perichromatin compartment, which contains the lowest density of DNA signal. Furthermore, histone modifications associated with active genes (H3K4me3/H3K36me3) are also heavily enriched in the low-density peri-chromatin compartment. Therefore, transcription likely occurs at the interface between the chromatin and interchromatin compartments. In contrast, repressive histone marks such as those associated with polycomb silenced genes (H3K27me3) and heterochromatin (H3K9me3) are highly enriched in the dense interior chromatin compartment (Li et al., 2019). Interestingly, CTCF and the cohesin complex both localize in the inter and perichromatin compartments, suggesting that loop extrusion may be due to chromatin passing through the complex rather than the complex itself moving through chromatin.

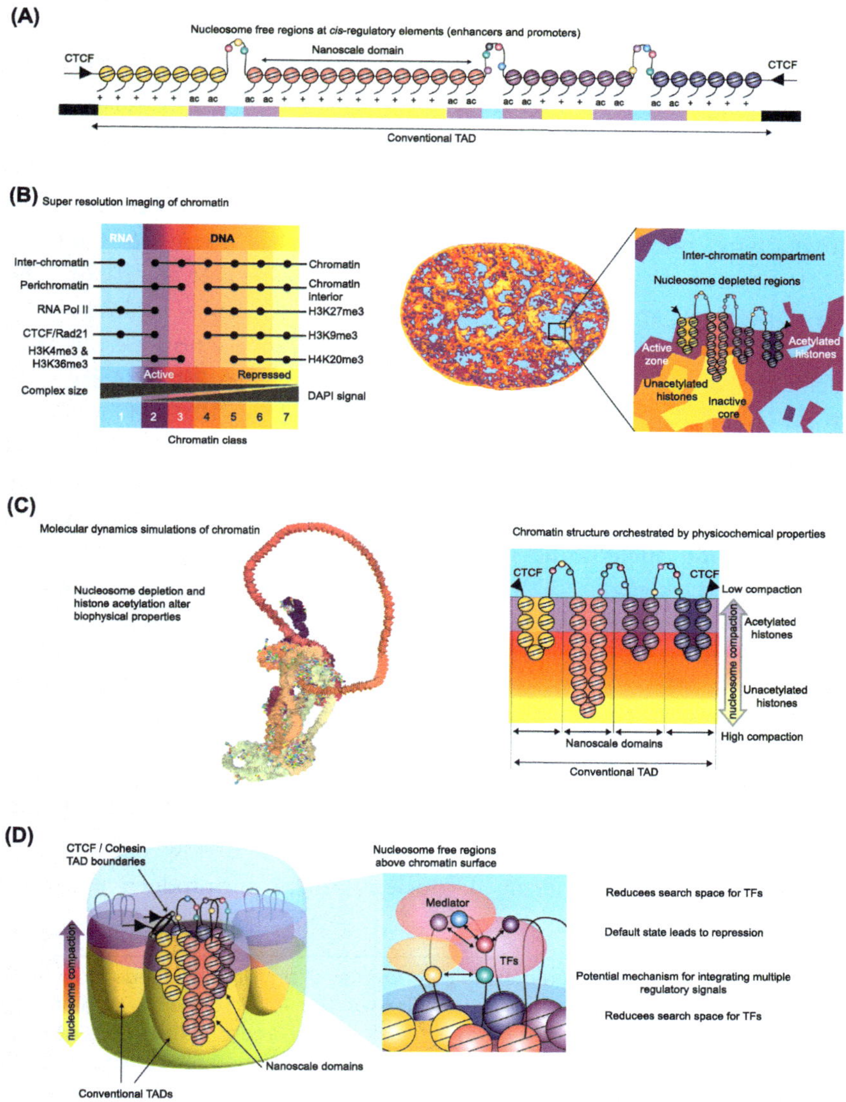

Fig. 3 Integration of 3C data with microscopy and molecular dynamics simulations to propose a unified model of chromatin structure. (A) Schematic of a simplified regulatory region comprising CTCF boundary elements, nucleosome-depleted regions (cis-regulatory elements including enhancers and promoters), and adjacent regions enriched in histone acetylation. (B) Classification of chromatin by super-resolution imaging. The nucleus can be partitioned into seven classes based on DNA signal intensity. This identifies an interchromatin compartment, rich in nascent RNA and polymerase; a perichromatin compartment, which contains active genes and promoters

(Continued)

7.2 Molecular dynamics simulations allow the physicochemical properties of chromatin to be studied

In silico approaches such as molecular dynamics (MD) simulations can complement experimental methods by providing mechanistic, high-resolution interpretations of chromatin behavior. To date, MD has been used extensively to study resolve DNA dynamics such as breathing and unwrapping, nucleosome sliding, and disassembly (Huertas & Cojocaru, 2021; Ettig et al., 2011; Rychkov et al., 2017; Zhang et al., 2016). They have also been applied to the conformational dynamics of histone tails (Erler et al., 2014; Shaytan et al., 2016) and to quantify how perturbations such as histone PTMs (Fenley et al., 2018; Li & Kono, 2018; Rajagopalan et al., 2017), histone variants (Bowerman et al., 2019) and underlying DNA sequence modulate nucleosome stability and dynamics (Sun et al., 2019). Additionally, MD has examined interactions between oligonucleosomes (Collepardo-Guevara et al., 2015) and the role of linker histone H1 in shaping nucleosome–nucleosome contacts and higher-order organization (Ozturk et al., 2020). However, due to computational constraints, current insights remain limited by model approximations and by the finite simulation timescales that can be practically achieved. Multidimensional high-resolution targeted 3C strategies such as MCCu make it possible to bring these perspectives together with molecular dynamics simulations.

7.3 Working towards a unified model of chromatin structure

A key question that 3C approaches aim to address is how enhancers and promoters find one another in 3D nuclear space, form productive contacts,

Fig. 3—Cont'd and a central repressed core. Overlay of the simplified regulatory region onto a nuclear image showing how folding of the linear genome separates the nucleosome-depleted regions from an acetylated layer and a central core of highly compacted nucleosomes. The nucleosome-depleted regions potentially coalesce on the surface of the aggregated nucleosomes. (C) Molecular dynamics simulations of the physicochemical properties of chromatin (including every base pair of DNA and every amino acid in the nucleosome) suggest that nanoscale domains observed in MCCu data can arise from alterations in the biophysical properties of chromatin following nucleosome depletion or histone acetylation. (D) Cohesin-mediated loop extrusion promotes strong interactions between CTCF boundary sites, leading to the formation of topologically associating domains (TADs) and insulation of nucleosome clusters from surrounding regions. Within these CTCF-bounded domains, nucleosome-depleted regions coalesce at the surface of nucleosome-rich compartments. These regions form complex fine-scale structures driven by transcription factor binding.

and thereby regulate gene expression (Yang & Hansen, 2024). Despite substantial progress in both super-resolution imaging and 3C-based methods, the mechanisms underlying enhancer–promoter interactions remain incompletely resolved. Current models are often built on (i) specific interactions mediated by DNA-binding proteins (Meisterernst & Roeder, 1991), (ii) loop extrusion (Fudenberg et al., 2016; Sanborn et al., 2015) and (iii) condensates driven by liquid–liquid phase separation (Du et al., 2024a; Shrinivas et al., 2019). A more comprehensive understanding, therefore, depends on integrating 3C with orthogonal approaches. High-resolution 3C measurements enable integration with chemically specific molecular dynamics simulations of chromatin, providing a route to test mechanistic hypotheses in a quantitatively constrained manner. Observations across complementary approaches are beginning to converge on a more unified model.

Multiple studies indicate that chromatin condensate as dense but dynamic. Specifically, chromatin folds stochastically and exhibits liquid-like behavior (Farr et al., 2021; Maeshima et al., 2016), consistent with phase separation models (Krietenstein & Rando, 2020; Mirny et al., 2019; Sabari et al., 2018; Du et al., 2024b; Lu et al., 2018). However, inside the condensate, the chromatin fiber may be non-uniform, with a central condensed core of nucleosomes forming a solid hydrogel (Strickfaden et al., 2020). Supporting this idea, super-resolution microscopy suggests that the innermost core of chromatin is strongly self-associating and largely repressive, in part because it excludes large macromolecular complexes such as RNA polymerase.

Surrounding the dense core, chromatin transitions into a more active layer that is enriched in acetylated histones. Both polymer modeling and 3C contact data indicate that acetylated nucleosome regions tend to segregate away from the denser, unacetylated core (MacPherson et al., 2018; Gibson et al., 2019; Nuebler et al., 2018). Specifically, charge neutralization reduces the affinity of histones for the negatively charged DNA backbone, thereby weakening internucleosomal interactions and reducing chromatin compaction. This aligns with decades of evidence that acetylation increases DNA accessibility and fosters a chromatin state that is more permissive to transcription and other DNA-based processes (Hong et al., 1993; Kouzarides, 2007). These differences in nucleosomal protein biophysics drive the self-assembly of the nucleosome chain into nanoscale domains, which are ordered by histone acetylation.

This model helps reconcile why experiments that force enhancer–promoter contact show that proximity can promote transcription (Deng et al., 2012;

Morgan et al., 2017). In this framework, localized proximity signals between adjacent nucleosome-depleted regions can arise because diffusion of nucleosome-depleted regions is effectively constrained to two dimensions along the chromatin surface, increasing the probability of physical contact. The chromatin surface may therefore provide a platform for specific interactions among transcription factors and the general transcription machinery that drive gene expression. Constraint by a two-dimensional surface may also potentially provides a mechanism by which multiple signals from different cis-regulatory elements can be summated to determine transcriptional output. This may be analogous to the kinetic co-segregation model of T cell activation in which signals from multiple receptors are summated by diffusion into the immune synapse (Davis & van der Merwe, 2006). Large protein complexes such as Mediator and the preinitiation complex potentially further retain nucleosome-depleted regions at this surface by limiting their diffusion back into the nucleosome-dense compartment (Maeshima et al., 2015).

Finally, this model aligns with and extends the prevalent loop extrusion model of genome organization (Fudenberg et al., 2016; Sanborn et al., 2015). Loop extrusion by cohesin is known to drive the formation of TADs and structural loops demarcated by CTCF sites but does not fully account for fine-scale enhancer–promoter interactions. Data from recent high-resolution 3C study suggested that cohesin-mediated loop extrusion provides a mesoscale framework, while the biophysical properties of chromatin play a more decisive role in enhancer–promoter engagement at the finer scale.

In eukaryotes, cohesin extrudes genomic DNA into loops, with CTCF constraining this process to form TADs (Fudenberg et al., 2016; Sanborn et al., 2015). Many TADs contain multiple nucleosome-depleted enhancers and promoters, typically flanked by regions of histone acetylation (McArthur & Capra, 2021). In this context, loop extrusion would be expected to increase the overall probability that active *cis*-regulatory elements encounter one another within the same TAD, while limiting crosstalk between domains.

Within a TAD, however, local chromatin decompaction and transcription factor clustering may contribute more directly to the structures that enable functional enhancer–promoter communication. This perspective is consistent with observations that cohesin depletion often produces only modest reductions in enhancer–promoter contact frequency and relatively subtle transcriptional changes for many genes (Nora et al., 2017; Rao et al., 2017; Schwarzer et al., 2017; Aljahani et al., 2022; Wutz et al., 2017; Kubo et al., 2021; Hsieh et al., 2022). Interestingly, studies show the cohesin loader NIPBL is enriched at active enhancers and promoters (Hua et al., 2021, 2021; Busslinger et al., 2017), and microscopy positions CTCF and cohesin in the inter- and perichromatin compartments (Miron et al., 2020). Together, these observations further support a model in which cohesin loading and loop extrusion preferentially occur at the surface of the nucleosome compartment, alongside RNA polymerase II and other large regulatory complexes.

References

Aljahani, A., Hua, P., Karpinska, M. A., Quililan, K., Davies, J. O. J., & Oudelaar, A. M. (2022). Analysis of sub-kilobase chromatin topology reveals nano-scale regulatory interactions with variable dependence on cohesin and CTCF. *Nature Communications, 13*, 2139. https://doi.org/10.1038/s41467-022-29696-5.

Andrey, G., Montavon, T., Mascrez, B., Gonzalez, F., Noordermeer, D., Leleu, M., ... Duboule, D. (2013). A switch between topological domains underlies HoxD genes collinearity in mouse limbs. *Science (New York, N. Y.), 340*, 1234167. https://doi.org/10.1126/science.1234167.

Banerji, J., Olson, L., & Schaffner, W. (1983). A lymphocyte-specific cellular enhancer is located downstream of the joining region in immunoglobulin heavy chain genes. *Cell, 33*, 729–740. https://doi.org/10.1016/0092-8674(83)90015-6.

Banerji, J., Rusconi, S., & Schaffner, W. (1981). Expression of a beta-globin gene is enhanced by remote SV40 DNA sequences. *Cell, 27*, 299–308. https://doi.org/10.1016/0092-8674(81)90413-x.

Bau, D., Sanyal, A., Lajoie, B. R., Capriotti, E., Byron, M., Lawrence, J. B., ... Marti-Renom, M. A. (2011). The three-dimensional folding of the alpha-globin gene domain reveals formation of chromatin globules. *Nature Structural & Molecular Biology, 18*, 107–114. https://doi.org/10.1038/nsmb.1936.

Beagrie, R. A., Scialdone, A., Schueler, M., Kraemer, D. C., Chotalia, M., Xie, S. Q., ... Branco, M. R. (2017). Complex multi-enhancer contacts captured by genome architecture mapping. *Nature, 543*, 519–524. https://doi.org/10.1038/nature21411.

Belkadi, A., Bolze, A., Itan, Y., Cobat, A., Vincent, Q. B., Antipenko, A., ... Abel, L. (2015). Whole-genome sequencing is more powerful than whole-exome sequencing for detecting exome variants. *Proceedings of the National Academy of Sciences of the United States of America, 112*, 5473–5478. https://doi.org/10.1073/pnas.1418631112.

Bowerman, S., Hickok, R. J., & Wereszczynski, J. (2019). Unique dynamics in asymmetric macroH2A-H2A hybrid nucleosomes result in increased complex stability. *The Journal of Physical Chemistry. B, 123*, 419–427. https://doi.org/10.1021/acs.jpcb.8b10668.

Busslinger, G. A., Stocsits, R. R., van der Lelij, P., Axelsson, E., Tedeschi, A., Galjart, N., & Peters, J. M. (2017). Cohesin is positioned in mammalian genomes by transcription, CTCF and Wapl. *Nature, 544*, 503–507. https://doi.org/10.1038/nature22063.

Cader, M. Z., Boroviak, K., Zhang, Q., Assadi, G., Kempster, S. L., Sewell, G. W., ... Mukhopadhyay, S. (2016). C13orf31 (FAMIN) is a central regulator of immunometabolic function. *Nature Immunology, 17*, 1046–1056. https://doi.org/10.1038/ni.3532.

Chen, C., Yu, W., Tober, J., Gao, P., He, B., Lee, K., ... Tan, K. (2019). Spatial genome re-organization between fetal and adult hematopoietic stem cells. *Cell Reports, 29*, 4200–4211.e4207. https://doi.org/10.1016/j.celrep.2019.11.065.

Chin, I. M., Gardell, Z. A., & Corces, M. R. (2024). Decoding polygenic diseases: Advances in noncoding variant prioritization and validation. *Trends in Cell Biology, 34*, 465–483. https://doi.org/10.1016/j.tcb.2024.03.005.

Chung, J. H., Whiteley, M., & Felsenfeld, G. (1993). A 5′ element of the chicken beta-globin domain serves as an insulator in human erythroid cells and protects against position effect in Drosophila. *Cell, 74*, 505–514. https://doi.org/10.1016/0092-8674(93)80052-g.

Claussnitzer, M., Cho, J. H., Collins, R., Cox, N. J., Dermitzakis, E. T., Hurles, M. E., ... MacArthur, D. G. (2020). A brief history of human disease genetics. *Nature, 577*, 179–189. https://doi.org/10.1038/s41586-019-1879-7.

Collepardo-Guevara, R., Portella, G., Vendruscolo, M., Frenkel, D., Schlick, T., & Orozco, M. (2015). Chromatin unfolding by epigenetic modifications explained by dramatic impairment of internucleosome interactions: A multiscale computational study. *Journal of the American Chemical Society, 137*, 10205–10215. https://doi.org/10.1021/jacs.5b04086.

Consortium, E. P., Moore, J. E., Purcaro, M. J., Pratt, H. E., Epstein, C. B., Shoresh, N., ... Dobin, A. (2022). Author correction: Expanded encyclopaedias of DNA elements in the human and mouse genomes. *Nature, 605*, E3. https://doi.org/10.1038/s41586-021-04226-3.

Cremer, T., & Cremer, C. (2001). Chromosome territories, nuclear architecture and gene regulation in mammalian cells. *Nature Reviews. Genetics, 2*, 292–301. https://doi.org/10.1038/35066075.

Cremer, T., & Cremer, M. (2010). Chromosome territories. *Cold Spring Harbor Perspectives in Biology, 2*, a003889. https://doi.org/10.1101/cshperspect.a003889.

Cuatrecasas, P., Fuchs, S., & Anfinsen, C. B. (1967). Catalytic properties and specificity of the extracellular nuclease of Staphylococcus aureus. *The Journal of Biological Chemistry, 242*, 1541–1547.

Cullen, K. E., Kladde, M. P., & Seyfred, M. A. (1993). Interaction between transcription regulatory regions of prolactin chromatin. *Science (New York, N. Y.), 261*, 203–206.

Davies, J. O., Oudelaar, A. M., Higgs, D. R., & Hughes, J. R. (2017). How best to identify chromosomal interactions: A comparison of approaches. *Nature Methods, 14*, 125–134. https://doi.org/10.1038/nmeth.4146.

Davies, J. O., Telenius, J. M., McGowan, S. J., Roberts, N. A., Taylor, S., Higgs, D. R., & Hughes, J. R. (2016). Multiplexed analysis of chromosome conformation at vastly improved sensitivity. *Nature Methods, 13*, 74–80. https://doi.org/10.1038/nmeth.3664.

Davis, S. J., & van der Merwe, P. A. (2006). The kinetic-segregation model: TCR triggering and beyond. *Nature Immunology, 7*, 803–809. https://doi.org/10.1038/ni1369.

de Wit, E., Bouwman, B. A., Zhu, Y., Klous, P., Splinter, E., Verstegen, M. J., ... Welling, M. (2013). The pluripotent genome in three dimensions is shaped around pluripotency factors. *Nature, 501*, 227–231. https://doi.org/10.1038/nature12420.

Dekker, J., Rippe, K., Dekker, M., & Kleckner, N. (2002). Capturing chromosome conformation. *Science (New York, N. Y.), 295*, 1306–1311. https://doi.org/10.1126/science.1067799.

Deng, W., Lee, J., Wang, H., Miller, J., Reik, A., Gregory, P. D., Dean, A., & Blobel, G. A. (2012). Controlling long-range genomic interactions at a native locus by targeted tethering of a looping factor. *Cell, 149*, 1233–1244. https://doi.org/10.1016/j.cell.2012.03.051.

Dixon, J. R., Jung, I., Selvaraj, S., Shen, Y., Antosiewicz-Bourget, J. E., Lee, A. Y., ... Xie, W. (2015). Chromatin architecture reorganization during stem cell differentiation. *Nature, 518*, 331–336. https://doi.org/10.1038/nature14222.

Dixon, J. R., Selvaraj, S., Yue, F., Kim, A., Li, Y., Shen, Y., ... Ren, B. (2012). Topological domains in mammalian genomes identified by analysis of chromatin interactions. *Nature, 485*, 376–380. https://doi.org/10.1038/nature11082.

Dostie, J., Richmond, T. A., Arnaout, R. A., Selzer, R. R., Lee, W. L., Honan, T. A., ... Nusbaum, C. (2006). Chromosome conformation capture carbon copy (5C): A massively parallel solution for mapping interactions between genomic elements. *Genome Research, 16*, 1299–1309. https://doi.org/10.1101/gr.5571506.

Du, M., Stitzinger, S. H., Spille, J. H., Cho, W. K., Lee, C., Hijaz, M., ... Cisse, I. I. (2024a). Direct observation of a condensate effect on super-enhancer controlled gene bursting. *Cell, 187*, 331–344.e317. https://doi.org/10.1016/j.cell.2023.12.005.

Du, M., Stitzinger, S. H., Spille, J. H., Cho, W. K., Lee, C., Hijaz, M., ... Cisse, I. I. (2024b). Direct observation of a condensate effect on super-enhancer controlled gene bursting. *Cell, 187*, 2595–2598. https://doi.org/10.1016/j.cell.2024.04.001.

Erler, J., Zhang, R., Petridis, L., Cheng, X., Smith, J. C., & Langowski, J. (2014). The role of histone tails in the nucleosome: A computational study. *Biophysical Journal, 107*, 2911–2922. https://doi.org/10.1016/j.bpj.2014.10.065.

Ettig, R., Kepper, N., Stehr, R., Wedemann, G., & Rippe, K. (2011). Dissecting DNA-histone interactions in the nucleosome by molecular dynamics simulations of DNA unwrapping. *Biophysical Journal, 101*, 1999–2008. https://doi.org/10.1016/j.bpj.2011.07.057.

Fang, R., Yu, M., Li, G., Chee, S., Liu, T., Schmitt, A. D., & Ren, B. (2016). Mapping of long-range chromatin interactions by proximity ligation-assisted ChIP-seq. *Cell Research, 26*, 1345–1348. https://doi.org/10.1038/cr.2016.137.

Farr, S. E., Woods, E. J., Joseph, J. A., Garaizar, A., & Collepardo-Guevara, R. (2021). Nucleosome plasticity is a critical element of chromatin liquid-liquid phase separation and multivalent nucleosome interactions. *Nature Communications, 12*, 2883. https://doi.org/10.1038/s41467-021-23090-3.

Fenley, A. T., Anandakrishnan, R., Kidane, Y. H., & Onufriev, A. V. (2018). Modulation of nucleosomal DNA accessibility via charge-altering post-translational modifications in histone core. *Epigenetics Chromatin, 11*, 11. https://doi.org/10.1186/s13072-018-0181-5.

Fudenberg, G., Imakaev, M., Lu, C., Goloborodko, A., Abdennur, N., & Mirny, L. A. (2016). Formation of chromosomal domains by loop extrusion. *Cell Reports, 15*, 2038–2049. https://doi.org/10.1016/j.celrep.2016.04.085.

Fullwood, M. J., Liu, M. H., Pan, Y. F., Liu, J., Xu, H., Mohamed, Y. B., ... Mei, P. H. (2009). An oestrogen-receptor-alpha-bound human chromatin interactome. *Nature, 462*, 58–64. https://doi.org/10.1038/nature08497.

Furey, T. S. (2012). ChIP-seq and beyond: New and improved methodologies to detect and characterize protein-DNA interactions. *Nature Reviews. Genetics, 13*, 840–852. https://doi.org/10.1038/nrg3306.

Galitsyna, A. A., & Gelfand, M. S. (2021). Single-cell Hi-C data analysis: Safety in numbers. *Briefings in Bioinformatics, 22*. https://doi.org/10.1093/bib/bbab316.

Gibson, B. A., Doolittle, L. K., Schneider, M. W. G., Jensen, L. E., Gamarra, N., Henry, L., ... Rosen, M. K. (2019). Organization of chromatin by intrinsic and regulated phase separation. *Cell, 179*, 470–484.e421. https://doi.org/10.1016/j.cell.2019.08.037.

Goel, V. Y., Huseyin, M. K., & Hansen, A. S. (2023). Region capture micro-C reveals coalescence of enhancers and promoters into nested microcompartments. *Nature Genetics, 55*, 1048–1056. https://doi.org/10.1038/s41588-023-01391-1.

Gong, H., Ma, F., & Zhang, X. (2023). [Advances in methods and applications of single-cell Hi-C data analysis]. *Sheng Wu Yi Xue Gong Cheng xue Za Zhi = Journal of Biomedical Engineering = Shengwu Yixue Gongchengxue Zazhi, 40*, 1033–1039. https://doi.org/10.7507/1001-5515.202303046.

Greenwald, W. W., Li, H., Benaglio, P., Jakubosky, D., Matsui, H., Schmitt, A., ... Frazer, K. A. (2019). Subtle changes in chromatin loop contact propensity are associated with differential gene regulation and expression. *Nature Communications, 10*, 1054. https://doi.org/10.1038/s41467-019-08940-5.

Groschel, S., Sanders, M. A., Hoogenboezem, R., de Wit, E., Bouwman, B. A., Erpelinck, C., ... van Lom, K. (2014). A single oncogenic enhancer rearrangement causes concomitant EVI1 and GATA2 deregulation in leukemia. *Cell, 157*, 369–381. https://doi.org/10.1016/j.cell.2014.02.019.

Guelen, L., Pagie, L., Brasset, E., Meuleman, W., Faza, M. B., Talhout, W., ... van Steensel, B. (2008). Domain organization of human chromosomes revealed by mapping of nuclear lamina interactions. *Nature, 453*, 948–951. https://doi.org/10.1038/nature06947.

Guo, Y., Xu, Q., Canzio, D., Shou, J., Li, J., Gorkin, D. U., ... Tang, Y. (2015). CRISPR inversion of CTCF sites alters genome topology and enhancer/promoter function. *Cell, 162*, 900–910. https://doi.org/10.1016/j.cell.2015.07.038.

Haarhuis, J. H. I., van der Weide, R. H., Blomen, V. A., Yanez-Cuna, J. O., Amendola, M., van Ruiten, M. S., ... van Steensel, B. (2017). The cohesin release factor WAPL restricts chromatin loop extension. *Cell, 169*, 693–707.e614. https://doi.org/10.1016/j.cell.2017.04.013.

Hong, L., Schroth, G. P., Matthews, H. R., Yau, P., & Bradbury, E. M. (1993). Studies of the DNA binding properties of histone H4 amino terminus. Thermal denaturation studies reveal that acetylation markedly reduces the binding constant of the H4 "tail" to DNA. *The Journal of Biological Chemistry, 268*, 305–314.

Hsieh, T. H., Weiner, A., Lajoie, B., Dekker, J., Friedman, N., & Rando, O. J. (2015). Mapping nucleosome resolution chromosome folding in yeast by micro-C. *Cell, 162*, 108–119. https://doi.org/10.1016/j.cell.2015.05.048.

Hsieh, T. S., Cattoglio, C., Slobodyanyuk, E., Hansen, A. S., Darzacq, X., & Tjian, R. (2022). Enhancer-promoter interactions and transcription are largely maintained upon acute loss of CTCF, cohesin, WAPL or YY1. *Nature Genetics, 54*, 1919–1932. https://doi.org/10.1038/s41588-022-01223-8.

Hsieh, T. S., Cattoglio, C., Slobodyanyuk, E., Hansen, A. S., Rando, O. J., Tjian, R., & Darzacq, X. (2020). Resolving the 3D landscape of transcription-linked mammalian chromatin folding. *Molecular Cell, 78*, 539–553.e538. https://doi.org/10.1016/j.molcel.2020.03.002.

Hua, P., Badat, M., Hanssen, L. L. P., Hentges, L. D., Crump, N., Downes, D. J., ... Taylor, S. (2021). Defining genome architecture at base-pair resolution. *Nature, 595*, 125–129. https://doi.org/10.1038/s41586-021-03639-4.

Huertas, J., & Cojocaru, V. (2021). Breaths, twists, and turns of atomistic nucleosomes. *Journal of Molecular Biology, 433*, 166744. https://doi.org/10.1016/j.jmb.2020.166744.

Hughes, J. R., & Davies, J. O. J. (2024). Single-cell technologies meet Hi-C. *Nature Genetics, 56*, 1542–1543. https://doi.org/10.1038/s41588-024-01855-y.

Hughes, J. R., Roberts, N., McGowan, S., Hay, D., Giannoulatou, E., Lynch, M., ... Higgs, D. R. (2014). Analysis of hundreds of cis-regulatory landscapes at high resolution in a single, high-throughput experiment. *Nature Genetics, 46*, 205–212. https://doi.org/10.1038/ng.2871.

Javierre, B. M., Burren, O. S., Wilder, S. P., Kreuzhuber, R., Hill, S. M., Sewitz, S., ... Thiecke, M. J. (2016). Lineage-specific genome architecture links enhancers and non-coding disease variants to target gene promoters. *Cell, 167*, 1369–1384. https://doi.org/10.1016/j.cell.2016.09.037 e1319.

Jin, F., Li, Y., Dixon, J. R., Selvaraj, S., Ye, Z., Lee, A. Y., ... Ren, B. (2013). A high-resolution map of the three-dimensional chromatin interactome in human cells. *Nature, 503*, 290–294. https://doi.org/10.1038/nature12644.

Kagey, M. H., Newman, J. J., Bilodeau, S., Zhan, Y., Orlando, D. A., van Berkum, N. L., ... Levine, S. S. (2010). Mediator and cohesin connect gene expression and chromatin architecture. *Nature, 467*, 430–435. https://doi.org/10.1038/nature09380.

Kieffer-Kwon, K. R., Tang, Z., Mathe, E., Qian, J., Sung, M. H., Li, G., ... Grontved, L. (2013). Interactome maps of mouse gene regulatory domains reveal basic principles of transcriptional regulation. *Cell, 155*, 1507–1520. https://doi.org/10.1016/j.cell.2013.11.039.

Kouzarides, T. (2007). Chromatin modifications and their function. *Cell, 128*, 693–705. https://doi.org/10.1016/j.cell.2007.02.005.

Krietenstein, N., & Rando, O. J. (2020). Mesoscale organization of the chromatin fiber. *Current Opinion in Genetics & Development, 61*, 32–36. https://doi.org/10.1016/j.gde.2020.02.022.

Krivega, I., & Dean, A. (2012). Enhancer and promoter interactions-long distance calls. *Current Opinion in Genetics & Development, 22*, 79–85. https://doi.org/10.1016/j.gde.2011.11.001.

Kubo, N., Ishii, H., Xiong, X., Bianco, S., Meitinger, F., Hu, R., ... Yu, M. (2021). Promoter-proximal CTCF binding promotes distal enhancer-dependent gene activation. *Nature Structural & Molecular Biology, 28*, 152–161. https://doi.org/10.1038/s41594-020-00539-5.

Leung, D., Jung, I., Rajagopal, N., Schmitt, A., Selvaraj, S., Lee, A. Y., ... Qiu, Y. (2015). Integrative analysis of haplotype-resolved epigenomes across human tissues. *Nature, 518*, 350–354. https://doi.org/10.1038/nature14217.

Li, G., Liu, Y., Zhang, Y., Kubo, N., Yu, M., Fang, R., ... Ren, B. (2019). Joint profiling of DNA methylation and chromatin architecture in single cells. *Nature Methods, 16*, 991–993. https://doi.org/10.1038/s41592-019-0502-z.

Li, H., Dalgleish, J. L. T., Lister, G., Maristany, M. J., Huertas, J., Dopico-Fernandez, A. M., ... Zhang, W. (2025). Mapping chromatin structure at base-pair resolution unveils a unified model of cis-regulatory element interactions. *Cell, 188*, 7175–7193.e7119. https://doi.org/10.1016/j.cell.2025.10.013.

Li, Z., & Kono, H. (2018). Investigating the influence of arginine dimethylation on nucleosome dynamics using all-atom simulations and kinetic analysis. *The Journal of Physical Chemistry. B, 122*, 9625–9634. https://doi.org/10.1021/acs.jpcb.8b05067.

Lieberman-Aiden, E., van Berkum, N. L., Williams, L., Imakaev, M., Ragoczy, T., Telling, A., ... Dorschner, M. O. (2009). Comprehensive mapping of long-range interactions reveals folding principles of the human genome. *Science (New York, N. Y.), 326*, 289–293. https://doi.org/10.1126/science.1181369.

Liu, Z., & Garrard, W. T. (2005). Long-range interactions between three transcriptional enhancers, active Vkappa gene promoters, and a 3′ boundary sequence spanning 46 kilobases. *Molecular and Cellular Biology, 25*, 3220–3231. https://doi.org/10.1128/MCB.25.8.3220-3231.2005.

Liu, Z., Chen, Y., Xia, Q., Liu, M., Xu, H., Chi, Y., Deng, Y., & Xing, D. (2023). Linking genome structures to functions by simultaneous single-cell Hi-C and RNA-seq. *Science (New York, N. Y.), 380*, 1070–1076. https://doi.org/10.1126/science.adg3797.

Lu, H., Yu, D., Hansen, A. S., Ganguly, S., Liu, R., Heckert, A., Darzacq, X., & Zhou, Q. (2018). Phase-separation mechanism for C-terminal hyperphosphorylation of RNA polymerase II. *Nature, 558*, 318–323. https://doi.org/10.1038/s41586-018-0174-3.

MacPherson, Q., Beltran, B., & Spakowitz, A. J. (2018). Bottom-up modeling of chromatin segregation due to epigenetic modifications. *Proceedings of the National Academy of Sciences of the United States of America, 115*, 12739–12744. https://doi.org/10.1073/pnas.1812268115.

Maeshima, K., Ide, S., Hibino, K., & Sasai, M. (2016). Liquid-like behavior of chromatin. *Current Opinion in Genetics & Development, 37*, 36–45. https://doi.org/10.1016/j.gde.2015.11.006.

Maeshima, K., Kaizu, K., Tamura, S., Nozaki, T., Kokubo, T., & Takahashi, K. (2015). The physical size of transcription factors is key to transcriptional regulation in chromatin domains. *Journal of Physics: Condensed Matter, 27*, 064116. https://doi.org/10.1088/0953-8984/27/6/064116.

McArthur, E., & Capra, J. A. (2021). Topologically associating domain boundaries that are stable across diverse cell types are evolutionarily constrained and enriched for heritability. *American Journal of Human Genetics, 108*, 269–283. https://doi.org/10.1016/j.ajhg.2021.01.001.

Meisterernst, M., & Roeder, R. G. (1991). Family of proteins that interact with TFIID and regulate promoter activity. *Cell, 67*, 557–567. https://doi.org/10.1016/0092-8674(91)90530-c.

Mifsud, B., Tavares-Cadete, F., Young, A. N., Sugar, R., Schoenfelder, S., Ferreira, L., ... Ewels, P. A. (2015). Mapping long-range promoter contacts in human cells with high-resolution capture Hi-C. *Nature Genetics, 47*, 598–606. https://doi.org/10.1038/ng.3286.

Minikel, E. V., Painter, J. L., Dong, C. C., & Nelson, M. R. (2024). Refining the impact of genetic evidence on clinical success. *Nature, 629*, 624–629. https://doi.org/10.1038/s41586-024-07316-0.

Mirny, L. A., Imakaev, M., & Abdennur, N. (2019). Two major mechanisms of chromosome organization. *Current Opinion in Cell Biology, 58*, 142–152. https://doi.org/10.1016/j.ceb.2019.05.001.

Miron, E., Oldenkamp, R., Brown, J. M., Pinto, D. M. S., Xu, C. S., Faria, A. R., ... de Ornellas, S. (2020). Chromatin arranges in chains of mesoscale domains with nanoscale functional topography independent of cohesin. *Science Advances, 6*. https://doi.org/10.1126/sciadv.aba8811.

Morgan, S. L., Mariano, N. C., Bermudez, A., Arruda, N. L., Wu, F., Luo, Y., ... Hu, J. F. (2017). Manipulation of nuclear architecture through CRISPR-mediated chromosomal looping. *Nature Communications, 8*, 15993. https://doi.org/10.1038/ncomms15993.

Morris, J. A., Caragine, C., Daniloski, Z., Domingo, J., Barry, T., Lu, L., ... Hao, S. (2023). Discovery of target genes and pathways at GWAS loci by pooled single-cell CRISPR screens. *Science (New York, N. Y.), 380*, eadh7699. https://doi.org/10.1126/science.adh7699.

Mukherjee, S., Erickson, H., & Bastia, D. (1988). Detection of DNA looping due to simultaneous interaction of a DNA-binding protein with two spatially separated binding sites on DNA. *Proceedings of the National Academy of Sciences of the United States of America, 85*, 6287–6291.

Mumbach, M. R., Rubin, A. J., Flynn, R. A., Dai, C., Khavari, P. A., Greenleaf, W. J., & Chang, H. Y. (2016). HiChIP: Efficient and sensitive analysis of protein-directed genome architecture. *Nature Methods, 13*, 919–922. https://doi.org/10.1038/nmeth.3999.

Murthy, A., Li, Y., Peng, I., Reichelt, M., Katakam, A. K., Noubade, R., ... van Lookeren Campagne, M. (2014). A Crohn's disease variant in Atg16l1 enhances its degradation by caspase 3. *Nature, 506*, 456–462. https://doi.org/10.1038/nature13044.

Nagano, T., Lubling, Y., Stevens, T. J., Schoenfelder, S., Yaffe, E., Dean, W., ... Fraser, P. (2013). Single-cell Hi-C reveals cell-to-cell variability in chromosome structure. *Nature, 502*, 59–64. https://doi.org/10.1038/nature12593.

Nelson, M. R., Tipney, H., Painter, J. L., Shen, J., Nicoletti, P., Shen, Y., ... Wang, J. (2015). The support of human genetic evidence for approved drug indications. *Nature Genetics, 47*, 856–860. https://doi.org/10.1038/ng.3314.

Ng, S. B., Turner, E. H., Robertson, P. D., Flygare, S. D., Bigham, A. W., Lee, C., ... Eichler, E. E. (2009). Targeted capture and massively parallel sequencing of 12 human exomes. *Nature, 461*, 272–276. https://doi.org/10.1038/nature08250.

Nora, E. P., Goloborodko, A., Valton, A. L., Gibcus, J. H., Uebersohn, A., Abdennur, N., ... Bruneau, B. G. (2017). Targeted degradation of CTCF decouples local insulation of chromosome domains from genomic compartmentalization. *Cell, 169*, 930–944.e922. https://doi.org/10.1016/j.cell.2017.05.004.

Nora, E. P., Lajoie, B. R., Schulz, E. G., Giorgetti, L., Okamoto, I., Servant, N., ... Sedat, J. (2012). Spatial partitioning of the regulatory landscape of the X-inactivation centre. *Nature, 485*, 381–385. https://doi.org/10.1038/nature11049.

Nuebler, J., Fudenberg, G., Imakaev, M., Abdennur, N., & Mirny, L. A. (2018). Chromatin organization by an interplay of loop extrusion and compartmental segregation. *Proceedings of the National Academy of Sciences of the United States of America, 115*, E6697–E6706. https://doi.org/10.1073/pnas.1717730115.

Oudelaar, A. M., & Higgs, D. R. (2021). The relationship between genome structure and function. *Nature Reviews. Genetics, 22*, 154–168. https://doi.org/10.1038/s41576-020-00303-x.

Oudelaar, A. M., Beagrie, R. A., Gosden, M., de Ornellas, S., Georgiades, E., Kerry, J., ... El-Sagheer, A. H. (2020). Dynamics of the 4D genome during in vivo lineage specification and differentiation. *Nature Communications, 11*, 2722. https://doi.org/10.1038/s41467-020-16598-7.

Ozturk, M. A., De, M., Cojocaru, V., & Wade, R. C. (2020). Chromatosome structure and dynamics from molecular simulations. *Annual Review of Physical Chemistry, 71*, 101–119. https://doi.org/10.1146/annurev-physchem-071119-040043.

Palstra, R. J., Tolhuis, B., Splinter, E., Nijmeijer, R., Grosveld, F., & de Laat, W. (2003). The beta-globin nuclear compartment in development and erythroid differentiation. *Nature Genetics, 35*, 190–194. https://doi.org/10.1038/ng1244.

Pasquali, L., Gaulton, K. J., Rodriguez-Segui, S. A., Mularoni, L., Miguel-Escalada, I., Akerman, I., ... van de Bunt, M. (2014). Pancreatic islet enhancer clusters enriched in type 2 diabetes risk-associated variants. *Nature Genetics, 46*, 136–143. https://doi.org/10.1038/ng.2870.

Phillips-Cremins, J. E., Sauria, M. E., Sanyal, A., Gerasimova, T. I., Lajoie, B. R., Bell, J. S., ... Sun, Y. (2013). Architectural protein subclasses shape 3D organization of genomes during lineage commitment. *Cell, 153*, 1281–1295. https://doi.org/10.1016/j.cell.2013.04.053.

Poleshko, A., Shah, P. P., Gupta, M., Babu, A., Morley, M. P., Manderfield, L. J., ... Sierra-Pagan, J. E. (2017). Genome-nuclear lamina interactions regulate cardiac stem cell lineage restriction. *Cell, 171*, 573–587.e514. https://doi.org/10.1016/j.cell.2017.09.018.

Pomerantz, M. M., Ahmadiyeh, N., Jia, L., Herman, P., Verzi, M. P., Doddapaneni, H., ... Davis, M. (2009). The 8q24 cancer risk variant rs6983267 shows long-range interaction with MYC in colorectal cancer. *Nature Genetics, 41*, 882–884. https://doi.org/10.1038/ng.403.

Qi, T., Song, L., Guo, Y., Chen, C., & Yang, J. (2024). From genetic associations to genes: Methods, applications, and challenges. *Trends in Genetics: TIG, 40*, 642–667. https://doi.org/10.1016/j.tig.2024.04.008.

Rajagopalan, M., Balasubramanian, S., & Ramaswamy, A. (2017). Structural dynamics of wild type and mutated forms of human L1 endonuclease and insights into its sequence specific nucleic acid binding mechanism: A molecular dynamics study. *Journal of Molecular Graphics & Modelling, 76*, 43–55. https://doi.org/10.1016/j.jmgm.2017.07.002.

Rao, S. S., Huntley, M. H., Durand, N. C., Stamenova, E. K., Bochkov, I. D., Robinson, J. T., ... Aiden, E. L. (2014). A 3D map of the human genome at kilobase resolution reveals principles of chromatin looping. *Cell, 159*, 1665–1680. https://doi.org/10.1016/j.cell.2014.11.021.

Rao, S. S. P., Huang, S. C., Glenn St Hilaire, B., Engreitz, J. M., Perez, E. M., Kieffer-Kwon, K. R., ... Bochkov, I. D. (2017). Cohesin loss eliminates all loop domains. *Cell, 171*, 305–320.e324. https://doi.org/10.1016/j.cell.2017.09.026.

Rychkov, G. N., Ilatovskiy, A. V., Nazarov, I. B., Shvetsov, A. V., Lebedev, D. V., Konev, A. Y., ... Onufriev, A. V. (2017). Partially assembled nucleosome structures at atomic detail. *Biophysical Journal, 112*, 460–472. https://doi.org/10.1016/j.bpj.2016.10.041.

Sabari, B. R., Dall'Agnese, A., Boija, A., Klein, I. A., Coffey, E. L., Shrinivas, K., ... Manteiga, J. C. (2018). Coactivator condensation at super-enhancers links phase separation and gene control. *Science (New York, N. Y.), 361*. https://doi.org/10.1126/science.aar3958.

Sanborn, A. L., Rao, S. S., Huang, S. C., Durand, N. C., Huntley, M. H., Jewett, A. I., ... Li, J. (2015). Chromatin extrusion explains key features of loop and domain formation in wild-type and engineered genomes. *Proceedings of the National Academy of Sciences of the United States of America, 112*, E6456–E6465. https://doi.org/10.1073/pnas.1518552112.

Sanyal, A., Lajoie, B. R., Jain, G., & Dekker, J. (2012). The long-range interaction landscape of gene promoters. *Nature, 489*, 109–113. https://doi.org/10.1038/nature11279.

Saunders, C. J., Miller, N. A., Soden, S. E., Dinwiddie, D. L., Noll, A., Alnadi, N. A., ... Fellis, J. (2012). Rapid whole-genome sequencing for genetic disease diagnosis in neonatal intensive care units. *Science Translational Medicine, 4*, 154ra135. https://doi.org/10.1126/scitranslmed.3004041.

Schoenfelder, S., Furlan-Magaril, M., Mifsud, B., Tavares-Cadete, F., Sugar, R., Javierre, B. M., ... Wingett, S. W. (2015). The pluripotent regulatory circuitry connecting promoters to their long-range interacting elements. *Genome Research, 25*, 582–597. https://doi.org/10.1101/gr.185272.114.

Schwarzer, W., Abdennur, N., Goloborodko, A., Pekowska, A., Fudenberg, G., Loe-Mie, Y., ... Spitz, F. (2017). Two independent modes of chromatin organization revealed by cohesin removal. *Nature, 551*, 51–56. https://doi.org/10.1038/nature24281.

Selvaraj, S., Jesse, R. D., Bansal, V., & Ren, B. (2013). Whole-genome haplotype reconstruction using proximity-ligation and shotgun sequencing. *Nature Biotechnology, 31*, 1111–1118. https://doi.org/10.1038/nbt.2728.

Sexton, T., Yaffe, E., Kenigsberg, E., Bantignies, F., Leblanc, B., Hoichman, M., ... Cavalli, G. (2012). Three-dimensional folding and functional organization principles of the Drosophila genome. *Cell, 148*, 458–472. https://doi.org/10.1016/j.cell.2012.01.010.

Shaytan, A. K., Armeev, G. A., Goncearenco, A., Zhurkin, V. B., Landsman, D., & Panchenko, A. R. (2016). Coupling between histone conformations and DNA geometry in nucleosomes on a microsecond timescale: Atomistic insights into nucleosome functions. *Journal of Molecular Biology, 428*, 221–237. https://doi.org/10.1016/j.jmb.2015.12.004.

Shrinivas, K., Sabari, B. R., Coffey, E. L., Klein, I. A., Boija, A., Zamudio, A. V., ... Chakraborty, A. K. (2019). Enhancer features that drive formation of transcriptional condensates. *Molecular Cell, 75*, 549–561.e547. https://doi.org/10.1016/j.molcel.2019.07.009.

Simonis, M., Klous, P., Homminga, I., Galjaard, R. J., Rijkers, E. J., Grosveld, F., ... de Laat, W. (2009). High-resolution identification of balanced and complex chromosomal rearrangements by 4C technology. *Nature Methods, 6*, 837–842. https://doi.org/10.1038/nmeth.1391.

Simonis, M., Klous, P., Splinter, E., Moshkin, Y., Willemsen, R., de Wit, E., ... de Laat, W. (2006). Nuclear organization of active and inactive chromatin domains uncovered by chromosome conformation capture-on-chip (4C). *Nature Genetics, 38*, 1348–1354. https://doi.org/10.1038/ng1896.

Smemo, S., Tena, J. J., Kim, K. H., Gamazon, E. R., Sakabe, N. J., Gomez-Marin, C., ... Wasserman, N. F. (2014). Obesity-associated variants within FTO form long-range functional connections with IRX3. *Nature, 507*, 371–375. https://doi.org/10.1038/nature13138.

Spilianakis, C. G., & Flavell, R. A. (2004). Long-range intrachromosomal interactions in the T helper type 2 cytokine locus. *Nature Immunology, 5*, 1017–1027. https://doi.org/10.1038/ni1115.

Stevens, T. J., Lando, D., Basu, S., Atkinson, L. P., Cao, Y., Lee, S. F., ... O'Shaughnessy-Kirwan, A. (2017). 3D structures of individual mammalian genomes studied by single-cell Hi-C. *Nature, 544*, 59–64. https://doi.org/10.1038/nature21429.

Strickfaden, H., Tolsma, T. O., Sharma, A., Underhill, D. A., Hansen, J. C., & Hendzel, M. J. (2020). Condensed chromatin behaves like a solid on the mesoscale in vitro and in living cells. *Cell, 183*, 1772–1784.e1713. https://doi.org/10.1016/j.cell.2020.11.027.

Sun, R., Li, Z., & Bishop, T. C. (2019). TMB library of nucleosome simulations. *Journal of Chemical Information and Modeling, 59*, 4289–4299. https://doi.org/10.1021/acs.jcim.9b00252.

Takayama, N., Murison, A., Takayanagi, S. I., Arlidge, C., Zhou, S., Garcia-Prat, L., ... Boutzen, H. (2021). The transition from quiescent to activated states in human hematopoietic stem cells is governed by dynamic 3D genome reorganization. *Cell Stem Cell, 28*, 488–501.e410. https://doi.org/10.1016/j.stem.2020.11.001.

Tan, L., Xing, D., Chang, C. H., Li, H., & Xie, X. S. (2018). Three-dimensional genome structures of single diploid human cells. *Science (New York, N. Y.), 361*, 924–928. https://doi.org/10.1126/science.aat5641.

Tolhuis, B., Palstra, R. J., Splinter, E., Grosveld, F., & de Laat, W. (2002). Looping and interaction between hypersensitive sites in the active beta-globin locus. *Molecular Cell, 10*, 1453–1465. https://doi.org/10.1016/s1097-2765(02)00781-5.

Uffelmann, E., & Posthuma, D. (2021). Emerging methods and resources for biological interrogation of neuropsychiatric polygenic signal. *Biological Psychiatry, 89*, 41–53. https://doi.org/10.1016/j.biopsych.2020.05.022.

van Steensel, B., & Belmont, A. S. (2017). Lamina-associated domains: Links with chromosome architecture, heterochromatin, and gene repression. *Cell, 169*, 780–791. https://doi.org/10.1016/j.cell.2017.04.022.

Vernimmen, D., De Gobbi, M., Sloane-Stanley, J. A., Wood, W. G., & Higgs, D. R. (2007). Long-range chromosomal interactions regulate the timing of the transition between poised and active gene expression. *The EMBO Journal, 26*, 2041–2051. https://doi.org/10.1038/sj.emboj.7601654.

Vernimmen, D., Marques-Kranc, F., Sharpe, J. A., Sloane-Stanley, J. A., Wood, W. G., Wallace, H. A., ... Higgs, D. R. (2009). Chromosome looping at the human alpha-globin locus is mediated via the major upstream regulatory element (HS -40). *Blood, 114*, 4253–4260. https://doi.org/10.1182/blood-2009-03-213439.

Vietri Rudan, M., Barrington, C., Henderson, S., Ernst, C., Odom, D. T., Tanay, A., & Hadjur, S. (2015). Comparative Hi-C reveals that CTCF underlies evolution of chromosomal domain architecture. *Cell Reports, 10*, 1297–1309. https://doi.org/10.1016/j.celrep.2015.02.004.

Wu, H., Zhang, J., Tan, L., & Xie, X. S. (2025). Single-cell Micro-C profiles 3D genome structures at high resolution and characterizes multi-enhancer hubs. *Nature Genetics, 57*, 1777–1786. https://doi.org/10.1038/s41588-025-02247-6.

Wutz, G., Varnai, C., Nagasaka, K., Cisneros, D. A., Stocsits, R. R., Tang, W., ... Hossain, M. J. (2017). Topologically associating domains and chromatin loops depend on cohesin and are regulated by CTCF, WAPL, and PDS5 proteins. *The EMBO Journal, 36*, 3573–3599. https://doi.org/10.15252/embj.201798004.

Yang, J. H., & Hansen, A. S. (2024). Enhancer selectivity in space and time: From enhancer-promoter interactions to promoter activation. *Nature Reviews. Molecular Cell Biology, 25*, 574–591. https://doi.org/10.1038/s41580-024-00710-6.

Zhang, B., Zheng, W., Papoian, G. A., & Wolynes, P. G. (2016). Exploring the free energy landscape of nucleosomes. *Journal of the American Chemical Society, 138*, 8126–8133. https://doi.org/10.1021/jacs.6b02893.

Zhao, Z., Tavoosidana, G., Sjolinder, M., Gondor, A., Mariano, P., Wang, S., ... Singh, U. (2006). Circular chromosome conformation capture (4C) uncovers extensive networks of epigenetically regulated intra- and interchromosomal interactions. *Nature Genetics, 38*, 1341–1347. https://doi.org/10.1038/ng1891.

Zhou, G. L., Xin, L., Song, W., Di, L. J., Liu, G., Wu, X. S., ... Liang, C. C. (2006). Active chromatin hub of the mouse alpha-globin locus forms in a transcription factory of clustered housekeeping genes. *Molecular and Cellular Biology, 26*, 5096–5105. https://doi.org/10.1128/MCB.02454-05.

Zhou, T., Zhang, R., Jia, D., Doty, R. T., Munday, A. D., Gao, D., ... Ma, J. (2024). GAGE-seq concurrently profiles multiscale 3D genome organization and gene expression in single cells. *Nature Genetics, 56*, 1701–1711. https://doi.org/10.1038/s41588-024-01745-3.

CHAPTER SEVEN

Post-transcriptional regulation of erythropoiesis

Merlin Nithya Gnanapragasam[a,b,c,*,1], Raja Sundari Meenakshi Sundaram[b], Sarah E. Adams[a,b], Rachael A. White[a,b], Anita R. Dhara[a,b], Parina N. Patel[a,b], Sumayea B. Mahmood Meem[a,b], and Mahesh Ramamoorthy[b]

[a]Department of Biological, Geological, and Environmental Sciences, at Cleveland State University, Cleveland, OH, United States
[b]Center for Gene Regulation in Health and Disease, at Cleveland State University, Cleveland, OH, United States
[c]Center for RNA Science and Therapeutics, Case Western Reserve University, Cleveland, OH, United States
*Corresponding author. e-mail address: m.gnanapragasam@csuohio.edu

Contents

[1] (Senior Author)

Current Topics in Developmental Biology, Volume 170
ISSN 0070-2153, https://doi.org/10.1016/bs.ctdb.2026.06.006

Abstract

Post-transcriptional mechanisms underlie various aspects of the erythropoiesis process, with a growing body of literature demonstrating its importance in the regulation of abundance, stability, and translational efficiency of the erythroid transcriptome. As the concerted actions of RNA binding proteins, alternative splicing, micro RNAs, long non-coding RNAs, and epitranscriptomic changes are increasingly uncovered, their importance in normal erythropoiesis and erythroid disorders are becoming better understood. Here, we systematically summarize the known post-transcriptional mechanisms in erythropoiesis, their perturbations in disease, and their potential in therapeutic applications.

1. Introduction

Red blood cells comprise 84 % of cells in the human body (Sender et al., 2016). Human adults produce around 2 million red blood cells per second (Sackmann, 1995; Socolovsky, 2022), amounting to approximately 200 billion red blood cells generated daily (Muckenthaler et al., 2017). Each red blood cell carries around 250 million hemoglobin molecules (Moore & von Lindern, 2018), which account for 95 % of protein content in red blood cells (Suresh et al., 2020). Therefore, erythropoiesis is a process of red cell formation that evolved to meet the mammalian tissue demand for oxygen supply. The erythropoiesis process involves precisely orchestrated steps undertaken by a pluripotent hematopoietic stem cell (HSC) as it traverses the differentiation pathway to culminate into a fully mature erythrocyte or a red blood cell.

Mammalian erythropoiesis involves two broadly divided phases: early erythropoiesis and terminal erythroid differentiation (Socolovsky, 2022). Early erythropoiesis encompasses the development of erythroid lineage committed progenitors from multipotent progenitors, and expansion of the erythroid progenitor pool. These lineage committed progenitors are characterized by their colony forming potential, beginning with early rare progenitors termed burst forming units-erythroid (BFU-Es), which give rise to more abundant colony forming units-erythroid (CFU-Es) progenitors. During terminal erythroid differentiation, CFU-Es lose their self-renewal ability and differentiate into proerythroblasts (Pro-E), erythroblasts (Baso-E), polychromatic erythroblasts (Poly-E), orthochromatic erythroblasts (Ortho-E) and finally an enucleated reticulocyte. During terminal erythroid differentiation, erythroblasts undergo 3–5 terminal cell divisions, reduce cell and nuclear size, undergo significant nuclear condensation, express very high levels of globins while suppressing non-erythroid genes, and finally enucleate in the bone marrow (Chen et al., 2009; Hattangadi et al., 2011; Murphy et al., 2021; Palis, 2014). These enucleated reticulocytes enter circulation where they further mature into discoid shaped red blood cells, which circulate in peripheral blood for around 120 days before undergoing senescence (Chen et al., 2009; Palis, 2014; Socolovsky, 2022).

There is a gamut of information on the mechanisms underlying the regulation of erythropoiesis at the transcriptional front, with studies demonstrating the roles of epigenetic factors, the chromatin landscape, genome organization and erythroid master-transcription factors. However, our understanding of post-transcriptional mechanisms that regulate erythropoiesis is relatively sparse.

Recent work has shown that there exists a poor correlation between transcriptional expression of a gene and its protein levels (Moore & von Lindern, 2018). This has led to the identification of mechanisms besides transcriptional regulation to modulate erythroid gene expression. Post-transcriptional events that affect transcript abundance, function, and structure, such as alternative splicing of RNA, RNA modifications, mRNA stability and translational efficiency, together contribute to proper erythropoietic progression. Various studies have shown that these regulatory pathways are crucial for normal erythropoiesis, and dysregulation can lead to erythroid-disorders (Li et al., 2023). Much of this post-transcriptional regulation of gene expression occurs through the concerted action of RNA binding proteins, long non-coding, micro RNAs, and RNA modifications (Fig. 1). A consolidation of available literature and what we know about the post-transcriptional regulation of erythropoiesis is presented below.

1.1 RNA binding proteins (RBPs)

RNA binding proteins (RBPs) form ribonucleoprotein particles (RNPs) after binding with RNA molecules, and regulate various RNA processes including transcription, splicing, modification, translation, and decay. Classic RBPs are characterized by the presence of one or more conserved and well defined classical RNA-binding domains (RBDs) such as RNA recognition motif (RRM), K-homology (KH), DEAD/DEAH helicase, and zinc finger domains, along with around 30 other domains with lower abundance, such as Pumilio homology domain (PUM-HD) (Gebauer et al., 2021). Unconventional RBPs have also been identified and these lack well defined classical RNA binding domains. Studies have shown that

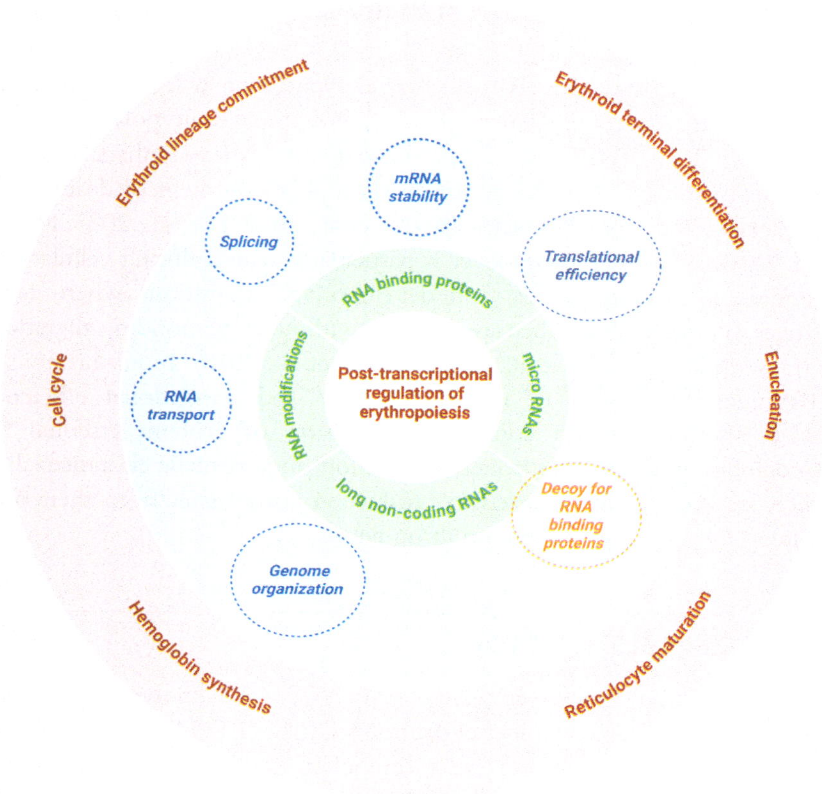

Fig. 1 Post-transcriptional regulation of erythropoiesis: post-transcriptional regulation is mediated by RNA binding proteins, microRNAs, long non-coding RNAs, and RNA editors, which via decoy functions and protein interactions, influence mRNA stability, translational efficiency, splicing, RNA transport, and genome organization. This regulation influences all facets of erythroid differentiation.

RBPs are key players in regulating erythropoiesis at the post-transcriptional level (Li et al., 2023; Moore et al., 2018). The interactions between RBPs and RNA molecules are known to be dynamic in nature, and RBPs can bind to specific regions on an RNA strand depending on the functional role of the RBP (He et al., 2023). To facilitate these stage-specific interactions with RNA molecules, RBPs themselves undergo differential regulation during erythropoiesis, examples of which are given below. These examples illustrate the involvement of some of the classical RNA binding proteins in the regulation of RNA splicing, transcript stability, and translational efficiency during erythropoiesis.

1.2 RBPs mediating mRNA stability and translational efficiency

Through interactions with cis-regulatory elements such as AU-rich elements (AREs), iron-responsive elements (IREs), and cytoplasmic polyadenylation elements (CPEs), RBPs can either stabilize mRNAs to enhance protein synthesis or promote their degradation through recruitment of deadenylase complexes and decapping machinery (He et al., 2023; Li et al., 2023; Moore et al., 2018). This regulatory layer is particularly critical during cellular differentiation, stress responses, and developmental transitions where rapid changes in protein abundance are required without corresponding alterations in transcription rates (Li et al., 2023; Moore et al., 2018). In erythropoiesis, RBP-mediated control of mRNA stability and translational efficiency ensures the precise and controlled expression of proteins essential for hemoglobin synthesis, cytoskeletal remodeling, and organelle clearance, all of which are collectively required to produce erythroid cells from their progenitors during the process of erythropoiesis.

2. ZFP36L2 (TIS11D)

Zinc Finger Protein 36 Like 2 (ZFP36L2), also known as TTP-like protein 2 (TIS11D), is a member of the tristetraprolin (TTP) family of RBPs. As a post-transcriptional regulator of erythropoiesis, it specifically binds to the AU-rich elements (AREs) in the 3′-UTR of mRNA that encodes tumor necrosis factor-alpha (TNF-alpha), destabilizing the mRNA (Carballo et al., 1998). After binding the target mRNA, it recruits mRNA decay machinery, such as deadenylases and exonucleases to the transcripts, thereby causing mRNA destabilization (Lai et al., 2003). Glucocorticoid stimulation of BFU-E erythroid progenitors leads to increase in expression of ZFP36L2, which then binds and negatively regulates the expression of several mRNAs that are induced during erythroid differentiation such as EPB41 gene (Erythrocyte Membrane Protein Band 4.1) (Zhang et al., 2013). Therefore, under stress conditions, this regulation by glucocorticoids can allow self-renewal of BFU-Es, and upon erythroid differentiation the levels of ZFP36L2 decrease. Under stress conditions, depletion of ZFP36L2 causes decreased BFU-E self-renewal, which in turn decreases CFU-E output and impairs erythroid lineage expansion. Thus, ZFP36L2 RBP facilitates self-renewal of BFU-Es while delaying differentiation by post-transcriptional regulation of mRNAs critical for erythroid differentiation (Stumpo et al., 2009; Zhang et al., 2013).

3. CSDE1

CSDE1 (Cold shock domain protein E1), also known as Upstream of N-Ras (UNR), is a highly conserved RBP which shows elevated expression during erythropoiesis compared to other hematopoietic cells (Moore et al., 2018). In erythroid cells, CSDE1 is known to bind to mRNAs involved in ribosome biogenesis, protein degradation, mitochondrial function, mitosis, and mRNA translation and exert broad post-transcriptional control over gene expression (Moore et al., 2018). At the mechanistic level, CSDE1 regulates mRNA stability and translation efficiency via interactions with co-factors including DEAD-box helicase 6 (DDX6), Poly(A)-binding protein (PABP), and hnRNPs (Moore & von Lindern, 2018).

In erythropoiesis, deletion of N-terminal cold shock domain of CSDE1 causes increase in Poly A binding protein c1 (PABPC1) protein levels, while the encoding mRNA levels was decreased (Horos et al., 2012). PABPC1 binds to polyA tail of transcripts and protects them from deadenylation and degradation, and can regulate translational efficiency. In addition, the deletion mutants also showed reduced protein expression of nucleolar ribogenesis factors. This modulation ensures balanced ribosome production and efficient protein synthesis, processes essential for erythroid proliferation and differentiation. In Diamond-Blackfan anemia (DBA), reduced CSDE1 activity is observed, which significantly impairs pathways such as ribosome biogenesis, protein homeostasis, and mitosis (Horos et al., 2012). Collectively, CSDE1 acts as a post-transcriptional regulator that coordinates ribosome production, translation, and proteostasis to ensure balanced protein synthesis during erythroid maturation, and its dysfunction contributes to the defective erythropoiesis characteristic of ribosomopathies such as Diamond-Blackfan Anemia (Horos et al., 2012).

QKI5 (targets **miR-124**, included in the miRNA section).

RNA-binding protein Quaking 5 (QKI5) plays a key role in regulating the biogenesis of miR-124 post-transcriptionally from its initial transcript primary miR-124–1 (pri-124–1) during erythropoiesis (Wang et al., 2017). Study by Wang et al., showed that QKI5 recognizes its QKI response element (QRE)- which is about 300 nucleotides upstream of pri-124–1 stem loops- binds to it and activates its processing by recruiting microprocessors during erythropoiesis (Wang et al., 2017). The job of microprocessors- which, in this case, is a trimeric protein complex made of DROSHA (drosha ribonuclease III) and DGCR8 (DGCR8 Microprocessor Complex Subunit)- is to perform the first step of miRNA maturation by recognizing and cutting hairpin-containing pri-124–1. This occurs by RNA looping, where QKI5 interacts directly with DGCR8, which acts as the bridge between QKI5 and the microprocessor. This study further showed that, in early erythropoiesis, QKI5 is abundant for this processing to occur and produce more mature miR-142. As erythroid differentiation progresses, QKI5 level decreases, resulting in reduced processing of pri-124–1 into mature miR-124. miR-124 targets TAL1 (T-cell acute lymphocytic leukemia 1) and c-Myb (MYB proto-oncogene, transcription factor), transcription factors involved in erythropoiesis. Therefore, when erythropoiesis proceeds, QK15 levels decrease, and leads to a decrease in mature miR-124 levels, allowing erythroid differentiation (Wang et al., 2017).

4. IRP1 and IRP2

Iron regulatory proteins (IRP1 and IRP2) are RBPs that are post-transcriptionally regulated to control appropriate expression of transcripts encoding iron, oxygen and energy metabolism to ensure appropriate homeostasis of erythropoiesis (Tsiftsoglou, 2021; Wilkinson & Pantopoulos, 2014). IRPs recognize and bind to the short RNA sequences called Iron-Responsive Elements (IREs) in the 5′ or 3′-UTR of the target mRNAs such as transferrin receptor 1 (TfR1/TFRC) mRNA, which controls iron uptake into the cell, and ALAS2 (5′-aminolevulinate synthase 2) mRNA, which encodes a key enzyme in heme biosynthesis, and regulates the stability and translational efficiency (Sanchez et al., 2011; Schranzhofer et al., 2006). Thereby, IRPs ensures that developing erythroid cells have enough iron to make heme and not end up making toxic amounts of heme (Schranzhofer et al., 2006).

IRP 1 can act both as an enzyme, called cytosolic aconitase, in the TCA cycle when iron is sufficient, and as an RBP to regulate mRNAs needed to ensure proper erythropoiesis (Tsiftsoglou, 2021; Wilkinson & Pantopoulos, 2014). Additionally, it regulates iron absorption by binding to the hypoxia inducible factor 2α (HIF2α) mRNA and regulating at the level of translation (Wilkinson & Pantopoulos, 2013). HIF2α is a transcription factor that modulates molecular responses to hypoxia, where it increases the production of erythropoietin (EPO) for enhanced erythropoiesis in hypoxic conditions (Kapitsinou et al., 2010). EPO is a major hematopoietic growth factor that binds to the surface of erythroid progenitor cells in the bone marrow and prevents apoptosis of BFU-E and CFU-E cells and stimulates their maturation into hemoglobin-producing erythroblasts (D'Andrea et al., 1989; Stephenson et al., 1971; Tusi et al., 2018). In low iron or normal oxygen condition, IRP1 binds and reduces the HIF2α mRNA levels, producing less HIF2α protein, so that adequate levels of EPO are maintained at physiological levels (Tsiftsoglou, 2021; Wilkinson & Pantopoulos, 2014).

5. AGO2

Agronaute-2 (AGO2) mediates miRNA-directed gene silencing via the RNA-induced silencing complex (RISC). In erythroid cells, the presence of AGO2 is crucial for normal erythropoiesis (O'Carroll et al., 2007). Conditional knockdown of AGO2 in the hematopoietic system of mice led to abnormal development of erythrocytes (O'Carroll et al., 2007). The knockdown also caused excessive accumulation of basophilic erythroblasts in the bone marrow and spleen of these mice which ultimately led to erythroid hyperplasia and splenomegaly. This was coupled with a severe reduction in the number of late-stage erythroblasts and functional red blood cells which gave rise to severe anemia (O'Carroll et al., 2007).

6. ELAVL1

RNA-binding protein ELAV-like RNA binding protein 1 (ELAVL1), also known as Human antigen R (HuR), post-transcriptionally regulates GATA-binding protein 1 (GATA1), an erythroid master transcription factor, to regulate and promote embryonic erythropoiesis in zebrafish (Li et al., 2014). Its isoform ELAVL1a binds to the 3′-UTR of GATA1 mRNA and stabilizes the GATA1 transcript or promotes its efficient translation. Overall, loss of ELAVL1a decreases the level of GATA1 mRNA and protein levels, which impairs erythropoiesis, and ectopic expression of GATA1 partially rescues the erythropoiesis defects (Li et al., 2014).

7. hnRNPs

hnRNP E1/E2 (Poly(C)-binding proteins; PCBP1/2) and hnRNP K (Heterogeneous nuclear ribonucleoprotein K/HNRNPK) belong to a subfamily of heterogenous nuclear ribonucleoproteins that act as important post-transcriptional regulators during hematopoietic differentiation via the shared hnRNP K homology (KH) domains. These domains enable the proteins to have sequence-specific RNA binding via CU-rich elements in the 3'UTR of their target mRNAs which in turn allows them to regulate cellular signaling pathways (Naarmann-de Vries et al., 2016).

hnRNP E1 and E2 function as crucial components of the α-complex that is assembled at various CU-rich sequences in the α-globin mRNA 3′UTR, where they maintain the α-globin mRNA stability by interacting with AU-rich element RNA-binding protein 1 (AUF1/hnRNP D) and cytoplasmic poly(A)-binding protein (PABP) to protect the transcript from undergoing deadenylation and cleavage (Chkheidze et al., 1999; Weiss & Liebhaber, 1994). These proteins therefore serve as versatile post-transcriptional regulators whose specific functions are determined by binding to distinct CU-rich sequences in 3′ UTRs and interactions with different protein partners, ultimately controlling critical developmental transitions in erythropoiesis via coordinated regulation of mRNA fate.

Another study in mice demonstrated that the two paralogous poly(C) binding proteins (hnRNP E1/E2; PCBP1/2), regulate mRNA abundance and stability through binding to C-rich 3′-UTR motifs of target mRNAs (Ji et al., 2021). This study by Ji et al., showed that these RNA-binding proteins are important in committed erythroid progenitor stage and during the early phases of terminal erythroid differentiation. Single knockout of PCBP1 or PCBP2 showed mild effects on red blood cell maturation, while double knockout mice showed severe anemia and defective erythroid differentiation. This indicated that both have overlapping functions to maintain the stability and splicing of mRNAs expressing adult hemoglobin genes (HBB, HBA) along with genes involved in heme synthesis and cell cycle regulation (Ji et al., 2021).

hnRNP K has multiple well characterized targets in erythroid cells. It forms a silencing complex at the differentiation control element in the 3′UTR of Arachidonate 15-Lipoxygenase (ALOX15) mRNA along with hnRNP E1 and DEAD-box helicase 6 (DDX6), which prevents the assembly of the 60 s and 40 s ribosomal subunits at the initiator AUG thereby inhibiting translation initiation (Naarmann-De Vries et al., 2021). Timely silencing and activation of ALOX15 is crucial for initiating mitochondria degradation in enucleated erythrocytes (Naarmann et al., 2008). Tyrosine phosphorylation of hnRNP K by c-Src kinase leads to translational activation in mature reticulocytes which in turn enhances ALOX15 synthesis to initiate mitochondrial degradation (Naarmann et al., 2008).

HnRNP K also binds to a specific sequence element in the 3′UTR of MYH9 (Myosin Heavy Chain 9) mRNA and mediates its translational repression (Naarmann-de Vries et al., 2016). This study by Naarmann-de Vries et al., also showed that MYH9 is involved in the process of erythrocyte enucleation, and an increase in MYH9 expression resulted in erythroid maturation independent enucleation.

8. PABPC1, AUF-1 and YB-1

Studies by van Zalen et al., showed that post-transcriptional events in erythroid cells are critical to ensure the high expression of β globin RNA (van Zalen et al., 2015). Poly(A) Binding Protein Cytoplasmic 1 (PABPC1) is an RNA binding protein that binds to and stabilizes the HBB mRNA by inhibiting its deadenylation. AU-rich element RNA-binding protein 1 (AUF-1/HNRNPD) and Y-box binding protein 1 (YB-1) are a part of an erythroid specific cytoplasmic messenger ribonucleoprotein (mRNP) complex that plays a critical role in maintaining the stability of HBB during erythroid differentiation. This complex specifically assembles on a CUGGG motif present on the 3'UTR of HBB mRNA and it exclusively forms only in erythroid cells. These studies further showed that the assembly of a messenger ribonucleoprotein β-complex on the β-globin 3′ UTR by RBPs, AUF1 and YB-1 enhances the binding of PABPC1 to the poly(A) tail thereby inhibiting the mRNA deadenylation of HBB, which in turn enhances the HBB mRNA stability in erythroid progenitor cells (van Zalen et al., 2015).

9. PABPC4

Poly(A) Binding Protein Cytoplasmic 4 (PABPC4) is another protein that binds to and stabilizes various erythroid targets like HBA1/2, HBB, BTG2, and SLC4A1 (Kini et al., 2014). Studies by Kini et al., showed that the mRNAs impacted by PABPC4 contain an AU-rich motif that enhances the interaction between PABPC4 and its target mRNAs. PABPC4 binds to mRNAs which have a critically shortened poly(A) tail and protects its targets from decay in erythroid cells (Kini et al., 2014).

10. LIN28B

Lin-28 Homolog B (LIN28B) is an RBP that drives a fetal-like erythroid program by acting post-transcriptionally. Studies by Lee et al., showed that it binds to precursor lethal-7 (let-7) RNA and blocks them from maturing into let-7 miRNA in erythroid cells (Lee et al., 2013). These studies further showed that let-7 miRNA upregulates B-cell CLL/lymphoma 11A (BCL11A), which is a known repressor of gamma globin. Reduced levels of mature let-7 miRNA cause repression of BCL11A, results in increased fetal hemoglobin and a shift in erythroid cells to a fetal phenotype (Lee et al., 2013). Further, LIN28B has also been shown to directly bind BCL11A transcript and inhibit its translational efficiency (Basak et al., 2020).

11. IGF2BP1

Another protein, Insulin-like growth factor 2 RNA-binding protein 1 (IGF2BP1), that belongs to the IGF2BP family, and acts as a post-transcriptional regulator of gamma globin expression (Coyne et al., 2025; Huang et al., 2024). Studies by Huang et al., demonstrated that this RBP binds to Hypermethylated in Cancer 2 (HIC2) transcript and increases the protein levels of HIC2, which is a transcriptional repressor of BCL11A (Huang et al., 2022). Increase in HIC2 leads to the decrease in the BCL11A levels, thereby promoting fetal hemoglobin expression (Huang et al., 2022, 2024).

Beyond its effect on BCL11A, a recent study deposited in bioRxiv preprint server demonstrates that IGF1BP1 also directly binds to fetal globin mRNAs HBG1 and HBG2, enhancing their translation and leading to a rise in fetal hemoglobin levels (Coyne et al., 2025). Hence, this dual mechanism makes this RBP an attractive therapeutic strategy to treat hemoglobinopathies such as sickle cell anemia and β-thalassemia (Coyne et al., 2025; Huang et al., 2024).

12. PUM1

A recent study from our laboratory described for the first time a role for Pumilio homologue 1 (PUM1) in erythropoiesis (Elagooz et al., 2022). PUM1 belongs to the eukaryotic PUF family of RNA binding proteins that show sequence specific binding to 3′UTRs of its targets via highly conserved Pumilio homology domain (PUM-HD) (Goldstrohm et al., 2018). Our study showed that PUM1 is expressed at higher levels during terminal erythroid differentiation, and its expression is maintained by master erythroid transcription factor Krüppel-like factor 1 (EKLF/KLF1). PUM1 binds to gamma globin mRNA and represses its stability and translational efficiency, thereby reducing the level of gamma globin protein and mediating the post-transcriptional regulation of fetal hemoglobin in humans. Knock down of PUM1 robustly increases fetal hemoglobin levels without affecting β-globin levels in human erythrocytes or limiting the progression of erythropoiesis. Further, an individual with heterozygous mutation in PUM1 RNA binding domain exhibited elevated fetal hemoglobin levels and increased percentage of fetal hemoglobin containing cells (F-cells), in the absence of anemia. Overall, these studies posit a potentially safe and effective treatment strategy for sickle cell anemia and β-thalassemia (Elagooz et al., 2022).

12.1 RBP mediated regulation of RNA splicing

Alternative splicing gives rise to distinct splice isoforms with unique temporal and spatial roles from a single mRNA. Mechanically, AS of pre-mRNA is undertaken by a macromolecular complex known as the spliceosome, that comprises of five small nuclear RNAs (U1, U2, U4, U5, and U6) and hundreds of proteins, collectively known as small nuclear ribonucleoproteins (snRNPs) (Tao et al., 2024). Besides the spliceosome, there are factors termed splicing regulatory elements (SREs), located in the enhancer or silencer regions of gene introns and exons known as exon or intron splicing enhancers (ESE or ISE) or silencers (ISS or ESS), which regulate AS by binding to trans-acting factors (Tao et al., 2024). Further, RNA binding proteins (RBP) serve as crucial regulators of AS by interacting with RNA to form ribonucleoprotein complexes that regulate various aspects of post-transcriptional regulation including mRNA splicing (Tao et al., 2024).

Terminal erythroid differentiation is regulated by an extensive and dynamic alternative splicing program that remodels the transcriptome as cells progress through their final divisions before enucleation (Pimentel et al., 2016). Alternative splicing switches play key roles during this process by selectively upregulating or downregulating individual exons of the same transcript, which alters the function of the encoded protein(s). Alternative splicing is mediated by a combination of RNA-binding proteins that include spicing factors (SR or serine/arginine rich proteins and hnRNPs) and tissue-specific regulators that coordinate stage-specific splicing events during erythropoiesis (Li et al., 2023; Moore et al., 2018).

13. U2AF1

U2 small nuclear RNA auxiliary factor 1 (U2AF1) is the small subunit of the U2 auxiliary factor heterodimer and constitutes the U2 snRNP of the spliceosome (Wu et al., 1999). Expression analysis of various erythroid cell populations revealed that U2AF1 is highly expressed in the burst forming unit erythroid cells (BFU-E) during early erythropoiesis (Zhang et al., 2019). Knocking down U2AF1 in human CD34+ hematopoietic stem cells led to severely impaired erythroid cell development which was characterized by failure to form BFU-E and CFU-E cells. Further, knockdown also delayed terminal erythroid differentiation. U2AF1 depletion altered the splicing patterns of various genes involved in apoptosis such as MTA1, HNRNPC, EIF3B, HNRNPD, BAX and SNGH1 (Zhang et al., 2019).

14. SF3B1

Splicing Factor 3b Subunit 1 (SF3B1) is a core component of the splicing machinery, and mutations of this protein are often found in myelodysplastic syndromes (Huang et al., 2018). Knocking down SF3B1 in CD34+ hematopoietic stem cells led to cell cycle arrest, apoptosis, delayed erythroid differentiation, and generation of polychromatic and orthochromatic erythroblasts with abnormal nuclei (Huang et al., 2018). Further, knockdown of SF3B1 in erythroid progenitor CFU-E cells caused altered splicing of E3 ligase Makorin Ring Finger Protien1 (MKRN1), which caused decreased expression of the large isoform of MKRN1, and increased expression of the smaller isoform, which further led to the activation of the p53 pathway (Huang et al., 2018).

Interestingly, a study by Pimentel et al., identified SF3BP1 among the prominent spliceosome associated factors that showed dynamically increased intron retention events during terminal erythroid differentiation. This intron retention could cause reduction in splicing capacity of erythroid cells as they prepare for enucleation (Pimentel et al., 2016).

15. RBM38

One key regulator of alternative splicing during late erythroid differentiation is RNA-binding motif protein 38 (RBM38/RNCP1), which is highly upregulated in CD71+ erythroid cells (Heinicke et al., 2013). RBM38 binds to GU-rich intronic sequence elements and activates splicing when it is recruited to downstream introns. One of its targets is Protein 4.1R (EPB41), a well characterized target that undergoes alternative splicing during erythropoiesis, and RBM38 mediates inclusion of exon 16 of this protein in later stages of erythropoiesis (Heinicke et al., 2013). This region encodes for a part of the spectrin-actin binding domain that is essential for membrane skeleton assembly and failure to include this exon causes the development of mechanically unstable red cells and elliptocytic red cells with anemia. Zinc finger DHHC-type palmitoyltransferase 16 (ZDHHC16) and Isochorismatase Domain Containing 2 (ISOC2) are additional targets of RBM38 regulation that exhibit differentiation-associated splicing switches consistent with RBM38 expression upregulation in mature erythroid cells (Heinicke et al., 2013).

16. MBNL1

Muscleblind-like 1 (MBNL1) is a critical sequence-specific pre-mRNA splicing factor which recognizes YGCY (Y = C or U) motifs through pairs of highly conserved zinc fingers. Work by Chen et al., showed that MBNL motifs are enriched in erythropoietic cells of mouse fetal liver, as MBNL1 is a major regulator of alternative splicing during terminal erythropoiesis (Cheng et al., 2014). MBNL1 is required for several steps of terminal differentiation such as enucleation, cell size maintenance and hemoglobin synthesis, and causes downregulation of developmentally repressed genes like CD44, Myb, and c-kit. Further, the knockdown also inhibited alternative splicing of MBNL1 target genes including NDEL1 (an important regulator of cytoskeletal dynamics) (Cheng et al., 2014).

17. PTBP1

Polypyrimidine tract binding protein 1 (PTBP1) is an RBP that belongs to the hnRNP family, and functions as a splicing factor regulating the post-transcriptional regulation of erythropoiesis (Rehn et al., 2022).

Studies by Rehn et al., showed that it binds to its target pre-mRNA Cell Division Cycle 42 (CDC42) and includes or skips exons, aiding in splicing regulatory activities. CDC42 is a factor reported to be important for translational initiation. Further, PTBP1 ensures correct processing of pre-rRNA transcripts and translation of ribosome proteins, preventing ribosomopathy (Rehn et al., 2022). In addition, studies in mice have shown that loss of PTBP1 causes mis-splicing that leads to the formation of alternative CDC42 gene isoforms required for erythroid differentiation, which results in reduced protein synthesis rate in erythropoiesis. Further, decreased erythroid progenitors were observed in the bone marrow of PTBP1 knockout mice. Interestingly, proteomics data revealed that hemoglobin proteins HBA and HBB, along with crucial erythropoietic factors such as Alpha Hemoglobin Stabilizing Protein (AHSP) and Ferritin had differential expression uncoupled from any changes in mRNA expression or alternative splicing. Overall, a loss of PTBP1 in mice led to anemia and decreased long-time survival (Rehn et al., 2022).

Although several RBPs with defined functions in erythropoiesis have been identified, the broader post-transcriptional regulatory landscape governed by RBPs is only beginning to be systematically characterized. The post-transcriptional regulation mediated by RNA binding proteins are essential to fine tune erythropoiesis and ensure a balance between proliferation, differentiation and hemoglobin synthesis (Fig. 2). Table 1 shows a compilation of the role of RNA binding proteins in erythropoiesis.

17.1 Micro RNAs

MicroRNAs (miRNAs) are small (~19–24 nucleotides), non-coding RNAs that regulate gene expression post-transcriptionally. Following their identification in Caenorhabditis elegans, miRNAs have been acknowledged as crucial regulators of various cellular processes, such as development, differentiation, proliferation, apoptosis, and stress responses (Bartel, 2004; He & Hannon, 2004). Despite their limited abundance, it is estimated that between 10–30 % of human genes are regulated by miRNAs (Lewis et al., 2005). They perform their function by binding to complementary sequences in target messenger RNAs (mRNAs), typically within the 3′ untranslated region, although there is increasing evidence for their binding in the 5′UTR and coding regions, resulting in translational repression or mRNA degradation (Byon & Papayannopoulou, 2012).

While there are several miRNAs that have been implicated in the post-transcriptional regulation of erythroid differentiation, challenges still exist

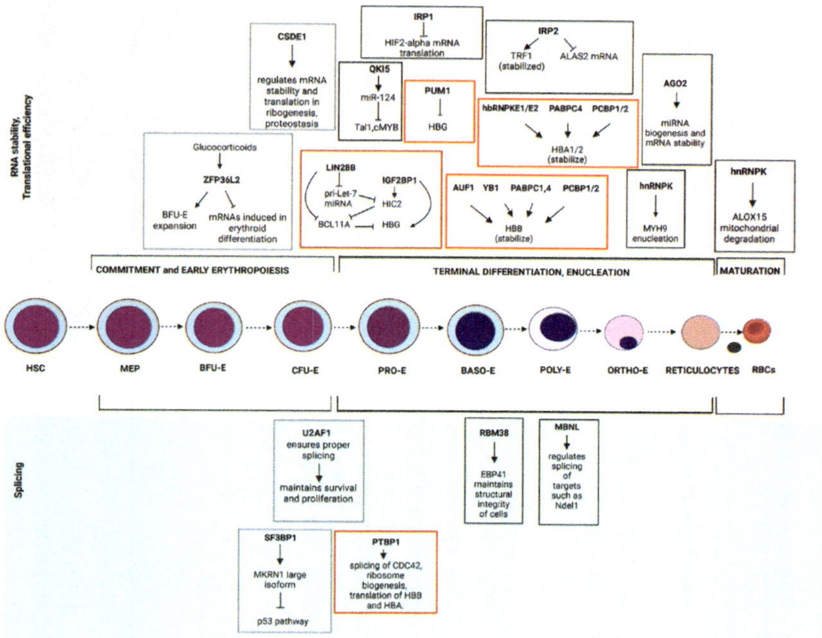

Fig. 2 RNA binding proteins in mammalian erythropoiesis: RNA binding proteins that regulate RNA stabilty, translational efficiency and splicing during mammalian erythropoiesis are shown. In grey boxes are those involved in early erythropoiesis. In black are those invovled in terminal erythroid differentiation. Red boxes show regulation of globin gene expression by RNA binding proteins.

in validating the targets of these miRNAs and in elucidating their involvement in this process either directly, or indirectly. Here, we focus on some of the well characterized miRNAs involved in erythropoiesis, beginning with early erythropoiesis, that includes erythroid lineage commitment from multipotent progenitors, and erythroid progenitor (BFU-e and CFU-e) proliferation and homeostasis, and terminal erythroid differentiation, that involves CFU-e yielding erythroid precursors that undergo 3–5 terminal cell divisions, accumulate hemoglobin, decrease in cell size, undergo nuclear condensation and polarization to yield orthochromatic erythroblasts, which enucleate in the bone marrow, and mature in circulation into red blood cells (Chen et al., 2009; Gnanapragasam et al., 2016; Socolovsky, 2022).

Table 1 Role of RNA binding proteins in erythropoiesis.

RBP name	Role in normal erythropoiesis	Role in aberrant erythropoiesis	References
Regulation of RNA stability and translational efficiency			
ZFP36L2	Binds AU-rich elements (AREs) in TNF-α mRNA to promote decay and maintain normal BFU-E self-renewal and differentiation balance.	Knockout leads to TNF-α accumulation, KLF1 and ALAS2 upregulation, premature differentiation, and defective BFU-E/CFU-E output.	Stumpo et al. (2009), Zhang et al. (2013)
CSDE1	Controls mRNA stability and translation of genes involved in ribosome biogenesis, mitochondrial function, mitosis, and protein degradation.	Reduced CSDE1 activity (e.g., in Diamond–Blackfan anemia) disrupts ribosome assembly and mRNA stability, leading to defective erythropoiesis and anemia.	Moore and von Lindern (2018), Horos et al. (2012)
QKI5	Binds QKI response element (QRE) in pri-miR-124-1 and recruits Drosha/ DGCR8 to promote miR-124 maturation, which represses erythroid transcription factors (TAL1, c-MYB). QK15 levels decrease during erythropoiesis to allow differentiation.	Dysregulation of QKI5 affects miR-124 processing, leading to mis regulation of transcription factors and altered red cell maturation.	Wang et al. (2017)
IRP1 / IRP2	Bind iron-responsive elements (IREs) in target mRNAs (TfR1, ALAS2, HIF2α) to maintain iron and heme balance for erythroid differentiation.	Dysregulation leads to iron imbalance, reduced heme synthesis, and defective erythropoiesis.	Tsiftsoglou (2021), Wilkinson and Pantopoulos (2014)

hnRNP K	Regulates translation and mRNA stability of ALOX15 and MYH9 and coordinates mitochondrial clearance and enucleation during late erythropoiesis.	Dysregulation impairs ALOX15 activation and mitochondrial degradation, leading to failed enucleation and abnormal erythroid maturation.	Naarmann et al. (2008), Naarmann-de Vries et al. (2016)
AGO2	Integrates miRNA biogenesis and mRNA stability control while also regulating oxidative stress response, terminal differentiation, and cytoskeletal remodelling.	Loss of AGO2 leads to erythroid hyperplasia, splenomegaly, and anaemia due to maturation arrest and defective red cell formation.	O'Carroll et al. (2007)
ELAVL1 (HuR)	Binds 3′-UTR of GATA1 mRNA to stabilize transcript and promote translation, supporting erythroid maturation.	Knockdown decreases GATA1 levels and impairs erythropoiesis in zebrafish; rescued by GATA1 overexpression.	Li et al. (2014)
hnRNP E1/ E2(PCBP1/ PCBP2)	Form α-complex on α-globin mRNA 3′UTR to prevent deadenylation; maintain α-globin mRNA stability during erythroid differentiation.	Loss or dysfunction destabilizes α-globin mRNA, disrupting globin chain balance and causing ineffective erythropoiesis. Double knockout causes severe anemia, defective erythroid differentiation, and loss of mRNA stability/splicing control.	Chkheidze et al. (1999), Weiss and Liebhaber (1994), Ji et al. (2021)

(continued)

Table 1 Role of RNA binding proteins in erythropoiesis. *(cont'd)*

RBP name	Role in normal erythropoiesis	Role in aberrant erythropoiesis	References
PABPC1	Stabilizes β-globin mRNA by preventing deadenylation; promotes efficient hemoglobin synthesis and erythroid maturation.	Loss or impaired function decreases β-globin mRNA stability and translation, leading to reduced hemoglobin and defective red cell development.	van Zalen et al. (2015)
AUF1 (HNRNPD)	Cooperates with YB-1 and PABPC1 in β-globin mRNA stabilization through assembly of erythroid-specific mRNP complexes.	Dual loss of AUF1 and YB-1 destabilizes β-globin mRNA, reducing hemoglobin synthesis and impairing erythroid differentiation.	van Zalen et al. (2015)
YB-1	Functions with AUF1 and PABPC1 to stabilize β-globin mRNA. Promotes erythroid-specific mRNP assembly and mRNA protection.	Knockdown disrupts β-globin mRNA stability and reduces erythroid transcript abundance, impairing red cell production.	van Zalen et al. (2015)
PABPC4	Binds and stabilizes erythroid mRNAs (HBB, HBA1/2, BTG2, SLC4A1) with shortened poly(A) tails, protecting them from decay.	Dysregulation impairs stabilization of key erythroid transcripts, leading to reduced globin synthesis and abnormal erythrocyte development.	Kini et al. (2014)
LIN28B	Binds pri-let-7 miRNA, preventing its maturation. Reduced let-7 increases γ-globin via suppression of BCL11A, promoting fetal-like erythroid phenotype. Can also repress translation of BCL11A.	Overexpression or dysregulation shifts erythroid program toward fetal stage and disrupts adult globin balance.	Basak et al. (2020), Lee et al. (2013)

IGF2BP1	Binds to HIC2 (a BCL11A repressor) transcript and increases its levels; and directly binds and stabilizes HBG1/HBG2 mRNAs to enhance fetal hemoglobin translation.	Dysregulation affects γ-globin and BCL11A levels.	Coyne et al. (2025), Huang et al. (2024)
PUM1	Represses γ-globin mRNA stability and translation during erythroid differentiation to regulate fetal hemoglobin levels.	Knockdown increases fetal hemoglobin (HbF) without affecting β-globin and without deleterious effects on terminal differentiation; Individual with heterozygous mutation in PUM1 RNA binding domain has elevated percentage of fetal hemoglobin and F cells (fetal hemoglobin containing cells), in the absence of anemia.	Elagooz et al. (2022)
RBP mediated regulation of RNA Splicing			
U2AF1	Functions as part of the U2 snRNP complex to regulate pre-mRNA splicing during early erythropoiesis and ensures proper differentiation of BFU-E and CFU-E cells.	Knockdown causes defective erythroid progenitor formation, increased apoptosis, and abnormal splicing of key apoptotic regulators (MTA1, BAX, EIF3B).	Zhang et al. (2019)

(continued)

Table 1 Role of RNA binding proteins in erythropoiesis. *(cont'd)*

RBP name	Role in normal erythropoiesis	Role in aberrant erythropoiesis	References
SF3B1	Core spliceosome component that ensures accurate pre-mRNA splicing and supports erythroid differentiation.	Knockdown causes mis-splicing of MLRN1 E3 ligase, leading to activation of the p53 pathway, apoptosis, cell cycle arrest in erythroid cells.	Zhang et al. (2019)
RBM38	Regulates alternative splicing during terminal erythroid differentiation, enables exon inclusion in EPB41 and other targets to ensure membrane stability and cytoskeletal assembly.	RBM38 deficiency disrupts exon inclusion in EPB41, which leads to mechanically unstable erythroid cells, elliptocytosis, and defective erythroid maturation.	Heinicke et al. (2013)
MBNL1	Controls stage-specific alternative splicing of genes involved in chromatin remodeling, RNA processing, and cytoskeletal organization. Also promotes terminal erythroid differentiation.	Knockdown leads to impaired enucleation, larger erythroblast size, reduced hemoglobin synthesis, persistence of developmentally repressed genes (CD44, Myb, c-kit), and mis-splicing of NDEL1, an important regulator of cytoskeletal dynamics.	Cheng et al. (2014)
PTBP1	Functions as a splicing factor; binds CDC42 pre-mRNA to ensure proper exon inclusion which is important to maintain appropriate protein synthesis rates. Also important for ribosome biogenesis.	Loss leads to CDC42 mis-splicing, reduced protein synthesis rates, defective ribosome biogenesis, anemia, reduced HBB, HBA proteins, and decreased erythroid progenitor survival.	Rehn et al. (2022)

17.2 Role in erythroid lineage commitment and early erythropoiesis

17.2.1 miR-150 and miR-15a

miR-150 and miR-15a have been shown to have a significant impact on the expression of MYB, a transcription factor and proto-oncogene that is known to regulate lineage fate and cell proliferation/maturation in many different tissues and is essential for normal erythropoiesis (Byon & Papayannopoulou, 2012). MYB is a key regulatory molecule in the 'decision-making' process of MEP commitment to either the megakaryocyte or erythroid lineages, where high MYB expression is observed in erythroid committed MEPs, and low MYB expression is observed in megakaryocyte committed MEPs (Lu et al., 2008; Zhang et al., 2012; Zhao et al., 2009). This point of bifurcation between the two lineages has been well characterized, and thus miR-150 and miR-15a have been shown to have a significant impact on the lineage fate of MEPs, as MYB is a direct target of both miRNAs (Lu et al., 2008; Zhao et al., 2009). miR-150 expression is highest at the HSPC stage and is quickly downregulated in MEPs destined for erythroid commitment resulting in increased MYB levels whereas maintained expression of miR-150 at the MEP stage promotes megakaryopoiesis (Lu et al., 2008).

Similarly, miR-15a expression is high at the HSPC stage and sharply decreases in MEPs. However, at the transition between the MEP and BFU-E stages there is a sharp increase in miR-15a expression (and thus a decrease in MYB) that was shown to enhance BFU-E formation (Zhao et al., 2009). After BFU-E establishment, miR-15a expression continues to decline to encourage proper CFU-E colony formation and further maturation. It has also been shown that MYB is an effector of miR-15a expression, forming an auto-regulatory feedback loop to carefully temper MYB expression in a stage specific manner (Zhao et al., 2009). Together, miR-150 and miR-15a provide an example of how multiple miRNAs can influence the expression patterns of a single gene to form a regulatory network that modulates the expression of their common target at sequential developmental stages, thus contributing to lineage specification and differentiation.

18. miR-125b-2 and miR-145

miR-125b-2 and miR−145 also play a role in the balance of the commitment of MEPs to either the megakaryocytic or erythroid lineages.

DICER1 and ST18 are direct targets of miR-125b-2. High levels of miR-125b-2 was shown to induce the proliferation and self-renewal of MEPs and contribute to lineage commitment trajectory of MEPs, where an increase in miR-125b-2 promotes megakaryopoiesis at the expense of erythropoiesis (Klusmann et al., 2010).

Friend leukemia integration 1 (FLI-1) is a transcription factor directly involved in MEP cell fate to either the megakaryocyte or erythroid lineage (Frontelo et al., 2019; Palii et al., 2019). FLI-1 is a direct target of miR-145, and miR-145 upregulation in early erythropoiesis (particularly at the MEP stage) results in the downregulation of FLI-1, driving the MEP population toward erythroid lineage commitment (Kumar et al., 2011).

19. miR-26a

miR-26a promotes erythroid differentiation by targeting Nemo-like kinase (NLK), a negative regulator of erythropoiesis (Wilkes et al., 2020). In patients with Diamond Blackman anemia, metformin-induced upregulation of miR-26a restores erythroid maturation. By suppressing NLK, it alleviates differentiation block and enhances red blood cell production (Wilkes et al., 2020). This highlights miR-26a as a mediator of erythroid maturation and a potential target for interventions in stress or pathological erythropoiesis.

20. miR-221/222 and miR-320

Among the first miRNAs discovered to have a functional role in erythropoiesis were miR-221/222 (Gabbianelli et al., 2010). They form a miR-cluster on the X chromosome and share the same seed sequence (Felli et al., 2009; Gabbianelli et al., 2010). miR-221/222 are highly expressed in HSPCs and directly target cKIT (receptor for Stem Cell Factor/SCF), resulting in repression of this stem cell factor receptor. Progressive downregulation of miR-221/222 throughout early erythropoiesis promotes cKIT expression, which is a crucial protein for the proliferation of erythroid progenitors (Felli et al., 2005; Hattangadi et al., 2011; Lee et al., 2018).

This developmental downregulation of miR-221/222 is in part affected by yet another miRNA – miR-320. In erythroid precursor cells, miR-320 levels are low and this results in increased expression of its target Scaffold/Matrix Attachment Region-binding protein 1 (SMAR1), which in turn inhibits expression of miR-221/222 (Mittal et al., 2013).

21. miR-223

miRNA 223 has been studied extensively in granulopoiesis and its expression declines during erythroid differentiation (Felli et al., 2009). miR-223 is expressed at reduced levels on the umbilical cord blood (UCB) $CD34^+$ progenitor cells, with its expression further decreasing throughout erythroid development. Moreover, factors such as different uremic toxins, erythropoietin levels, and dialysis can modulate this miRNA expression (Brisot et al., 2025). Ectopic expression of miR-223 impairs colony formation and normal erythroid differentiation, resulting in the accumulation of immature erythroid cells and a reduction in orthochromatic erythroblasts. LIM-domain only 2 (LMO2), a transcriptional co-regulator crucial for erythropoiesis, is recognized as a key functional target of miR-223 (Felli et al., 2009). The depletion of LMO2 results in embryonic mortality due to inadequate yolk sac erythropoiesis (Felli et al., 2009), emphasizing the importance of miR-223 downregulation for appropriate erythroid development.

22. miR-126

miR-126 directly targets Protein Tyrosine Phosphatase Non-Receptor Type 9 (PTPN9), a non-receptor tyrosine phosphatase that plays a critical role in erythropoiesis (Bu et al., 2014; Huang et al., 2011). PTPN9 contributes to erythropoiesis by dephosphorylating STAT3 and maintaining optimal phosphorylation of GATA1, a crucial erythroid specific transcription factor, thereby providing a necessary balance between early-stage proliferation and induction of essential RBC genes such as hemoglobin (Bu et al., 2014). Downregulation of miR-126 increases the expression of PTPN9, which promotes CFU-E colony formation/establishment, and hemoglobin accumulation, features that are important for proper erythropoiesis (Huang et al., 2011).

23. miR-24

miR-24 is highly expressed in human CD34+ cells and decreases throughout erythroid differentiation (Wang et al., 2008). The direct target of miR-24 is the Activin A receptor type 1B (ALK4), which binds Activin, an EPO cooperative cytokine that regulates erythropoiesis (Wang et al., 2008). High expression of miR-24 attenuates Activin signaling and promotes the necessary cell proliferation of early erythroid precursors, as high levels of Activin produces an anti-proliferative effect. Decreasing miR-24 levels in CD34+ cells results in an increase in BFU-E and CFU-E formation (Wang et al., 2008). As Activin and EPO work cooperatively to promote erythroid maturation and hemoglobin accumulation (Zhang et al., 2012), it follows that miR-24 downregulation throughout this process is essential for proper differentiation.

23.1 Role in terminal differentiation, enucleation and maturation

23.1.1 miR-144/451

miR-144 and miR-451 are among the most well studied miRNAs in erythropoiesis (Xu et al., 2020; Yang et al., 2010). miRNA-144 and miRNA 451 constitute a cluster of miRNAs derived from a single transcript and possess an identical seed sequence. These miRNAs are increased during terminal differentiation under GATA1 transcriptional regulation (Yang et al., 2010). They play a central role in coordinating erythroid differentiation, enucleation, and red cell maturation (Dore et al., 2008). miR-451 constitutes approximately 60 % of the miRNA content in mature erythrocytes, and processing is dependent on AGO2 (Argonaute RISC catalytic component 2) activity, and is independent of DICER (Dicer 1, ribonuclease III) (Rasmussen et al., 2010; Yang et al., 2010). During terminal erythropoiesis, miR-144/451 expression increases markedly, ensuring proper balance between proliferation, differentiation, and survival of erythroid progenitors. miRNA-451 also plays a major role in the regulation of erythropoiesis by repressing GATA binding protein 2 (GATA2), a negative regulator of this process (Jin et al., 2012).

This erythroid-specific miR-144/451 cluster, activated by GATA1, is crucial for erythropoietin (EPO)-induced maturation. The absence of miR-144 and/or miR-451 disrupts erythropoiesis by derepressing many target genes, such as RAB14 (member RAS oncogene family; influences vesicular trafficking), COX10 (cytochrome c oxidase assembly factor heme A:farnesyltransferase; an essential component of mitochondrial respiration) and CAP1 (Cyclase Associated Actin Cytoskeleton Regulatory Protein 1; a negative modulator of erythroid differentiation and enucleation) (Huang et al., 2021; Kim et al., 2015; Xu et al., 2019). Inhibition or loss of miR-144/451 results in upregulation of RAB14, leading to impaired EPO-driven erythropoiesis— demonstrating that miR-144/451 fine-tunes vesicular transport pathways critical for RBC maturation (Kim et al., 2015). By targeting COX10, miR-144/451 suppresses mitochondrial respiration during terminal differentiation, facilitating the metabolic shift toward glycolysis in mature erythrocytes (Xu et al., 2019). Lastly, miR-144/451 also target CAP1, and by repressing CAP1, miR-144/451 supports Actin dynamics necessary for nuclear extrusion (Huang et al., 2021).

A direct target of miR-144/451 is MYC, expression of which is carefully modulated throughout erythropoiesis. Conversely, MYC downregulates miR-144/451, creating a GATA1–miR-144/451–MYC feedback loop that sustains the requisite MYC levels for optimal erythroid proliferation and differentiation (Xu et al., 2020). Moreover, miR-144/451 regulates the LKB1/AMPK/mTOR signaling pathway, facilitating the survival of erythroid precursors during stress erythropoiesis, especially during recovery from acute anemia (Fang et al., 2018).

Another target of miR-451 is 14–3-3ζ (YWHAZ), a phospho-serine/threonine-binding protein that influences nuclear translocation of the transcription factor Forkhead box O3 (FOXO3) (Patrick et al., 2010; Yu et al., 2010). FOXO3 typically induces genes associated with antioxidant defense mechanisms. In miR-144/451 knockout erythroblasts, an excess of 14–3-3ζ induces the translocation of FOXO3 from the nucleus to the cytoplasm, thereby decreasing the expression of FOXO3-dependent antioxidant genes, notably CAT (catalase) and GPX1 (glutathione peroxidase 1) (Yu et al., 2010). This disruption impairs the erythroid antioxidant response (Yu et al., 2010). Thus, miR-451 targeting of 14–3-3ζ modulates the activity of FOXO3, enhancing the cell's ability to handle oxidative stress.

miR-144/451 also plays a vital role in erythropoiesis through the regulation of chromatin condensation (Kretov et al., 2024).

Studies by Kretov et a., showed that it primarily accomplishes this effect by targeting High Mobility Group Nucleosomal Binding Domain 2 (HMGN2), a crucial chromatin-associated protein.The depletion of miR-144/451 results in impaired chromatin condensation and inadequate erythrocyte maturation. Further, reducing HMGN2 levels can somewhat ameliorates this condition, highlighting a conserved regulatory axis from zebrafish to humans (Kretov et al., 2024).

24. miR-let-7d

miR-let-7d is a microRNA that has been shown to regulate iron uptake during late erythroid development (Byon & Papayannopoulou, 2012). It targets the non-IRE DMT1 (DMT -IRE) isoform, inhibiting iron transfer into erythroid precursors (Andolfo et al., 2010). Adequate iron import is required for hemoglobin synthesis and appropriate chromatin condensation as erythroblasts differentiate into reticulocytes. Overexpression of miR-let-7d at this stage can cause functional iron restriction, delaying hemoglobinization. Therefore, miR-let-7d functions as an effector of the iron transport machinery in erythroid cells. Its dysregulation in late-stage erythroblasts negatively impacts erythropoiesis due to insufficient iron for hemoglobin synthesis, emphasizing the necessity for tightly controlled miRNA expression for effective RBC maturation (Andolfo et al., 2010).

25. miR-218

miR-218 regulates iron metabolism during erythropoiesis by targeting 5′-aminolevulinate synthase 2 (ALAS2), the enzyme that catalyzes the first step in heme biosynthesis (Li et al., 2015). Overexpression inhibits erythroid differentiation and disrupts iron utilization. This results in mitochondrial iron accumulation and impaired heme production. Thus, miR-218 links iron homeostasis to red blood cell maturation (Li et al., 2015).

miR-124 (targeted by QKI5, discussed in RBP section).

During erythropoiesis, mature miR-124 is generated from the binding of RNA-binding protein QKI5 to the distal RNA motif of pri-miR-124–1, which in turn recruits the microprocessor complex, consisting of DiGeorge Syndrome Critical Region 8 (DGCR8) and Drosha ribonuclease III (DROSHA) (Wang et al., 2017). This complex cleaves the primary transcript resulting in pre-miR-124, which is then processed into mature miR-124. Mature miR-124 temporally inhibits the transcription factors TAL1 and MYB, subsequently limiting early and mid-stage erythroid proliferation and differentiation. As QKI5 and miR-124 levels decrease in the later stages, suppression of these proteins is reduced, allowing the progression of terminal erythroid maturation. This characterizes miR-124 as a temporal regulator in erythropoiesis and further demonstrates the propensity for multiple miRNAs to target multiple genes that contribute to extensive regulatory networks (Wang et al., 2017).

26. miR-669

During the last stages of erythroblast differentiation several genes, notably XK (X-linked Kx blood group protein) and AKAP7 (A-kinase anchoring protein 7), exhibit increased expression levels, each contributing to critical functions in RBCs. XK encodes a membrane-associated protein that participates in cell-surface organization, and AKAP7 functions as a scaffold protein for protein kinase A, promoting heme synthesis (Kotaki et al., 2020). miR-669 negatively modulates XK and AKAP7 by binding to their 3′UTRs, thereby inhibiting production. In in vivo models, overexpression of miR-669 was demonstrated to inhibit erythroblast maturation by restricting the transition from late basophilic to orthochromatic erythroblasts (Kotaki et al., 2020). Therefore, appropriate downregulation of miR-669 (as observed in normal erythropoiesis) is essential for optimal XK and AKAP7 expression to promote later stages of terminal erythroid differentiation.

27. miR-181a and miR-142

Both miR-181a and miR-142 play crucial roles in the last stages of erythroid differentiation by regulating the timing of nuclear condensation and enucleation (Figueroa et al., 2018; Rivkin et al., 2017). In the active nuclear transcription phase of erythroblasts, miR-181a is highly abundant and directly binds with the 3′-UTR of XPO7 (Exportin 7), inhibiting both its transcript and protein levels. As differentiation continues miR-181a expression decreases, facilitating an increase in XPO7 levels, which subsequently promotes chromatin condensation and promotes enucleation. Further, overexpression of miR-181a reduces XPO7 levels and inhibits enucleation (Figueroa et al., 2018).

miR-142 has been shown to maintain erythropoiesis by promoting hematopoietic progenitor proliferation and facilitating enucleation by directly targeting Ras-related C3 botulinum toxin substrate 1 (RAC1) (Rivkin et al., 2017). The deletion of miR-142 generates erythrocytes displaying aberrant biconcave morphology, impaired membrane elasticity, dysregulated reactive oxygen species (ROS) metabolism, and a reduced lifespan. miR-142 targeting of RAC1 in the late stages of erythropoiesis, namely in the Ortho-E and reticulocyte stages, promotes the enucleation process that is essential for proper RBC formation and functionality (Rivkin et al., 2017).

28. miR-320

miR-320 inhibits erythroid development by targeting the transferrin receptor CD71/TFRC (Byon & Papayannopoulou, 2012; Chen et al., 2008). During the final maturation stage of reticulocytes, miR-320 expression rises and CD71 levels decrease. The reduction in CD71 inhibits iron uptake and proliferation signals, allowing reticulocytes to mature into erythrocytes. miR-320 regulation contributes to iron import and proliferative activity promoting hemoglobinization, nuclear condensation, and terminal differentiation. miR-320 dysregulation, whether over- or under-expression, can disrupt these processes (Chen et al., 2008), underscoring its role as a fine-tuning "brake" that coordinates erythroid maturation.

28.1 Globin gene regulation

28.1.1 miR-let7

LIN28B RNA binding protein that is expressed at high levels in fetal erythroid cells binds and represses miR-let7 (Lee et al., 2013), which post-transcriptionally inhibits HIC2 (HIC ZBTB Transcriptional Repressor 2) (Huang et al., 2024). HIC2 in turn inhibits BCL11A (B-cell CLL/lymphoma 11A), a key repressor of γ-globin gene expression. This hierarchical regulatory cascade — let-7a/f inhibiting HIC2 to thereby allow BCL11A to inhibit γ-globin (HBG), exemplifies a miRNA-mediated inhibitory mechanism that regulates fetal hemoglobin silencing in adult erythroid cells (Huang et al., 2024), and reveals miR-let7 as a prospective target for diseases such as β-thalassemia and sickle cell anemia.

28.1.2 miR-326

miR-326 is a microRNA that modulates fetal hemoglobin synthesis by directly targeting the transcription factor Krüppel-like factor 1 (EKLF/KLF1), a key regulator of globin gene switching (Li et al., 2018). Studies by Li et al, showed that miR-326 binds to the 3′-untranslated region of EKLF mRNA and suppresses its translation, resulting in reduced KLF1 protein levels in human erythroid cells. In K562 cells and primary human $CD34^+$ progenitors, miR-326 overexpression decreases KLF1 protein and this is accompanied by elevated γ-globin (HbF) expression and altered β-globin and BCL11A levels, consistent with a shift toward fetal hemoglobin production. Conversely, inhibition of miR-326 increases KLF1 levels and reduces γ-globin expression, indicating that endogenous miR-326 regulates globin gene balance in erythropoiesis. Furthermore, miR-326 expression in reticulocytes from β-thalassemia major patients correlates positively with HbF levels, underscoring its potential role in HbF regulation and therapeutic relevance for hemoglobinopathies (Li et al., 2018).

28.1.3 miR-23a and miR-27a

miR-23a and miR-27a are involved in the control of β-like globin gene expression during erythropoiesis. These microRNAs directly target and suppress the transcriptional repressors Krüppel-like factor 3 (KLF3) and Sp1 transcription factor (SP1), leading to enhanced β-globin expression (Ma et al., 2013). Conversely, KLF3 and SP1 regulate the transcription of the miR-23a/27a cluster. This regulatory circuit maintains balanced globin gene expression during red blood cell maturation. This highlights the importance of microRNA–transcription factor interactions in fine-tuning hemoglobin production (Ma et al., 2013)·

28.1.4 miR-34a

miR-34a is a microRNA that promotes fetal hemoglobin (γ-globin) expression in erythroid cells. Stable overexpression of miR-34a in K562 cells increases γ-globin mRNA and protein levels, demonstrating its ability to reactivate fetal hemoglobin in an erythroid model system (Ward et al., 2016). Interestingly, in silico analyses identified a binding site for miR-34a on γ-globin mRNA. Since miR-34a primarily functions as a repressor, Ward and colleagues performed a luciferase assay to evaluate its direct effect on γ-globin mRNA (Ward et al., 2016). Intriguingly, miR-34a activated HBG1 and not HBG2 γ-globin mRNA, and the activation suggested an indirect mechanism involving a repressor protein. Mechanistically, this induction is associated with a reduction in total and phosphorylated STAT3 (Signal Transducer And Activator Of Transcription 3) protein, a transcription factor known to repress γ-globin expression, while other repressors such as YY1 (YY1 Transcription Factor) and HDAC1 (Histone Deacetylase 1) remain unaffected. These findings indicate that miR-34a enhances γ-globin expression at least in part through STAT3 down-regulation (Ward et al., 2016).

28.1.5 miR-2355-5p

miR-2355–5p regulates γ-globin expression in human erythroid cells by directly targeting Krüppel-like factor 6 (KLF6), a transcription factor important for erythroid differentiation (Cheng et al., 2021). By inhibiting KLF6, miR-2355–5p relieves repression of the γ-globin gene, resulting in increased fetal hemoglobin (HbF) production. Overexpression of miR-2355–5p enhances γ-globin levels, while its inhibition reduces HbF, confirming its regulatory role (Cheng et al., 2021).

28.1.6 miR-96

Argonaute RISC Catalytic Component 2 (AGO2) protein directly interacts with gamma globin mRNA at a higher extent in adult blood reticulocytes compared to cord blood reticulocytes which suggested direct regulation by miRNA (Azzouzi et al., 2011). Studies by Azzouzi et al., showed that miR-96 enriched more significantly in AGO2 complexes in adult blood compared to cord blood. miR-96 knockdown increased gamma globin, and over-expression decreased gamma globin, without altering β-globin in cord blood derived erythroid cells. Reporter assays were used to show that that miR-96 directly targets the open reading frame (ORF) of γ-globin at the predicted target site (Azzouzi et al., 2011).

28.1.7 miR-210-3p

miR-210–3p is induced under hypoxic conditions and plays a central role in stress-adapted erythropoiesis (Hu et al., 2021; Sarakul et al., 2013). It is regulated by GATA1 and promotes erythroid differentiation through direct targeting of SMAD2 (SMAD family member 2) (Hu et al., 2021). In addition, it indirectly enhances γ-globin expression via modulation of the PLC-β1/PKCα signaling pathway (Hu et al., 2021; Sarakul et al., 2013). Supporting this mechanism, it was demonstrated that PLC-β1 directly regulates miR-210 expression during mithramycin-induced erythroid differentiation in K562 cells, highlighting the functional importance of this signaling–microRNA axis in γ-globin upregulation (Bavelloni et al., 2014). Through these interconnected pathways, miR-210–3p links oxygen sensing, intracellular signaling, and erythroid maturation, underscoring its potential as a therapeutic target in hemoglobinopathies (Sarakul et al., 2013).

28.1.8 miR-144

In zebrafish development, miR-144 progressively elevates and inhibits embryonic α-globin by targeting the transcription factor KLFD (Fu et al., 2009). KLFD subsequently induces the transcription of miR-144, creating a negative feedback loop that regulates KLFD levels with precision. In mammals, the most likely ortholog is KLF1, and KLF1 −/− embryos display reduced ζ-globin levels. This feedback assures efficient regulation of embryonic α-globin during erythropoiesis (Fu et al., 2009).

As previously discussed, hundreds of miRNAs have been implicated to be contributing factors in the process of erythropoiesis. While we have focused on a fraction of the major contributors here, there are sure to be many more whose functions have yet to be characterized. The present challenges that exist in elucidating these functions are in part due to the nature of miRNAs, given that multiple miRNAs can influence multiple genes and conversely, multiple mRNAs impact the function of many miRNAs. This creates a vast regulatory network of cooperativity that demonstrates the important role miRNAs play in the fine tuning and temporal control of the expression of key regulatory factors that contribute to steady-state and dysregulated erythropoiesis (Fig. 3). In conclusion, while extensive knowledge has been gained over the last few decades about how miRNAs influence erythropoiesis, much is yet to be discovered about their contribution to this complex process. Table 2 shows a compilation of the role of micro RNAs in erythropoiesis.

28.2 Long non-coding RNAs (lncRNAs)

While 75 % of the human genome is transcribed, only 2 % represent protein coding genes. The rest of the transcription contributes to non-coding RNAs (Djebali et al., 2012). Long non-coding RNA (lncRNA) is defined as a subclass of noncoding RNAs (ncRNAs) that surpass 200 base pairs (bp) (Chen & Kim, 2024). According to the analysis of human genome and transcriptome data, there are more than 20,000 lncRNAs identified, with some studies estimating in excess of 100,000 lncRNAs in humans (Chen & Kim, 2024; Mattick et al., 2023). Around 50 % of lncRNAs are polyadenylated, and similar to mRNAs, around 98 % are spliced (Chen & Kim, 2024; Tsagakis et al., 2020). LncRNAs are highly diverse, vastly abundant, poorly evolutionarily conserved among species, and without a protein-coding potential despite a small open reading frame (Chen & Kim, 2024). LncRNA expression is tightly regulated in a tissue and developmental stage-specific manner, and is frequently dysregulated in

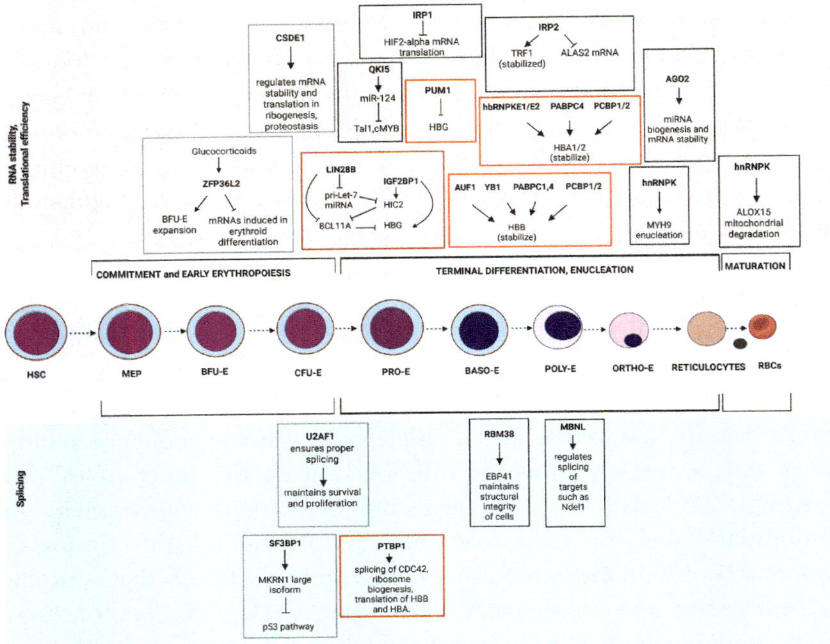

Fig. 3 miRNAs in mammalian erythropoiesis: miRNAs that play a role in post-transcriptional regulation of both early erythorpoiesi and terminal differentiation and maturation, are shown. Red boxes shown miRNAs invovled in globin gene regulation.

various disease states, underscoring its precise control (Paralkar et al., 2014). In erythropoiesis, regulatory control by lncRNAs fine-tunes the gene expression profiles of cells as they undergo differentiation from a hematopoietic stem cell into a mature, enucleated red blood cell.

Several classes of lncRNA have been identified in recent years including intergenic RNAs (located between protein-coding genes), intronic RNAs or ilncRNAs (transcribed from within the introns of protein-coding genes), enhancer RNAs or elncRNAs (transcribed from enhancer regions, antisense to other genes), and pseudogene-derived RNAs (plncRNAs) derived from pseudogenes (Hon et al., 2017), which play pivotal roles in eukaryotic development by diverse modes of regulation of gene expression via transcriptional and post-transcriptional mechanisms (Paralkar et al., 2014). Diverse functions of lncRNAs are associated with their nuclear or cytoplasmic cellular location (Ji Heon Noh, 2016; Miao et al., 2019). Overall, lncRNAs can modulate gene expression at the transcriptional level

Table 2 Role of microRNAs in erythropoiesis.

miRNA	Direct/indirect target (s)	Function in erythropoiesis	References
Role in erythroid lineage commitment and early erythropoiesis			
miR-150/15a	MYB	Forms regulatory feedback loop to fine-tune MYB expression at sequential developmental stages for lineage specification and differentiation.	Lu et al. (2008), Zhao et al. (2009)
miR-125b-2	DICER/ST18	Induces proliferation/self-renewal of MEPs, and lineage decision.	Klusmann et al. (2010)
miR-145	FLI-1	Downregulates Fli-1, promotes MEP commitment to erythroid lineage.	Klusmann et al. (2010), Kumar et al. (2011)
miR-26a	NLK	By targeting NLK, a negative regulator of erythropoiesis, miR-26a promotes erythroid differentiation.	Wilkes et al. (2020)
miR-221/222	cKIT	Progressively downregulated through early erythropoiesis to modulate cKIT levels to promote progenitor proliferation.	Felli et al. (2005), Lee et al. (2018), Mittal et al. (2013)
miR-223	LMO2	Progressively downregulated during erythropoiesis to promote erythroid progenitor commitment, differentiation, and maturation.	Brisot et al. (2025), Felli et al. (2009)
miR-126	PTPN9	Downregulation promotes colony formation.	Bu et al. (2014), Huang et al. (2011)
miR-24	ALK4	Stage specific up/down-regulation provides fine-tuning of activin signaling to balance proliferation and differentiation.	Wang et al. (2008), Zhang et al. (2012)

(continued)

Table 2 Role of microRNAs in erythropoiesis. (*cont'd*)

miRNA	Direct/indirect target (s)	Function in erythropoiesis	References
Role in terminal differentiation, enucleation and maturation			
miR-144/451	GATA2, NRF2, 14-3-3ζ, RAB14, COX10, CAP10, Hmgn2, MYC	Progressively upregulated during erythropoiesis to modulate expression of erythroid specific transcription factors to promote erythroblast formation, proliferation, hemoglobinization, protection against oxidative stress, chromatin condensation, and enucleation.	Dore et al. (2008), Fang et al. (2018), Huang et al. (2021), Kim et al. (2015), Kretov et al. (2024), Patrick et al. (2010), Xu et al. (2020), Yang et al. (2010), Yu et al. (2010)
miR-let7d	DMT -IRE	Downregulation promotes DMT-IRE expression to promote maturation and iron homeostasis.	Andolfo et al. (2010)
miR-218	ALAS2	Overexpression inhibits differentiation and leads to iron accumulation.	Li et al. (2015)
miR-124	TAL1, MYB	Promotes erythroid differentiation and maturation through TAL and MYB modulation.	Wang et al. (2017)
miR-669	AKAP7, XK	Downregulation increases expression of AKAP7/XK promoting terminal differentiation.	Kotaki et al. (2020)
miR-181a	XPO7	Downregulation prompts rise in XPO7 levels, essential for enucleation.	Figueroa et al. (2018)
miR-142	RAC1	Upregulation leads to repression of RAC1, and modulates enucleation	Rivkin et al. (2017)

miR-320	SMAR1, CD71/TFRC	Downregulation in early stages increases SMAR1 which results decrease of miR-221/222 leading to progenitor expansion. Upregulation in later stages inhibits CD71 expression to promote erythroid maturation.	Chen et al. (2008)
miR-let7	HIC2	Upregulation inhibits HIC2, a repressor of BCL11A, resulting in downregulation of γ-globin.	Huang et al. (2024), Lee et al. (2013)
miR-326	KLF1	Binds to 3′UTR of KLF1 mRNA and suppresses its translation; decreases in γ-globin.	Li et al. (2018)
miR-23a/-27a	KLF3/SP1	Suppresses expression of KLF3/SP1, thereby promoting β-globin transcription.	Ma et al. (2013)
miR-34a	STAT3	miR-34a expression leads to a reduction in STAT3 phosphorylation, activating γ-globin expression; overexpression activates HBG1 and not HBG2.	Ward et al. (2016)
miR-2355-5p	KLF6	Inhibits KLF6, thereby relieves repression of γ-globin.	Cheng et al. (2021)
Mir-96	Gamma globin	Directly interacts with γ-globin mRNA, and overexpression decreases γ-globin without altering β-globin.	Azzouzi et al. (2011)
miR-210-3p	SMAD2	Upregulated under hypoxic condition, and is regulated by GATA1. Promotes erythroid differentiation via SMAD2 and indirectly enhances γ-globin expression.	Hu et al. (2021), Sarakul et al. (2013)

(eg., by interacting with and recruiting chromatin modifying proteins and altering chromatin looping, either at cis or trans), or at the post-transcriptional level (eg., by acting as decoys for RNA binding proteins and microRNAs and impacting RNA stability and translational efficiency) (Ji Heon Noh, 2016). LncRNAs discussed below illustrate their roles in post-transcriptional and chromatin regulation during early erythropoiesis, terminal erythroid differentiation, and hemoglobin gene regulation.

28.3 LncRNAs involved in post-transcriptional regulation

28.3.1 HBBP1

Hemoglobin subunit beta pseudogene 1 (HBBP1), is a lncRNA expressed from a pseudogene in the β-globin locus (between γ- and δ-globin genes) (Huang et al., 2017; Kiefer et al., 2011), and is predominantly localized in the cytoplasm (Ma et al., 2021). Studies by Ma et al., showed that HBBP1 is expressed at higher levels during later stages of human erythroid differentiation (Ma et al., 2021). These studies further demonstrated that HBBP1 interacts directly with hnRNPA1 (heterogeneous nuclear ribonucleoprotein A1), an RNA-binding protein that is known to mediate RNA decay. HBBP1 thereby acts as a molecular decoy for hnRNAP1, and upregulates TAL1 (T cell acute lymphocytic leukemia 1), a master transcriptional regulator of erythropoiesis, by preventing hnRNPA1's degradation of TAL1 mRNA. Further, knockdown of HBBP1 lncRNA led to fewer erythroid progenitors and erythroblasts along with impaired hemoglobin induction, in both ex vivo and xenograft models of human erythroid differentiation. This study also found a positive correlation between the levels of HBBP1 and TAL1, and fetal hemoglobin. Another study documented a positive correlation between HBBP1, ELK1 (ETS transcription factor ELK1), and fetal hemoglobin (Ma et al., 2021). Notably, HBBP1 harbors variants are associated with the severity of β-thalassemia and high fetal hemoglobin levels (Galarneau et al., 2010; Giannopoulou et al., 2012; Nuinoon et al., 2010), and deletion of HBBP1 DNA sequence led to an increase in LCR–γ-globin chromatin looping (P. Huang et al., 2017).

28.3.2 UCA1

Urothelial carcinoma-associated 1 (UCA1) is an intergenic non-coding RNA that is predominantly expressed in the cytoplasm of erythroid cells (Liu et al., 2018). UCA1 was first discovered in bladder cancer and is well studied in different types of cancers (Wang et al., 2006; Xue et al., 2016). Studies by Liu et al., showed that in erythroid cells, UCA1 is dynamically expressed with maximal expression at the proerythroblasts stage (Liu et al., 2018). Their studies further showed that UCA1 transcription is controlled by erythroid transcriptional regulator GATA1. When UCA1 was knocked down it led to impairment of heme biosynthesis and inhibition of erythroid differentiation. UCA1 interacts with the RNA binding protein poly-pyrimidine tract-binding protein 1 (PTBP1) to bind to 5'-aminolevulinate synthase 2 (ALAS2) mRNA, an important protein for heme synthesis, to stabilize the mRNA and exert post-transcriptional regulation in heme metabolism (Liu et al., 2018).

28.3.3 LncRNA Fas-AS1/Saf

LncRNA Fas-antisense 1 (LncRNA Fas-AS1/Saf), is transcribed from the antisense strand of the first intron of the human Fas receptor (APO-1/CD95) gene and is not polyadenylated (Villamizar, Chambers, Mo, et al., 2016; Villamizar, Chambers, Riberdy, et al., 2016). Its expression increases during terminal erythroid differentiation, and the expression is induced by the binding of master transcription factors GATA1 and KLF1 to the Saf promoter (Villamizar, Chambers, Mo, et al., 2016). Studies by Villamizar et al., showed that overexpression of the lncRNA Saf renders cells less susceptible to Fas-mediated apoptosis. Mechanistically, lncRNA Saf binds to Fas pre-mRNA and SPF45 (Splicing factor 45) protein, which promotes alternative splicing to generate a soluble Fas protein isoform. Consequently, lncRNA Saf expression reduces surface Fas receptor levels, shielding maturing erythroblasts from apoptotic signals and ensuring their survival in the final stages of erythropoiesis (Villamizar, Chambers, Mo, et al., 2016).

28.3.4 DLEU2/ ShlncRNA-EC6

Deleted in Lymphocytic Leukemia 2 (DLEU2)/shlncRNA-EC6 shows dynamic expression during murine erythroid differentiation (Alvarez-Dominguez et al., 2014). Alvarez-Dominguez et al., showed that this lncRNA is broadly expressed in different cell types but in erythroid cells it is transcribed via a different promoter that is regulated by GATA1 and generates poly-adenylated and non-polyadenylated transcripts that are erythroid restricted (Alvarez-Dominguez et al., 2014). These studies further showed that both these erythroid restricted transcripts show similar stability, and microRNAs 16–11 and 15a are released from the polyadenylated precursor. In addition, DLEU2 non-polyadenylated variant was shown to act in cis to upregulate SRYD7/CLLD5 (SPRY domain-containing protein 7/Chronic Lymphocytic Leukemia Deletion

Region Gene 6), a gene located 46 kb away, which codes for a potential RNA binding protein whose functions in erythropoiesis are yet uncharacterized (Kulczyńska & Siatecka, 2016; Wang et al., 2015). DLEU2 levels increase during terminal erythroid differentiation, and knockdown of DLUE2 impairs enucleation without affecting earlier stages of proliferation or differentiation (Wang et al., 2015).

Another study showed that, DLEU2 post-transcriptionally suppresses RAC1 (Ras-related C3 botulinum toxin substrate 1) expression by binding to the 3′-untranslated region (3′-UTR) of RAC1 mRNA (Wang et al., 2015). These studies by Wang et al., demonstrated that RAC1 acts as a positive regulator of PIP5K gene expression, which prevents enucleation (Wang et al., 2015). Thus, by suppressing the inhibitory effects of the RAC1/PIP5K signaling pathway, DLEU2 mediates enucleation (Alvarez-Dominguez et al., 2014; Wang et al., 2015).

28.4 LncRNAs involved in chromatin regulation

28.4.1 LincRNA-EPS

In a study by Hu et al, aimed at identifying lncRNAs expressed during mouse erythropoiesis, 427 putative lncRNAs were identified, out of which 163 had a gradual increase in expression and 42 had a gradual decrease in expression (Hu et al., 2011). Among them, LincRNA erythroid pro-survival (LincRNA-EPS), an intergenic lncRNA that is localized in the nucleus, showed high expression during terminal differentiation of erythropoiesis. When LincRNA-EPS was inhibited, it led to an impairment in terminal differentiation and enucleation along with an increase in apoptosis. In macrophages, LincRNA-EPS interacts with hnRNPL (Heterogeneous nuclear ribonucleoprotein L) protein to alter nucleosome positioning and repress immune regulatory genes (Atianand et al., 2016), while protein interaction partner in erythroid cells is not known. In erythroid cells, lincRNA-EPS prevents apoptosis by repressing PYCARD (PYD and CARD domain containing), a proapoptotic gene that activates caspase (Hu et al., 2011).

28.4.2 LncEry

Another lncRNA that goes up in expression dramatically during murine erythroid differentiation is the mouse long non-coding RNA lncEry (Yang et al., 2022). When IncEry is knocked down it leads to impaired erythroid differentiation and maturation. LncEry interacts with WD repeat-containing protein 82 (WDR82) and promotes the transcription of globin genes, and KLF1, a master transcriptional regulator of erythropoiesis, thereby regulating the progression of erythroid differentiation (Yang et al., 2022).

28.4.3 ncRNA-a3

The ncRNA-a3 is an enhancer RNA that is transcribed from 50-kb downstream of TAL1, a master transcription factor regulating hematopoiesis (Matur et al., 2025). This eRNA regulates erythroid-specific TAL1 enhancer/promoter loop and cofactor recruitment (p300/BRG1) to the TAL1 locus, to induce TAL1 expression and transcription program in erythropoiesis. Depletion of ncRNA-a3 leads to reduced erythroid progenitor colonies, and impaired terminal erythroid differentiation (Matur et al., 2025).

28.4.4 Bloodlinc/alncRNA-EC7

Bloodlinc or alncRNA-EC7 is a long non-coding RNA that is conserved in both mouse and human, and is critical for the red blood cell development (Alvarez-Dominguez et al., 2017). Alvarez-Dominguez and colleagues estimated that it exists at 5–50 copies per cell in the late stage erythorblasts and is around 10-fold greater in expression than the median lncRNA levels in tissues, suggesting that high level Bloodlinc expression is specific to late-stage differentiating erythroblasts (Alvarez-Dominguez et al., 2017). These studies further showed that Bloodlinc is transcribed from the super enhancer for Band 3 anion transport protein (also known as Solute Carrier Family 4 Member 1/SLC4A1), an important membrane transporter for erythroid cells. When Bloodlinc is knocked down in murine erythroid cells, it leads to a decrease in enucleated cells and an increase in immature erythroblasts. In addition, these studies demonstrated that Bloodlinc can interact with the heterogeneous nuclear ribonucleoprotein U (HNRNPU) protein to form a ribonucleoprotein complex that regulates genes essential for terminal differentiation through trans-chromosomal interactions (Alvarez-Dominguez et al., 2017).

28.4.5 GATA2AS

GATA2AS is a 2080 bp lncRNA that is localized in the nucleus, and is expressed in erythroid progenitor cells but decreases in levels during differentiation (Liu et al., 2024). It is transcribed opposite to the transcription factor GATA2, a regulator of erythropoiesis (Liu et al., 2024; Oh et al., 2005). Liu and colleagues showed that GATA2AS interacts with erythroid transcription factors, ZBTB7A (Zinc finger and BTB domain-containing protein 7A) and KLF1 (Krüppel-like factor 1), and knockout of GATA2AS decreased their binding genome wide (Liu et al., 2024). Further, knockout or knockdown of GATA2S in human erythroid cells caused downregulation of GATA2, dysregulation of genes important for terminal differentiation, and downregulation of fetal globin genes (HBG1 and HBG2). This leads to decreased erythroid cell proliferation and accelerated differentiation. Overall, these studies established GATA2AS as an essential regulator of human erythroid differentiation (Liu et al., 2024).

28.4.6 SRA

The long non-coding RNA steroid receptor RNA activator (SRA) is an intergenic lncRNA that functions as an RNA coactivator and can regulate cell fate program including the erythroid transcriptional program (Lanz et al., 1999; Sawaengdee et al., 2020; Sheng et al., 2018). Sawaengdee and colleagues showed that SRA can directly interact with chromatin regulators such as CTCF (CCCTC-binding factor), WDR5 (WD repeat domain 5) and EZH2 (Enhancer of Zeste Homolog 2), and can bind to the chromatin of genes belonging to proliferation, Wnt signaling, NF-κB signaling, regulation of protein phosphorylation, cell differentiation, and metabolism (Sawaengdee et al., 2020). Further, SRA was also shown to bind to the globin genes and their locus control regions. Knockdown of SRA caused a decrease in the expression of globin genes and other erythroid genes in proerythrobasts, suggesting that SRA facilitates the erythroid transcriptional program (Sawaengdee et al., 2020).

28.4.7 BGLT3

BGLT3 is a 1.6 kb enhancer RNA (eRNA) that is localized in the nucleus, and is cotranscribed with gamma globin from the same DNA strand as the γ-globin gene and the other β-type globin genes (Huang et al., 2017; Kiefer et al., 2011). Ivaldi and colleagues showed that BGLT3 locus functions as an enhancer, and the locus is enriched with CBP/p300, LDB1 complex, and histone marks associated with enhancer activity (Ivaldi et al., 2018). Their studies further showed that both BGLT3 transcription and transcript are required for positively regulating γ-globin genes. Transcription through BGLT3 locus allows for looping of the locus to the γ-globin genes, and BGLT3 lncRNA transcript interacts with MED12 (mediator complex subunit 12) mediator subunit. Both together, contribute to supporting γ-globin transcription by increasing RNA Pol II density on the gene beyond the level contributed by the β-globin locus control region (LCR) enhancer

(Ivaldi et al., 2018). In addition, HIDALGO, a long non coding RNA that encompasses the HBG1 and BGLT3 locus has been described, which when knocked down reduced HBG1 but not HBG2 γ-globin transcription (Werner et al., 2017).

28.4.8 Myrlin

Myb enhancer long intergenic non-coding RNA termed Myrlin, is a long non-coding RNA that is transcribed from murine −81 kb MYB (MYB proto-oncogene) enhancer (Kim et al., 2024). Kim et al., showed that Myrlin is primarily nuclear localized, and similar to MYB, its expression decreases during erythroid differentiation (Kim et al., 2024). Further, dissection of the lcnRNA transcript versus transcriptional activity at the locus demonstrated a delineation of functions. Myrlin TSS deletion led to reduction of Myrlin and MYB levels along with disruption of MYB enhancer hub, and reduction of LDB1 (LIM domain binding 1) complex occupancy across MYB enhancers. On the other hand, reduction of Myrlin transcript levels (without altering the sequence at the −81kb MYB enhancer) led to the reduction of MYB levels without causing the loss of the MYB enhancer hub. However, this caused a decrease in KMT2A (Lysine Methyltransferase 2A) and WDR5 (WD repeat domain 5) enrichment at the MYB locus mediated by direct interaction of Myrlin transcript with KMT2A complex components including WDR5, which deposit H3K4me3 mark of active transcription. Through this interaction, Myrlin helps recruit CDK9 (Cyclin Dependent Kinase 9), the protein kinase submit of pTEFb (Positive Transcription Elongation Factor b) that confers RNA Pol II serine 2 phosphorylation and allows efficient RNA Pol II elongation at the MYB gene (Kim et al., 2024).

29. LncRNA PCED1B-AS1 and DANCR

A study by Ren et al., investigated the interaction between lncRNAs and chromatin accessibility during hematopoiesis, and observed stage specific changes through erythropoiesis, and these changes were more dramatic during terminal erythroid differentiation (Ren et al., 2021). Two of these erythroid stage-specific lncRNAs, were functionally characterized:

LncRNA PCED1B-AS1 (PC-esterase domain containing 1B anti-sense RNA 1) was significantly decreased in KLF1 null neonatal anemia and in sickle cell disease (Demirci et al., 2020; Magor et al., 2015; Ren et al., 2021). Knockdown of this lncRNA caused decreased erythroid progenitor colonies, and a delay in terminal erythroid differentiation (Ren et al., 2021). Further overexpression of this lncRNA promoted the expression of globin genes during terminal erythroid differentiation. Its functions in terminal erythroid differentiation is mediated by interaction and cooperation with GATA1 in regulating chromatin accessibility (Zhu et al., 2019).

DANCR (Differentiation Antagonizing Non-Protein Coding RNA) is a long non-coding RNA whose expression peaks at the MEP stage (Ren et al., 2021). Depletion of DANCR leads to a decrease in the production of erythroid progenitors and in the expression of erythroid specific genes including HBG and HBB globin genes. Overexpression of DANCR promotes erythroid differentiation while inhibiting megakaryocyte differentiation, suggesting that it regulates the balance between the two lineages by coordinating accessible chromatin, and contributes to lineage decisions (Ren et al., 2021).

30. CpoxeRNA

Coproporphyrinogen oxidase enhancer-associated RNA (CpoxeRNA) is an erythroid-specific enhancer RNA identified by studying RNA-seq and promoter Capture Hi-C data from mouse E14.5 fetal liver (Xie & Dean, 2025). The study by B. Xie and A. Dean showed that CpoxeRNA is transcribed on the antisense stand within a super enhancer upstream from the coproporphyrinogen oxidase (CPOX) gene that codes for CPOX, an enzyme important for heme synthesis (Xie & Dean, 2025). It is predominantly located in the nucleus and interacts with CTCF/cohesion complex components to regulate chromatin looping both in cis and trans. When CpoxeRNA is knocked down, it leads to a decrease in CPOX enzyme that is important for heme synthesis, and also a decrease in other erythroid genes bound by CpoxeRNA, such as ERCC1 (ERCC excision repair 1). In addition, CpoxeRNA depletion leads to impaired proliferation and enucleation during terminal erythroid differentiation (Xie & Dean, 2025).

31. HIKER

Hypoxia-Induced Kinase-mediated Erythropoietic Regulator (HIKER/ LINC02228), is a nuclear localized lncRNA, that is upregulated under hypoxic conditions and contributes to excessive erythrocytosis in Common Mountain sickness (CMS/Monge's disease) (Azad et al., 2023). Studies by Azad and colleagues showed that the downregulation of HIKER in iPSC derived CD34+ showed a significant reduction of burst forming unit (BFU-e) numbers in CMS cells under hypoxia (Azad et al., 2023). This reduction in BFU-e colony numbers was rescued by the expression of CSNK2B Casein Kinase 2 beta, the regulatory subunit of Casein Kinase 2 in HIKER knock-down cells, suggesting that CSNK2B mediates the functions of HIKER. Further, these studies showed that CSNK2B plays an evolutionarily conserved role in erythropoiesis, and the overexpression of CSNK2B causes excessive erythropoiesis under hypoxia (Azad et al., 2023).

32. Lockd lncRNA

Lockd (lncRNA downstream of CDKN1B), is a 434 bp poly-adenylated lncRNA that originates 4 kb downstream of the cell cycle regulator CDKN1B gene (Paralkar et al., 2016). Paralkar and colleagues showed that Lockd has higher abundance in mouse erythroblasts compared to other tissues. Further, deletion of the Lockd gene caused a reduction of both Lockd and CDKN1B transcript levels. In addition, insertion of poly-adenylation (polyA) cassette 80 bp downstream of the transcription start site while preserving the cis elements in the genomic locus, caused a reduction in Lockd lncRNA levels without altering CDKN1B transcript levels. Therefore, while cis element at the promoter of Lockd locus positively regulates CDKB1B, the Lockd lncRNA transcript is dispensable for this function (Paralkar et al., 2016).

As described above, lncRNAs modulate erythropoiesis through post-transcriptional regulation and chromatin remodeling. While lncRNA functions in erythropoiesis are emerging, their impact on post-transcriptional regulation remains understudied. Overall, lncRNAs fine-tune gene expression and modulate function as multipotent progenitors commit to erythropoiesis and as erythroid progenitors undergo terminal differentiation and mature into an enucleated red blood cell (Fig. 4). Table 3 shows a compilation of the role of long non-coding RNAs in erythropoiesis.

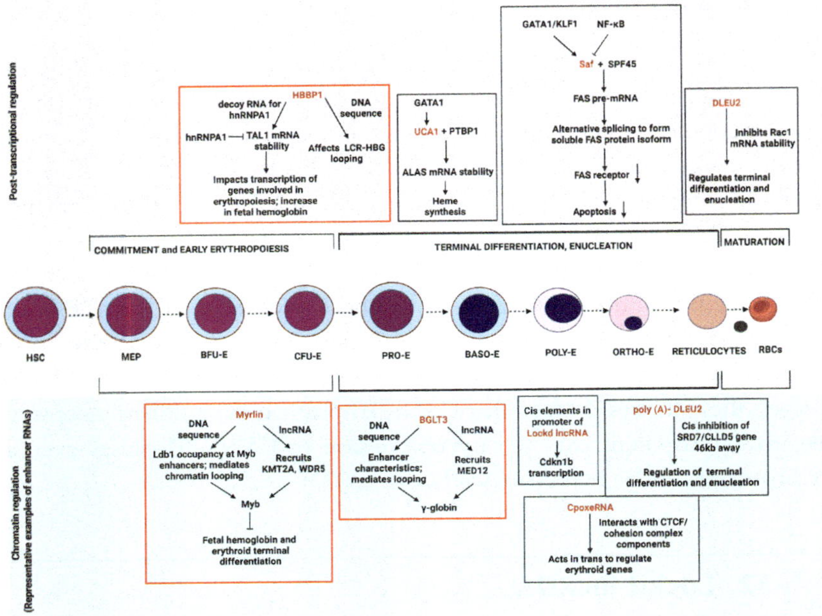

Fig. 4 LncRNAs in mammalian erythropoiesis: LncRNAs involved in post-transcriptional regulation and examples of lncRNAs that modulate chromatin organization during mammalian erythropoiesis are shown. Red boxes indicate lncRNAs involved in globin gene regulation.

32.1 RNA modifications

Recent years have seen an increase in the number of RNA chemical marks, adding an additional layer of gene regulation known as epitranscriptomics or the study of RNA modifications (Cantara et al., 2011). The field of epitranscriptomics has close ties to the field of erythropoiesis as the one of the first reported mRNAs with methylated derivatives was the globin mRNA from immature duck erythrocytes (Perry & Scherrer, 1975). In eukaryotic mRNA, over 150 RNA modifications apart from the modification at the 5′ cap structure N^7-methylguanosine (m^{7}G) (Rottman et al., 1974; Wei et al., 1975) have been described. Among these only a handful have been studied for their functions in cell fate regulation (Gunage & Zon, 2024). These include (Fig. 5): N6-methyladenosine (m6A) (Wei et al., 1976), N1-methyladenosine (m1A) (Dunn, 1961), 2′-O-methylated nucleosides (Dong et al., 2012), 5-methylcytosine (m^{5}C) (Squires et al., 2012), and internal N^6-methyladenosine (m^{6}A) (Dong et al., 2012; Squires et al., 2012;

Table 3 Role of long non-coding RNAs in erythropoiesis.

lncRNAs	Protein Interaction (s) in erythroid cells	Function	References
Role in post-transcriptional regulation			
HBBP1	hnRNPA1	It acts as a molecular decoy for hnRNPA1 and prevents the post-transcriptional degradation of TAL1 mRNA by hnRNPA1, thereby stabilizing TAL1 transcripts.	Ma et al. (2021), Giannopoulou et al. (2012), Nuinoon et al. (2010), Huang et al. (2017)
UCA1	PTBP1	Regulates ALAS2 mRNA stability Downregulation disrupts heme synthesis and terminal erythroid differentiation	Liu et al. (2018)
Fas-AS1/Saf	SPF45	LncRNA Saf expression reduces surface Fas receptor levels, shielding maturing erythroblasts from apoptotic signals and ensuring their survival in the final stages of erythropoiesis.	Villamizar et al. (2016)
EC6/DLEU2	RAC1	Promotes enucleation	Alvarez-Dominguez et al. (2014), Wang et al. (2015)
Role in chromatin organization			
LincRNA-EPS	Unknown	Prevents apoptosis by repressing PYCARD during terminal erythroid differentiation.	Hu et al. (2011)
LncEry	WDR82	Promotes transcription of KLF1 and globin genes.	Yang et al. (2022)
ncRNA-a3	TAL1	Regulates erythroid specific TAL1 enhancer/promoter loop. Depletion leads to impaired terminal erythroid differentiation.	Matur et al. (2025)

(continued)

Table 3 Role of long non-coding RNAs in erythropoiesis. (*cont'd*)

lncRNAs	Protein Interaction (s) in erythroid cells	Function	References
Bloodlinc/alncRNA-EC7	hnRNPU	Ectopic expression leads to an increase in enucleation, a decrease in immature erythroblasts, and stimulates erythroid gene expression.	Alvarez-Dominguez et al. (2017)
GATA2AS	Interacts with ZBTB7A and KLF1	Knockout or knockdown decreases GATA2, HBG1 and HBG2 levels, decreases proliferation and accelerates erythroid differentiation due to dysregulation of genes important for erythroid terminal differentiation.	Liu et al. (2023)
SRA	Interacts with CTCF, WDR5, and EZH2.	Binds erythroid genes, and globin genes and their locus control regions, and stimulates their expression.	Sawaengdee et al. (2020)
BGLT3	MED12	BGLT3 transcription and transcript are required for positively regulating γ-globin genes. Transcription through BGLT3 locus allows for looping of the locus to the γ-globin genes, and transcript interacts with MED12.	Ivaldi et al. (2018), Werner et al. (2017)
Myrlin	Interacts with KMT2A complex components including WDR5	Both Myrlin transcript and transcriptional activity contribute to MYB expression. Myrlin transcript interacts with KMT2A complex components and recruits CDK9 to allow RNA Pol II elongation at the MYB gene, and the locus promotes MYB enhancer hub.	Kim et al. (2024)

LncRNA PCED1B-AS1	GATA1	Decreased in KLF1 null neonatnal anemia and in sickle cell disease. Knockdown causes decreased erythroid progenitor colonies, and a delay in terminal erythroid differentiation.	Ren et al. (2021), Demirci et al. (2020), Magor et al. (2015), Ren et al. (2021), Zhu et al. (2019)
DANCR	Unknown	Knockdown decreases expression of erythroid specific genes including HBG and HBB globin genes. Overexpression promotes erythroid differentiation while inhibiting megakaryocyte differentiation.	Ren et al. (2021)
CpoxeRNA	Interacts with CTCF/ cohesion complex components	Regulates CPOX levels. Depletion leads to impaired proliferation and enucleation during terminal erythroid differentiation.	Xie and Dean (2025)
HIKER/ LINC02228	Unknown	Contributes to excessive erythrocytosis in Common Mountain sickness (CMS/ Monge's disease); CSNK2B mediates the functions of HIKER.	Azad et al. (2023)
Lockd lncRNA	Unknown	Cis element in Lockd promoter positively regulates CDKN1B, but Lockd lcnRNA transcript is dispensable for this regulation.	Paralkar et al. (2016)

Wei et al., 1976), pseudouridine (Carlile et al., 2014; Schwartz et al., 2014), addition of terminal uridines (Connell et al., 1997), and conversion from adenosine to inosine (Bass & Weintraub, 1988). Following is a brief description of these epitanscriptomics marks.

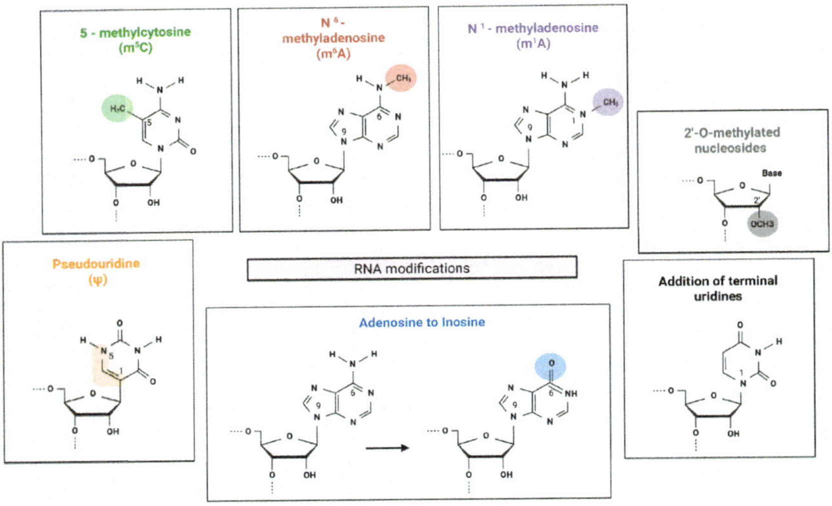

Fig. 5 The chemical structures of RNA modifications discussed in this chapter are shown.

33. N6-methyladenosine (m6A)

The N6-methyladenosine (m6A) mark is the most abundant non-cap internal modification present in the messenger RNA of all higher eukaryotes whose function has remained elusive (Jia et al., 2013). The abundance of m^6A has been shown to be ~0.1–0.4 % of total adenosine residues in cellular mRNA, and the average content of m^6A has been estimated to be 3–5 residues per mammalian mRNA (Meyer et al., 2015; Schibler et al., 1977; Wei et al., 1976). Marks of m6A occurring in adenines of RRACH sequence (R=G or A; H=A, C, or U) are in abundance in the 3′ untranslated regions (3′ UTRs), in long internal exons, and around stop codons (Dominissini et al., 2012). The location of these m6A marks within the 3′ UTRs or near stop codons usually promotes the transcript turnover. However, enrichment of m6A in the 5′ UTR region facilitates cap-independent translation initiation by recruiting eukaryotic initiation factor 3 (eIF3), a mechanism particularly important under conditions of cellular stress (Du et al., 2016; Meyer et al., 2015; Wang et al., 2014).

Details of proteins that are involved as writers, structural stabilizers and erasers of m6A proteins are also discussed in the RNA binding proteins section. The m6A marks are deposited in the nucleus by the m6A methyltransferase (MTase) complex (writers), consisting of METTL3 (methyltransferase-like 3), the catalytic subunit, METTL14 (methyltransferase-like 14), structural scaffold protein, WTAP (Wilm's tumor 1-associating protein), RNA-binding protein that stabilizes target mRNA, and other cofactors (Meyer et al., 2015; Yan et al., 2023).

Reduced METTL3 protein, and hence reduced m6A levels in zebrafish embryos, has been shown to lead to the disruption and complete loss of the transient population of primitive blood cells in the early embryo generated from the endothelium (Zhang et al., 2017). The epitranscriptome mapping revealed that the lack of the primitive HSC emergence was caused by reduced m6A mRNA methylation of NOTCH1 (Notch homolog protein 1) and of other arterial endothelial genes (Lv et al., 2018). There also exists an interplay between epigenetic and epitranscriptomic mechanisms in controlling m6A levels via modulating METTL3 binding to mRNAs during development (Aguilo et al., 2015; Chen et al., 2015).

m6A modification and its regulators are key participants in HSC homeostasis since loss of YTHDF2 (YTH N6-methyladenosine RNA binding protein F2), a protein that mediates targeted m6A-mRNA degradation resulted in an increased proliferation of HSC without altering lineage bias (Wang et al., 2018). Additionally, after 5-FU (5-Fluorouracil) treatment (to induce hematopoietic failure), the regenerative capacity of HSCs increased after YTHDF2 loss (Wang et al., 2018). m6A mapping revealed that mRNAs of transcription factors important for stem cell self-renewal, such as TAL1 and GATA2 are m6A-labeled and recognized by YTHDF2, which drives them to decay sites suggesting that YTHDF2 normally suppresses HSC self-renewal through RNA degradation (Li et al., 2018). YTHDF3 (YTH N6-methyladenosine RNA binding protein F3) is another YTH domain containing m6A reader, that cooperates with METTL3 to recognize and promote the translation of CCND1 (Cyclin D1) mRNA, the m6A-modified regulator of cell cycle progression, thereby supporting HSC self-renewal (Zhang et al., 2022).

The m6A writers METTL3 and METTL14 (Methyltransferase Like 3 and 14) control HSC self-renewal, and METTL3/METTL14-deficient mice displayed the accumulation of HSCs and a reduction in their self-renewal capacity (Yao et al., 2018). METTL3 enhances the expression of genes related to HSC quiescence, maintenance, and self-renewal, such as NR4A2 (Nuclear receptor subfamily 4 group A member 2), CDK1A (Cyclin-dependent kinase inhibitor 1A), BMI1 (BMI proto-oncogene, polycomb ring finger), and PRDM16 (PR/SET domain-containing 16) (Yao et al., 2018). Further, m6A is essential for HSC quiescence and self-renewal, since METTL3 knock-out HSCs are less quiescent and have increased metabolism, as well as mitochondrial activity (Cheng et al., 2019).

During erythropoiesis, m6A sites have been identified on the mRNA molecules of transcription factors that coordinate erythroid differentiation, such as KLF1, GATA1, and GATA2 (Kuppers et al., 2019). These genes function as principal regulators of erythropoiesis. The requirement of m6A during normal erythroid differentiation is further revealed in CRISPR-Cas knockdown studies of the MTase complex subunits METTL3, METTL14, and WTAP; all of which result in impairment of erythroid differentiation (Kuppers et al., 2019). Specifically, m^6A promotes translation of enzymes that modify histones and affect gene activation, such as SET Domain Containing (SETD) histone methyltransferases, which enhances Histone H3 Lysine 4 trimethylation (H3K4me3) histone marks and KLF1 recruitment at erythroid gene promoters—thus helping maintain erythroid transcriptional identity (Kuppers et al., 2019). In addition, m6A methyl transferase function promotes selective translation of m6A marked mRNAs such as those coding for ribosomal components and polyA RNA binding proteins (Kuppers et al., 2019). Overall, these studies show that m6A methyltransferase function is critical just after MEP stage of erythroid lineage specification, and reveal a physiological role for m6A dependent regulation through promoting translation of mRNAs that are required for erythroid gene expression and lineage specification.

METTL16 is another m6A methyltransferase with distinct targets relevant to erythroid homeostasis (Yoshinaga et al., 2022). METTL16 depletion in erythroblasts causes loss of m6A on DNA-repair transcripts, accumulation of DNA damage, and increased apoptosis (Flaherty et al., 2025; Yoshinaga et al., 2022). METTL16 targets the methionine adenosyltransferase 2A (MAT2A) transcript, responsible for producing S-Adenosylmethionine (SAM), a universal methyl donor essential for various methylation reactions in cells (Flaherty et al., 2025). METTL16 indirectly influences SAM levels by modulating MAT2A expression, thereby impacting global methylation capacity and cellular methylation-dependent processes which indirectly influence both RNA methylation and chromatin states crucial for erythroid gene expression. Supplementing cells with SAM can rescue some defects observed upon METTL16 depletion, demonstrating that METTL16's control of SAM contributes functionally to erythropoiesis (Yoshinaga et al., 2022).

There is a fine balance that exists in terms of the m6A levels on RNA since persistence of m6A marks also impacts the regulation of HSC self-renewal and differentiation. This was identified from studies of the m6A eraser ALKBH5 (AlkB homolog 5, RNA demethylase) (Gao et al., 2023). Loss of ALKBH5 results in destabilization of the OGDH (oxoglutarate dehydrogenase) mRNA, impaired mitochondrial oxidative phosphorylation and ATP production; and consequently, reduced HSC metabolic activity, self-renewal, and long-term repopulation potential (Gao et al., 2023).

Further, a recent study by Gunage et al., deposited in bioRxiv, identified ALKBH5 as a direct regulator of ATXN2 mRNA methylation. The study showed that ATXN2 overexpression accelerated the erythroid differentiation of HSPCs and rescued the erythroid differentiation defect in ALKBH5 mutant cells. Mechanistically, ATXN2 facilitated the release of stress granule stored m6A-modified RNAs including erythroid-specific and SG-enriched RNAs, which then are loaded onto functional ribosomes, allowing better translation and accelerated erythroid differentiation during stress (Gunage et al., 2024).

34. N1-methyladenosine (m1A)

N1-methyladenosine (m1A) is found in high abundance in rRNAs and tRNAs, while it has also been found in other types of RNAs such as mRNA, mitochondrial RNA and lncRNAs (Xiong et al., 2023). The methyl moiety is provided by S-adenosylmethionine (SAM) (Dominissini et al., 2016), and TRMT6/61A tRNA methyltransferases (TRMT), serve as writers of m1A on tRNAs and mRNAs (Anderson et al., 1998; Bar-Yaacov et al., 2016), while NML (Nucleomethylin) and BMT2 (Base Methyltransferase of 25S rRNA 2 Homolog/ S-adenosylmethionine sensor upstream of mTORC1) serve as writers of rRNA (Ozanick et al., 2005; Waku et al., 2016). Similar to m6A marks, AlkB family members and YTH domain protein family members are involved in erasing and reading m1A marks: ALKBH1/3/7 (AlkB homolog 1/3/7) (Liu et al., 2016; Ougland et al., 2004; Zhang et al., 2021) and FTO (Fat mass and obesity-associated protein/ALKBH9) (Wei et al., 2018) serve as erasers of these marks, and YTHDF1/2/3 (YTH N6-methyladenosine RNA binding protein F1/2/3) serve as readers of these marks (Dai et al., 2018; Seo & Kleiner, 2020; Zheng et al., 2020).

This modification plays a role in development and disease, including stress responses and tumorigenesis. Upon stress, m1A levels increase in mRNAs, and this contributes to stress granule formation (Alriquet et al., 2020). In the context of erythropoiesis, deletion of Nucleomethylin (NML) that catalyzes m1A formation in rRNA, causes severe defects in embryonic erythropoiesis leading to lethality in mice (Murakami et al., 2018). In these mice, while rRNA levels were not altered, m1A level was significantly reduced in 28S rRNA. Further, while m6A is enriched in 3′UTRs, m1A is enriched in 5'UTRs and dramatically around the start codon, and this promotes translation of these transcripts (Dominissini et al., 2012; Safra et al., 2017; Xiong et al., 2023).

35. 5-methylcytosine (m5C)

5-methylcytosine (m5C) is a common RNA modification in eukaryotic cells, observed in mRNAs, tRNAs, rRNAs and ncRNAs (Chen et al., 2021; Hussain et al., 2013). The m5C regulation depends on 'writers' or methyltransferases (examples: NOP2, NSUN2, NSUN3, NSUN4, NSUN5, NSUN6, NSUN7, DNMT1, DNMT3A, DNMT3B, and TRDMT1), 'erasers' or demethylases (examples: TET1, TET2, TET3, and ALKBH1) and 'readers' or m5C binding proteins that recognize and bind to the m5C modification sites (examples: ALYREF and YBX1) (Chen et al., 2021; Hussain et al., 2013; Nombela et al., 2021; Zhang et al., 2021). RNA m5C modification affects multiple aspects of biological processes, including RNA export, RNA stability, and translation (Cui et al., 2022). Despite their demonstrated roles in cancer progression, there is not much insight into their roles in erythropoiesis, which remains elusive.

36. ψ Synthesis and effectors

Another RNA modification is the inclusion of Ψ or pseudouridine in the RNA sequence. Ψ is the fifth nucleoside and substitutes for nearly ~0.2–0.6 % of the total uridines in human mRNA and has been detected in almost all known RNA species (Jalan et al., 2024). Ψ stabilizes the RNA backbone and promotes transcript stability, and this modification is rarely seen in 5′UTRs but is enriched in CDS and 3′ UTRs (Jalan et al., 2024). Functionally, ψ regulates splicing and it is enriched in alternative splicing-associated elements and binding sites for RNA-binding proteins (RBPs) that function in splicing and RNA higher-order structure (Gilbert & Nachtergaele, 2023; Gunage & Zon, 2024). Ψ amounts to 1.4% of all bases in human ribosomal RNAs and has been shown to modulate rRNA conformational dynamics and defects in rRNA pseudouridylation decrease ribosome-ligand interactions and translational fidelity (Barozzi et al., 2022). Ψ also enhances the affinity for tRNA binding to the A and P sites (Jack et al., 2011).

Not much is known about the role of ψ in erythropoiesis, but Ψ is known to have an important role in HSC function. Dysfunction of Ψ is implicated in cancer with Ψ significantly increased in serum and urine samples from patients with various hematologic malignancies, including leukemia (Carlile et al., 2019). An example of how Ψ impacts hematopoeisis comes from studies of a ψ synthesis enzyme, dyskerin. Dyskerin or dyskeratosis congenita 1 enzyme is one of several enzymes that catalyze the conversion of uridine to ψ by bonding C5 of the heterocyclic ring to the C1′ atom of the pentose sugar. The dyskerin enzyme DKC1 (Dyskeratosis Congenita 1), along with small-nucleolar RNAs, guide rRNA pseudouridylation. The DKC1 gene, which encodes dyskerin, is associated with X-linked dyskeratosis congenita (DC), a rare inherited bone marrow failure syndrome with patients demonstrating an increased risk of developing leukemia, particularly AML. Dyskerin is also a component of the telomerase complex and is involved in the pseudouridylation of ribosomal RNA (rRNA). Dysregulation of dyskerin can lead to abnormal rRNA pseudouridylation patterns, affecting ribosome function and potentially contributing to leukemogenesis (Garus & Autexier, 2021). CD34+ cells from patients with the DKC1 mutation show improper dyskerin function that impairs the in vitro growth properties and the in vivo repopulating ability of human HSPCs (Carrascoso-Rubio et al., 2021).

Another example of ψ synthesis enzymes is the PUS family which are 13 demonstrated members with diverse targets. RNA-independent PUS enzymes edit a variety of RNA substrates, including tRNAs, mRNAs, and others. Ψ can be added cotranscriptionally to modify mRNA precursors (pre-mRNAs) in humans by PUS1 (Pseudouridine Synthase 1), PUS7 (Pseudouridine Synthase 7), and RPUSD4 (RNA pseudouridine synthase D4) (Gilbert & Nachtergaele, 2023; Gunage & Zon, 2024). The PUS7 enzyme is reported to modulate tRNA-derived small fragments (tRFs). PUS7 binds distinct tRNAs and controls biogenesis of tRFs, and PUS7-mediated Ψ deploys specific tRFs to repress translation in stem cells. Thus, PUS7 loss impairs early embryogenesis and impacts hematopoietic commitment (Guzzi et al., 2018).

37. 2′-O-methyl (2′OMe)

2′-O-methyl (2'OMe) denotes the methylation of the 2′-hydroxyl group of ribose sugar from any one of the four ribonucleotides. It is found in tRNAs, mRNAs, snoRNAs, small non-coding RNA and is the most abundant rRNA modification (Ayadi et al., 2019). 2'Ome protects RNA from nuclease activity, enhances RNA-RNA interactions, destabilizes RNA-protein interactions and changes RNA tertiary structures (Dimitrova et al., 2019). During erythroid differentiation, rRNA 2'OMe dynamics followed a distinct trajectory and genetic perturbation of individual 2'OMe sites on HSC rRNAs demonstrated their distinct roles in modulating proliferation and differentiation. Specifically, 28S-Gm4588, a specific 2′OMe site, is progressively lost during differentiation, and was identified as a key regulator of HSC self-renewal. Low methylation at this site led to translational skewing, which promoted differentiation. Functionally, HSCs with diminished 28S 2′OMe exhibited impaired self-renewal capacity ex-vivo, and loss of fitness in-vivo in bone marrow transplantations (Rabany et al., 2026).

38. A-to-I editing

The existence in higher eukaryotes of adenosine deaminases (ADAR) that deaminate genomically encoded adenosine to inosine (A-to-I editing) in double-stranded RNA, forms another common form of RNA modification that can be introduced both co-transcriptionally and post-transcriptionally (Hartner et al., 2009). Interestingly, A-I editing can occur on all forms of RNA such as pre-mRNAs, mRNAs, and noncoding RNAs, including microRNAs (miRNAs), long non-coding RNAs (lncRNAs), and tRNAs (Liu et al., 2022). Of the three ADAR proteins present in mammals, ADAR1 (ADAR/Adenosine Deaminase Acting On RNA), ADAR2 (ADARB1/Adenosine Deaminase RNA Specific B1), and ADAR3 (ADARB2/Adenosine Deaminase RNA Specific B2), only ADAR1 is expressed broadly throughout mammalian cell types (Liddicoat et al., 2016). ADAR1 gene KO mice are embryonic lethal by E12.5 and have massive failure in hematopoietic activity and this was attributed to its enzymatic functions (XuFeng et al., 2009). Selective knockout of ADAR1 in the hematopoietic progenitor cell (HPC) leads to their increased apoptosis, when they underwent differentiation (XuFeng et al., 2009). ADAR1 gene mutations are associated with disorders of the immune system and blood development and their dysfunction leading to malignant reprogramming of myeloid progenitors seen in chronic myeloid leukemia and multiple myeloma (Gunage & Zon, 2024; Liu et al., 2022).

39. ZCCHC6/DIS3L2 RNA editors

The terminal uridylyltransferase (TUTase/TUT) enzyme ZCCHC6 (TUT7/Terminal uridylyltransferase 7) adds uridines to the 3′ end of target RNAs (terminal uridines), which serves as a recognition signal for the 3′to 5′ exonuclease DIS3L2 (DIS3-like 3′−5′ exoribonuclease 2), leading to the degradation of target RNAs, especially during cellular differentiation (Faehnle et al., 2014; Pirouz et al., 2016; Thomas et al., 2015). During erythropoiesis, a recent finding shows that deficiency of ZCCHC6 and DIS3L2 in mouse models led to fetal and perinatal reticulocytosis, an accumulation of RNA-rich precursors of terminal erythroid cells, suggesting their crucial roles in terminal red cell differentiation (Han et al., 2025). This was not the case when a ZCCHC6 paralogue, ZCCHC11 (Zinc Finger CCHC-Type Containing 11) which also functions as a TUTase, was deleted from the mouse, indicating the specificity of the enzyme for this function. Interestingly, the study also showed that the degree of reticulocytosis attenuated with age suggesting that there may be differential RNA regulation in the bone marrow and adult peripheral blood compared to the embryonic spleen and blood (Han et al., 2025). Further, knockout of ZCCHC6 and DIS3L2 led to the increase in human fetal hemoglobin transcript and protein levels, and also the levels of mouse Hbb-bh1 globin (mouse embryonic β-globin), by prolonging the mRNA half-life, adding to the evidence for post-transcriptional regulation of hemoglobin gene switching (Han et al., 2025).

39.1 Post-transcriptional regulation of hemoglobin switching-a paradigm for post-transcriptional mechanisms regulating normal development and physiology

We discuss below mechanisms identified thus far in the post-transcriptional regulation of hemoglobin switching mediated through RNA binding proteins, miRNAs, lncRNAs and RNA modifications. While these factors are described in more detail in previous sections, a summary of their roles in hemoglobin switching is presented below (Fig. 6).

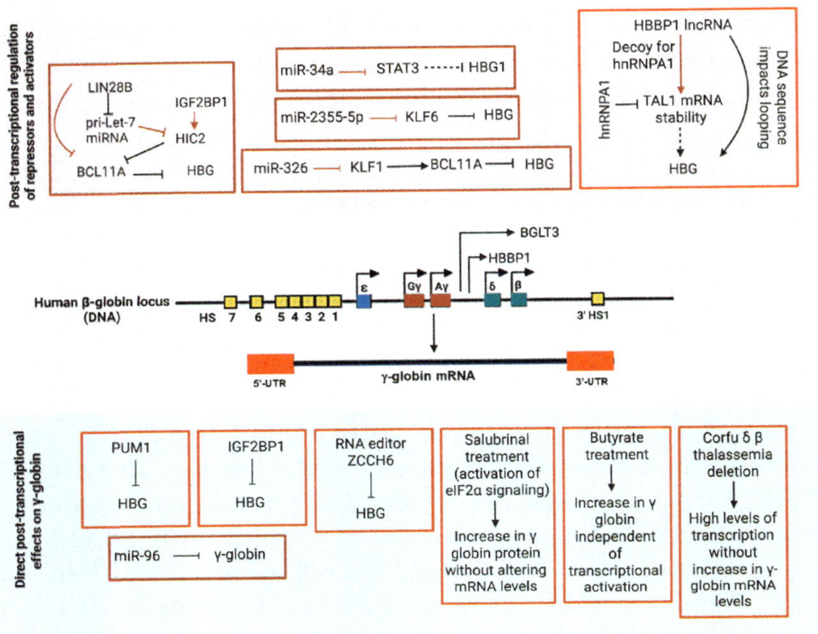

Fig. 6 Post-transcriptional regulation of β-globin switching: RNA binding proteins, lncRNAs, miRNAs and RNA editors involved in the post-transcriptional regulation of hemoglobin switching, both directly and indirectly, are shown. For indirect regulators, red arrows show the post-transcriptional regulation aspect of the cascade. Additionally, post-transcriptional regulation of hemoglobin switching observed in disease and upon chemical treatments are also shown.

Hemoglobin is a tetramer of two α-globin chains and two β-globin chains and each chain has an embedded heme group (Oilman et al., 1988; Shen et al., 1981). The human alpha globin locus is on chromosome 16 and encodes α-chains of hemoglobin: ζ- and α-globin, and the human β- globin locus on chromosome 11 contains the genes that encode β-chains of hemoglobin. They are located in the order of their expression during development: 5′ε, γ, δ, and β 3′ (Stamatoyannopoulos, 2005). Two switches in globin expression occur during development (Stamatoyannopoulos, 2005). The first one takes place at 5 weeks postconception, whereby γ-globin replaces ε globin as the major β chain subunit, and α-globin replaces ζ-globin as the major α chain subunit. The second switch takes place at the time of birth, whereby β globin and a small amount of δ-globin replaces γ- globin. At this time, hemoglobin switches from fetal hemoglobin (HbF) to adult hemoglobin (HbA). In adult erythroid cells, HbF is less than 1 %.

Amelioration of the symptoms and increased life expectancy has been documented in β-thalassemia patients carrying a diverse group of genetic mutations which cause persistent synthesis of γ-globin after birth, known collectively as hereditary persistence of fetal hemoglobin (HPFH). The elevated levels of γ-globin chains combine with the excess α chains leading to reduced dyserythropoiesis and lower blood transfusion requirement in β thalassemia (Stamatoyannopoulos, 2005; Vinjamur et al., 2018). Due to these observations, treatment strategies aimed at preserving or reactivating fetal hemoglobin production have been pursued. Various cis elements, trans factors and DNA methylation are implicated in the γ-globin silencing in adult erythrocytes (Blobel & Khandros, 2024; Vinjamur et al., 2018).

Globin mRNAs increase from less than 1 % to over 95 % of the total transcripts in differentiated RBCs (Bastos et al., 1977; Bastos & Aviv, 1977; Weiss & Liebhaber, 1994). This increase arises from robust transcriptional mechanisms and is also supported by long half-lives (24–60 h) (Aviv et al., 1976; Weiss & Liebhaber, 1994, 1995). While transcriptional regulation of hemoglobin gene regulation and its therapeutic relevance in fetal hemoglobin induction has been studied extensively, post-transcriptional regulation of globin genes has been understudied, and its relevance in fetal hemoglobin induction is under-appreciated (Blobel & Khandros, 2024; Elagooz et al., 2022; Liebhaber, 1997).

Early studies on post-transcriptional regulation of globin genes were fueled by the identification of a naturally occurring α-thalassemia mutation, α constant spring, which leads to a destabilized α globin mRNA (Clegg et al., 1971; Liebhaber & Kan, 1981; Russell et al., 1997). These studies led to the identification of cis elements in the 3′UTR, and RNA binding proteins that interact with these elements to regulate the stability of the α globin transcript.

As an extension of these studies, comparisons were made with the 3′UTR of β globin, and the importance of this UTR in the maintenance of the β globin stability was documented (Russell & Liebhaber, 1996). Although the mechanisms involved in the regulation of α globin mRNA stability have been investigated extensively (Russell et al., 1997), similar studies on the β globin genes are sparse. Specifically, post-transcriptional mechanisms that modulate β globin switching have not been thoroughly explored, although the following few studies point to substantial evidence for this regulation:

Studies performed in the early 1990s showed that K562 cells when induced towards megakaryocytic lineage, underwent a dramatic reduction in the cytoplasmic levels of γ globin mRNA potentially by post-transcriptional mechanisms (Lumelsky & Forget, 1991). However, it was only in the early 2000s that the clinical relevance of this regulation was documented. Firstly, studies on Corfu δ β thalassemia deletion confirmed that even though this deletion resulted in high level transcription of the γ globin genes, this did not cause an accumulation of the γ globin mRNA, indicating that these genes can be post-transcriptionally regulated (Chakalova et al., 2005). Further, it was noted that sickle cell patients who were treated with butyrate demonstrated high increases in γ globin levels that were independent of its transcriptional activation (Weinberg et al., 2005). In addition, it was recently demonstrated that a chemical compound called Salubrinal, an activator of eIF2α signaling, elicited a dramatic post-transcriptional induction of γ globin without altering its mRNA levels, erythroid differentiation, or the total hemoglobin content (Hahn & Lowrey, 2013).

Spurred by evidence from these studies, further investigations probed for factors that can mediate the post-transcriptional regulation of fetal hemoglobin. γ globin mRNA isolated from human adult reticulocytes has been shown to be bound by Argonaute2, a member of miRNA induced silencing complex (RISC) (Azzouzi et al., 2011). In addition, miRNA96 has been shown to target γ globin ORF to mediate its post-transcriptional regulation (Azzouzi et al., 2011). On the other hand miR34a has been shown to target γ globin indirectly by reducing STAT3 levels, and stabilizes HBG1 but not the HBG2 homolog (Ward et al., 2016). Other miRNAs such as miR-326 and miR-2355–5p inhibit transcription factors KLF1 and KLF6 respectively to modulate γ globin levels (Cheng et al., 2021; Li et al., 2018).

Further, let7 miRNAs are repressed by LIN28B RNA binding proteins that are highly expressed in fetal erythroid cells. Let7 miRNAs post-transcriptionally inhibit HIC2, which further inhibits BCL11A, a γ globin repressor (Huang et al., 2024; Lee et al., 2013). Therefore, high LIN28B levels and low let7 miRNAs lead to high γ globin levels.

Our lab has shown that another RNA binding protein, PUM1, with hitherto unknown functions in erythropoiesis, is a post-transcriptional repressor of γ globin (Elagooz et al., 2022). PUM1 can directly bind to γ globin and impair transcript stability and translational efficiency. This is to our knowledge the first RNA binding protein that has been shown to directly bind to γ globin transcript. PUM1 knockdown led to induction of fetal hemoglobin without impairing the progression of terminal erythroid differentiation. We also identified a novel human heterozygous mutation in PUM1 RNA binding domain that causes higher than baseline levels of fetal hemoglobin in the absence of anemia in the individual.

A recent study by Coyne et al., deposited in the bioRXiv preprint server, demonstrates that another RNA binding protein, IGF2BP1, can directly bind and stabilize γ globin transcript, while also binding HIC2 transcript and activating HIC2 post-transcriptionally, thereby decreasing BCL11A and increasing γ globin levels (Coyne et al., 2025).

As discussed earlier (see Long non-coding RNAs section), LncRNAs have been shown to modulate γ-globin expression, both via the DNA sequence at the lncRNA locus by influencing chromatin looping in cis and/or trans by interacting and influencing the recruitment of chromatin regulators. BGLT3 is one such lncRNA that positively influences transcription of HBG genes both through its DNA sequence mediated looping and by cis interactions of lncRNA with the HBG genes (Ivaldi et al., 2018). In addition, knockdown of HIDALGO, a long non coding RNA that encompasses the HBG1 and BGLT3 locus, reduced the transcription of HBG1 but not HBG2 (Werner et al., 2017). On the other hand, HBBP1, an intergenic lncRNA in the β-globin locus, acts as a decoy for hnRNPA1 thereby stabilizing TAL1 transcript and modulating transcription of genes involved in erythropoiesis along with increasing fetal hemoglobin levels (Ma et al., 2021). In addition, HBBP1 DNA sequence harbors variants associated with severity of β-thalassemia, and high fetal hemoglobin levels (Galarneau et al., 2010; Giannopoulou et al., 2012; Nuinoon et al., 2010), and deleting HBBP1 increased looping between LCR and γ-globin (Huang et al., 2017).

Further, a recent study in BioRXiv showed that erythroid specific gene transcripts including globin genes are m6A methylated and this methylation at the 3'end decreases as the differentiation progresses (Gunage et al., 2024). Further, terminal uridylyl transferase enzyme, ZCCH6 (Zinc Finger CCHC-Type Containing 6) that adds uridines to 3'ends of its target RNAs to prime for degradation, was shown to be involved in regulating the half-life of human fetal γ globin and mouse Hbb-bh1 globin (ortholog of human gamma globin) (Han et al., 2025). Knockout of ZCCHC6 or DISL2 (DIS3-like 3′−5′ exoribonuclease 2) an exonuclease that recognizes the terminal uridylyl signal, leads to the increase of these developmentally silenced globins, adding further evidence for post-transcriptional regulation of hemoglobin switching (Han et al., 2025).

39.2 Dysregulation of post-transcriptional mechanisms leading to disease

Myeloproliferative neoplasms (MPNs), characterized by rapid proliferation of myeloid progenitors are clonal myeloid malignancies caused by abnormal response to hematopoietic growth factors (Hautin et al., 2020). Depending on the primarily affected myeloid lineage, MPNs are classified into three diseases, as provided by World Health Organization (WHO) and the International Consensus Classification (ICC): polycythemia vera where erythrocytes are overproduced, essential thrombocythemia where platelets are overproduced, and primary myelofibrosis that is characterized by bone marrow scarring. Further, both polycythemia vera and thrombocythemia can evolve into secondary myelofibrosis. Additionally, all MPNs may evolve towards acute myeloblastic leukemia, a condition with a very poor prognosis. Besides the driver mutations causing MPNs, recent data mainly obtained by Next Generation Sequencing, indicate that 'additional mutations' are recurrently found, particularly in genes involved in epigenetic and splicing regulation (Arber et al., 2016; Vainchenker & Kralovics, 2017). RNA splicing factors SF3B1 (Splicing Factor 3b Subunit 1), U2AF1 (U2 small nuclear RNA auxiliary factor 1), SRSF2 (serine and arginine rich splicing factor 2), ZRSR2 (zinc finger CCCH-type, RNA binding motif and serine/arginine rich 2) and PRPF8 (Pre-mRNA Processing Factor 8) are frequent 'additional mutations' in MPN. In profibrotic or overt primary myelofibrosis, SRSF2 and U2AF1 mutations were the most frequent: 12.2 % and 13.8 % respectively, followed by SF3B1 (7.7 %) and ZRSR2 (5.8 %) mutations, whereas PRPF8 (2.1 %) mutations were rarer. SRSF2 mutations are also found in early stages of primary myelofibrosis at a frequency of 8.5 %. Splicing factor mutations were less frequent in polycythemia vera and essential thrombocythemia than in primary myelofibrosis (Hautin et al., 2020).

SF3B1 mutations are frequently associated with myelodysplastic/myeloproliferative neoplasms, MDS/MPN with ring sideroblasts and thrombocytosis (formerly refractory anemia with ring sideroblasts associated with marked thrombocytosis [RARS-T]) (Malcovati et al., 2011; Patnaik et al., 2016). SRSF2 mutations are also seen in 33 % of chronic myelomonocytic leukemia (CMML) patients; other splicing mutations in CMML include U2AF1 (8 %), ZRSR2 (9 %) and SF3B1 (5 %) (Yoshida et al., 2011).

Splicing factor U2AF1 is an U2 auxiliary factor protein involved in recognizing the 3′ boundary of introns at the AG splice site acceptor dinucleotide (Taylor & Lee, 2019). U2AF1 mutations are correlated with anemia and thrombocytopenia (Smith et al., 2019; Tefferi et al., 2016). Mutations in U2AF1 in MDS and AML leads to the inclusion of exon 4 in IRAK4, a serine/threonine kinase mediating TLR signaling and NF-κB and MAPK activation (Smith et al., 2019).

Ribosomopathies are diseases resulting from mutations in genes encoding either ribosome biogenesis factors or ribosomal proteins, overall leading to an altered translational profile (Danilova & Gazda, 2015). Diamond Blackfan Anemia (DBA) is one such ribosomopathy and a sporadic heterogeneous genetic disorder characterized by red blood cell aplasia in association with skeletal anomalies and short stature that classically appear soon after birth (Wlodarski et al., 2024). DBA has been shown to be associated with both ribosomal and non-ribosomal mutations in genes located across more than 11 chromosomes which are responsible for encoding the ribosomal proteins (Draptchinskaia et al., 1999; Fujiwara, 2017). Although the molecular mechanisms between ribosomal proteins (RP) haploinsufficiency and anemia have not yet been clearly elucidated, there exists a hypothesis of a "ribosomal stress" mechanism, which implies that a defective ribosome biosynthesis leads to apoptosis in erythroid progenitors which in turn is leads to erythroid failure (Lipton & Ellis, 2009). DBA affects both the large and small subunits of ribosomal protein (RP) genes. Among these mutated genes are RPL5, RPL11, RPL35A, RPS7, RPS10, RPS17, RPS19, RPS24, and RPS26 among others (Arbiv et al., 2018; Errichiello et al., 2017; Gazda et al., 2008).

40. Therapeutic relevance and potential strategies for modulating post-transcriptional regulation in erythroid disorders

Strategies to modulate post-transcriptional regulation are being developed to treat hematological diseases and cancers. The strategies described below include use of small molecule inhibitors, antisense oligonucleotides, microRNAs, and decoy RNAs to target splicing, RNA stability, and translational efficiency.

41. Small-molecule inhibitors modulating splicing

Splicing factor mutations are attractive therapeutic targets in the treatment of cancers since they occur early during oncogenesis, are observed in dominant clones and can be effective treatment options in cancers with no other treatment targets (Taylor & Lee, 2019).

Small molecules that can target the splicing catalysis process and can bind directly to SF3b complex include drugs such as spliceostatin A, sudemycin, pladienolide, E7107 (analog of pladienolide), H3B-8800 (orally bioavailable analog of E7107) and herboxidiene (Fan et al., 2011; Hasegawa et al., 2011; Kaida et al., 2007; Kotake et al., 2007). However, none of these drugs have been proven safe for clinical trials.

Alternative targets of splicing machinery have focused on factors such as RBM39 and a phase II trial of the orally bioavailable RBM39 degrader E7820 is ongoing (Taylor & Lee, 2019). Splicing factors are the most abundant arginine-methylated substrates in cells, and this methylation is mediated by enzymes such as PRMT5, the arginine methyltransferase. PRMT5 inhibitor class of drugs are under clinical investigation, in advanced MDS/AML and advanced solid tumors/hematological malignancies (Fong et al., 2019).

42. Antisense oligonucleotide (ASO)

ASO therapy uses modified nucleic acids to base pair with pre-mRNA and change splicing by inhibiting RNA–RNA or splicing factor–RNA interactions.

Successful examples of antisense oligonucleotide (ASO) therapy include the Food and Drug Administration (FDA) approved nusinersen and eteplirsen (Finkel et al., 2017; McDonald et al., 2021) for the treatment of spinal muscular atrophy and Duchenne muscular dystrophy, respectively.

The use of ASO-directed blocking of cryptic splice sites caused by mutations in *HBB (β-globin)* in rescuing normal splicing in mouse models of β-thalassemia, shows the potential for a human therapeutic strategy (Lacerra et al., 2000; Suwanmanee et al., 2002; Svasti et al., 2009; Vacek et al., 2003; Xie et al., 2011).

Ribavirin is an inhibitor of viral replication, and its derivative TRX-201, acts as physical mimics of the m^7G cap (Assouline et al., 2009; Kentsis et al., 2004). It inhibits the interaction between eIF4E and its site of binding on target mRNAs, the m^7G cap structure at the 5′ end of transcripts. Thus, Ribavirin prevents the interaction of eIF4E with target mRNA and reduces mRNA export and protein expression, thereby decreasing cell proliferation and clonogenicity. Ribavirin showed initial clinical efficacy in a phase II clinical trial, as monotherapy for acute myeloid leukemia (AML) (Assouline et al., 2009).

Fat mass and obesity-associated protein (FTO) is an RNA N^6-methyladenosine (m^6A) demethylase that functions as an oncogene in various cancers including AML, glioblastoma, breast cancer and pancreatic cancer. A small molecule inhibitor called Bisantrene, interacts with FTO residues key for interactions with mRNA targets (K216, S229, H231 and E234). Bisantrene successfully concluded phase II clinical trials against AML (Su et al., 2020).

43. Molecular glues

The emergence of molecular glues, a novel class of compounds that function by enhancing the affinity between two proteins has allowed for targeting undruggable proteins such as RNA Binding Motif Protein 39 (RBM39), a multifunctional protein involved in the regulation of transcription, splicing and translation of tumor-related genes. REC1245 is the newest compound in this class of drugs and has cleared FDA approval for Phase I/II clinical trials in lymphoma and solid tumors. REC1245 provides the glue for the interaction of RBM39 with DCAF15 (DDB1 and CUL4-associated factor 15), a component of the E3 ubiquitin ligase complex, resulting in ubiquitination and degradation of RBM39 (Faust et al., 2020; Monteys et al., 2021).

44. Aptamers

Aptamers are short single stranded nucleic acids designed to bind to their targets with high affinity (Adachi & Nakamura, 2019). Thus far, five therapeutic aptamers have made it to clinical trials for the treatment of cancer, two of them (AS1411 and APTA-16) targeting nucleolin, an RNA and DNA binding protein in the nucleolus that is important for maturation of ribosomal RNA and ribosome biogenesis (Bates et al., 2009). AS1411 is being tested in combination with the anti-cancer drug cytarabine, where an improved response in AML patients was observed (Mongelard & Bouvet, 2010).

45. Micro RNAs

There are more than 552 studies found under the heading "miRNAs" for intervention/treatment in the ClinicalTrials.gov portal, which includes a few recently completed clinical trials for diagnostic miRNAs (Brillante et al., 2024).

TargomiRs are nonliving bacterial minicells (nanoparticles) that serve as drug delivery vehicles that function as leak resistant micro-reservoir carriers to allow efficient packaging of a range of different molecules, including nucleic acids (Jivrajani & Nivsarkar, 2016). TargomiRs loaded with miR-16-based mimic microRNA (miRNA) and targeted to EGFR have been designed to counteract the loss of the miR-15 and miR-16 family miRNAs, associated with unsuppressed tumor growth in preclinical models of malignant pleural mesothelioma (van Zandwijk et al., 2017).

Recently, a rAAV5-mediated delivery of a-miRNA (artificial miRNA) expression is ongoing in a clinical trial wherein, two patients with familial amyotrophic lateral sclerosis (ALS) and mutations in the gene encoding superoxide dismutase 1 (*SOD1*) were treated with a single intrathecal infusion of adeno-associated virus encoding a microRNA targeting SOD1 (Mueller et al., 2020).

Remlarsen is a microRNA-29 mimic that represses extracellular matrix expression and fibroplasia in the skin. In a clinical trial, remlarsen repressed collagen expression and the development of fibroplasia in incisional skin wounds (Gallant-Behm et al., 2019).

Another miR inhibitor in human study is cobomarsen (MRG-106), an inhibitor of miR-155, a microRNA with a strong link to cutaneous T cell lymphoma (CTCL) pathogenesis. The inhibitor is being used to study the efficacy of cobomarsen in mycosis fungoides (MF) patients. The study has been proposed to expand and include patients with other hematologic malignancies in which miR-155 is elevated and relevant, including chronic lymphocytic leukemia (CLL), diffuse large B cell lymphoma (DLBCL), and adult T cell lymphoma/leukemia (ATLL) (Seto et al., 2018).

46. Targeting RNA modifications

The link between deregulation of RNA modifications and human diseases suggests that targeting them may pave way for novel therapeutics.

A recent study identified an inhibitor STM2457 of METTL3 (Methyltransferase-like protein 3, which is the primary catalytic subunit of m^6A methyl transferase complex that writes m^6A on RNAs), as a potential treatment strategy against leukemia (Yankova et al., 2021). Many RNA modifications including m6A, Ψ or pseudouridine, m5C and m7G can improve translational efficiency while alleviating immunogenicity (Andries et al., 2015). The role of N1-pseudouridine is well documented as it is a key mutation incorporated into the COVID19 mRNA (Morais et al., 2021).

47. Conclusion

Continued advances in studies on RNA binding proteins, non-coding RNA, RNA stability, translational efficiency, and RNA modifications during erythropoiesis, will reveal additional mechanistic insights on how RNA metabolism and post-transcriptional regulation contribute to normal erythropoiesis and dyserythropoiesis. This will also provide novel ways of designing therapeutics to cure erythroid disorders.

Despite the potential, mechanisms involving post-transcriptional regulation have not been well tapped into for devising therapeutic strategies for erythroid disorders, which include defects in red blood cell production, function, and hemoglobin synthesis. Re-expression of the fetal form of hemoglobin in genetic diseases such as β-thalassemia and sickle cell anemia have been shown to ameliorate the disease (Galanello et al., 2009; Nuinoon et al., 2010; Uda et al., 2008). This has spurred extensive research on discovering inducers of HbF that can be used in therapeutics for curing β-thalassemia and sickle cell anemia. Further studies could contribute towards identification of drugs targeting RNA modifying enzymes, splicing, RNA binding proteins, and non-coding RNAs, to increase fetal hemoglobin production. As our understanding of post-transcriptional regulation in erythropoiesis and hemoglobin synthesis grows, successful therapies targeting these mechanisms could soon become a reality for curing erythroid disorders.

Acknowledgements

Studies by the authors are supported by NIH R01DK13493 and R15DK146338.

References

Adachi, T., & Nakamura, Y. (2019). Aptamers: A review of their chemical properties and modifications for therapeutic application. *Molecules (Basel, Switzerland), 24*(23), 1–14. https://doi.org/10.3390/molecules24234229.

Aguilo, F., Zhang, F., Sancho, A., Fidalgo, M., Di Cecilia, S., Vashisht, A., ... Walsh, M. J. (2015). Coordination of m6A mRNA methylation and gene transcription by ZFP217 regulates pluripotency and reprogramming. *Cell Stem Cell, 17*(6), 689–704. https://doi.org/10.1016/j.stem.2015.09.005.

Alriquet, M., Calloni, G., Martınez-Limon, A., Delli Pont, R., Hanspach, G., Hengesbach, M., ... Vabulas, M. (2020). The protective role of m1A during stress-induced granulation. *Journal of Molecular Cell Biology, 12*(11), 870–880. http://mc.manuscriptcentral.com/jmcb.

Alvarez-Dominguez, J. R., Hu, W., Yuan, B., Shi, J., Park, S. S., Gromatzky, A. A., ... Lodish, H. F. (2014). Global discovery of erythroid long noncoding RNAs reveals novel regulators of red cell maturation. *Blood, 123*(4), 570–581. https://doi.org/10.1182/blood-2013-10-530683.

Alvarez-Dominguez, J. R., Knoll, M., Gromatzky, A. A., & Lodish, H. F. (2017). The super-enhancer-derived alncRNA-EC7/bloodlinc potentiates red blood cell development in trans. *Cell Reports, 19*(12), 2503–2514. https://doi.org/10.1016/j.celrep.2017.05.082.

Anderson, J., Phan, L., Cuesta, R., Carlson, B. A., Pak, M., Asano, K., ... Hinnebusch, A. G. (1998). The essential Gcd10p-Gcd14p nuclear complex 15 required for 1- methyladenosine modification and maturation of initiator methionyl-tRNA. *Genes and Development, 12*(23), 3650–3662. https://doi.org/10.1101/gad.12.23.3650.

Andolfo, I., De Falco, L., Asci, R., Russo, R., Colucci, S., Gorrese, M., ... Iolascon, A. (2010). Regulation of divalent metal transporter 1 (DMT1) non-IRE isoform by the microRNA Let-7d in erythroid cells. *Haematologica, 95*(8), 1244–1252. https://doi.org/10.3324/haematol.2009.020685.

Andries, O., Mc Cafferty, S., De Smedt, S. C., Weiss, R., Sanders, N. N., & Kitada, T. (2015). N1-methylpseudouridine-incorporated mRNA outperforms pseudouridine-incorporated mRNA by providing enhanced protein expression and reduced immunogenicity in mammalian cell lines and mice. *Journal of Controlled Release, 217*, 337–344. https://doi.org/10.1016/j.jconrel.2015.08.051.

Arber, D. A., Orazi, A., Hasserjian, R., Thiele, J., Borowitz, M. J., Le Beau, M. M., ... Vardiman, J. W. (2016). The 2016 revision to the World Health Organization classification of myeloid neoplasms and acute leukemia. *Blood, 127*(20), 2391–2405. https://doi.org/10.1182/blood-2016-03-643544.

Arbiv, O. A., Cuvelier, G., Klaassen, R. J., Fernandez, C. V., Robitaille, N., Steele, M., ... Dror, Y. (2018). Molecular analysis and genotype-phenotype correlation of Diamond-Blackfan anemia. *Clinical Genetics, 93*(2), 320–328. https://doi.org/10.1111/cge.13158.

Assouline, S., Culjkovic, B., Cocolakis, E., Rousseau, C., Beslu, N., Amri, A., ... Borden, K. L. B. (2009). Molecular targeting of the oncogene eIF4E in acute myeloid leukemia (AML): A proof-of-principle clinical trial with ribavirin. *Blood, 114*(2), 257–260. https://doi.org/10.1182/blood-2009-02-205153.

Atianand, M. K., Hu, W., Satpathy, A. T., Shen, Y., Ricci, E. P., Alvarez-Dominguez, J. R., ... Fitzgerald, K. A. (2016). A long noncoding RNA lincRNA-EPS Acts as a transcriptional brake to restrain inflammation. *Cell, 165*(7), 1672–1685. https://doi.org/10.1016/j.cell.2016.05.075.

Aviv, H., Voloch, Z., Bastos, R., & Levy, S. (1976). Biosynthesis and stability of globin mRNA in cultured erythroleukemic friend cells. *Cell, 8*(4), 495–503. https://doi.org/10.1016/0092-8674(76)90217-8.

Ayadi, L., Galvanin, A., Pichot, F., Marchand, V., & Motorin, Y. (2019). RNA ribose methylation (2′-O-methylation): Occurrence, biosynthesis and biological functions. *Biochimica et Biophysica Acta - Gene Regulatory Mechanisms, 1862*(3), 253–269. https://doi.org/10.1016/j.bbagrm.2018.11.009.

Azad, P., Zhou, D., Tu, H. C., Villafuerte, F. C., Traver, D., Rana, T. M., & Haddad, G. G. (2023). Long noncoding RNA HIKER regulates erythropoiesis in Monge's disease via CSNK2B. *Journal of Clinical Investigation, 133*(11), https://doi.org/10.1172/JCI165831.

Azzouzi, I., Moest, H., Winkler, J., Fauchère, J. C., Gerber, A. P., Wollscheid, B., ... Speer, O. (2011). Microrna-96 directly inhibits γ-Globin expression in human erythropoiesis. *PLoS One, 6*(7), https://doi.org/10.1371/journal.pone.0022838.

Bar-Yaacov, D., Frumkin, I., Yashiro, Y., Chujo, T., Ishigami, Y., Chemla, Y., ... Mishmar, D. (2016). Mitochondrial 16S rRNA Is methylated by tRNA methyltransferase TRMT61B in all vertebrates. *PLoS Biology, 14*(9), 1–21. https://doi.org/10.1371/journal.pbio.1002557.

Barozzi, C., Zacchini, F., Asghar, S., & Montanaro, L. (2022). Ribosomal RNA Pseudouridylation: Will newly available methods finally define the contribution of this modification to human ribosome plasticity? *Frontiers in Genetics, 13*(June), 1–7. https://doi.org/10.3389/fgene.2022.920987.

Bartel, D. P. (2004). MicroRNAs: Genomics, biogenesis, mechanism, and function. *Cell, 116*(2), 281–297. https://doi.org/10.1016/S0092-8674(04)00045-5.

Basak, A., Munschauer, M., Lareau, C. A., Montbleau, K. E., Ulirsch, J. C., Hartigan, C. R., ... Sankaran, V. G. (2020). Control of human hemoglobin switching by LIN28B-mediated regulation of BCL11A translation. *Nature Genetics*. https://doi.org/10.1038/s41588-019-0568-7.

Bass, B. L., & Weintraub, H. (1988). An unwinding activity that covalently modifies its double-stranded RNA substrate. *Cell, 55*(6), 1089–1098. https://doi.org/10.1016/0092-8674(88)90253-X.

Bastos, R. N., & Aviv, H. (1977). Theoretical analysis of a model for globin messenger RNA accumulation during erythropoiesis. *Journal of Molecular Biology, 110*(2), 205–218. https://doi.org/10.1016/S0022-2836(77)80069-7.

Bastos, R. N., Volloch, Z., & Aviv, H. (1977). Messenger RNA population analysis during erythroid differentiation: A kinetical approach. *Journal of Molecular Biology, 110*(2), 191–203. https://doi.org/10.1016/S0022-2836(77)80068-5.

Bates, P. J., Laber, D. A., Miller, D. M., Thomas, S. D., & Trent, J. O. (2009). Discovery and development of the G-rich oligonucleotide AS1411 as a novel treatment for cancer. *Experimental and Molecular Pathology, 86*(3), 151–164. https://doi.org/10.1016/j.yexmp.2009.01.004.

Bavelloni, A., Poli, A., Fiume, R., Blalock, W., Matteucci, A., Ramazzotti, G., ... Faenza, I. (2014). PLC-beta 1 regulates the expression of miR-210 during mithramycin-mediated erythroid differentiation in K562 cells. *Oncotarget, 5*(12), 4222–4231. https://doi.org/10.18632/oncotarget.1972.

Blobel, G. A., & Khandros, E. (2024). Elevating fetal hemoglobin - recently discovered regulators and mechanisms. *Blood Journal*. https://doi.org/10.1182/blood.2023022190.

Brillante, S., Volpe, M., & Indrieri, A. (2024). Advances in MicroRNA therapeutics: From preclinical to clinical studies. *Human Gene Therapy, 35*(17–18), 628–648. https://doi.org/10.1089/hum.2024.113.

Brisot, E., Leprêtre, P. M., Hamza, E., Fourdinier, O., Brigant, B., Ouled-Haddou, H., ... Metzinger, L. (2025). Uremic toxins levels are associated with miR-223 in chronic kidney disease-associated anemia. *Non-Coding RNA Research, 13*(January), 121–130. https://doi.org/10.1016/j.ncrna.2025.04.009.

Bu, Y., Su, F., Wang, X., Gao, H., Lei, L., Chang, N., ... Xiong, J. W. (2014). Protein tyrosine phosphatase PTPN9 regulates erythroid cell development through STAT3 dephosphorylation in zebrafish. *Journal of Cell Science, 127*(12), 2761–2770. https://doi.org/10.1242/jcs.145367.

Byon, J. C. H., & Papayannopoulou, T. (2012). MicroRNAs: Allies or Foes in erythropoiesis? *Journal of Cellular Physiology, 227*(1), 7–13. https://doi.org/10.1002/jcp.22729.

Cantara, W. A., Crain, P. F., Rozenski, J., McCloskey, J. A., Harris, K. A., Zhang, X., ... Agris, P. F. (2011). The RNA modification database, RNAMDB: 2011 update. *Nucleic Acids Research, 39*(1), 195–201. https://doi.org/10.1093/nar/gkq1028.

Carballo, E., Lai, W. S., & Blackshear, P. J. (1998). Feedback inhibition of macrophage tumor necrosis factor-α production by tristetraprolin. *Science (New York, N. Y.), 281*(5379), 1001–1005. https://doi.org/10.1126/science.281.5379.1001.

Carlile, T. M., Martinez, N. M., Schaening, C., Su, A., Bell, T. A., Zinshteyn, B., & Gilbert, W. V. (2019). mRNA structure determines modification by pseudouridine synthase 1. *Nature Chemical Biology, 15*(10), 966–974. https://doi.org/10.1038/s41589-019-0353-z.

Carlile, T. M., Rojas-Duran, M. F., Zinshteyn, B., Shin, H., Bartoli, K. M., & Gilbert, W. V. (2014). Pseudouridine profiling reveals regulated mRNA pseudouridylation in yeast and human cells. *Nature, 515*(7525), 143–146. https://doi.org/10.1038/nature13802.

Carrascoso-Rubio, C., Zittersteijn, H. A., Pintado-Berninches, L., Fernández-Varas, B., Lozano, M. L., Manguan-Garcia, C., ... Guenechea, G. (2021). Generation of dyskeratosis congenita-like hematopoietic stem cells through the stable inhibition of DKC1. *Stem Cell Research and Therapy, 12*(1), 1–8. https://doi.org/10.1186/s13287-021-02145-8.

Chakalova, L., Osborne, C. S., Dai, Y. F., Goyenechea, B., Metaxotou-Mavromati, A., Kattamis, A., ... Fraser, P. (2005). The Corfu δβ thalassemia deletion disrupts γ-globin gene silencing and reveals post-transcriptional regulation of HbF expression. *Blood, 105*(5), 2154–2160. https://doi.org/10.1182/blood-2003-11-4069.

Chen, K., Liu, J., Heck, S., Chasis, J. A., An, X., & Mohandas, N. (2009). Resolving the distinct stages in erythroid differentiation based on dynamic changes in membrane protein expression during erythropoiesis. *Proceedings of the National Academy of Sciences of the United States of America, 106*(41), 17413–17418. https://doi.org/10.1073/pnas.0909296106.

Chen, L. L., & Kim, V. N. (2024). Small and long non-coding RNAs: Past, present, and future. *Cell, 187*(23), 6451–6485. https://doi.org/10.1016/j.cell.2024.10.024.

Chen, S. Y., Wang, Y., Telen, M. J., & Chi, J. T. (2008). The genomic analysis of erythrocyte microRNA expression in sickle cell diseases. *PLoS One, 3*(6), https://doi.org/10.1371/journal.pone.0002360.

Chen, T., Hao, Y. J., Zhang, Y., Li, M. M., Wang, M., Han, W., ... Zhou, Q. (2015). M6A RNA methylation is regulated by microRNAs and promotes reprogramming to pluripotency. *Cell Stem Cell, 16*(3), 289–301. https://doi.org/10.1016/j.stem.2015.01.016.

Chen, Y. S., Yang, W. L., Zhao, Y. L., & Yang, Y. G. (2021). Dynamic transcriptomic m5C and its regulatory role in RNA processing. *Wiley Interdisciplinary Reviews: RNA, 12*(4), 1–23. https://doi.org/10.1002/wrna.1639.

Cheng, A. W., Shi, J., Wong, P., Luo, K. L., Trepman, P., Wang, E. T., ... Lodish, H. F. (2014). Muscleblind-like 1 (Mbnl1) regulates pre-mRNA alternative splicing during terminal erythropoiesis. *Blood, 124*(4), 598–610. https://doi.org/10.1182/blood-2013-12-542209.

Cheng, Y., Luo, H., Izzo, F., Pickering, B. F., Nguyen, D., Myers, R., ... Kharas, M. G. (2019). m6A RNA methylation maintains hematopoietic stem cell identity and symmetric commitment. *Cell Reports, 28*(7), 1703–1716.e6. https://doi.org/10.1016/j.celrep.2019.07.032.

Cheng, Y., Shang, X., Chen, D., Pang, D., Zhao, C., & Xu, X. (2021). MicroRNA-2355-5p regulates γ-globin expression in human erythroid cells by inhibiting KLF6. *British Journal of Haematology, 193*(2), 401–405. https://doi.org/10.1111/bjh.17134.

Chkheidze, A. N., Lyakhov, D. L., Makeyev, A. V., Morales, J., Kong, J., & Liebhaber, S. A. (1999). Assembly of the α-Globin mRNA stability complex reflects binary interaction between the pyrimidine-rich 3′ untranslated region determinant and Poly(C) binding protein αCP. *Molecular and Cellular Biology, 19*(7), 4572–4581. https://doi.org/10.1128/mcb.19.7.4572.

Clegg, J. B., Weatherall, D. J., & Milner, P. F. (1971). Haemoglobin constant spring - A chain termination mutant? *Nature, 234*(5328), 337–340. https://doi.org/10.1038/234337a0.

Connell, G. J., Byrne, E. M., & Simpson, L. (1997). Guide RNA-independent and guide RNA-dependent uridine insertion into cytochrome b mRNA in a mitochondrial lysate from Leishmania tarentolae. *Role of RNA secondary structure. Journal of Biological Chemistry, 272*(7), 4212–4218. https://doi.org/10.1074/jbc.272.7.4212.

Coyne, S., Wu, T., Hossain, M., Vinjamur, D., Starrs, M., Zeng, J., ... Bauer, D. E. (2025). Direct and indirect regulation of fetal globin transcript by RNA-binding protein IGF2BP1. *BioRxiv, 2,* 2025. https://www.biorxiv.org/content/10.1101/2025.09.09.675143v1.abstract.

Cui, L., Ma, R., Cai, J., Guo, C., Chen, Z., Yao, L., ... Shi, Y. (2022). *RNA modifications: Importance in immune cell biology and related diseases. Signal Transduction and Targeted Therapy,* 7, Springer US. https://doi.org/10.1038/s41392-022-01175-9.

D'Andrea, A. D., Fasman, G. D., & Lodish, H. F. (1989). Erythropoietin receptor and interleukin-2 receptor β chain: A new receptor family. *Cell, 58*(6), 1023–1024. https://doi.org/10.1016/0092-8674(89)90499-6.

Dai, X., Wang, T., Gonzalez, G., & Wang, Y. (2018). Identification of YTH domain-containing proteins as the readers for N1-methyladenosine in RNA. *Analytical Chemistry, 90*(11), 6380–6384. https://doi.org/10.1021/acs.analchem.8b01703.

Danilova, N., & Gazda, H. T. (2015). Ribosomopathies: How a common root can cause a tree of pathologies. *DMM Disease Models and Mechanisms, 8*(9), 1013–1026. https://doi.org/10.1242/dmm.020529.

Demirci, S., Gudmundsdottir, B., Li, Q., Haro-Mora, J. J., Nassehi, T., Drysdale, C., ... Uchida, N. (2020). βT87Q-Globin gene therapy reduces sickle hemoglobin production, allowing for ex vivo anti-sickling activity in human erythroid cells. *Molecular Therapy Methods and Clinical Development, 17*(June), 912–921. https://doi.org/10.1016/j.omtm.2020.04.013.

Dimitrova, D. G., Teysset, L., & Carré, C. (2019). RNA 2′-O-methylation (Nm) modification in human diseases. *Genes, 10*(2), 1–23. https://doi.org/10.3390/genes10020117.

Djebali, S., Davis, C. A., Merkel, A., Dobin, A., Lassmann, T., Mortazavi, A., ... Gingeras, T. R. (2012). Landscape of transcription in human cells. *Nature, 489*(7414), 101–108. https://doi.org/10.1038/nature11233.

Dominissini, D., Moshitch-Moshkovitz, S., Schwartz, S., Salmon-Divon, M., Ungar, L., Osenberg, S., ... Rechavi, G. (2012). Topology of the human and mouse m6A RNA methylomes revealed by m6A-seq. *Nature, 485*(7397), 201–206. https://doi.org/10.1038/nature11112.

Dominissini, D., Nachtergaele, S., Moshitch-Moshkovitz, S., Peer, E., Kol, N., Ben-Haim, M. S., ... He, C. (2016). The dynamic N1 -methyladenosine methylome in eukaryotic messenger RNA. *Nature, 530*(7591), 441–446. https://doi.org/10.1038/nature16998.

Dong, H., Chang, D. C., Hua, M. H. C., Lim, S. P., Chionh, Y. H., Hia, F., ... Shi, P. Y. (2012). 2′-O Methylation of internal adenosine by flavivirus NS5 methyltransferase. *PLoS Pathogens, 8*(4), https://doi.org/10.1371/journal.ppat.1002642.

Dore, L. C., Amigo, J. D., Dos Santos, C. O., Zhang, Z., Gai, X., Tobias, J. W., ... Weiss, M. J. (2008). A GATA-1-regulated microRNA locus essential for erythropoiesis. *Proceedings of the National Academy of Sciences of the United States of America, 105*(9), 3333–3338. https://doi.org/10.1073/pnas.0712312105.

Draptchinskaia, N., Gustavsson, P., Andersson, B., Pettersson, M., Willig, T. N., Dianzani, I., ... Dahl, N. (1999). The gene encoding ribosomal protein S19 is mutated in Diamond-Blackfan anaemia. *Nature Genetics, 21*(2), 169–175. https://doi.org/10.1038/5951.

Du, H., Zhao, Y., He, J., Zhang, Y., Xi, H., Liu, M., ... Wu, L. (2016). YTHDF2 destabilizes m 6 A-containing RNA through direct recruitment of the CCR4-NOT deadenylase complex. *Nature Communications,* 7. https://doi.org/10.1038/ncomms12626.

Dunn, D. B. (1961). The occurence of 1-methyladenine in ribonucleic acid. *BBA - Biochimica et Biophysica Acta, 46*(1), 198–200. https://doi.org/10.1016/0006-3002(61)90668-0.

Elagooz, R., Dhara, A. R., Gott, R. M., Adams, S. E., White, R. A., Ghosh, A., ... Gnanapragasam, M. N. (2022). PUM1 mediates the posttranscriptional regulation of human fetal hemoglobin. *Blood Advances, 6*(23), 6016–6022. https://doi.org/10.1182/bloodadvances.2021006730.

Errichiello, E., Vetro, A., Mina, T., Wischmeijer, A., Berrino, E., Carella, M., ... Zuffardi, O. (2017). Whole exome sequencing in the differential diagnosis of Diamond-Blackfan anemiaClinical and molecular study of three patients with novel RPL5 and mosaic RPS19 mutations. *Blood Cells, Molecules, and Diseases, 64*, 38–44. https://doi.org/10.1016/j.bcmd.2017.03.002.

Faehnle, C. R., Walleshauser, J., & Joshua-Tor, L. (2014). Mechanism of Dis3l2 substrate recognition in the Lin28-let-7 pathway. *Nature, 514*(7521), 252–256. https://doi.org/10.1038/nature13553.

Fan, L., Lagisetti, C., Edwards, C. C., Webb, T. R., & Potter, P. M. (2011). Sudemycins, novel small molecule analogues of FR901464, induce alternative gene splicing. *ACS Chemical Biology, 6*(6), 582–589. https://doi.org/10.1021/cb100356k.

Fang, X., Shen, F., Lechauve, C., Xu, P., Zhao, G., Itkow, J., ... Yu, D. (2018). miR-144/451 represses the LKB1/AMPK/mTOR pathway to promote red cell precursor survival during recovery from acute anemia. *Haematologica, 103*(3), 406–416. https://doi.org/10.3324/haematol.2017.177394.

Faust, T. B., Yoon, H., Nowak, R. P., Donovan, K. A., Li, Z., Cai, Q., ... Fischer, E. S. (2020). Structural complementarity facilitates E7820-mediated degradation of RBM39 by DCAF15. *Nature Chemical Biology, 16*(1), 7–14. https://doi.org/10.1038/s41589-019-0378-3.

Felli, N., Fontana, L., Pelosi, E., Botta, R., Bonci, D., Facchiano, F., ... Peschle, C. (2005). MicroRNAs 221 and 222 inhibit normal erythropoiesis and erythroleukemic cell growth via kit receptor down-modulation. *Proceedings of the National Academy of Sciences of the United States of America, 102*(50), 18081–18086. https://doi.org/10.1073/pnas.0506216102.

Felli, N., Pedini, F., Romania, P., Biffoni, M., Morsilli, O., Castelli, G., ... Marziali, G. (2009). MicroRNA 223-dependent expression of LMO2 regulates normal erythropoiesis. *Haematologica, 94*(4), 479–486. https://doi.org/10.3324/haematol.2008.002345.

Figueroa, A. A., Fasano, J. D., Morilla, S. M., Venkatesan, S., Kupfer, G., & Hattangadi, S. M. (2018). MiR-181a regulates erythroid enucleation via the regulation of Xpo7 expression. *Haematologica, 103*(8), e341–e344. https://doi.org/10.3324/haematol.2017.171785.

Finkel, R. S., Mercuri, E., Darras, B. T., Connolly, A. M., Kuntz, N. L., Kirschner, J., ... De Vivo, D. C. (2017). Nusinersen versus sham control in infantile-onset spinal muscular atrophy. *New England Journal of Medicine, 377*(18), 1723–1732. https://doi.org/10.1056/nejmoa1702752.

Flaherty, J. N., Sivasudhan, E., Tegowski, M., Xing, Z., McGinnis, M. M., Hunter, O. V., ... Conrad, N. K. (2025). The catalytic efficiency of METTL16 affects cellular processes by governing the intracellular S-adenosylmethionine setpoint. *Cell Reports, 44*(7), https://doi.org/10.1016/j.celrep.2025.115966.

Fong, J. Y., Pignata, L., Goy, P. A., Kawabata, K. C., Lee, S. C. W., Koh, C. M., ... Guccione, E. (2019). Therapeutic targeting of RNA splicing catalysis through inhibition of protein arginine methylation. *Cancer Cell, 36*(2), 194–209.e9. https://doi.org/10.1016/j.ccell.2019.07.003.

Frontelo, P., Manwani, D., Galdass, M., Karsunky, H., Lohmann, F., Gallagher, P. G., & Bieker, J. J. (2019). Novel role for EKLF in megakaryocyte lineage commitment. *Blood, 110*(12), 3871–3881. https://doi.org/10.1182/blood-2007-03-082065.An.

Fu, Y. F., Du, T. T., Dong, M., Zhu, K. Y., Jing, C. B., Zhang, Y., ... Liu, T. X. (2009). Mir-144 selectively regulates embryonic ά-hemoglobin synthesis during primitive erythropoiesis. *Blood, 113*(6), 1340–1349. https://doi.org/10.1182/blood-2008-08-174854.

Fujiwara, T. (2017). GATA transcription factors: Basic principles and related human disorders. *Tohoku Journal of Experimental Medicine, 242*(2), 83–91. https://doi.org/10.1620/tjem.242.83.

Gabbianelli, M., Testa, U., Morsilli, O., Pelosi, E., Saulle, E., Petrucci, E., ... Peschle, C. (2010). Mechanism of human Hb switching: A possible role of the kit receptor/miR 221-222 complex. *Haematologica, 95*(8), 1253–1260. https://doi.org/10.3324/haematol.2009.018259.

Galanello, R., Sanna, S., Perseu, L., Sollaino, M. C., Satta, S., Lai, M. E., ... Cao, A. (2009). Amelioration of Sardinian β0 thalassemia by genetic modifiers. *Blood, 114*(18), 3935–3937. https://doi.org/10.1182/blood-2009-04-217901.

Galarneau, G., Palmer, C. D., Sankaran, V. G., Orkin, S. H., Hirschhorn, J. N., & Lettre, G. (2010). Fine-mapping at three loci known to affect fetal hemoglobin levels explains additional genetic variation. *Nature Genetics, 42*(12), 1049–1051. https://doi.org/10.1038/ng.707.

Gallant-Behm, C. L., Piper, J., Lynch, J. M., Seto, A. G., Hong, S. J., Mustoe, T. A., ... Marshall, W. S. (2019). A MicroRNA-29 Mimic (Remlarsen) represses extracellular matrix expression and fibroplasia in the skin. *Journal of Investigative Dermatology, 139*(5), 1073–1081. https://doi.org/10.1016/j.jid.2018.11.007.

Gao, Y., Zimmer, J. T., Vasic, R., Liu, C., Gbyli, R., Zheng, S. J., ... Halene, S. (2023). ALKBH5 modulates hematopoietic stem and progenitor cell energy metabolism through m6A modification-mediated RNA stability control. *Cell Reports, 42*(10), 113163. https://doi.org/10.1016/j.celrep.2023.113163.

Garus, A., & Autexier, C. (2021). Dyskerin: An essential pseudouridine synthase with multifaceted roles in ribosome biogenesis, splicing, and telomere maintenance. *RNA (New York, N. Y.), 27*(12), 1441–1458. https://doi.org/10.1261/rna.078953.121.

Gazda, H. T., Sheen, M. R., Vlachos, A., Choesmel, V., O'Donohue, M. F., Schneider, H., ... Beggs, A. H. (2008). Ribosomal protein L5 and L11 mutations are associated with cleft palate and abnormal thumbs in Diamond-Blackfan anemia patients. *American Journal of Human Genetics, 83*(6), 769–780. https://doi.org/10.1016/j.ajhg.2008.11.004.

Gebauer, F., Schwarzl, T., Valcárcel, J., & Hentze, M. W. (2021). RNA-binding proteins in human genetic disease. *Nature Reviews Genetics, 22*(3), 185–198. https://doi.org/10.1038/s41576-020-00302-y.

Giannopoulou, E., Bartsakoulia, M., Tafrali, C., Kourakli, A., Poulas, K., Stavrou, E. F., ... Patrinos, G. P. (2012). A single nucleotide polymorphism in the HBBP1 gene in the human β-globin locus is associated with a mild β-thalassemia disease phenotype. *Hemoglobin, 36*(5), 433–445. https://doi.org/10.3109/03630269.2012.717515.

Gilbert, W. V., & Nachtergaele, S. (2023). mRNA regulation by RNA modifications. *Annual Review of Biochemistry, 92*, 175–198. https://doi.org/10.1146/annurev-biochem-052521-035949.

Gnanapragasam, M. N., Mcgrath, K. E., Catherman, S., Xue, L., Palis, J., & Bieker, J. J. (2016). EKLF/KLF1-regulated cell cycle exit is essential for erythroblast. *Enucleation Key Points*. https://doi.org/10.1182/blood-2016-03.

Goldstrohm, A. C., Hall, T. M. T., & McKenney, K. M. (2018). Post-transcriptional regulatory functions of mammalian pumilio proteins. *Trends in genetics, 34*, Elsevier Ltd. 972–990. https://doi.org/10.1016/j.tig.2018.09.006.

Gunage, R., Lin, S., Wiley, D., Choudhuri, A., Smith, M., Han, T., ... Zon, L. I. (2024). RNA methylation controls stress granule formation and function during erythropoiesis. *BioRxiv* 2024.11.07.622459. http://biorxiv.org/content/early/2024/11/08/2024.11.07.622459.abstract.

Gunage, R., & Zon, L. I. (2024). Role of RNA modifications in blood development and regeneration. *Experimental hematology, 138*, 1–27. https://doi.org/10.1016/j.exphem.2024.104279.

Guzzi, N., Cieśla, M., Ngoc, P. C. T., Lang, S., Arora, S., Dimitriou, M., ... Bellodi, C. (2018). Pseudouridylation of tRNA-derived fragments steers translational control in stem cells. *Cell, 173*(5), 1204–1216.e26. https://doi.org/10.1016/j.cell.2018.03.008.

Hahn, C. K., & Lowrey, C. H. (2013). Eukaryotic initiation factor 2α phosphorylation mediates fetal hemoglobin induction through a post-transcriptional mechanism. *Blood, 122*(4), 477–485. https://doi.org/10.1182/blood-2013-03-491043.

Han, A., Yermalovich, A., Najia, M. A. T., Pearson, D. S., Fujiwara, Y., Bolgov, M., ... Daley, G. Q. (2025). RNA editors sculpt the transcriptome during terminal erythropoiesis. *BioRxiv, 2025* 04.03.647020. http://biorxiv.org/content/early/2025/04/15/2025.04.03.647020.abstract.

Hartner, J. C., Walkley, C. R., Lu, J., & Orkin, S. H. (2009). ADAR1 is essential for the maintenance of hematopoiesis and suppression of interferon signaling. *Nature Immunology, 10*(1), 109–115. https://doi.org/10.1038/ni.1680.

Hasegawa, M., Miura, T., Kuzuya, K., Inoue, A., Ki, S. W., Horinouchi, S., ... Mizukami, T. (2011). Identification of SAP155 as the target of GEX1A (Herboxidiene), an antitumor natural product. *ACS Chemical Biology, 6*(3), 229–233. https://doi.org/10.1021/cb100248e.

Hattangadi, S. M., Wong, P., Zhang, L., Flygare, J., & Lodish, H. F. (2011). From stem cell to red cell: Regulation of erythropoiesis at multiple levels by multiple proteins, RNAs, and chromatin modifications. *Blood, 118*(24), 6258–6268. https://doi.org/10.1182/blood-2011-07-356006.

Hautin, M., Mornet, C., Chauveau, A., Bernard, D. G., Corcos, L., & Lippert, E. (2020). Splicing anomalies in myeloproliferative neoplasms: Paving the way for new therapeutic venues. *Cancers, 12*(8), 1–22. https://doi.org/10.3390/cancers12082216.

He, L., & Hannon, G. J. (2004). MicroRNAs: Small RNAs with a big role in gene regulation. *Nature Reviews Genetics, 5*(7), 522–531. https://doi.org/10.1038/nrg1379.

He, S., Valkov, E., Cheloufi, S., & Murn, J. (2023). The nexus between RNA-binding proteins and their effectors. *Nature Reviews Genetics, 24*(5), 276–294. https://doi.org/10.1038/s41576-022-00550-0.

Heinicke, L. A., Nabet, B., Shen, S., Jiang, P., van Zalen, S., Cieply, B., ... Carstens, R. P. (2013). The RNA binding protein RBM38 (RNPC1) regulates splicing during late erythroid differentiation. *PLoS One, 8*(10), 1–14. https://doi.org/10.1371/journal.pone.0078031.

Hon, C. C., Ramilowski, J. A., Harshbarger, J., Bertin, N., Rackham, O. J. L., Gough, J., ... Forrest, A. R. R. (2017). An atlas of human long non-coding RNAs with accurate 5′ ends. *Nature, 543*(7644), 199–204. https://doi.org/10.1038/nature21374.

Horos, R., IJspeert, H., Pospisilova, D., Sendtner, R., Andrieu-Soler, C., Taskesen, E., ... Von Lindern, M. (2012). Ribosomal deficiencies in Diamond-Blackfan anemia impair translation of transcripts essential for differentiation of murine and human erythroblasts. *Blood, 119*(1), 262–272. https://doi.org/10.1182/blood-2011-06-358200.

Hu, C., Yan, Y., FU, C., Ding, J., Li, T., Wang, S., & Fang, L. (2021). Effects of miR-210-3p on the erythroid differentiation of K562 cells under hypoxia. *Molecular Medicine Reports, 24*(2), 1–12. https://doi.org/10.3892/mmr.2021.12202.

Hu, W., Yuan, B., Flygare, J., & Lodish, H. F. (2011). Long noncoding RNA-mediated anti-apoptotic activity in murine erythroid terminal differentiation. *Genes and Development, 25*(24), 2573–2578. https://doi.org/10.1101/gad.178780.111.

Huang, P., Keller, C. A., Giardine, B., Grevet, J. D., Davies, J. O. J., Hughes, J. R., Kurita, R., Nakamura, Y., Hardison, R. C., & Blobel, G. A. (2017). Comparative analysis of three-dimensional chromosomal architecture identifies a novel fetal hemoglobin regulatory element. *Genes and Development, 31*(16), 1704–1713. https://doi.org/10.1101/gad.303461.117.

Huang, P., Peslak, S. A., Ren, R., Khandros, E., Qin, K., Keller, C. A., Giardine, B., Bell, H. W., Lan, X., Sharma, M., Horton, J. R., Abdulmalik, O., Chou, S. T., Shi, J., Crossley, M., Hardison, R. C., Cheng, X., & Blobel, G. A. (2022). HIC2 controls developmental hemoglobin switching by repressing BCL11A transcription. *Nature Genetics, 54*(9), 1417–1426. https://doi.org/10.1038/s41588-022-01152-6.

Huang, P., Peslak, S. A., Shehu, V., Keller, C. A., Giardine, B., Shi, J., Hardison, R. C., Blobel, G. A., & Khandros, E. (2024). let-7 miRNAs repress HIC2 to regulate BCL11A transcription and hemoglobin switching. *Blood, 143*(19), 1980–1991. https://doi.org/10.1182/blood.2023023399.

Huang, X., Chao, R., Zhang, Y., Wang, P., Gong, X., Liang, D., & Wang, Y. (2021). CAP1, a target of miR-144/451, negatively regulates erythroid differentiation and enucleation. *Journal of Cellular and Molecular Medicine, 25*(5), 2377–2389. https://doi.org/10.1111/jcmm.16067.

Huang, X., Gschweng, E., Van Handel, B., Cheng, D., Mikkola, H. K. A., & Witte, O. N. (2011). Regulated expression of microRNAs-126/126* inhibits erythropoiesis from human embryonic stem cells. *Blood, 117*(7), 2157–2165. https://doi.org/10.1182/blood-2010-08-302711.

Huang, Y., Hale, J., Wang, Y., Li, W., Zhang, S., Zhang, J., Zhao, H., Guo, X., Liu, J., Yan, H., Yazdanbakhsh, K., Huang, G., Hillyer, C. D., Mohandas, N., Chen, L., Sun, L., & An, X. (2018). SF3B1 deficiency impairs human erythropoiesis via activation of p53 pathway: Implications for understanding of ineffective erythropoiesis in MDS. *Journal of Hematology and Oncology, 11*(1), 1–12. https://doi.org/10.1186/s13045-018-0558-8.

Hussain, S., Aleksic, J., Blanco, S., Dietmann, S., & Frye, M. (2013). Characterizing 5-methylcytosine in the mammalian epitranscriptome. *Genome Biology, 14*(11), 1–10. https://doi.org/10.1186/gb4143.

Ivaldi, M. S., Diaz, L. F., Chakalova, L., Lee, J., Krivega, I., & Dean, A. (2018). Fetal g-globin genes are regulated by the BGLT3 long noncoding RNA locus. *Blood, 132*(18). https://ashpublications.org/blood/article-pdf/132/18/1963/1469486/blood862003.pdf.

Jack, K., Bellodi, C., Landry, D. M., Niederer, R. O., Meskauskas, A., Musalgaonkar, S., Kopmar, N., Krasnykh, O., Dean, A. M., Thompson, S. R., Ruggero, D., & Dinman, J. D. (2011). RRNA Pseudouridylation Defects Affect Ribosomal Ligand Binding and Translational Fidelity from Yeast to Human Cells. *Molecular Cell, 44*(4), 660–666. https://doi.org/10.1016/j.molcel.2011.09.017.

Jalan, A., Jayasree, P. J., Karemore, P., Narayan, K. P., & Khandelia, P. (2024). Decoding the 'Fifth' Nucleotide: Impact of RNA Pseudouridylation on Gene Expression and Human Disease. *Molecular Biotechnology, 66*(7), 1581–1598. https://doi.org/10.1007/s12033-023-00792-1.

Ji Heon Noh, K. M. K. (2016). Cytoplasmic functions of lncRNAs. *Physiology & Behavior, 176*(1), 100–106. https://doi.org/10.1002/wrna.1471.Cytoplasmic.

Ji, X., Jha, A., Humenik, J., Ghanem, L. R., Kromer, A., Duncan-Lewis, C., Traxler, E., Weiss, M. J., Barash, Y., & Liebhaber, S. A. (2021). RNA-Binding Proteins PCBP1 and PCBP2 Are Critical Determinants of Murine Erythropoiesis. *Molecular and Cellular Biology, 41*(9), 1–15. https://doi.org/10.1128/mcb.00668-20.

Jia, G., Fu, Y., & He, C. (2013). Reversible RNA adenosine methylation in biological regulation. *Trends in Genetics, 29*(2), 108–115. https://doi.org/10.1016/j.tig.2012.11.003.

Jin, H. L., Kim, J. S., Kim, Y. J., Kim, S. J., Broxmeyer, H. E., & Kim, K. S. (2012). Dynamic expression of specific miRNAs during erythroid differentiation of human embryonic stem cells. *Molecules and Cells, 34*(2), 177–183. https://doi.org/10.1007/s10059-012-0090-6.

Jivrajani, M., & Nivsarkar, M. (2016). Ligand-targeted bacterial minicells: Futuristic nano-sized drug delivery system for the efficient and cost effective delivery of shRNA to cancer cells. *Nanomedicine: Nanotechnology, Biology, and Medicine, 12*(8), 2485–2498. https://doi.org/10.1016/j.nano.2016.06.004.

Kaida, D., Motoyoshi, H., Tashiro, E., Nojima, T., Hagiwara, M., Ishigami, K., Watanabe, H., Kitahara, T., Yoshida, T., Nakajima, H., Tani, T., Horinouchi, S., & Yoshida, M. (2007). Spliceostatin A targets SF3b and inhibits both splicing and nuclear retention of pre-mRNA. *Nature Chemical Biology, 3*(9), 576–583. https://doi.org/10.1038/nchembio.2007.18.

Kapitsinou, P. P., Liu, Q., Unger, T. L., Rha, J., Davidoff, O., Keith, B., Epstein, J. A., Moores, S. L., Erickson-Miller, C. L., & Haase, V. H. (2010). Hepatic HIF-2 regulates erythropoietic responses to hypoxia in renal anemia. *Blood, 116*(16), 3039–3048. https://doi.org/10.1182/blood-2010-02-270322.

Kentsis, A., Topisirovic, I., Culjkovic, B., Shao, L., & Borden, K. L. B. (2004). Ribavirin suppresses eIF4E-mediated oncogenic transformation by physical mimicry of the 7-methyl guanosine mRNA cap. *Proceedings of the National Academy of Sciences of the United States of America, 101*(52), 18105–18110. https://doi.org/10.1073/pnas.0406927102.

Kiefer, C. M., Lee, J., Hou, C., Dale, R. K., Lee, Y. T., Meier, E. R., Miller, J. L., & Dean, A. (2011). Distinct Ldb1/NLI complexes orchestrate γ-globin repression and reactivation through ETO2 in human adult erythroid cells. *Blood, 118*(23), 6200–6208. https://doi.org/10.1182/blood-2011-06-363101.

Kim, J., Diaz, L. F., Miller, M. J., Leadem, B., Krivega, I., & Dean, A. (2024). An enhancer RNA recruits KMT2A to regulate transcription of Myb. *Cell Reports, 43*(7), https://doi.org/10.1016/j.celrep.2024.114378.

Kim, M., Tan, Y. S., Cheng, W. C., Kingsbury, T. J., Heimfeld, S., & Civin, C. I. (2015). MIR144 and MIR451 regulate human erythropoiesis via RAB14. *British Journal of Haematology, 168*(4), 583–597. https://doi.org/10.1111/bjh.13164.

Kini, H. K., Kong, J., & Liebhaber, S. A. (2014). Cytoplasmic Poly(A) Binding Protein C4 Serves a Critical Role in Erythroid Differentiation. *Molecular and Cellular Biology, 34*(7), 1300–1309. https://doi.org/10.1128/mcb.01683-13.

Klusmann, J. H., Li, Z., Böhmer, K., Maroz, A., Koch, M. L., Emmrich, S., Godinho, F. J., Orkin, S. H., & Reinhardt, D. (2010). miR-125b-2 is a potential oncomiR on human chromosome 21 in megakaryoblastic leukemia. *Genes and Development, 24*(5), 478–490. https://doi.org/10.1101/gad.1856210.

Kotake, Y., Sagane, K., Owa, T., Mimori-Kiyosue, Y., Shimizu, H., Uesugi, M., Ishihama, Y., Iwata, M., & Mizui, Y. (2007). Splicing factor SF3b as a target of the antitumor natural product pladienolide. *Nature Chemical Biology, 3*(9), 570–575. https://doi.org/10.1038/nchembio.2007.16.

Kotaki, R., Kawashima, M., Yamaguchi, A., Suzuki, N., Koyama-Nasu, R., Ogiya, D., Okuyama, K., Yamamoto, Y., Takamatsu, M., Kurosaki, N., Ando, K., Murata, A., Ohtsuka, M., Nakagawa, S., Katagiri, K., & Kotani, A. (2020). Overexpression of miR-669m inhibits erythroblast differentiation. *Scientific Reports, 10*(1), 1–10. https://doi.org/10.1038/s41598-020-70442-y.

Kretov, D. A., Folkes, L., Mora-Martin, A., Walawalkar, I. A., Imrat, Syedah, N., Vanuytsel, K., Moxon, S., Murphy, G. J., & Cifuentes, D. (2024). The miR-144/Hmgn2 regulatory axis orchestrates chromatin organization during erythropoiesis. *Nature Communications, 15*(1), https://doi.org/10.1038/s41467-024-47982-2.

Kulczyńska, K., & Siatecka, M. (2016). A regulatory function of long non-coding RNAs in red blood cell development. *Acta Biochimica Polonica, 63*(4), 675–680. https://doi.org/10.18388/abp.2016_1351.

Kumar, M. S., Narla, A., Nonami, A., Mullally, A., Dimitrova, N., Ball, B., McAuley, J. R., Poveromo, L., Kutok, J. L., Galili, N., Raza, A., Attar, E., Gilliland, D. G., Jacks, T., & Ebert, B. L. (2011). Coordinate loss of a microRNA and protein-coding gene cooperate in the pathogenesis of 5q- syndrome. *Blood, 118*(17), 4666–4673. https://doi.org/10.1182/blood-2010-12-324715.

Kuppers, D. A., Arora, S., Lim, Y., Lim, A. R., Carter, L. M., Corrin, P. D., Plaisier, C. L., Basom, R., Delrow, J. J., Wang, S., Hansen He, H., Torok-Storb, B., Hsieh, A. C., & Paddison, P. J. (2019). N6-methyladenosine mRNA marking promotes selective translation of regulons required for human erythropoiesis. *Nature Communications, 10*(1), https://doi.org/10.1038/s41467-019-12518-6.

Lacerra, G., Sierakowska, H., Carestia, C., Fucharoen, S., Summerton, J., Weller, D., & Kole, R. (2000). Restoration of hemoglobin A synthesis in erythroid cells from peripheral blood of thalassemic patients. *Proceedings of the National Academy of Sciences of the United States of America, 97*(17), 9591–9596. https://doi.org/10.1073/pnas.97.17.9591.

Lai, W. S., Kennington, E. A., & Blackshear, P. J. (2003). Tristetraprolin and Its Family Members Can Promote the Cell-Free Deadenylation of AU-Rich Element-Containing mRNAs by Poly(A) Ribonuclease. *Molecular and Cellular Biology, 23*(11), 3798–3812. https://doi.org/10.1128/mcb.23.11.3798-3812.2003.

Lanz, R. B., McKenna, N. J., Onate, S. A., Albrecht, U., Wong, J., Tsai, S. Y., Tsai, M. J., & O'Malley, B. W. (1999). A steroid receptor coactivator, SRA, functions as an RNA and is present in an SRC-1 complex. *Cell, 97*(1), 17–27. https://doi.org/10.1016/S0092-8674(00)80711-4.

Lee, J. Y., Kim, M., Heo, H. R., Ha, K. S., Han, E. T., Park, W. S., Yang, S. R., & Hong, S. H. (2018). Inhibition of MicroRNA-221 and 222 enhances hematopoietic differentiation from human pluripotent stem cells via c-KIT upregulation. *Molecules and Cells, 41*(11), 971–978. https://doi.org/10.14348/molcells.2018.0244.

Lee, Y. T., De Vasconcellos, J. F., Yuan, J., Byrnes, C., Noh, S. J., Meier, E. R., Kim, K. S., Rabel, A., Kaushal, M., Muljo, S. A., & Miller, J. L. (2013). LIN28B-mediated expression of fetal hemoglobin and production of fetal-like erythrocytes from adult human erythroblasts ex vivo. *Blood, 122*(6), 1034–1041. https://doi.org/10.1182/blood-2012-12-472308.

Lewis, B. P., Burge, C. B., & Bartel, D. P. (2005). Conserved seed pairing, often flanked by adenosines, indicates that thousands of human genes are microRNA targets. *Cell, 120*, Cell Press,15–20. https://doi.org/10.1016/j.cell.2004.12.035.

Li, X., Lu, Y. C., Dai, K., Torregroza, I., Hla, T., & Evans, T. (2014). Elavl1a regulates zebrafish erythropoiesis via posttranscriptional control of gata1. *Blood, 123*(9), 1384–1392. https://doi.org/10.1182/blood-2013-09-526962.

Li, Y., Liu, D., Zhang, X., Li, Z., Ye, Y., Liu, Q., Shen, J., Chen, Z., Huang, H., Liang, Y., Han, X., Liu, J., An, X., Mohandas, N., & Xu, X. (2018). miR-326 regulates HbF synthesis by targeting EKLF in human erythroid cells. e2 *Experimental Hematology, 63*, 33–40. https://doi.org/10.1016/j.exphem.2018.03.004.

Li, Y., Liu, S., Sun, H., Yang, Y., Qi, H., Ding, N., Zheng, J., Dong, X., Qu, H., Zhang, Z., & Fang, X. (2015). miR-218 inhibits erythroid differentiation and alters iron metabolism by targeting ALAS2 in K562 cells. *International Journal of Molecular Sciences, 16*(12), 28156–28168. https://doi.org/10.3390/ijms161226088.

Li, Y., Zhang, H., Hu, B., Wang, P., Wang, W., & Liu, J. (2023). Post-transcriptional regulation of erythropoiesis. *Blood Science, 5*(3), 150–159. https://doi.org/10.1097/BS9.0000000000000159.

Li, Z., Qian, P., Shao, W., Shi, H., He, X. C., Gogol, M., Yu, Z., Wang, Y., Qi, M., Zhu, Y., Perry, J. M., Zhang, K., Tao, F., Zhou, K., Hu, D., Han, Y., Zhao, C., Alexander, R., Xu, H., & Li, L. (2018). Suppression of m6A reader Ythdf2 promotes hematopoietic stem cell expansion. *Cell Research, 28*(9), 904–917. https://doi.org/10.1038/s41422-018-0072-0.

Liddicoat, B. J., Hartner, J. C., Piskol, R., Ramaswami, G., Chalk, A. M., Kingsley, P. D., Sankaran, V. G., Wall, M., Purton, L. E., Seeburg, P. H., Palis, J., Orkin, S. H., Lu, J., Li, J. B., & Walkley, C. R. (2016). Adenosine-to-inosine RNA editing by ADAR1 is essential for normal murine erythropoiesis. *Experimental Hematology, 44*(10), 947–963. https://doi.org/10.1016/j.exphem.2016.06.250.

Liebhaber, S. A. (1997). mRNA stability and the control of gene expression. *Nucleic Acids Symposium Series, 36*, 29–32. https://doi.org/10.1038/npg.els.0005972.

Liebhaber, S. A., & Kan, W. (1981). Differentiation of the mRNA transcripts originating from the α1- and α2-globin loci in normals and α-thalassemics. *Journal of Clinical Investigation, 68*(2), 439–446. https://doi.org/10.1172/JCI110273.

Lipton, J. M., & Ellis, S. R. (2009). Diamond-Blackfan anemia: Diagnosis, treatment, and molecular pathogenesis. *Hematology/Oncology Clinics of North America, 23*(2), 261–282. https://doi.org/10.1016/j.hoc.2009.01.004.

Liu, F., Clark, W., Luo, G., Wang, X., Fu, Y., Wei, J., ... He, C. (2016). ALKBH1-mediated tRNA demethylation regulates translation. *Cell, 167*(3), 816–828.e16. https://doi.org/10.1016/j.cell.2016.09.038.

Liu, G., Kim, J., Nguyen, N., Zhou, L., & Dean, A. (2024). Long noncoding RNA GATA2AS influences human erythropoiesis by transcription factor and chromatin landscape modulation. *Blood, 143*(22), 2300–2313. https://doi.org/10.1182/blood.2023021287.

Liu, G., States, U., Kim, J., Institutes, N., States, U., Nguyen, N., …Statement, D.S. (2023). Long noncoding RNA GATA2AS influences human erythropoiesis by transcription factor and chromatin landscape regulation and chromatin landscape modulation Laboratory of Cellular and Developmental Biology, National Institute of Diabetes and Digestive Corres.

Liu, J., Li, Y., Tong, J., Gao, J., Guo, Q., Zhang, L., ... Shi, L. (2018). Long non-coding RNA-dependent mechanism to regulate heme biosynthesis and erythrocyte development. *Nature Communications, 9*(1), 1–15. https://doi.org/10.1038/s41467-018-06883-x.

Liu, J., Wang, F., Zhang, Y., Liu, J., & Zhao, B. (2022). ADAR1-mediated RNA editing and its role in cancer. *Frontiers in Cell and Developmental Biology, 10*(July), 1–12. https://doi.org/10.3389/fcell.2022.956649.

Lu, J., Guo, S., Ebert, B. L., Zhang, H., Peng, X., Bosco, J., ... Golub, T. R. (2008). MicroRNA-mediated control of cell fate in Megakaryocyte-Erythrocyte progenitors. *Developmental Cell, 14*(6), 843–853. https://doi.org/10.1016/j.devcel.2008.03.012.

Lumelsky, N. L., & Forget, B. G. (1991). Negative regulation of globin gene expression during megakaryocytic differentiation of a human erythroleukemic cell line. *Molecular and Cellular Biology, 11*(7), 3528–3536. https://doi.org/10.1128/mcb.11.7.3528.

Lv, J., Zhang, Y., Gao, S., Zhang, C., Chen, Y., Li, W., ... Liu, F. (2018). Endothelial-specific m6A modulates mouse hematopoietic stem and progenitor cell development via Notch signaling. *Cell Research, 28*(2), 249–252. https://doi.org/10.1038/cr.2017.143.

Ma, S. P., Xi, H. R., Gao, X. X., Yang, J. M., Kurita, R., Nakamura, Y., ... Lu, D. R. (2021). Long noncoding RNA HBBP1 enhances γ-globin expression through the ETS transcription factor ELK1. *Biochemical and Biophysical Research Communications, 552*, 157–163. https://doi.org/10.1016/j.bbrc.2021.03.051.

Ma, Y., Liu, S., Gao, J., Chen, C., Zhang, X., Yuan, H., ... Yu, J. (2021). Genome-wide analysis of pseudogenes reveals HBBP1's human-specific essentiality in erythropoiesis and implication in β-thalassemia. *Developmental Cell, 56*(4), 478–493. https://doi.org/10.1016/j.devcel.2020.12.019 e11.

Ma, Y., Wang, B., Jiang, F., Wang, D., Liu, H., Yan, Y., ... Yu, J. (2013). A feedback loop consisting of microRNA 23a/27a and the β-like globin suppressors KLF3 and SP1 regulates globin gene expression. *Molecular and Cellular Biology, 33*(20), 3994–4007. https://doi.org/10.1128/mcb.00623-13.

Magor, G. W., Tallack, M. R., Gillinder, K. R., Bell, C. C., Mccallum, N., Williams, B., & Perkins, A. C. (2015). KLF1 -Null neonates display hydrops fetalis and a deranged erythroid transcriptome. *Blood, 125*(15), 2405–2418. https://doi.org/10.1182/blood-2014-08-590968.Sequencing.

Malcovati, L., Papaemmanuil, E., Bowen, D. T., Boultwood, J., Della Porta, M. G., Pascutto, C., ... Cazzola, M. (2011). Clinical significance of SF3B1 mutations in myelodysplastic syndromes and myelodysplastic/myeloproliferative neoplasms. *Blood, 118*(24), 6239–6246. https://doi.org/10.1182/blood-2011-09-377275.

Mattick, J. S., Amaral, P. P., Carninci, P., Carpenter, S., Chang, H. Y., Chen, L. L., ... Wu, M. (2023). Long non-coding RNAs: Definitions, functions, challenges and recommendations. *Nature Reviews. Molecular Cell Biology, 24*(6), 430–447. https://doi.org/10.1038/s41580-022-00566-8.

Matur, M., Uzun, Y., Shi, X., Hamamoto, K., Qiu, Y., & Huang, S. (2025). Noncoding RNA, ncRNA-a3, epigenetically regulates TAL1 transcriptional program during erythropoiesis. *Molecular and Cellular Biology, 45*(4), 169–184. https://doi.org/10.1080/10985549.2025.2482079.

McDonald, C. M., Shieh, P. B., Abdel-Hamid, H. Z., Connolly, A. M., Ciafaloni, E., Wagner, K. R., ... Mendell, J. R. (2021). Open-label evaluation of eteplirsen in patients with duchenne muscular dystrophy amenable to exon 51 skipping: PROMOVI Trial. *Journal of Neuromuscular Diseases, 8*(6), 989–1001. https://doi.org/10.3233/JND-210643.

Meyer, K. D., Patil, D. P., Zhou, J., Zinoviev, A., Skabkin, M. A., Elemento, O., ... Jaffrey, S. R. (2015). 5′ UTR m6A promotes cap-independent translation. *Cell, 163*(4), 999–1010. https://doi.org/10.1016/j.cell.2015.10.012.

Miao, H., Wang, L., Zhan, H., Dai, J., Chang, Y., Wu, F., ... Song, X. (2019). A long noncoding RNA distributed in both nucleus and cytoplasm operates in the PYCARD-regulated apoptosis by coordinating the epigenetic and translational regulation. *PLoS Genetics, 15*(5), 1–24. https://doi.org/10.1371/journal.pgen.1008144.

Mittal, S. P. K., Mathai, J., Kulkarni, A. P., Pal, J. K., & Chattopadhyay, S. (2013). MiR-320a regulates erythroid differentiation through MAR binding protein SMAR1. *International Journal of Biochemistry and Cell Biology, 45*(11), 2519–2529. https://doi.org/10.1016/j.biocel.2013.07.006.

Mongelard, F., & Bouvet, P. (2010). AS-1411, a guanosine-rich oligonucleotide aptamer targeting nucleolin for the potential treatment of cancer, including acute myeloid leukemia. *Current Opinion in Molecular Therapeutics, 12*(1), 107–114.

Monteys, A. M., Hundley, A. A., Ranum, P. T., Tecedor, L., Muehlmatt, A., Lim, E., ... Davidson, B. L. (2021). Regulated control of gene therapies by drug-induced splicing. *Nature, 596*(7871), 291–295. https://doi.org/10.1038/s41586-021-03770-2.

Moore, K. S., Lindern, M., & Von (2018). RNA binding proteins and regulation of mRNA translation in erythropoiesis. *Frontiers in Physiology, 9*(July), 1–17. https://doi.org/10.3389/fphys.2018.00910.

Moore, K. S., & von Lindern, M. (2018). RNA binding proteins and regulation of mRNA translation in erythropoiesis. *Frontiers in Physiology, 9*(JUL), 1–17. https://doi.org/10.3389/fphys.2018.00910.

Moore, K. S., Yagci, N., Van Alphen, F., Paolini, N. A., Horos, R., Held, N. M., ... Von Lindern, M. (2018). Csde1 binds transcripts involved in protein homeostasis and controls their expression in an erythroid cell line. *Scientific Reports, 8*(1), 1–14. https://doi.org/10.1038/s41598-018-20518-7.

Morais, P., Adachi, H., & Yu, Y. T. (2021). The critical contribution of pseudouridine to mRNA COVID-19 vaccines. *Frontiers in Cell and Developmental Biology, 9*(November), 1–9. https://doi.org/10.3389/fcell.2021.789427.

Muckenthaler, M. U., Rivella, S., Hentze, M. W., & Galy, B. (2017). *A red carpet for iron metabolism. Cell, 168*, Elsevier Inc344–361. https://doi.org/10.1016/j.cell.2016.12.034.

Mueller, C., Berry, J. D., McKenna-Yasek, D. M., Gernoux, G., Owegi, M. A., Pothier, L. M., ... Brown, R. H. (2020). SOD1 suppression with adeno-associated virus and microRNA in familial ALS. *New England Journal of Medicine, 383*(2), 151–158. https://doi.org/10.1056/nejmoa2005056.

Murakami, S., Suzuki, T., Yokoyama, W., Yagi, S., Matsumura, K., Nakajima, Y., ... Motohashi, H. (2018). Nucleomethylin deficiency impairs embryonic erythropoiesis. *Journal of Biochemistry, 163*(5), 413–423. https://doi.org/10.1093/jb/mvx086.

Murphy, Z. C., Murphy, K., Myers, J., Getman, M., Couch, T., Schulz, V. P., ... Steiner, L. A. (2021). Regulation of RNA polymerase II activity is essential for terminal erythroid maturation. *Blood, 138*(18), 1740–1756. https://doi.org/10.1182/blood.2020009903.

Naarmann-de Vries, I. S., Brendle, A., Bähr-Ivacevic, T., Benes, V., Ostareck, D. H., & Ostareck-Lederer, A. (2016). Translational control mediated by hnRNP K links NMHC IIA to erythroid enucleation. *Journal of Cell Science, 129*(6), 1141–1154. https://doi.org/10.1242/jcs.174995.

Naarmann-De Vries, I. S., Senatore, R., Moritz, B., Marx, G., Urlaub, H., Niessing, D., ... Ostareck-Lederer, A. (2021). Methylated HNRNPK acts on RPS19 to regulate ALOX15 synthesis in erythropoiesis. *Nucleic Acids Research, 49*(6), 3507–3523. https://doi.org/10.1093/nar/gkab116.

Naarmann, I. S., Harnisch, C., Flach, N., Kremmer, E., Kühn, H., Ostareck, D. H., & Ostareck-Lederer, A. (2008). mRNA silencing in human erythroid cell maturation: Heterogeneous nuclear ribonucleoprotein K controls the expression of its regulator c-Src. *Journal of Biological Chemistry, 283*(26), 18461–18472. https://doi.org/10.1074/jbc.M710328200.

Nombela, P., Miguel-López, B., & Blanco, S. (2021). The role of m6A, m5C and ⊠ RNA modifications in cancer: Novel therapeutic opportunities. *Molecular Cancer, 20*, Molecular Cancer 1–30. https://doi.org/10.1186/s12943-020-01263-w.

Nuinoon, M., Makarasara, W., Mushiroda, T., Setianingsih, I., Wahidiyat, P. A., Sripichai, O., ... Fucharoen, S. (2010). A genome-wide association identified the common genetic variants influence disease severity in β0-thalassemia/hemoglobin E. *Human Genetics, 127*(3), 303–314. https://doi.org/10.1007/s00439-009-0770-2.

O'Carroll, D., Mecklenbrauker, I., Das, P. P., Santana, A., Koenig, U., Enright, A. J., ... Tarakhovsky, A. (2007). A slicer-independent role for Argonaute 2 in hematopoiesis and the microRNA pathway. *Genes and Development, 21*(16), 1999–2004. https://doi.org/10.1101/gad.1565607.

Oh, J. H., Jin, O. Y., Hahn, Y., Kim, M. R., Byun, S. S., Jeon, Y. J., ... Kim, N. S. (2005). Transcriptome analysis of human gastric cancer. *Mammalian Genome, 16*(12), 942–954. https://doi.org/10.1007/s00335-005-0075-2.

Oilman, J. G., Johnson, M. E., & Mishima, N. (1988). Four base-pair DNA deletion in human Aγ globin-gene promoter associated with low Aγ expression in adults. *British Journal of Haematology, 68*(4), 455–458. https://doi.org/10.1111/J.1365-2141.1988.TB04235.X.

Ougland, R., Zhang, C. M., Liiv, A., Johansen, R. F., Seeberg, E., Hou, Y. M., ... Falnes, P. (2004). AlkB restores the biological function of mRNA and tRNA inactivated by chemical methylation. *Molecular Cell, 16*(1), 107–116. https://doi.org/10.1016/j.molcel.2004.09.002.

Ozanick, S., Krecic, A., Andersland, J., & Anderson, J. T. (2005). The bipartite structure of the tRNA m1A58 methyltransferase from S. cerevisiae is conserved in humans. *RNA (New York, N. Y.), 11*(8), 1281–1290. https://doi.org/10.1261/rna.5040605.

Palii, C. G., Cheng, Q., Gillespie, M. A., Shannon, P., Mazurczyk, M., Napolitani, G., ... Brand, M. (2019). Single-cell proteomics reveal that quantitative changes in co-expressed lineage-specific transcription factors determine cell fate. *Cell Stem Cell, 24*(5), 812–820.e5. https://doi.org/10.1016/j.stem.2019.02.006.

Palis, J. (2014). Primitive and definitive erythropoiesis in mammals. *Frontiers in Physiology, 5*(JAN), https://doi.org/10.3389/fphys.2014.00003.

Paralkar, V. R., Mishra, T., Luan, J., Yao, Y., Kossenkov, A. V., Anderson, S. M., ... Weiss, M. J. (2014). Lineage and species-specific long noncoding RNAs during erythro-megakaryocytic development. *Blood, 123*(12), 1927–1937. https://doi.org/10.1182/blood-2013-12-544494.

Paralkar, V. R., Taborda, C. C., Huang, P., Yao, Y., Kossenkov, A. V., Prasad, R., ... Weiss, M. J. (2016). Unlinking an lncRNA from its associated cis element. *Molecular Cell, 62*(1), 104–110. https://doi.org/10.1016/j.molcel.2016.02.029.

Patnaik, M. M., Lasho, T. L., Finke, C. M., Hanson, C. A., King, R. L., Ketterling, R. P., ... Tefferi, A. (2016). Predictors of survival in refractory anemia with ring sideroblasts and thrombocytosis (RARS-T) and the role of next-generation sequencing. *American Journal of Hematology, 91*(5), 492–498. https://doi.org/10.1002/ajh.24332.

Patrick, D. M., Zhang, C. C., Tao, Y., Yao, H., Qi, X., Schwartz, R. J., ... Olson, E. N. (2010). Defective erythroid differentiation in miR-451 mutant mice mediated by 14-3-3ζ. *Genes and Development, 24*(15), 1614–1619. https://doi.org/10.1101/gad.1942810.

Perry, R. P., & Scherrer, K. (1975). The methylated constituents of globin mRNA. *FEBS Letters, 57*(1), 73–78. https://doi.org/10.1016/0014-5793(75)80155-4.

Pimentel, H., Parra, M., Gee, S. L., Mohandas, N., Pachter, L., & Conboy, J. G. (2016). A dynamic intron retention program enriched in RNA processing genes regulates gene expression during terminal erythropoiesis. *Nucleic Acids Research, 44*(2), 838–851. https://doi.org/10.1093/nar/gkv1168.

Pirouz, M., Du, P., Munafò, M., & Gregory, R. I. (2016). Dis3l2-mediated decay is a quality control pathway for noncoding RNAs. *Cell Reports, 16*(7), 1861–1873. https://doi.org/10.1016/j.celrep.2016.07.025.

Rabany, O., Ben Dror, S., Arafat, M., Aharoni levitanus, H., Halperin, Y., Marchand, V., ... Nachmani, D. (2026). Dynamic rRNA methylation regulates translation in the hematopoietic system and is essential for stem cell fitness. *Blood, 147*(5), https://doi.org/10.1182/blood.2024028300.

Rasmussen, K. D., Simmini, S., Abreu-Goodger, C., Bartonicek, N., Di Giacomo, M., Bilbao-Cortes, D., ... O'Carroll, D. (2010). The miR-144/451 locus is required for erythroid homeostasis. *Journal of Experimental Medicine, 207*(7), 1351–1358. https://doi.org/10.1084/jem.20100458.

Rehn, M., Wenzel, A., Frank, A. K., Schuster, M. B., Pundhir, S., Jørgensen, N., ... Porse, B. T. (2022). PTBP1 promotes hematopoietic stem cell maintenance and red blood cell development by ensuring sufficient availability of ribosomal constituents. *Cell Reports, 39*(6), 110793. https://doi.org/10.1016/j.celrep.2022.110793.

Ren, Y., Zhu, J., Han, Y., Li, P., Wu, J., Qu, H., ... Fang, X. (2021). Regulatory association of long noncoding RNAs and chromatin accessibility facilitates erythroid differentiation. *Blood Advances, 5*(23), 5396–5409. https://doi.org/10.1182/BLOODADVANCES.2021005167.

Rivkin, N., Chapnik, E., Mildner, A., Barshtein, G., Porat, Z., Kartvelishvily, E., ... Hornstein, E. (2017). Erythrocyte survival is controlled by microRNA-142. *Haematologica, 102*(4), 676–685. https://doi.org/10.3324/haematol.2016.156109.

Rottman, F., Shatkin, A. J., & Perry, R. P. (1974). Sequences containing methylated nucleotides at the 5′ termini of messenger RNAs: Possible implications for processing. *Cell, 3*(3), 197–199. https://doi.org/10.1016/0092-8674(74)90131-7.

Russell, J. E., & Liebhaber, S. A. (1996). The stability of human beta-globin mRNA is dependent on structural determinants positioned within its 3' untranslated region. *Blood, 87*(12), 5314–5323.

Russell, J. E., Morales, J., & Liebhaber, S. A. (1997). The role of mRNA stability in the control of globin gene expression. *Progress in Nucleic Acid Research and Molecular Biology, 57*, 249–287. https://doi.org/10.1016/S0079-6603(08)60283-4.

Sackmann, E. (1995). Physical basis of self-organization and function of membranes: Physics of vesicles. *Handbook of Biological Physics, 1*(C), 213–304. https://doi.org/10.1016/S1383-8121(06)80022-9.

Safra, M., Sas-Chen, A., Nir, R., Winkler, R., Nachshon, A., Bar-Yaacov, D., ... Schwartz, S. (2017). The m1A landscape on cytosolic and mitochondrial mRNA at single-base resolution. *Nature, 551*(7679), 251–255. https://doi.org/10.1038/nature24456.

Sanchez, M., Schwanhaeusser, B., Blake, J., Bähr-Ivacevic, T., Benes, V., Selbach, M., ... Hentze, M. W. (2011). Iron regulatory protein-1 and -2: Transcriptome-wide definition of binding mRNAs and shaping of the cellular proteome by iron regulatory proteins. *Blood, 118*(22), 1–3. https://doi.org/10.1182/blood-2011-04-343541.

Sarakul, O., Vattanaviboon, P., Tanaka, Y., Fucharoen, S., Abe, Y., Svasti, S., & Umemura, T. (2013). Enhanced erythroid cell differentiation in hypoxic condition is in part contributed by miR-210. *Blood Cells, Molecules, and Diseases, 51*(2), 98–103. https://doi.org/10.1016/j.bcmd.2013.03.005.

Sawaengdee, W., Cui, K., Zhao, K., Hongeng, S., Fucharoen, S., & Wongtrakoongate, P. (2020). Genome-wide transcriptional regulation of the long non-coding RNA steroid receptor RNA activator in human erythroblasts. *Frontiers in Genetics, 11*(August), https://doi.org/10.3389/fgene.2020.00850.

Schibler, U., Kelley, D. E., & Perry, R. P. (1977). Comparison of methylated sequences in messenger RNA and heterogeneous nuclear RNA from mouse L cells. *Journal of Molecular Biology, 115*(4), 695–714. https://doi.org/10.1016/0022-2836(77)90110-3.

Schranzhofer, M., Schifrer, M., Cabrera, J. A., Kopp, S., Chiba, P., Beug, H., & Müllner, E. W. (2006). Remodeling the regulation of iron metabolism during erythroid differentiation to ensure efficient heme biosynthesis. *Blood, 107*(10), 4159–4167. https://doi.org/10.1182/blood-2005-05-1809.

Schwartz, S., Bernstein, D. A., Mumbach, M. R., Jovanovic, M., Herbst, R. H., León-Ricardo, B. X., ... Regev, A. (2014). Transcriptome-wide mapping reveals widespread dynamic-regulated pseudouridylation of ncRNA and mRNA. *Cell, 159*(1), 148–162. https://doi.org/10.1016/j.cell.2014.08.028.

Sender, R., Fuchs, S., & Milo, R. (2016). Are we really vastly outnumbered? Revisiting the ratio of bacterial to host cells in humans. *Cell, 164*, 337–340. https://doi.org/10.1016/j.cell.2016.01.013 Cell Press.

Seo, K. W., & Kleiner, R. E. (2020). YTHDF2 recognition of N1-methyladenosine (m1A)-modified RNA Is associated with transcript destabilization. *ACS Chemical Biology, 15*(1), 132–139. https://doi.org/10.1021/acschembio.9b00655.

Seto, A. G., Beatty, X., Lynch, J. M., Hermreck, M., Tetzlaff, M., Duvic, M., & Jackson, A. L. (2018). Cobomarsen, an oligonucleotide inhibitor of miR-155, co-ordinately regulates multiple survival pathways to reduce cellular proliferation and survival in cutaneous T-cell lymphoma. *British Journal of Haematology, 183*(3), 428–444. https://doi.org/10.1111/bjh.15547.

Shen, S., Slightom, J. L., & Smithies, O. (1981). A history of the human fetal globin gene duplication. *Cell, 26*(2 PART 2), 191–203. https://doi.org/10.1016/0092-8674(81)90302-0.

Sheng, L., Ye, L., Zhang, D., Cawthorn, W. P., & Xu, B. (2018). New insights into the long non-coding RNA SRA: Physiological functions and mechanisms of action. *Frontiers in Medicine, 5*(SEP), 1–16. https://doi.org/10.3389/fmed.2018.00244.

Smith, M. A., Choudhary, G. S., Pellagatti, A., Choi, K., Bolanos, L. C., Bhagat, T. D., ... Starczynowski, D. T. (2019). U2AF1 mutations induce oncogenic IRAK4 isoforms and activate innate immune pathways in myeloid malignancies. *Nature Cell Biology, 21*(5), 640–650. https://doi.org/10.1038/s41556-019-0314-5.

Socolovsky, M. (2022). The role of specialized cell cycles during erythroid lineage development: Insights from single-cell RNA sequencing. *International Journal of Hematology, 116*(2), 163–173. https://doi.org/10.1007/s12185-022-03406-9.

Squires, J. E., Patel, H. R., Nousch, M., Sibbritt, T., Humphreys, D. T., Parker, B. J., ... Preiss, T. (2012). Widespread occurrence of 5-methylcytosine in human coding and non-coding RNA. *Nucleic Acids Research, 40*(11), 5023–5033. https://doi.org/10.1093/nar/gks144.

Stamatoyannopoulos, G. (2005). Control of globin gene expression during development and erythroid differentiation. *Experimental hematology, 33*, 259–271. https://doi.org/10.1016/j.exphem.2004.11.007 Elsevier Inc.

Stephenson, J. R., Axelrad, A. A., McLeod, D. L., & Shreeve, M. M. (1971). Induction of colonies of hemoglobin-synthesizing cells by erythropoietin in vitro. *Proceedings of the National Academy of Sciences of the United States of America, 68*(7), 1542–1546. https://doi.org/10.1073/pnas.68.7.1542.

Stumpo, D. J., Broxmeyer, H. E., Ward, T., Cooper, S., Hangoc, G., Yang, J. C., ... Blackshear, P. J. (2009). Targeted disruption of Zfp36l2, encoding a CCCH tandem zinc finger RNA-binding protein, results in defective hematopoiesis. *Blood, 114*(12), 2401–2410. https://doi.org/10.1182/blood-2009-04-214619.

Su, R., Dong, L., Li, Y., Gao, M., Han, L., Wunderlich, M., ... Chen, J. (2020). Targeting FTO suppresses cancer stem cell maintenance and immune evasion. *Cancer Cell, 38*(1), 79–96.e11. https://doi.org/10.1016/j.ccell.2020.04.017.

Suresh, S., Rajvanshi, P. K., & Noguchi, C. T. (2020). The many facets of erythropoietin physiologic and metabolic response. *Frontiers in Physiology, 10*(January), 1–20. https://doi.org/10.3389/fphys.2019.01534.

Suwanmanee, T., Sierakowska, H., Fucharoen, S., & Kole, R. (2002). Repair of a splicing defect in erythroid cells from patients with β-thalassemia/HbE disorder. *Molecular Therapy, 6*(6), 718–726. https://doi.org/10.1006/mthe.2002.0805.

Svasti, S., Suwanmanee, T., Fucharoen, S., Moulton, H. M., Nelson, M. H., Maeda, N., ... Kole, R. (2009). RNA repair restores hemoglobin expression in IVS2-654 thalassemic mice. *Proceedings of the National Academy of Sciences of the United States of America, 106*(4), 1205–1210. https://doi.org/10.1073/pnas.0812436106.

Tao, Y., Zhang, Q., Wang, H., Yang, X., & Mu, H. (2024). Alternative splicing and related RNA binding proteins in human health and disease. *Signal Transduction and Targeted Therapy, 9*Springer US. https://doi.org/10.1038/s41392-024-01734-2.

Taylor, J., & Lee, S. C. (2019). Mutations in spliceosome genes and therapeutic opportunities in myeloid malignancies. *Genes Chromosomes and Cancer, 58*(12), 889–902. https://doi.org/10.1002/gcc.22784.

Tefferi, A., Lasho, T. L., Finke, C. M., Elala, Y., Hanson, C. A., Ketterling, R. P., ... Pardanani, A. (2016). Targeted deep sequencing in primary myelofibrosis. *Blood Advances, 1*(2), 105–111. https://doi.org/10.1182/bloodadvances.2016000208.

Thomas, M. P., Liu, X., Whangbo, J., McCrossan, G., Sanborn, K. B., Basar, E., ... Lieberman, J. (2015). Apoptosis triggers specific, rapid, and global mRNA decay with 3' uridylated intermediates degraded by DIS3L2. *Cell Reports, 11*(7), 1079–1089. https://doi.org/10.1016/j.celrep.2015.04.026.

Tsagakis, I., Douka, K., Birds, I., & Aspden, J. L. (2020). Long non-coding RNAs in development and disease: Conservation to mechanisms. *Journal of Pathology, 250*(5), 480–495. https://doi.org/10.1002/path.5405.

Tsiftsoglou, A. S. (2021). Erythropoietin (EPO) as a key regulator of erythropoiesis, bone remodeling and endothelial transdifferentiation of multipotent mesenchymal stem cells (MSCs): Implications in regenerative medicine. *Cells, 10*(8), https://doi.org/10.3390/cells10082140.

Tusi, B. K., Wolock, S. L., Weinreb, C., Hwang, Y., Hidalgo, D., Zilionis, R., ... Socolovsky, M. (2018). Population snapshots predict early haematopoietic and erythroid hierarchies. *Nature Publishing Group*. https://doi.org/10.1038/nature25741.

Uda, M., Galanello, R., Sanna, S., Lettre, G., Sankaran, V. G., Chen, W., ... Cao, A. (2008). Genome-wide association study shows BCL11A associated with persistent fetal hemoglobin and amelioration of the phenotype of β-thalassemia. *Proceedings of the National Academy of Sciences of the United States of America, 105*(5), 1620–1625. https://doi.org/10.1073/pnas.0711566105.

Vacek, M. M., Ma, H., Gemignani, F., Lacerra, G., Kafri, T., & Kole, R. (2003). High-level expression of hemoglobin A in human thalassemic erythroid progenitor cells following lentiviral vector delivery of an antisense snRNA. *Blood, 101*(1), 104–111. https://doi.org/10.1182/blood-2002-06-1869.

Vainchenker, W., & Kralovics, R. (2017). Genetic basis and molecular pathophysiology of classical myeloproliferative neoplasms. *Blood, 129*(6), 667–679. https://doi.org/10.1182/blood-2016-10-695940.

van Zalen, S., Lombardi, A. A., Jeschke, G. R., Hexner, E. O., & Russell, J. E. (2015). AUF-1 and YB-1 independently regulate β-globin mRNA in developing erythroid cells through interactions with poly(A)-binding protein. *Mechanisms of Development, 136*(215), 40–52. https://doi.org/10.1016/j.mod.2015.02.003.

van Zandwijk, N., Pavlakis, N., Kao, S. C., Linton, A., Boyer, M. J., Clarke, S., ... Reid, G. (2017). Safety and activity of microRNA-loaded minicells in patients with recurrent malignant pleural mesothelioma: A first-in-man, phase 1, open-label, dose-escalation study. *The Lancet Oncology, 18*(10), 1386–1396. https://doi.org/10.1016/S1470-2045(17)30621-6.

Villamizar, O., Chambers, C. B., Mo, Y. Y., Torry, D. S., Hofstrand, R., Riberdy, J. M., ... Wilber, A. (2016). Fas-antisense long noncoding RNA is differentially expressed during maturation of human erythrocytes and confers resistance to Fas-mediated cell death. *Blood Cells, Molecules, and Diseases, 58*, 57–66. https://doi.org/10.1016/j.bcmd.2016.03.002.

Villamizar, O., Chambers, C. B., Riberdy, J. M., Persons, D. A., & Wilber, A. (2016). Long noncoding RNA Saf and splicing factor 45 increase soluble Fas and resistance to apoptosis. *Oncotarget*, 7(12), 13810–13826. https://doi.org/10.18632/oncotarget.7329.

Vinjamur, D. S., Bauer, D. E., & Orkin, S. H. (2018). Recent progress in understanding and manipulating haemoglobin switching for the haemoglobinopathies. *British Journal of Haematology, 180*(5), 630–643. https://doi.org/10.1111/BJH.15038.

Waku, T., Nakajima, Y., Yokoyama, W., Nomura, N., Kako, K., Kobayashi, A., ... Fukamizu, A. (2016). NML-mediated rRNA base methylation links ribosomal subunit formation to cell proliferation in a p53-dependent manner. *Journal of Cell Science, 129*(12), 2382–2393. https://doi.org/10.1242/jcs.183723.

Wang, C., Wu, X., Shen, F., Li, Y., Zhang, Y., & Yu, D. (2015). Shlnc-EC6 regulates murine erythroid enucleation by Rac1-PIP5K pathway. *Development Growth and Differentiation, 57*(6), 466–473. https://doi.org/10.1111/dgd.12225.

Wang, F., Song, W., Zhao, H., Ma, Y., Li, Y., Zhai, D., ... Yu, J. (2017). The RNA-binding protein QKI5 regulates primary miR-124-1 processing via a distal RNA motif during erythropoiesis. *Cell Research, 27*(3), 416–439. https://doi.org/10.1038/cr.2017.26.

Wang, H., Zuo, H., Liu, J., Wen, F., Gao, Y., Zhu, X., ... Ju, Z. (2018). Loss of YTHDF2-mediated m6A-dependent mRNA clearance facilitates hematopoietic stem cell regeneration. *Cell Research, 28*(10), 1035–1038. https://doi.org/10.1038/s41422-018-0082-y.

Wang, Q., Huang, Z., Xue, H., Jin, C., Ju, X. L., Han, J. D. J., & Chen, Y. G. (2008). MicroRNA miR-24 inhibits erythropoiesis by targeting activin type I receptor ALK4. *Blood, 111*(2), 588–595. https://doi.org/10.1182/blood-2007-05-092718.

Wang, X., Lu, Z., Gomez, A., Hon, G. C., Yue, Y., Han, D., ... He, C. (2014). N 6-methyladenosine-dependent regulation of messenger RNA stability. *Nature, 505*(7481), 117–120. https://doi.org/10.1038/nature12730.

Wang, X. S., Zhang, Z., Wang, H. C., Cai, J. L., Xu, Q. W., Li, M. Q., ... Chen, W. F. (2006). Rapid identification of UCA1 as a very sensitive and specific unique marker for human bladder carcinoma. *Clinical Cancer Research, 12*(16), 4851–4858. https://doi.org/10.1158/1078-0432.CCR-06-0134.

Ward, C. M., Li, B., & Pace, B. S. (2016). Stable expression of miR-34a mediates fetal hemoglobin induction in K562 cells. *Experimental Biology and Medicine, 241*(7), 719–729. https://doi.org/10.1177/1535370216636725.

Wei, C. M., Gershowitz, A., & Moss, B. (1975). Methylated nucleotides block 5′ terminus of HeLa cell messenger RNA. *Cell, 4*(4), 379–386. https://doi.org/10.1016/0092-8674(75)90158-0.

Wei, C. M., Gershowitz, A., & Moss, B. (1976). 5′-Terminal and internal methylated nucleotide sequences in hela cell mRNA. *Biochemistry, 15*(2), 397–401. https://doi.org/10.1021/bi00647a024.

Wei, J., Liu, F., Lu, Z., Fei, Q., Ai, Y., He, P. C., ... He, C. (2018). Differential m 6 A, m 6 A m, and m 1 A demethylation mediated by FTO in the cell nucleus and cytoplasm. *Molecular Cell, 71*(6), 973–985.e5. https://doi.org/10.1016/j.molcel.2018.08.011.

Weinberg, R. S., Ji, X., Sutton, M., Perrine, S., Galperin, Y., Li, Q., ... Atweh, G. F. (2005). Butyrate increases the efficiency of translation of γ-globin mRNA. *Blood, 105*(4), 1807–1809. https://doi.org/10.1182/blood-2004-02-0454.

Weiss, I. M., & Liebhaber, S. A. (1994). Erythroid cell-specific determinants of α-globin mRNA stability. *Molecular and Cellular Biology, 14*(12), 8123–8132. https://doi.org/10.1128/mcb.14.12.8123-8132.1994.

Weiss, I. M., & Liebhaber, S. A. (1995). Erythroid cell-specific mRNA stability elements in the α2-globin 39 nontranslated region. *Molecular and Cellular Biology, 15*(5), 2457–2465. https://doi.org/10.1128/mcb.15.5.2457.

Werner, M. S., Sullivan, M. A., Shah, R. N., Nadadur, R. D., Grzybowski, A. T., Galat, V., ... Ruthenburg, A. J. (2017). Chromatin-enriched lncRNAs can act as cell-type specific activators of proximal gene transcription. *Nature Structural and Molecular Biology, 24*(7), 596–603. https://doi.org/10.1038/nsmb.3424.

Wilkes, M. C., et al. (2020). Metformin-induced suppression of Nemo-like kinase improves erythropoiesis in preclinical models of Diamond–Blackfan anemia through induction of miR-26a. *Exp. Hematol. 91*, 65–77.

Wilkinson, N., & Pantopoulos, K. (2013). IRP1 regulates erythropoiesis and systemic iron homeostasis by controlling HIF2α mRNA translation. *Blood, 122*(9), 1658–1669. https://doi.org/10.1182/blood-2013-03-492454.

Wilkinson, N., & Pantopoulos, K. (2014). The IRP/IRE system in vivo: Insights from mouse models. *Frontiers in pharmacology, 5*(July), 1–15. https://doi.org/10.3389/fphar.2014.00176.

Wlodarski, M. W., Vlachos, A., Farrar, J. E., Da Costa, L. M., Kattamis, A., Dianzani, I., ... Leblanc, T. M. (2024). Diagnosis, treatment, and surveillance of Diamond-Blackfan anaemia syndrome: International consensus statement. *The Lancet Haematology, 11*(5), e368–e382. https://doi.org/10.1016/S2352-3026(24)00063-2.

Wu, S., Romfo, C. M., Nilsen, T. W., Green, M. R., & Page, S. D. S. (1999). Functional recognition of the39 splice site AG by the splicing factor U2AF35. *Nature, 402*(December), 832–835.

Xie, B., & Dean, A. (2025). A super enhancer-derived enhancer RNA acts together with CTCF/Cohesin in trans to regulate erythropoiesis. *Genes, 16*(4), https://doi.org/10.3390/genes16040389.

Xie, S. Y., Li, W., Ren, Z. R., Huang, S. Z., Zeng, F., & Zeng, Y. T. (2011). Correction of β654-thalassaemia mice using direct intravenous injection of siRNA and antisense RNA vectors. *International Journal of Hematology, 93*(3), 301–310. https://doi.org/10.1007/s12185-010-0727-1.

Xiong, W., Zhao, Y., Wei, Z., Li, C., Zhao, R., Ge, J., & Shi, B. (2023). N1-methyladenosine formation, gene regulation, biological functions, and clinical relevance. *Molecular Therapy, 31*(2), 308–330. https://doi.org/10.1016/j.ymthe.2022.10.015.

Xu, L., Wu, F., Yang, L., Wang, F., Zhang, T., Deng, X., ... Yu, D. (2020). miR-144/451 inhibits c-Myc to promote erythroid differentiation. *FASEB Journal, 34*(10), 13194–13210. https://doi.org/10.1096/fj.202000941R.

Xu, P., Palmer, L. E., Lechauve, C., Zhao, G., Yao, Y., Luan, J., ... Paralkar, V. R. (2019). Regulation of gene expression by miR-144/451 during mouse erythropoiesis. *Blood, 133*(23), 2518–2528. https://doi.org/10.1182/blood.2018854604.

Xue, M., Chen, W., & Li, X. (2016). Urothelial cancer associated 1: A long noncoding RNA with a crucial role in cancer. *Journal of Cancer Research and Clinical Oncology, 142*(7), 1407–1419. https://doi.org/10.1007/s00432-015-2042-y.

XuFeng, R., Boyer, M. J., Shen, H., Li, Y., Yu, H., Gao, Y., ... Cheng, T. (2009). ADAR1 is required for hematopoietic progenitor cell survival via RNA editing. *Proceedings of the National Academy of Sciences of the United States of America, 106*(42), 17763–17768. https://doi.org/10.1073/pnas.0903324106.

Yan, X., Liu, F., Yan, J., Hou, M., Sun, M., Zhang, D., ... Yin, P. (2023). WTAP–VIRMA counteracts dsDNA binding of the m6A writer METTL3–METTL14 complex and maintains N 6-adenosine methylation activity. *Cell Discovery, 9*, Springer US4–8. https://doi.org/10.1038/s41421-023-00604-5.

Yang, S., Maurin, T., Robine, N., Rasmussen, K. D., Jeffrey, K. L., Chandwani, R., ... Lai, E. C. (2010). Conserved vertebrate mir-451 provides a platform for Dicer-independent, Ago2-mediated microRNA biogenesis. *Proceedings of the National Academy of Sciences of the United States of America, 107*(34), 15163–15168. https://doi.org/10.1073/pnas.1006432107.

Yang, S., Sun, G., Wu, P., Chen, C., Kuang, Y., Liu, L., ... Cheng, H. (2022). WDR82-binding long noncoding RNA lncEry controls mouse erythroid differentiation and maturation. *Journal of Experimental Medicine, 219*(4), https://doi.org/10.1084/jem.20211688.

Yankova, E., Blackaby, W., Albertella, M., Rak, J., De Braekeleer, E., Tsagkogeorga, G., ... Kouzarides, T. (2021). Small-molecule inhibition of METTL3 as a strategy against myeloid leukaemia. *Nature, 593*(7860), 597–601. https://doi.org/10.1038/s41586-021-03536-w.

Yao, Q. J., Sang, L., Lin, M., Yin, X., Dong, W., Gong, Y., & Zhou, B. O. (2018). Mettl3–Mettl14 methyltransferase complex regulates the quiescence of adult hematopoietic stem cells. *Cell Research, 28*(9), 952–954. https://doi.org/10.1038/s41422-018-0062-2.

Yoshida, K., Sanada, M., Shiraishi, Y., Nowak, D., Nagata, Y., Yamamoto, R., ... Ogawa, S. (2011). Frequent pathway mutations of splicing machinery in myelodysplasia. *Nature, 478*(7367), 64–69. https://doi.org/10.1038/nature10496.

Yoshinaga, M., Han, K., Morgens, D. W., Horii, T., Kobayashi, R., Tsuruyama, T., ... Takeuchi, O. (2022). The N 6-methyladenosine methyltransferase METTL16 enables erythropoiesis through safeguarding genome integrity. *Nature Communications, 13*(1), https://doi.org/10.1038/s41467-022-34078-y.

Yu, D., Dos Santos, C. O., Zhao, G., Jiang, J., Amigo, J. D., Khandros, E., ... Weiss, M. J. (2010). miR-451 protects against erythroid oxidant stress by repressing 14-3-3ζ. *Genes and Development, 24*(15), 1620–1633. https://doi.org/10.1101/gad.1942110.

Zhang, C., Chen, Y., Sun, B., Wang, L., Yang, Y., Ma, D., ... Liu, F. (2017). M6A modulates haematopoietic stem and progenitor cell specification. *Nature, 549*(7671), 273–276. https://doi.org/10.1038/nature23883.

Zhang, J., Zhao, H., Wu, K., Peng, Y., Han, X., Zhang, H., ... Liu, J. (2019). Knockdown of spliceosome U2AF1 significantly inhibits the development of human erythroid cells. *Journal of Cellular and Molecular Medicine, 23*(8), 5076–5086. https://doi.org/10.1111/jcmm.14370.

Zhang, L., Prak, L., Rayon-Estrada, V., Thiru, P., Flygare, J., Lim, B., & Lodish, H. F. (2013). ZFP36L2 is required for self-renewal of early burst-forming unit erythroid progenitors. *Nature, 499*(7456), 92–96. https://doi.org/10.1038/nature12215.

Zhang, L. S., Xiong, Q. P., Peña Perez, S., Liu, C., Wei, J., Le, C., ... He, C. (2021). ALKBH7-mediated demethylation regulates mitochondrial polycistronic RNA processing. *Nature Cell Biology, 23*(7), 684–691. https://doi.org/10.1038/s41556-021-00709-7.

Zhang, L., Sankaran, V. G., & Lodish, H. F. (2012). MicroRNAs in erythroid and megakaryocytic differentiation and megakaryocyte-erythroid progenitor lineage commitment. *Leukemia, 26*, Nature Publishing Group. 2310–2316. https://doi.org/10.1038/leu.2012.137.

Zhang, Q., Liu, F., Chen, W., Miao, H., Liang, H., Liao, Z., ... Zhang, B. (2021). The role of rna m5c modification in cancer metastasis. *International Journal of Biological Sciences, 17*(13), 3369–3380. https://doi.org/10.7150/IJBS.61439.

Zhang, X., Cong, T., Wei, L., Zhong, B., Wang, X., Sun, J., ... Wang, J. (2022). YTHDF3 modulates hematopoietic stem cells by recognizing RNA m6A modification on Ccnd1. *Haematologica, 107*(10), 2381–2394. https://doi.org/10.3324/haematol.2021.279739.

Zhao, H., Kalota, A., Jin, S., & Gewirtz, A. M. (2009). Autoregulatory feedback loop in human hematopoietic cells the c-myb proto-oncogene and microRNA-15a comprise an active. *Blood, 113*(3), 505–516. https://doi.org/10.1182/blood-2008-01-136218.

Zheng, Q., Gan, H., Yang, F., Yao, Y., Hao, F., Hong, L., & Jin, L. (2020). Cytoplasmic m1A reader YTHDF3 inhibits trophoblast invasion by downregulation of m1A-methylated IGF1R. *Cell Discovery, 6*(1), 1–12. https://doi.org/10.1038/s41421-020-0144-4.

Zhu, J., Ren, Y., Han, Y., Jin, T., Li, Y., Ruan, X., ... Fang, X. (2019). Long noncoding RNA PCED1B-AS1 promotes erythroid differentiation coordinating with GATA1 and chromatin remodeling. *Blood Science, 1*(2), 161–167. https://doi.org/10.1097/BS9.0000000000000031.

CHAPTER EIGHT

Decoding the erythroid niche: From single-cell programs to spatially resolved architecture in mouse and human hematopoiesis

Pan Wang[a,*], Kehan Ren[b], and Peng Ji[a,*]
[a]Department of Pathology, Feinberg School of Medicine, Northwestern University, Chicago, IL, United States
[b]Department of Medicine, Renaissance School of Medicine, Stony Brook University, Stony Brook, NY, United States
*Corresponding authors. e-mail address: pan.wang@northwestern.edu; peng-ji@fsm.northwestern.edu

Contents

Abstract

Erythropoiesis is a complex, spatially constrained morphogenetic process that transforms hematopoietic progenitors into specialized, enucleated red blood cells. While decades of research have defined the cell-intrinsic transcriptional and epigenetic circuits, governed by master regulators such as GATA1 and KLF1, the critical role of the microenvironmental niche remains less understood. Recent advances in single-cell omics have revealed a multi-layered landscape in which transcriptional potential is often uncoupled from proteomic and metabolic output. The transition from dissociative single-cell sequencing to spatially resolved omics further shaped our understanding of how these molecular programs are executed in situ. This chapter summarizes the broader landscape of omics profiling while focusing on how spatial transcriptomics and proteomics are uncovering the anatomical locations of erythropoiesis within the hematopoietic organs. We highlight a significant evolutionary divergence in niche architecture: while murine terminal erythropoiesis is organized with C1q + macrophage-centered erythroblastic islands, human erythropoiesis

Current Topics in Developmental Biology, Volume 170
ISSN 0070-2153, https://doi.org/10.1016/bs.ctdb.2026.06.009

exhibits a unique, self-sustaining capacity characterized by autonomous erythroblast clustering independent of macrophages. We further discuss the technical hurdles of spatial profiling in mineralized bone and the potential for multimodal spatial omics to link chromatin accessibility, RNA, and protein abundance within intact tissues. By integrating molecular trajectories with microanatomical context, these technologies provide a new framework for understanding ineffective erythropoiesis and identifying therapeutic targets in hematologic diseases.

1. Introduction

Erythropoiesis is a highly organized hierarchical progression of erythroid progenitor cells. It is fundamentally a spatially constrained morphogenetic program executed with extraordinary fidelity. During differentiation, erythroid precursors experience an extraordinary physical and morphogenetic challenge, including cell shrinkage, chromatin condensation, hemoglobinization, enucleation, and reticulocyte remodeling (Chen et al., 2009; Dzierzak & Philipsen, 2013; Ji et al., 2011; Koury et al., 2007; Liu et al., 2021; Mei et al., 2021; Palis, 2014; Zhao et al., 2016) (Fig. 1). This physical reorganization creates an obligatory dependence on the microenvironment, in which the niche must serve as both a mechanical scaffold and a metabolic fueling station to ensure the successful generation of a biconcave red blood cell.

Erythropoiesis cannot be understood solely as a cell-autonomous trajectory. The most distinctive late events of erythroid differentiation, particularly enucleation and the assembly of a mature membrane and cytoskeletal architecture, occur in specialized niches where direct cell-cell contact, local metabolite availability, and physical microanatomy are difficult to treat as incidental (Chasis & Mohandas, 2008; May & Forrester, 2020). In bone marrow, perivascular stromal and endothelial niches provide factors such as SCF and CXCL12 that shape progenitor maintenance and lineage output (Comazzetto et al., 2019; Comazzetto et al., 2021). Terminal differentiation, by contrast, has long been known to be associated with erythroblastic islands, which are macrophage-centered niches that integrate adhesion, iron and heme handling, cytokine support, metabolic buffering, and pyrenocyte engulfment to promote erythroid maturation (Bessis & Breton-Gorius, 1962; Chasis & Mohandas, 2008; Chow et al., 2013; Comazzetto et al., 2021; Hom et al., 2015; Li et al., 2020; Manwani & Bieker, 2008; Niu & Zhang, 2024; Zheng et al., 2025).

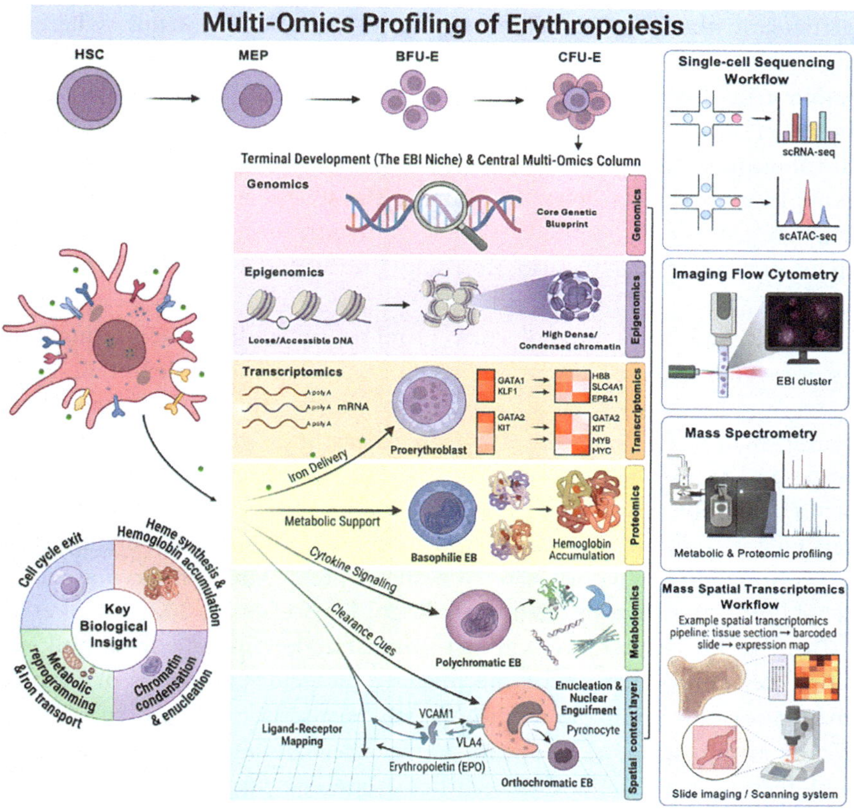

Fig. 1 Multi-omics profiling of erythropoiesis. This conceptual overview illustrates the integration of genomic, epigenomic, transcriptomic, proteomic, and metabolomic layers across successive stages of erythroid differentiation. Spatially organized erythroblastic islands provide a niche in which erythroblasts interact with macrophages and stromal cells, either physically or via paracrine signaling, to coordinate iron handling, metabolic and transcriptional reprogramming, cytokine signaling, and nuclear extrusion, ultimately enabling efficient hemoglobin accumulation and terminal maturation. The figure was created with reference to AI-generated visual elements and subsequently refined by the authors.

In the tissue setting, the spatial perspective of erythropoiesis is also species dependent. In mice, the spleen becomes a major site of stress erythropoiesis, whereas in rats and humans, splenic erythropoiesis is generally limited (Palis, 2014; Paulson et al., 2011; Zhang et al., 2019). Nevertheless, erythropoiesis occurs predominantly in the bone marrow under physiologic conditions in adult mammals. Historically, studies of erythropoiesis have been predominantly using mouse models. This is particularly the case when investigating erythroblastic islands, where ex vivo reconstitution of macrophages and erythroid progenitor cells has traditionally been used (Chasis & Mohandas, 2008; Li et al., 2019; May & Forrester, 2020). The field lacks a clear understanding of the in vivo spatial distribution and frequency of these islands in hematopoietic organs, such as bone marrow or fetal liver. In humans, the study of erythroblastic islands remains particularly limited, with few investigations directly examining their in vivo spatial organization and prevalence within hematopoietic tissues.

Recent advances in molecular profiling of erythropoiesis using bulk analyses and single-cell approaches (An et al., 2014; Kingsley et al., 2013; Merryweather-Clarke et al., 2011; Stuart & Satija, 2019) have been transformative for our understanding of erythroid developmental trajectories and regulatory programs, but there remains a gap in how molecular states are coupled to position, contact, and local microenvironment during terminal execution. Spatially resolved omics now enables linking gene expression and regulatory state to tissue context while preserving cellular neighborhood structure (Han et al., 2026; Larsson et al., 2021; Ståhl et al., 2016). In this chapter, we will first summarize key insights from bulk and single-cell omics that define the intrinsic molecular programs of terminal maturation and then discuss how spatial transcriptomic and spatial proteomic approaches are reshaping our understanding of erythroid niche architecture in mouse and human hematopoietic tissues.

2. Multi-omics studies in erythropoiesis

The molecular understanding of erythropoiesis has shifted from identifying individual regulators to resolving a multi-layered, discordant landscape. While transcriptional programs define the cell's potential, they are frequently uncoupled from the actual proteomic and metabolic execution. This chapter explores how integrative omics have exposed the molecular hand-offs and bottlenecks that occur as a cell translates its genomic blueprint into a specialized physical reality (Fig. 1).

2.1 Transcriptional circuitry and the epigenetic foundation of commitment

Early bulk transcriptomic studies provided important insight into the molecular basis of erythroid lineage commitment by systematically characterizing gene expression programs in human and murine erythropoiesis. A central principle arising from these studies is the remarkable conservation of the erythroid program across developmental stages and anatomical niches. Developmentally resolved profiling across the yolk sac, fetal liver, and bone marrow demonstrated that erythroid gene expression unfolds in synchronized, development-specific waves. Importantly, these studies showed that although the anatomical niches differ, erythroid maturation follows a conserved sequence of functional programs, with early emphasis on heme biosynthesis (e.g., Alas2) and iron transport (e.g., Slc11a2), followed by cell-cycle exit and membrane assembly (Kingsley et al., 2013; Mei et al., 2021). For human erythroid progenitor subpopulations, early transcriptomic profiling identified coordinated gene programs associated with membrane biogenesis, mitochondrial clearance, and enucleation, revealing previously unrecognized structure within erythroid maturation (Merryweather-Clarke et al., 2011). Subsequent cross-species analyses refined these observations by defining a maturation clock and showed that the most pronounced transcriptional changes occur during the transition from basophilic to polychromatophilic erythroblasts, when organelle clearance pathways are activated well before enucleation (An et al., 2014; An et al., 2015; Huang et al., 2020).

This developmental program is governed by a core transcriptional network, with GATA1 and KLF1 acting as central regulators. Seminal work by Welch et al. using GATA1-null and rescued erythroblasts revealed that Gata1 does not merely activate erythroid identity; it functions as a global genomic switch that simultaneously orchestrates the induction of structural and metabolic genes (e.g., *Hbb-b1*, *Slc4a1*) while actively repressing alternative-lineage programs (e.g., *Gata2*, *Kit*) (Welch et al., 2004). This provided the first evidence that erythroid commitment requires the simultaneous repression of stemness to prevent lineage reversal. Hodge et al. complemented this by demonstrating that Klf1 acts as a critical integrator for terminal maturation, directly coordinating the high-level expression of β-globin with the synthesis of membrane and cytoskeleton proteins, thereby linking metabolic readiness with structural integrity (Hodge et al., 2006). These findings established that lineage commitment involves not only gene activation but also the active repression of alternative programs.

The stability of these transcriptional waves is fundamentally rooted in a multi-tier epigenetic foundation. This stability is maintained through coordinated changes in chromatin state during erythropoiesis. Epigenetic profiling during commitment revealed a clear dichotomy: activation of erythroid genes is accompanied by rapid gains of H3K9ac and H3K4me3 at enhancers, whereas alternative lineage programs (myeloid and lymphoid) are silenced through promoter compaction and H3K27me3 deposition. This ensures that once a cell crosses the pro-erythroblast threshold, its chromatin landscape becomes physically refractory to non-erythroid signals (Wong et al., 2011). The precision of this circuitry is further shaped by the sequential assembly of modular enhancers over differentiation. Using high-resolution ChIP-seq and ATAC-seq, studies have shown that GATA1, TAL1, and KLF1 do not bind their targets in a stochastic manner. Instead, they follow a chronological assembly hierarchy in which priming factors, such as TAL1, first occupy closed chromatin regions to recruit chromatin remodeling complexes, which then create the accessibility required for GATA1 and KLF1 to bind and execute the final transcriptional program (Schulz et al., 2019; Xu et al., 2012). This stepwise activation effectively locks the cell into its specialized fate, providing a reference map for interpreting the regulatory failures seen in pathological erythropoiesis.

Beyond local enhancer activation and chromatin accessibility, erythroid maturation is accompanied by large-scale reorganization of three-dimensional genome architecture. Hi-C studies show dynamic rewiring of enhancer–promoter contacts, with stage-specific loops enriched for erythroid regulators such as GATA1 that link distal elements to their target genes (Bi et al., 2022; Li et al., 2023a). Although overall compartments and many topological domains are retained, a substantial fraction of regulatory loops is remodeled across maturation, adding a structural layer to transcriptional control. During late terminal differentiation, this is coupled to global chromatin compaction and heterochromatin relocalization, with weakening of many topologically associating domain (TAD) boundaries but selective preservation of active erythroid domains (Li et al., 2023b). Together, these findings indicate that terminal erythropoiesis undergoes progressive nuclear reorganization, preserving essential erythroid regulatory contacts within an increasingly condensed genome.

2.2 Post-transcriptional re-engineering: The proteomic and metabolic pivot

A central feature of terminal erythropoiesis is that phenotypic execution intensifies as genomic access becomes progressively restricted, highlighting that the mature red cell state is not simply a transcriptomic outcome but a product of extensive post-transcriptional remodeling (Mei et al., 2021; Ren et al., 2022). Quantitative proteomics has delineated this transition with high resolution. Gautier et al. mapped human erythroid proteomes across seven differentiation stages, which revealed massive accumulation of globins and membrane-cytoskeletal proteins (SLC4A1, ANK1, SPTA1, EPB41) and concurrent depletion of mitochondrial proteins, RNA-processing factors, and ribosomal subunits as mitophagy and ribosome clearance proceed (Gautier et al., 2016). Parallel work in mice confirmed these dynamics, demonstrating conserved proteomic waves aligned with enucleation and reticulocyte maturation (Gautier et al., 2020). Notably, comparative transcriptome–proteome analyses uncovered substantial discordance between mRNA and protein abundance, particularly for mitochondrial enzymes, ribosomal subunits, and cytoskeletal regulators. These findings indicate that erythroid fate and maturation are shaped not only by transcriptional programs but also by changes in translational capacity, as ribosome abundance constrains translation of key regulators such as GATA1 (Khajuria et al., 2018), while later stages rely on selective translation of membrane and metabolic proteins (Alvarez-Dominguez et al., 2017).

As transcriptional and translational capacity becomes increasingly constrained, terminal erythropoiesis depends on coordinated metabolic and membrane remodeling to complete differentiation. Prior to enucleation, proteomics and metabolomics studies reveal a coordinated shift in enzymatic machinery, including PKLR, HK1, and G6PD, that redirects metabolic flux away from oxidative phosphorylation toward glycolysis and the pentose phosphate pathway (Gonzalez-Menendez et al., 2021; Huang et al., 2018; Nemkov et al., 2022). This redirection is essential for maintaining redox homeostasis during the sudden surge in heme biosynthesis. Simultaneously, lipidomic and membrane proteome analyses reveal a substantial reorganization of the bilayer. The redistribution of phosphatidylserine, sphingomyelin, and cholesterol drives a progressive reduction in membrane surface area and a strengthening of vertical membrane cytoskeletal interactions (Liu et al., 2010; Minetti et al., 2025; Mohandas & Gallagher, 2008). Collectively, these post-transcriptional programs complete erythroid maturation by converting a committed cellular state into the deformable, biconcave structure of the mature erythrocyte.

2.3 Single-cell resolution: Resolving asynchronous maturation and bottlenecks

Single-cell technologies have transformed how we understand erythropoiesis, shifting the field away from a rigid stepwise model toward a more continuous view of maturation states resolved at high resolution. Early landmark scRNA-seq analyses reconstructed pseudotemporal trajectories and identified stage-specific transcriptional programs governed by canonical regulators such as GATA1, KLF1, and TAL1 (Paul et al., 2016; Tusi et al., 2018). These studies reveal that erythroid maturation is inherently asynchronous; transcriptionally distinct intermediate states frequently coexist within a single morphologically defined population, demonstrating that molecular identity often precedes or diverges from classical immunophenotypic markers. Importantly, high-resolution profiling has revealed discrete quality-control checkpoints during terminal erythropoiesis. Among these, a key checkpoint occurs at the pro-erythroblast-to-basophilic erythroblast transition, where cell-cycle exit must coincide with a surge in heme biosynthesis via ALAS2 and FECH and with adequate globin production to prevent free-heme toxicity (Suragani et al., 2012; Zhang et al., 2018). Disruption of this coordination, often due to iron limitation, activates the HRI stress pathway, inducing inflammatory and p53-dependent apoptotic programs such as S100A8/A9 and diverting cells away from terminal maturation, thereby contributing to ineffective erythropoiesis in stress and disease states (Dutt et al., 2011; Schneider et al., 2016).

The integration of single-cell chromatin accessibility (scATAC-seq) has added a structural layer to our understanding of erythropoiesis. Through this approach, Corces et al. revealed that erythroid lineage commitment is a progressive restriction of regulatory potential. At the single-cell level, this manifests as the selective opening of erythroid-specific enhancers, such as the β-globin locus control region, coupled with the active closure of myeloid-specific regulatory elements, including SPI1/PU.1-associated enhancers (Corces et al., 2016). Building on this, Ludwig et al. demonstrated that these chromatin accessibility changes are distributed continuously across individual cells, with erythroid enhancers gradually gaining accessibility cell by cell rather than switching on synchronously at a defined differentiation stage. As a result, cells traditionally assigned to the same progenitor compartment occupy distinct positions along an accessibility gradient (Ludwig et al., 2019). These studies suggest that the transition from a progenitor to an erythroblast is a graded process, with enhancer activation occurring in a priming phase that precedes overt transcriptional induction. Consistent with this model, Ranzoni et al. showed that hematopoietic progenitors undergo an early, coordinated remodeling of regulatory landscapes. This involves the systematic silencing of multipotent programs and the priming of erythroid enhancers enriched for GATA and KLF motifs (Ranzoni et al., 2021). Interestingly, integrated scRNA-seq and scATAC-seq analyses in human developmental hematopoiesis show that this chromatin priming often precedes overt transcriptional commitment. This creates an intermediate state of molecular competence in which enhancers are accessible, but lineage transcripts remain only partially upregulated, demonstrating that chromatin potential precedes transcriptional execution. Multi-omics studies in malignant contexts reinforce this concept: chromatin-defined lineage bias can become decoupled from transcriptional output (Granja et al., 2019), highlighting that accessibility encodes potential, not destiny.

How these primed chromatin states are resolved into irreversible fate decisions depends on protein-level regulation. Single-cell proteomic analyses indicate that lineage execution is governed by quantitative shifts in co-expressed transcription factors, with the relative abundance of GATA1 compared with competing factors such as TAL1 and KLF1 serving as a key instructive signal (Palii et al., 2019). These stoichiometric relationships link chromatin priming to lineage commitment by helping set the threshold of GATA1 dominance required to activate the full erythroid gene expression program.

3. Spatial omics studies in erythropoiesis

High-resolution single-cell omics have clarified the cell-intrinsic requirements for erythropoiesis, but in doing so, they remove the process from the architectural context of hematopoietic tissues. In vivo, erythropoiesis is not a cell-autonomous program but a spatially organized process that unfolds within defined anatomical niches. Within these niches, developing erythroid cells encounter local physical and cellular cues that shape maturation. The spatial positioning of erythroid progenitors and precursors relative to stromal cells, vascular structures, and nursing cells influences the timing of differentiation, the efficiency of organelle clearance, and the mechanical priming of the cytoskeleton. The move from dissociative to spatial omics, therefore, represents a shift from cataloging cellular potential to understanding how erythropoiesis is executed within its native physiological setting (Fig. 2).

3.1 Spatial transcriptomic technologies and the hematopoietic landscape

Spatial transcriptomics (ST) encompasses a range of approaches that quantify gene expression while preserving the spatial information of transcripts within

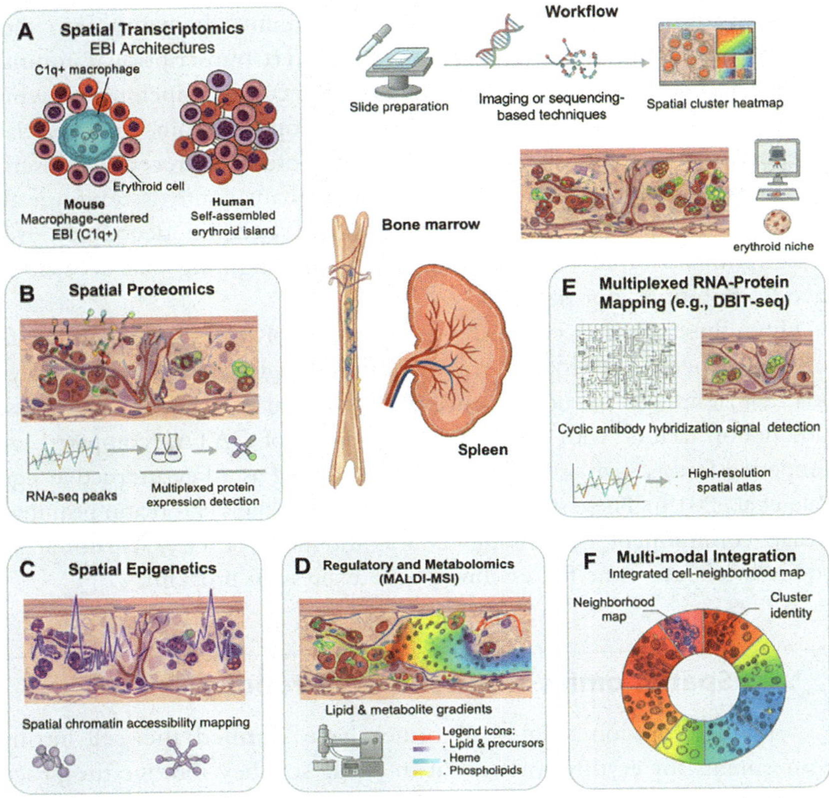

Fig. 2 Spatial omics technologies for resolving erythroid niche organization across species. Major spatial omics modalities used to study erythropoietic microenvironments are illustrated. (A) Spatial transcriptomics distinguishes macrophage-centered EBIs in mice and self-assembled erythroid islands in humans. (B) Spatial proteomic platforms offer the capacity to localize protein expression within intact marrow and may eventually resolve protein enrichment at erythroblast interfaces. (C) Emerging spatial epigenomic methods, including spatial ATAC-seq, provide opportunities to map regulatory landscapes associated with erythroid maturation. (D) Spatial metabolomic imaging (e.g., MALDI-MSI) can detect lipid and metabolite gradients and may reveal metabolic features relevant to terminal differentiation. (E) Multiplexed RNA–protein co-mapping platforms (such as DBiT-seq) align transcriptomic and protein signals within the same tissue grid. (F) Multi-modal integration frameworks conceptually unify these layers, enabling future reconstruction of cell-neighborhood structure and niche organization in situ. The figure was created with reference to AI-generated visual elements and subsequently refined by the authors.

tissue sections. Unlike dissociative single-cell sequencing, which loses the "anatomical address" of a cell, ST allows researchers to map the molecular heterogeneity of the bone marrow directly onto its histological framework. Current spatial transcriptomic platforms generally fall into two categories. Imaging-based methods (e.g., MERFISH, seqFISH, Xenium) achieve subcellular resolution through sequential fluorescence in situ hybridization, whereas sequencing-based approaches (e.g., Visium, Slide-seq, GeoMx) use spatially barcoded capture probes to register transcripts to defined locations (Chen et al., 2015; Eng et al., 2019; Moffitt et al., 2016; Rodriques et al., 2019; Ståhl et al., 2016).

Applying these integrated approaches to the bone marrow has transformed our understanding of niche geography. Earlier work by Baccin et al. combined droplet-based scRNA-seq with high-resolution laser-capture microdissection (LCM-seq) to create a comprehensive spatial map of the murine bone marrow. Their work identified distinct specialized cytokine-secreting cells, specifically Adipo-CAR and Osteo-CAR reticular cells, which differentially localize to sinusoidal and arteriolar surfaces. By employing the RNA-Magnet algorithm, they demonstrated that these stromal subsets create specialized peri-vascular micro-niches where localized ligand-receptor interactions, such as Cxcl12-Cxcr4 and Kitl-Kit, are spatially concentrated (Baccin et al., 2020). These findings suggest that the bone marrow is not a homogenous mixture of factors, but a highly organized collection of neighborhoods that provide spatially restricted cues for hematopoietic maintenance and maturation.

Building on these foundational insights, more recent studies have successfully applied advanced technologies to the challenging environment of mineralized bone. Xiao et al. overcame technical hurdles in tissue decalcification and mRNA degradation by using the 10xGenomics Visium platform to analyze fully mineralized murine femurs. By integrating spatial data with scRNA-seq through deconvolution, they successfully mapped skeletal stem and progenitor cell subtypes to distinct anatomical sites. Furthermore, they identified spatially restricted signaling gradients of WNT, BMP, and TGFβ and verified niche-specific interactions between smooth muscle cells, endothelial cells, and macrophages (Xiao et al., 2023). This work addresses a long-standing technical barrier, establishing an approach for studying hematopoietic and erythroid maintenance within the endogenous, calcified architecture of the bone marrow.

Despite these advancements, the unique physiology of the bone marrow continues to pose significant obstacles for spatial analysis. Beyond the primary hurdle of decalcification, the tissue is highly susceptible to sampling bias and mechanical artifacts, such as the "crush" effects often seen in clinical biopsies. Cooper et al. addressed these limitations by advocating for a pathology-led integration of spatial data. They emphasize that because spatial transcriptomic outputs can mask underlying architectural damage, high-dimensional molecular maps must be cross-validated with traditional H&E morphology to ensure that gene expression profiles correspond to intact, healthy tissue rather than processing artifacts (Cooper et al., 2025). This rigorous quality control framework is essential for translating murine spatial discoveries into reliable human hematopoietic models.

Building on the technical solutions for mineralized tissue, Wang provided a comprehensive framework for optimizing bone processing specifically for spatial omics (Wang, 2025). This perspective addresses a critical bottleneck: the fundamental conflict between the rapid decalcification required for clinical turnaround and the delicate preservation of nucleic acids and tissue architecture. While traditional strong acids, such as hydrochloric acid, are effective for softening bone, they irreversibly degrade fragile RNA, rendering samples unsuitable for high-throughput spatial assays. To bridge this gap, Wang advocated standardized EDTA-based protocols maintained at neutral pH (7.4–8.0), which protect RNA from hydrolysis while effectively chelating calcium. By aligning rigorous quality metrics from molecular pathology with spatial transcriptomics, this work establishes a baseline for achieving the molecularly robust processing required to accurately map the bone marrow's native signaling landscape.

Building on these technical optimizations, our recent work (Han et al., 2023; Han et al., 2026) provides a detailed spatial characterization of the erythroblastic island (EBI) across species and developmental stages. To move beyond traditional in vitro reconstitution models, we employed a multimodal approach combining 10xGenomics Visium, scRNA-seq deconvolution, and cyclic immunofluorescence (CyCIF) to map erythropoiesis in the mouse fetal liver, newborn bone marrow, and adult spleen. This spatial mapping revealed a significant conceptual advance: C1q+ macrophages serve as the conserved central nursing cells of the murine EBI during terminal erythropoiesis. These C1q+ macrophages showed a strong positive spatial correlation with erythroid cells and highly expressed genes critical for EBI function, a finding that remained consistent across different hematopoietic organs and under both physiologic and stress conditions.

However, applying these same technologies to human tissues revealed a striking evolutionary divergence in erythroid niche architecture. Unlike the macrophage-dependent clusters identified in mice, human fetal liver and bone marrow lacked a strong spatial correlation between erythroid cells and macrophages. Instead, we observed a robust association between erythroid progenitors and maturing erythroid cells, suggesting that human erythropoiesis possesses a unique self-sustaining capacity. This self-association becomes even more pronounced in regenerating human bone marrow following chemotherapy and was further validated in a human bone marrow organoid model (Han et al., 2023; Han et al., 2026). Notably, this erythroid-autonomous clustering is also detectable in mice, where erythroid cells and their progenitors show strong mutual association alongside the canonical macrophage-centered islands, indicating that erythroid self-organization is a conserved feature of terminal erythropoiesis that becomes the dominant organizing principle in humans. This study underscores that the regulatory logic of the erythroid niche is not universal, highlighting a critical shift from macrophage-centric support in mice to autonomous cell-cell interactions in humans.

Despite these advances, the application of spatial transcriptomics to erythropoiesis remains limited. While emerging platforms can achieve the resolution needed to resolve individual cells and complex tissue architectures, these approaches have not yet been widely applied or systematically validated in erythroid systems. As a result, key dynamic processes, such as erythroid cell-cell and erythroid-niche interactions, and erythroid transcriptomic changes across different anatomic locations, remain underexplored in situ. Addressing these gaps will require broader adoption of high-resolution and multimodal spatial technologies, coupled with integrative proteomic and imaging approaches, to directly link transcriptional states with the structural and functional changes that define terminal erythroid maturation.

3.2 Emerging spatial proteomic approaches and implications for erythropoiesis

While spatial transcriptomics provides a high-dimensional view of the molecular landscape, the functional execution of these programs is ultimately mediated by proteins. The transition from transcript-based mapping to spatial proteomics is essential for validating cell phenotypes and capturing the physical architecture of the niche at subcellular resolution. Among current high-plex imaging technologies, CODEX (CO-Detection by indEXing) has emerged as a particularly powerful platform for hematopoietic research. Unlike traditional immunohistochemistry and immunofluorescence, CODEX utilizes DNA-barcoded antibodies and iterative cycles of fluorescent imaging to simultaneously detect dozens of protein markers. This enables precise identification of rare cell populations and quantification of their spatial proximity within intact tissue sections.

One influential application of this approach is a spatial atlas of human bone marrow, generated by integrating single-cell transcriptomics with highly multiplexed CODEX imaging (Bandyopadhyay et al., 2024). By profiling millions of cells across diverse clinical samples, this work resolved the marrow into distinct spatial neighborhoods—functional units composed of specific stromal, hematopoietic, and immune cell populations. Although the analysis focused primarily on global hematopoietic organization and immune architecture, it also delineated discrete perivascular and endosteal microenvironments enriched for specialized niche cells. This protein-resolved spatial framework provides an important reference for studying erythropoiesis in situ, enabling direct assessment of physical interactions between erythroid precursors and supporting nursing cells within intact marrow architecture.

Building on the protein-resolved atlases generated by CODEX, recent work by Wu et al. (2024) further expanded our understanding of the marrow's physical architecture by defining the anatomy of lineage-specific production sites (Wu et al., 2024). While platforms like CODEX identify broad cellular neighborhoods, this study developed an analytical pipeline to visualize stepwise erythropoiesis in situ, identifying discrete microanatomical structures that serve as specialized factories for red blood cell production. By mapping the spatial distribution of terminal erythroid progenitors (CFU-E) and their offspring in mice, the researchers discovered that erythropoiesis is not randomly distributed but occurs in monoclonal clusters that decorate the surface of specific sinusoids.

A key conceptual advancement of this study is the characterization of the marrow's resilient anatomy and its capacity for local plasticity. The work demonstrates that these erythroid production sites are remarkably durable, maintaining their core organizational principles even under acute insults such as phlebotomy or systemic infection. In response to erythropoietic stress, the marrow exhibits plasticity by synchronously increasing both the total number of these production sites and their individual cellular output through the recruitment of upstream progenitors. Interestingly, the study also revealed that these stress responses are highly heterogeneous across the skeleton; for example, the skull showed a notable lack of increased erythropoiesis after hemorrhage compared to long bones. This research provides a crucial framework for understanding how the spatial organization of the marrow dictates its functional response to physiological demand, establishing that dedicated microanatomical sites are the fundamental units of erythroid maintenance and regeneration in mice. Extending this murine spatial paradigm to humans remains a critical open direction. Translating these insights will require high-resolution spatial proteomic approaches (such as CODEX) to map lineage markers, microanatomical niches, and localized signalosomes in intact human biopsy tissues under steady-state and stress conditions.

Looking ahead, integrating spatial proteomics with spatial transcriptomics and emerging multimodal assays, such as those that jointly measure RNA and protein (Liu et al., 2023; Liu et al., 2020), offers a way to connect transcriptional priming with phenotypic execution during erythropoiesis. By mapping protein localization and abundance at cell-cell interfaces within intact erythroid niches, these strategies can clarify how erythroid programs are assembled and executed in situ.

3.3 Other spatial omics and integrated approaches

Beyond transcriptomic and protein-based spatial profiling, emerging spatial multi-omics technologies now enable chromatin, RNA, protein, and metabolic layers to be interrogated within intact tissues. Spatial ATAC-seq and related approaches adapt chromatin-accessibility profiling to tissue sections, revealing region-specific regulatory landscapes in developing and adult organs (Deng et al., 2022; Guo et al., 2025). An additional layer comes from spatial metabolomics and lipidomics, where imaging mass spectrometry approaches such as MALDI and nano-DESI map metabolites, redox intermediates, and membrane lipids directly within tissue sections at cellular or near-cellular resolution (Alexandrov, 2020; Rappez et al., 2021). Spatial genomics adds yet another axis of information by enabling the mapping of DNA sequences, clonal structures, and somatic lesions directly within tissue architecture. In situ genome sequencing and Slide-DNA-seq preserve cellular and neighborhood relationships while resolving genomic variation at high spatial resolution (Payne et al., 2021; Zhao et al., 2022). Building on these single-modality advances, multimodal approaches such as DBiT-seq enable joint mapping of mRNA and antibody-derived protein tags within the same tissue section at near-single-cell resolution. In parallel, workflows that integrate spatial transcriptomics with multiplexed or hyperplex immunofluorescence extend this RNA-protein alignment to larger protein panels and broader tissue contexts (Liu et al., 2020; Zhang et al., 2025). Although these integrative platforms have been applied primarily to neural, tumor, and developmental tissues, they provide a conceptual and technical template for future studies in hematopoietic organs.

For erythropoiesis, such spatial multi-omics strategies have not yet been systematically implemented, but they hold considerable promise. Approaches capable of mapping metabolic and genomic features in situ will be particularly valuable, given the central roles of metabolic state, lipid remodeling, and clonal dynamics in terminal maturation and stress responses. More broadly, integrating regulatory, transcriptional, proteomic, metabolic, and genomic layers within intact erythroid niches will clarify how spatially organized programs coordinate and execute terminal maturation. As these technologies mature, multi-layered spatial omics will be essential for resolving how erythroid programs are specified, assembled, and functionally realized within their native microenvironments.

4. Conclusions and future perspectives

Spatially resolved analyses indicate that terminal erythropoiesis does not follow a single architectural blueprint, but instead reflects species-specific strategies shaped by the mechanical and metabolic demands of red cell production. In mice, erythropoiesis is organized around macrophage-centered islands together with erythroid self-organized clusters close to the sinusoids, whereas in humans, maturation occurs predominantly within erythroblast clusters that persist independently of a central nursing cell. Advances in spatial transcriptomic technologies now enable an unbiased and comprehensive assessment of undisturbed cellular organization that does not depend on antibody-based detection, thereby reducing concerns about nonspecific staining. Together, these findings prompt a reassessment of long-standing assumptions about the physical organization of erythroid differentiation in situ.

This reframed view of the niche raises critical questions about the divergent regulatory logics across species. In the macrophage-dependent murine model, although the structural roles of the central nursing cell are well documented, we still lack a clear understanding of how macrophages dynamically coordinate their phagocytic and metabolic support across varying physiological states. Conversely, the discovery of a self-sustaining, adhesion-driven human niche suggests that human erythropoiesis may be governed by intrinsic cluster dynamics rather than external myeloid scaffolding.

Interestingly, the diminished role of physical macrophage-erythroblast contact in humans suggests that human niche macrophages may exert their influence through non-canonical, long-range mechanisms. Rather than serving as a tethered anchor, these macrophages may function through paracrine signaling or by secreting specialized extracellular matrix components that stabilize erythroid clusters from a distance. This "remote nursing" model may indicate that the human niche is more fluid and spatially flexible than its murine counterpart. More broadly, it will be important to determine whether ineffective erythropoiesis in diseases such as myelodysplastic syndromes and congenital dyserythropoietic anemias reflects defects in autonomous erythroid clustering, disrupted macrophage-derived paracrine signaling, or a more general breakdown of the spatial niche, as each scenario implies distinct therapeutic opportunities.

These questions also expose conceptual limitations in current spatial approaches. Most platforms capture static snapshots of a highly dynamic process, making it difficult to resolve how erythroblastic islands assemble, mature, and ultimately disperse over time. Addressing this will require strategies that integrate spatial context with inferred temporal progression. It also remains unclear whether human erythroid niches maintain a stable architecture or adapt in response to stress. For instance, macrophages may be recruited under certain conditions to support clearance or metabolic buffering. The species differences observed here carry important translational implications: if human terminal erythroid maturation is largely independent of macrophages, then the impact of immunomodulatory or macrophage-targeted therapies on erythropoiesis may need to be reconsidered, particularly given our limited understanding of stromal-erythroid interactions in human bone marrow.

As various omics technologies continue to improve in resolution and multimodal integration, they will increasingly enable the examination of transcriptional states, protein localization, and tissue architecture within a unified framework. Such integrated maps will be critical for refining mechanistic models of human erythropoiesis and for identifying new strategies to restore effective red cell production in disease.

References

Alexandrov, T. (2020). *Annual Review of Biomedical Data Science, 3*, 61–87.
Alvarez-Dominguez, J. R., Zhang, X., & Hu, W. (2017). *Blood, 129*, 619–629.
An, X., Schulz, V. P., Li, J., Wu, K., Liu, J., Xue, F., ... Gallagher, P. G. (2014). *Blood, 123*, 3466–3477.
An, X., Schulz, V. P., Mohandas, N., & Gallagher, P. G. (2015). *Current Opinion in Hematology, 22*, 206–211.
Baccin, C., Al-Sabah, J., Velten, L., Helbling, P. M., Grunschlager, F., Hernandez-Malmierca, P., ... Haas, S. (2020). *Nature Cell Biology, 22*, 38–48.
Bandyopadhyay, S., Duffy, M. P., Ahn, K. J., Sussman, J. H., Pang, M., Smith, D., ... Tan, K. (2024). *Cell, 187*, 3120–3140. e29.
Bessis, M. C., & Breton-Gorius, J. (1962). *Blood, 19*, 635–663.
Bi, H., Hou, Y., Wang, J., Xia, Z., Wang, D., Liu, Y., ... Ji, P. (2022). *iScience, 25*, 105554.
Chasis, J. A., & Mohandas, N. (2008). *Blood, 112*, 470–478.
Chen, K., Liu, J., Heck, S., Chasis, J. A., An, X., & Mohandas, N. (2009). *Proceedings of the National Academy of Sciences of the United States of America, 106*, 17413–17418.
Chen, K. H., Boettiger, A. N., Moffitt, J. R., Wang, S., & Zhuang, X. (2015). *Science (New York, N. Y.), 348*, aaa6090.
Chow, A., Huggins, M., Ahmed, J., Hashimoto, D., Lucas, D., Kunisaki, Y., ... Frenette, P. S. (2013). *Nature Medicine, 19*, 429–436.
Comazzetto, S., Murphy, M. M., Berto, S., Jeffery, E., Zhao, Z., & Morrison, S. J. (2019). *Cell Stem Cell, 24*, 477–486 e6.
Comazzetto, S., Shen, B., & Morrison, S. J. (2021). *Developmental Cell, 56*, 1848–1860.
Cooper, R. A., Thomas, E., Sozanska, A. M., Pescia, C., & Royston, D. J. (2025). *Leukemia: Official Journal of the Leukemia Society of America, Leukemia Research Fund, UK, 39*, 291–295.
Corces, M. R., Buenrostro, J. D., Wu, B., Greenside, P. G., Chan, S. M., Koenig, J. L., ... Chang, H. Y. (2016). *Nature Genetics, 48*, 1193–1203.
Deng, Y., Bartosovic, M., Ma, S., Zhang, D., Kukanja, P., Xiao, Y., ... Fan, R. (2022). *Nature, 609*, 375–383.
Dutt, S., Narla, A., Lin, K., Mullally, A., Abayasekara, N., Megerdichian, C., ... Ebert, B. L. (2011). *Blood, 117*, 2567–2576.
Dzierzak, E., & Philipsen, S. (2013). *Cold Spring Harbor Perspectives in Medicine, 3*, a011601.
Eng, C. L., Lawson, M., Zhu, Q., Dries, R., Koulena, N., Takei, Y., ... Cai, L. (2019). *Nature, 568*, 235–239.
Gautier, E. F., Ducamp, S., Leduc, M., Salnot, V., Guillonneau, F., Dussiot, M., ... Mayeux, P. (2016). *Cell Reports, 16*, 1470–1484.
Gautier, E. F., Leduc, M., Ladli, M., Schulz, V. P., Lefèvre, C., Boussaid, I., ... Mayeux, P. (2020). *Blood Advances, 4*, 1464–1477.
Gonzalez-Menendez, P., Romano, M., Yan, H., Deshmukh, R., Papoin, J., Oburoglu, L., ... Kinet, S. (2021). *Cell Reports, 34*, 108723.
Granja, J. M., Klemm, S., McGinnis, L. M., Kathiria, A. S., Mezger, A., Corces, M. R., ... Greenleaf, W. J. (2019). *Nature Biotechnology, 37*, 1458–1465.
Guo, P., Chen, Y., Mao, L., Cardilla, A., Lee, C. N., Cui, Y., ... Deng, Y. (2025). *Nature Communications, 16*, 5945.
Han, X., Ren, K., Wang, P., Bi, H., Aydemir, I., Li, E., & Ji, P. (2023). *Blood, 142*, 140.
Han, X., Ren, K., Wang, P., Bi, H., Li, E., Aydemir, I., ... Ji, P. (2026). *Nature Genetics, 58*, 1620–1631.
Hodge, D., Coghill, E., Keys, J., Maguire, T., Hartmann, B., McDowall, A., ... Perkins, A. (2006). *Blood, 107*, 3359–3370.
Hom, J., Dulmovits, B. M., Mohandas, N., & Blanc, L. (2015). *Immunologic Research, 63*, 75–89.

Huang, N. J., Lin, Y. C., Lin, C. Y., Pishesha, N., Lewis, C. A., Freinkman, E., ... Lodish, H. (2018). *Blood, 131*, 2955–2966.
Huang, P., Zhao, Y., Zhong, J., Zhang, X., Liu, Q., Qiu, X., ... Xu, X. (2020). *Proceedings of the National Academy of Sciences of the United States of America, 117*, 12868–12876.
Ji, P., Murata-Hori, M., & Lodish, H. F. (2011). *Trends in Cell Biology, 21*, 409–415.
Khajuria, R. K., Munschauer, M., Ulirsch, J. C., Fiorini, C., Ludwig, L. S., McFarland, S. K., ... Sankaran, V. G. (2018). *Cell, 173*, 90–103. e19.
Kingsley, P. D., Greenfest-Allen, E., Frame, J. M., Bushnell, T. P., Malik, J., McGrath, K. E., ... Palis, J. (2013). *Blood, 121*, e5–e13.
Koury, S., Yarlagadda, S., Moskalik-Liermo, K., Popli, N., Kim, N., Apolito, C., ... Bofinger, D. (2007). *Genomics, 90*, 574–582.
Larsson, L., Frisén, J., & Lundeberg, J. (2021). *Nature Methods, 18*, 15–18.
Li, D., Wu, F., Zhou, S., Huang, X. J., & Lee, H. Y. (2023a). *Nature Structural & Molecular Biology, 30*, 463–474.
Li, D., Zhao, X. Y., Zhou, S., Hu, Q., Wu, F., & Lee, H. Y. (2023b). *Nucleic Acids Research, 51*, 6634–6653.
Li, W., Guo, R., Song, Y., & Jiang, Z. (2020). *Frontiers in Cell and Developmental Biology, 8*, 613885.
Li, W., Wang, Y., Zhao, H., Zhang, H., Xu, Y., Wang, S., ... An, X. (2019). *Blood, 134*, 480–491.
Liu, J., Guo, X., Mohandas, N., Chasis, J. A., & An, X. (2010). *Blood, 115*, 2021–2027.
Liu, Y., DiStasio, M., Su, G., Asashima, H., Enninful, A., Qin, X., ... Fan, R. (2023). *Nature Biotechnology, 41*, 1405–1409.
Liu, Y., Mei, Y., Han, X., Korobova, F. V., Prado, M. A., Yang, J., ... Ji, P. (2021). *Blood, 137*, 398–409.
Liu, Y., Yang, M., Deng, Y., Su, G., Enninful, A., Guo, C. C., ... Fan, R. (2020). *Cell, 183*, 1665–1681. e18.
Ludwig, L. S., Lareau, C. A., Bao, E. L., Nandakumar, S. K., Muus, C., Ulirsch, J. C., ... Sankaran, V. G. (2019). *Cell Reports, 27*, 3228–3240. e7.
Manwani, D., & Bieker, J. J. (2008). *Current Topics in Developmental Biology, 82*, 23–53.
May, A., & Forrester, L. M. (2020). *Experimental Hematology, 91*, 10–21.
Mei, Y., Liu, Y., & Ji, P. (2021). *Blood Reviews, 46*, 100740.
Merryweather-Clarke, A. T., Atzberger, A., Soneji, S., Gray, N., Clark, K., Waugh, C., ... Robson, K. J. (2011). *Blood, 117*, e96–e108.
Minetti, G., Dorn, I., Köfeler, H., Perotti, C., & Kaestner, L. (2025). *Cell Death Discovery, 11*, 79.
Moffitt, J. R., Hao, J., Bambah-Mukku, D., Lu, T., Dulac, C., & Zhuang, X. (2016). *Proceedings of the National Academy of Sciences of the United States of America, 113*, 14456–14461.
Mohandas, N., & Gallagher, P. G. (2008). *Blood, 112*, 3939–3948.
Nemkov, T., Yoshida, T., Nikulina, M., & D'Alessandro, A. (2022). *Frontiers in Physiology, 13*, 833242.
Niu, C., & Zhang, J. (2024). *Frontiers in Immunology, 15*, 1466669.
Palii, C. G., Cheng, Q., Gillespie, M. A., Shannon, P., Mazurczyk, M., Napolitani, G., ... Brand, M. (2019). *Cell Stem Cell, 24*, 812–820 e5.
Palis, J. (2014). *Frontiers in Physiology, 5*, 3.
Paul, F., Arkin, Y., Giladi, A., Jaitin, D. A., Kenigsberg, E., Keren-Shaul, H., ... Amit, I. (2016). *Cell, 164*, 325.
Paulson, R. F., Shi, L., & Wu, D. C. (2011). *Current Opinion in Hematology, 18*, 139–145.
Payne, A. C., Chiang, Z. D., Reginato, P. L., Mangiameli, S. M., Murray, E. M., Yao, C. C., ... Chen, F. (2021). *Science (New York, N. Y.), 371*.

Ranzoni, A. M., Tangherloni, A., Berest, I., Riva, S. G., Myers, B., Strzelecka, P. M., ... Cvejic, A. (2021). *Cell Stem Cell, 28*, 472–487. e7.

Rappez, L., Stadler, M., Triana, S., Gathungu, R. M., Ovchinnikova, K., Phapale, P., ... Alexandrov, T. (2021). *Nature Methods, 18*, 799–805.

Ren, K., Li, E., & Ji, P. (2022). *Current Opinion in Hematology, 29*, 137–143.

Rodriques, S. G., Stickels, R. R., Goeva, A., Martin, C. A., Murray, E., Vanderburg, C. R., ... Macosko, E. Z. (2019). *Science (New York, N. Y.), 363*, 1463–1467.

Schneider, R. K., Schenone, M., Ferreira, M. V., Kramann, R., Joyce, C. E., Hartigan, C., ... Ebert, B. L. (2016). *Nature Medicine, 22*, 288–297.

Schulz, V. P., Yan, H., Lezon-Geyda, K., An, X., Hale, J., Hillyer, C. D., ... Gallagher, P. G. (2019). *Cell Reports, 28*, 2996–3009.e7.

Ståhl, P. L., Salmén, F., Vickovic, S., Lundmark, A., Navarro, J. F., Magnusson, J., ... Frisén, J. (2016). *Science (New York, N. Y.), 353*, 78–82.

Stuart, T., & Satija, R. (2019). *Nature Reviews. Genetics, 20*, 257–272.

Suragani, R. N., Zachariah, R. S., Velazquez, J. G., Liu, S., Sun, C. W., Townes, T. M., & Chen, J. J. (2012). *Blood, 119*, 5276–5284.

Tusi, B. K., Wolock, S. L., Weinreb, C., Hwang, Y., Hidalgo, D., Zilionis, R., ... Socolovsky, M. (2018). *Nature, 555*, 54–60.

Wang, S. (2025). *Journal of Histotechnology, 48*, 68–77.

Welch, J. J., Watts, J. A., Vakoc, C. R., Yao, Y., Wang, H., Hardison, R. C., ... Weiss, M. J. (2004). *Blood, 104*, 3136–3147.

Wong, P., Hattangadi, S. M., Cheng, A. W., Frampton, G. M., Young, R. A., & Lodish, H. F. (2011). *Blood, 118*, e128–e138.

Wu, Q., Zhang, J., Kumar, S., Shen, S., Kincaid, M., Johnson, C. B., ... Lucas, D. (2024). *Nature, 627*, 839–846.

Xiao, X., Juan, C., Drennon, T., Uytingco, C. R., Vishlaghi, N., Sokolowskei, D., ... Tower, R. J. (2023). *Bone Research, 11*, 59.

Xu, J., Shao, Z., Glass, K., Bauer, D. E., Pinello, L., Van Handel, B., ... Orkin, S. H. (2012). *Developmental Cell, 23*, 796–811.

Zhang, J., Liu, Y., Han, X., Mei, Y., Yang, J., Zhang, Z. J., ... Ji, P. (2019). *Experimental Hematology, 78*, 21–34 e3.

Zhang, S., Macias-Garcia, A., Velazquez, J., Paltrinieri, E., Kaufman, R. J., & Chen, J. J. (2018). *Blood, 131*, 450–461.

Zhang, X., Zhang, M., Xu, Y., Song, Y., Yang, X., Wu, Y., ... Zhang, J. (2025). *Genome Biology, 26*, 428.

Zhao, B., Yang, J., & Ji, P. (2016). *Nucleus, 7*, 425–429.

Zhao, T., Chiang, Z. D., Morriss, J. W., LaFave, L. M., Murray, E. M., Del Priore, I., ... Chen, F. (2022). *Nature, 601*, 85–91.

Zheng, L., Wang, J., Jin, X., Cheng, Q., Zhang, X., Li, Y., ... Shi, L. (2025). *Blood Science, 7*, e00228.

CHAPTER NINE

Transcription elongation mechanisms that control gene expression in erythropoiesis

Kaustav Mukherjee* **and James J. Bieker***
Department of Stem Cell Biology & Regenerative Medicine, Mount Sinai School of Medicine, New York, NY, United States
*Corresponding authors. e-mail address: kaustav.mukherjee@mssm.edu; james.bieker@mssm.edu

Contents

Abstract

The regulation of RNA transcript abundance relies on the control of two key early steps in gene expression: transcription initiation and productive elongation. Although DNA-directed RNA polymerase provides the intrinsic catalytic activity required for RNA synthesis, its ability to effectively produce a transcript relies on accurate recruitment to specific DNA sites to initiate RNA polymerization, followed by release from promoter-proximal pausing into a productive elongation machine. These steps are governed by accessory factors that not only facilitate the assembly of a functional pre-initiation complex but also exert both negative and positive control over polymerase pause/elongation status. We describe experiments that have elucidated these mechanisms in the context of erythropoiesis.

Current Topics in Developmental Biology, Volume 170
ISSN 0070-2153, https://doi.org/10.1016/bs.ctdb.2026.06.004

1. Transcription by RNA polymerase II: A historical layout

An essential tenet of the classic central dogma of molecular biology is that information flows from DNA to RNA (Crick, 1970). Although the discovery of reverse transcriptase put a wrinkle in this idea, the process of transcription remains a critical control point in relaying information from the genome to a functional moiety that enables cellular regulation to occur (Jacob & Monod, 1961).

Given this model of information flow established shortly after the discovery of the double stranded helical structure of DNA, early studies searched for an enzymatic activity that could copy DNA into RNA; more specifically, a DNA-dependent RNA polymerase. Conceptually, this was modeled on the groundbreaking studies from the Kornberg lab that discovered DNA-dependent DNA polymerase (reviewed in (Kornberg, 1969)). The dependence on template DNA was critical, as polynucleotide phosphorylase activity, an enzyme capable of synthesizing nucleic acid polymers, had been isolated but was not necessarily dependent on a template. Historically, a DNA-dependent activity was first identified in mammalian cells (Weiss, 1960), but then was quickly followed by its discovery in bacteria (*E. coli*), when it was extensively characterized from that more biochemically amenable source (Chamberlin & Berg, 1962; Furth et al., 1962).

Progress from an in vitro activity to actual isolation of the protein was accomplished by Burgess, who found that bacterial RNA polymerase consists of multiple subunits (Burgess et al., 1969). Importantly, these studies led to identification of a more moderately associated subunit (σ) that plays a critical role in promoter-directed regulation of activity, foreshadowing observations to be seen with the eukaryotic enzymes. These experiments set the stage for the ground-breaking identification of three mammalian RNA polymerases by Roeder (Roeder & Rutter, 1969).

Although bacterial RNA polymerase could accurately initiate transcription from an included DNA template in vitro, this was not the case with the mammalian homologues (reviewed in (Roeder, 2005; Struhl, 1999)). Persistent yet elegant biochemical studies ultimately reproduced accurate transcription in vitro with semi-purified components for all three mammalian RNA polymerases (Grummt, 1981; Ng et al., 1979; Weil et al., 1979). Related to our focus here on pre-mRNA transcription by RNA polymerase II (originally called 'heterogenous nuclear RNA'), these investigations led to the identification of 'general transcription initiation factors' that together form a pre-initiation complex of proteins that enable proper recognition and localization of the polymerase to a target sequence and efficient elongation (Dignam et al., 1983; Matsui et al., 1980).

Given this history, it was not surprising that transcription initiation was looked on as the rate-limiting step in the process of generating an RNA transcript. Such a model continued to direct in vitro studies after chromatinized rather than naked DNA templates were used for reconstitution in vitro, albeit now requiring additional activators and cofactors to reverse the repressive ground state (Struhl, 1999). But there were hints that the regulatory aspect of eukaryotic pre-mRNA transcription might well be more complex than a simple binary decision between 'on' and 'off'. For example, early analyses of transcripts in isolated nuclei showed an uneven abundance of transcripts and polymerase that were biased toward the 5′-end of the gene (Fraser et al., 1978; Gariglio et al., 1981; Tamm & Kikuchi, 1979).

An approximation to polymerase pausing in vitro came from defining differentially sarkosyl-sensitive steps that separated 'template committed' from 'rapid start' steps in transcription, as monitored by accumulation of a 186-nucleotide transcript from the Adenovirus-2 major late promoter (Hawley & Roeder, 1985; Hawley & Roeder, 1987). Still, a confounding issue for establishing the existence of a truly paused polymerase came from finding that 'premature termination' could be a distinct form of expression control, where the polymerase is not paused but rather is no longer associated with the template, as seen in regulation of the c-myc gene (Bentley & Groudine, 1986; Bentley & Groudine, 1988).

However at about the same time, the Lis lab determined that the Drosophila heat shock genes were already bound by RNA polymerase II prior to induction in vivo (Gilmour & Lis, 1986), and that it was paused ~25 bp downstream of the transcription initiation site (Rougvie & Lis, 1988). This led to a fundamental reassessment of potential genetic control mechanisms of the level of RNA transcript generation and accumulation. Additional support for this conceptual change came from identification of pausing in both viral and mammalian DNA templates and at multiple gene targets and by methodological advances that increased the sensitivity and resolution of protein/DNA interactions in vivo (reviewed in (Adelman & Lis, 2012; Lis, 2019)). As a result, a complete analysis of transcriptional regulation must include analyses of initiation and elongation mechanisms that, coupled to degradation and termination (not discussed here), coordinate the final output level of the transcript.

2. Poised RNA polymerase II at inactive genes: A genome-wide perspective

Soon after it was established that RNA Polymerase II elongation is a rate-limiting transcription regulatory step, multiple groups working on β-globin gene regulation showed that the globin distal enhancer or locus control region (LCR) drives a switch from paused to elongating RNA Polymerase II at the β-globin gene leading to its activation (Johnson et al., 2001; Johnson et al., 2003; Sawado et al., 2003; Schubeler et al., 2001). A common finding in these studies was that even though transcription is initiated by the successful establishment of DNase-accessible chromatin, a pre-initiation complex, and RNA Polymerase II at the β-globin promoter, expression is high only when elongation-competent RNA Polymerase II engages with the LCR. This indicates that the LCR-drives β-globin expression by triggering the release of paused RNA Polymerase II into productive elongation, rather than by merely establishing a transcription pre-initiation complex (general schematic in Fig. 1). These focused studies on globin regulation inspired later studies using improved genomics techniques to show a global pattern of erythroid gene activation by regulated RNA Polymerase II pause-release. We discuss below the findings from genome-wide studies on RNA Polymerase II pausing and elongation in general, before moving on to the specific studies on erythropoiesis and erythroid gene regulation.

Earlier genome-wide studies used chromatin immunoprecipitation followed by microarray (ChIP-on-chip) to show that both RNA Polymerase II and activation-associated histone marks like H3K4me3 were enriched at the promoters of genes that were inactive in human embryonic stem (ES) cells and primary cells like hepatocytes and B-cells (Guenther et al., 2007). The inactive genes in these cells produced little to no mRNA and had low levels of the RNA Polymerase II elongation associated histone mark H3K36me3, suggesting that their mRNA level correlated with elongating RNA Polymerase II at the gene bodies rather than with paused RNA Polymerase II at the transcription start site (TSS). This study was complemented by GRO-seq (gene run-on sequencing) data that examined nascent transcription by RNA Polymerase II and found a similar pattern where RNA Polymerase II was present in a paused state at most active and inactive genes (Core & Lis, 2008). However, higher expression of active genes was driven by higher levels of elongating RNA Polymerase II. Further, independent groups working on *Drosophila* observed that around

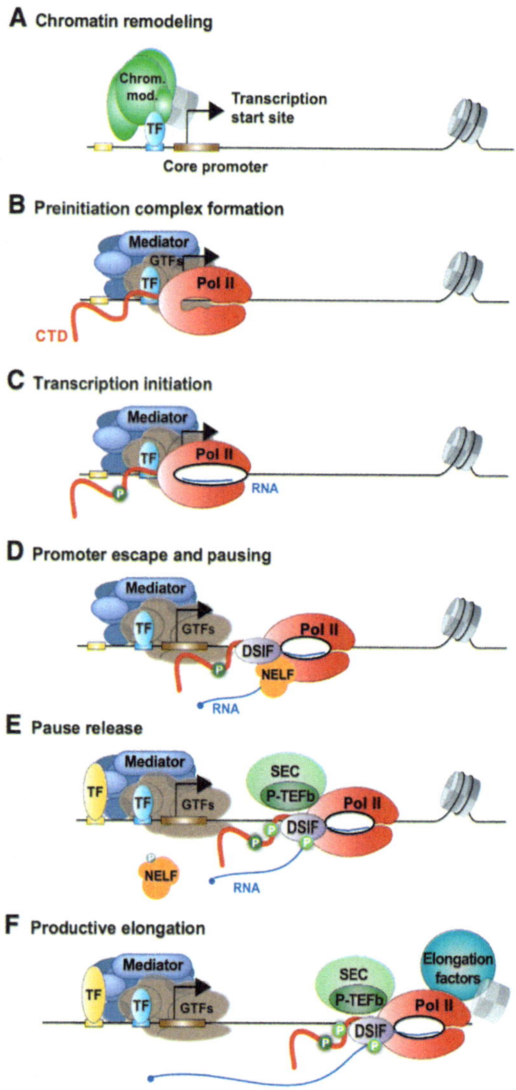

Fig. 1 Schematic of the key steps of mammalian transcription. (A) Chromatin modifiers recruited by "pioneer" transcription factors (TF–blue) create accessible chromatin near the core promoter. (B) Preinitiation complex is formed by the recruitment of general transcription factors (GTFs), Mediator complex, and RNA Polymerase II (Pol II). The C-terminal domain containing multiple hepta-peptide repeats is marked as CTD. (C) RNA Polymerase II transcribes a short distance generating a promoter-proximal small RNA and subsequently pauses near the transcription start site (TSS), where it is phosphorylated at Ser5 of its CTD, and associates with elongation factors DSIF (DRB-sensitivity inducing factor) and NELF (negative elongation factor) to remain

(Continued)

ten percent of all genes had RNA Polymerase II bound at the promoter but not at the gene bodies indicating that the RNA Polymerase II was in a "poised" state at these genes (Muse et al., 2007; Zeitlinger et al., 2007). Many of these genes are important for developmental processes in the *Drosophila* embryo, thus indicating that one of the final targets of signaling mechanisms that drive development are regulators of RNA Polymerase II elongation.

This was followed by comprehensive genomic studies on RNA Polymerase II pausing and elongation patterns during cell growth and development and their correlation with global binding of elongation factors and histone modifications (Kininis et al., 2009; Rahl et al., 2010; Welboren et al., 2009). Kininis et. al. used ChIP-chip and Welboren et. al. used ChIP-sequencing to study the effect of estrogen signaling on human MCF7 breast cancer cells while Rahl et. al. used ChIP-seq on mouse embryonic stem cells. Their data independently confirmed the findings of the above studies that the vast majority of active and inactive genes have paused RNA Polymerase II at their promoters, whereas only active genes have elongating RNA Polymerase II at their gene bodies (Fig. 2A). This correlates with the gene activation-associated histone mark H3K4me3 at both inactive and active genes, but H3K79me2 is present only at the active genes with elongating RNA Polymerase II. Both the Welboren and Rahl studies introduced the concept of RNA Polymerase II Traveling Ratio (also used interchangeably with the term "Pausing Index") defined as the ratio of RNA Polymerase II occupancy at the TSS (paused RNA Polymerase II) to RNA Polymerase II occupancy at gene bodies (elongating RNA Polymerase II), to measure the effective rate of RNA Polymerase II pause-release into elongation (Fig. 2B). Thus, genes with lower Traveling Ratio have higher RNA Polymerase II elongation and genes with higher Traveling Ratio have higher promoter-proximal pausing.

Fig. 1—Cont'd stably paused. (D) The binding of additional transcription factors at enhancers (TF – yellow) helps recruit P-TEFb (positive transcription elongation factor b) through various co-factor complexes such as the super-elongation complex (SEC) shown in the schematic. (E) The CDK9 kinase subunit of P-TEFb phosphorylates RNA Polymerase II CTD at Ser2, DSIF, and NELF, following which, NELF dissociates from the complex and RNA Polymerase II is released from its paused state into elongation. (F) RNA Polymerase II progresses into productive elongation aided by the association of yet other elongation factors that modify downstream nucleosomes enabling passage of the elongation machinery. *Adapted from Core and Adelman (2019).*

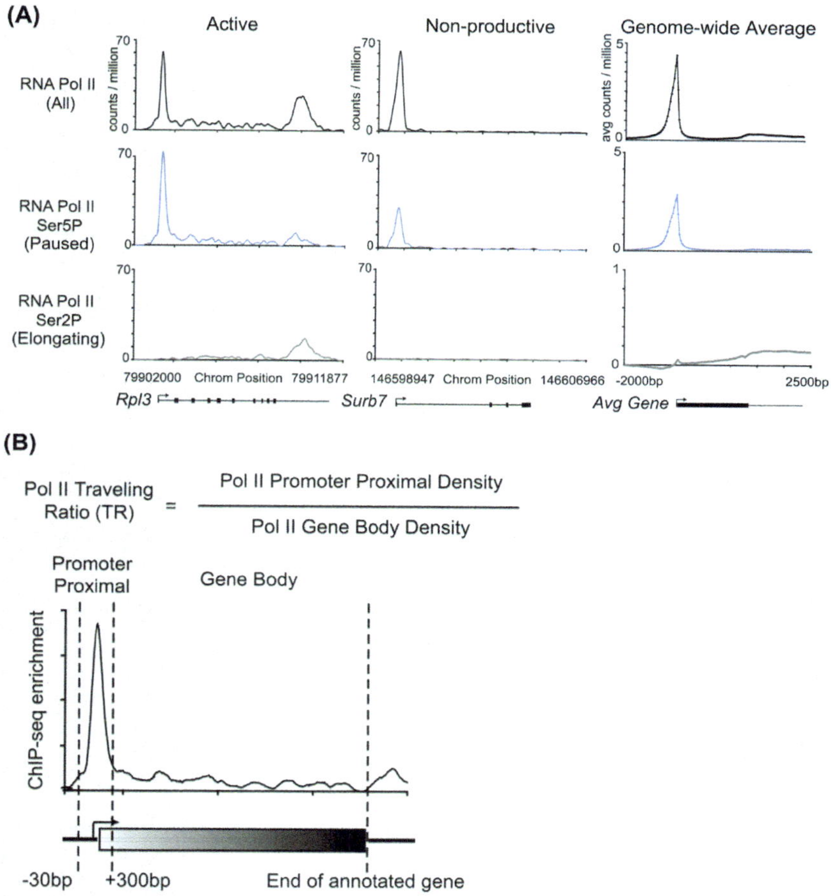

Fig. 2 RNA Polymerase II occupancy profiles and traveling ratio definition. (A). RNA Polymerase II occupancy determined by ChIP-seq at the active gene *Rpl3,* the inactive (non-productive) gene *Surb7* and the genome-wide average occupancy at all genes. (B). Schematic depicting the definition and calculation of RNA Polymerase II traveling ratio (TR) that is used interchangeably with pausing index. *Modified from Rahl et al. (2010).*

Further, most genes' promoters with TSS-paused RNA Polymerase II also have high binding of negative elongation factors DSIF (Spt4/5) and NELF at the TSS. Whereas depleting DSIF caused increased RNA Polymerase II elongation globally, NELF depletion only affected RNA Polymerase II elongation at specific genes (Sun et al., 2011). Cellular stimulation or differentiation causes higher recruitment of P-TEFb by the bromodomain protein BRD4, both positive elongation factors (Hargreaves et al., 2009), and the

Cdk9 subunit of P-TEFb phosphorylates Ser2 of the RNA Polymerase II C-terminal domain, which triggers elongation and gene expression. P-TEFb counteracts the effect of NELF and DSIF by phosphorylating both these factors, leading to NELF dissociation, whereas the phosphorylated form of DSIF is a positive elongation factor and travels into the gene bodies with RNA Polymerase II (Fig. 1) (Fujinaga et al., 2004; Kim & Sharp, 2001; Yamada et al., 2006). A summary of key elongation factors is shown in Fig. 3.

Overall, these molecular and early genome-wide studies provide concrete evidence that at genes whose expression is turned on or off based on specific cellular requirements, RNA Polymerase II elongation is a key rate-limiting step of global transcriptional control. These findings spurred further investigations into various aspects of RNA Polymerase II elongation control in mouse and human erythropoiesis as discussed below, with many of the studies co-evolving with advances in knowledge of fundamental

Protein	Function
P-TEF-b/Cdk9/CycT1/ (S.c.:Bur1/Bur2)	Cyclin-dependent kinase, releases Pol II from pausing and phosphorylates Pol II, NELF, and DSIF
DSIF (S.c.:Spt4-Spt5)	Stabilizes Pol II pausing, recruits elongation factors, stimulates elongation
NELF	Negative elongation factor
PARP-1	Poly[adenosine diphosphate (ADP-ribose)] polymerase, transfers ADP-ribose to NELF to inhibit NELF
BRD4	Recruits and activates P-TEFb
TFIIS	Resumes Pol II elongation from backtracked-arrest
TFIIF	Prevents transient pausing of Pol II
Elongin	Increases transcription elongation rate
Super elongation complex	Recruits P-TEFb, increases Pol II elongation rate
Little elongation complex	Required for snRNA expression in metazoans
Paf1C	Associated with Pol II and Spt4-Spt5, recruits chromatin remodelers, histone chaperones, and modifiers
GDOWN1	Tightly associated with Pol II, stabilizes pausing
RNA processing factors	CCR4-NOT, THO/TREX, Xrn2: mRNA processing and export factors that also regulate elongation

Fig. 3 Table summarizing the key elongation factors and complexes in mammalian cells and their molecular functions. "S.c." indicates *Saccharomyces cerevisiae* homolog.

mechanisms of RNA Pol II control in eukaryotes and mammals (Adelman & Lis, 2012; Chen et al., 2018; Core & Adelman, 2019; Lis, 2019). Concurrently, elegant cryo and EM studies elucidated structural changes of the multiprotein components bound to template as they occur at initiation, paused, and elongation stages (reviewed in (Osman & Cramer, 2020)).

3. RNA Polymerase II pausing and elongation control is crucial for optimal stage-specific erythroid gene expression

The first study that demonstrated global transcription elongation control of erythroid genes used a *Zebrafish* model and erythroid cells derived by *ex vivo* differentiation of primary human CD34+ hematopoietic stem cells (HSCs) (Bai et al., 2010). They showed that *Tif1γ* (TRIM33 in mammals) is a part of positive transcription elongation complexes that includes P-TEFb (Cdk9-Cyclin T1), PAF, and FACT, which together were important for activating the Zebrafish erythroid genes (Bai et al., 2010). This was based on their earlier observations that the *Tif1γ* mutant *moonshine (mon)* Zebrafish had severe anemia from early embryonic stages, but *Tif1γ* function was unclear and not directly implicated in controlling RNA Pol II elongation at the time (Ransom et al., 2004). They followed this up with studies on mouse using conditional Tif1-γ (*Trim33*) knockout (KO) in hematopoietic lineages and found that Tif1-γ KO negatively affects erythroid development, and B-cell development while enhancing myeloid development (Bai et al., 2013). This indicates that Tif1-γ is a regulator of erythroid-myeloid lineage skewing as also seen by the lower expression of important erythroid genes and higher expression of myeloid genes in the Tif1-γ KO. Finally, whereas global RNA Polymerase II levels remained unchanged in Tif1-γ KO, elongation competent RNA Polymerase II phospho-Ser2 levels were low suggesting a global loss of efficient transcription elongation due to the absence of Tif1-γ.

Simultaneously, genome-wide studies examining the status of RNA Polymerase II pausing and elongation in erythropoiesis were published (Ghamari et al., 2013; Stadhouders et al., 2012; Wong et al., 2011). To delineate the impact of RNA Polymerase II pausing and elongation on erythroid gene activation during the transition from early to late erythroid stages, the Lodish group separated early (Ter119-) and late terminally maturing erythroblasts (Ter119+) from E14.5 mouse embryonic fetal liver for profiling the genome-wide occupancy of RNA Polymerase II and various histone modifications. At erythroid genes activated during the transition from early erythroid fate specification to late erythroid maturation, RNA Polymerase II showed a paused profile near the promoters in early stage erythroblasts. However, the RNA Polymerase II elongation rate was higher at these genes (lower Traveling Ratio) in late-stage erythroblasts compared to early-stage, and the histone mark for active elongation, H3K79me2, correlated directly with RNA Polymerase II occupancy and elongation rate (Fig. 4). This directly demonstrates that erythroid genes are induced during terminal maturation by triggering the release of paused RNA Polymerase II into productive elongation (Wong et al., 2011). This is likely due to the action of the positive elongation factor P-TEFb, which was found to be globally co-localized with paused RNA Polymerase II phospho-Ser5 at erythroid gene promoters in mouse erythroleukemia (MEL) cells (Ghamari et al., 2013). Erythroid transcription factor activity is also important for triggering elongation as seen at the *Myb* gene in MEL cells, where transcription elongation of *Myb* was regulated by long-range enhancer-promoter interactions mediated by the erythroid transcription factors Klf1, Gata1, Tal1 and Ldb1 and involved cooperation with chromatin looping factors such as CTCF (Stadhouders et al., 2012).

To directly analyze genome-wide RNA Polymerase II pausing and elongation during human primary erythroid differentiation, the Steiner lab performed Cut&Tag (Kaya-Okur et al., 2019) for the phospho-Ser5

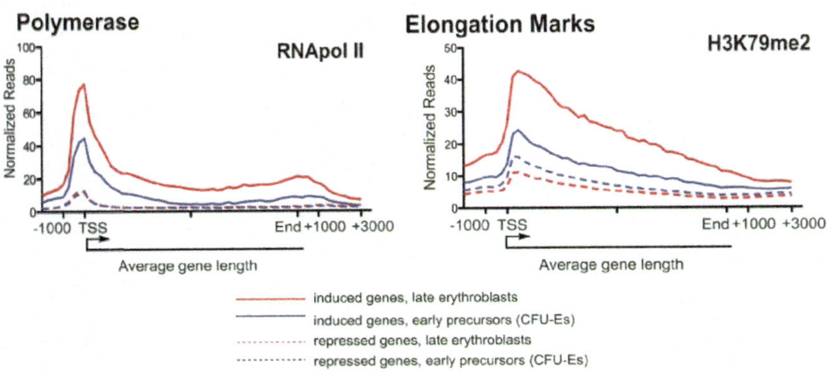

Fig. 4 RNA Polymerase II and Histone H3 Lysine 79 dimethyl (H3K79me2) occupancy during erythropoiesis. ChIP-seq was performed to detect RNA Polymerase II and H3K79me2 in primary E14.5 fetal liver mouse erythroblasts sorted into early (Ter119-) and late (Ter119) stage cells. *Adapted from Wong et al. (2011).*

and phospho-Ser2 modified Polymerases in intermediate and late-stage erythroblasts derived from erythroid differentiation of healthy human peripheral blood hematopoietic stem and progenitor cells (HSPCs) (Murphy et al., 2021). There was an overall decrease in the protein level of RNA Polymerase II (and its modified forms) during human erythroid maturation also reflected at most erythroid genes where RNA Polymerase II occupancy was decreased at later stages. However, at specific erythroid genes that are highly expressed in mature erythroblasts such as HBA (α-globin genes) and the mature erythroid marker *GYPA* (Glycophorin A), RNA Polymerase II occupancy was maintained during maturation. This reveals a remarkable capability of erythroid cells to maintain high RNA Polymerase II occupancy at specific genes whose functions are required in terminally mature erythroid cells, while removing RNA Polymerase II from all other genes. They also found that the traveling ratio of RNA Polymerase II Ser5p at mitochondrial and ribosomal genes was higher in intermediate erythroblasts than in terminal erythroblasts, and that these genes were repressed in intermediate erythroblasts. Thus, as erythroid cells mature, non-erythroid genes are repressed by first preventing efficient RNA Polymerase II elongation, followed by overall reduced RNA Polymerase II occupancy at these genes (Fig. 5A). However, it is difficult to distinguish the contribution of RNA Polymerase II initiation and termination from ChIP-seq or Cut&Tag data. Meanwhile, the Sankaran and Churchman groups used a different approach called NET-seq (native elongating transcript sequencing) (Mayer et al., 2015) that harnesses the stability of the complex formed between RNA, RNA Polymerase II and DNA during transcription to immunoprecipitate RNA Polymerase II and sequence the nascent RNA rather than the DNA (as in ChIP or Cut&Tag). They assayed cells harvested each day until Day 12 of human HSPC erythroid differentiation thereby capturing the majority of molecular events governing erythropoiesis (Martell et al., 2023). They then calculated the RNA Polymerase II Pausing Index or Traveling Ratio and showed that the overall expressed protein-coding genes had higher RNA Polymerase II pausing in the early stages of erythropoiesis (days 0–5) compared with later stages (days 6–11). There was a dip in traveling ratio from days 6–9 indicating higher elongation and coinciding with the high cell cycle phases of erythroid commitment when erythroid progenitors transition to mature precursors. This was also seen from their gene ontology analysis where genes with dynamic traveling ratio changes from high-low-high during erythroid differentiation were mostly

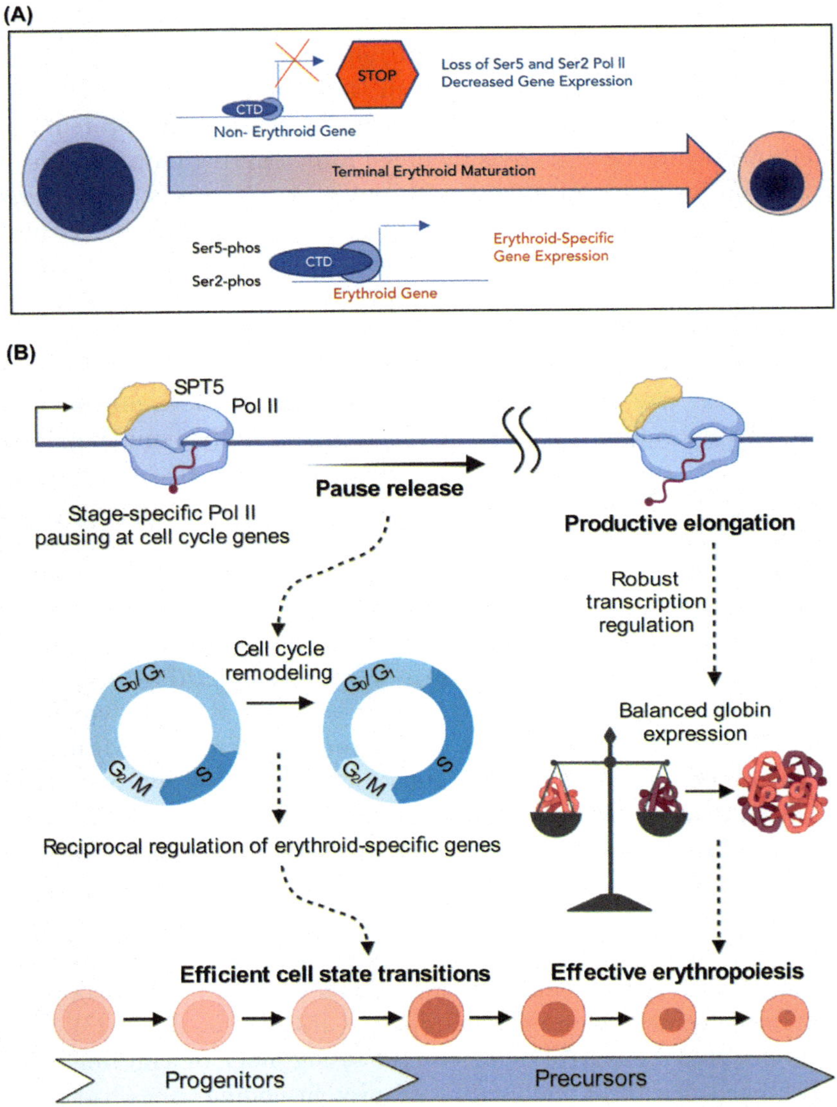

Fig. 5 Regulated release of RNA Polymerase II into productive elongation is necessary for proper erythroid fate specification and terminal maturation in human erythropoiesis. (A) Showing that during human terminal erythroid maturation of primary CD34+ HSPCs, both the paused (RNA Polymerase II Ser5p) and elongating (RNA Polymerase II Ser2p) forms are maintained specifically on erythroid genes, whereas non-erythroid genes lose RNA Polymerase II. (B) showing the specific contributions of regulated RNA Polymerase II pause-release in controlling cell cycle and globin expression during erythroid fate specification and terminal erythroid maturation of primary human CD34+ HSPCs. *(A) Schematic adapted from Murphy et al. (2021). (B) Model from Martell et al. (2023).*

involved in cell cycle processes. These genome-wide studies demonstrate that RNA Polymerase II pausing and elongation controls govern global gene expression of processes critical for erythropoiesis such as protein translation, mitochondria function, and cell division (Fig. 5B).

Studies on avian erythropoiesis (Penagos-Puig et al., 2023) also confirmed and further extended the above findings from human primary cells. This study compared erythrocytes from embryonic and adult chicken to erythroblasts and performed a technique called ATAC-see (Chen et al., 2016) that enables visualization of open chromatin regions. They found as expected that overall open chromatin was reduced in mature erythrocytes compared to erythroblasts, but the open chromatin was mostly present on the nuclear periphery. Nascent RNA labeling showed that most of the active transcription in mature erythrocytes also occurred at the nuclear periphery, correlating with these open chromatin domains. Importantly, they also showed using immunofluorescence that RNA Polymerase II Ser5p (paused) also colocalized with these open chromatin domains at the nuclear periphery in mature erythrocytes, whereas RNA Polymerase II Ser2p (elongating) levels were lower and not fully localized to the nuclear periphery. Using ChIP-seq they showed that the overall level of paused RNA Polymerase II was comparable in mature adult erythrocytes and erythroblasts, whereas elongating RNA Polymerase II was significantly lower. This resembles the findings from mouse and human erythroid cells during maturation. Thus, they concluded that paused RNA Polymerase II localizes into specific accessible chromatin compartments during erythroid maturation in avian models of erythropoiesis. They also performed Hi-C to evaluate chromosome compartments and interactions and found that in contrast to erythroblasts that have long-range distinct topologically associated domains (TADs), the mature erythrocytes lacked most of these TADs and instead had short range interactions that they termed "minidomains". Interestingly, these short-range interacting minidomains in mature erythrocytes were enriched for genes with high RNA Polymerase II pausing, and these genes were involved in various processes related to transcription elongation and RNA transport. This correlates well with the overall reduced gene expression in mature erythrocytes, but why the mature erythrocytes maintain paused RNA Polymerase II at specific genes, and restricted to specific spatial compartments in the nucleus, remains an open question.

4. RNA Polymerase II recruitment and pause-release at erythroid genes occur during stochastic transcription "bursts"

Although general depictions of transcription portray continuous mRNA production by RNA Polymerase II, single cell visualization of eukaryotic and prokaryotic transcription revealed discontinuous mRNA production by a process referred to as transcriptional bursting (Chubb et al., 2006; Dar et al., 2012; Suter et al., 2011); reviewed in (Bahar Halpern et al., 2015; Rodriguez & Larson, 2020). Gene expression can be fine-tuned by modulating burst size and burst frequency (Dar et al., 2012) and these features are regulated by cis-regulatory sequences such as promoters and enhancers and epigenetic processes such as histone modifications (Larsson et al., 2019; Nicolas et al., 2018). These studies show that cell-type transcript specificity depends largely on regulation of burst frequency by histone acetylation of enhancers whereas promoter elements mostly control burst size. This was also found to be true for the murine β-globin locus, where interaction with the LCR caused increased transcription burst frequency that activates or elevates its expression (Bartman et al., 2016; Bender et al., 2012). Similarly, changes in transcriptional bursting dynamics were shown to be important for regulating other mouse erythroid genes by using a combination of nascent RNA FISH and RNA Polymerase II ChIP-seq (Bartman et al., 2019). Since the readout of most transcriptional bursting studies is the nascently produced mRNA and the duration of the burst is longer than the time required for RNA Polymerase II recruitment or pause-release, two likely scenarios are possible. *First,* burst initiation is followed by RNA Polymerase II recruitment leading to elongation and mRNA production. *Second,* the observed transcription burst correlates with high nascent mRNA production and occurs after RNA Polymerase II recruitment, coinciding with release of paused RNA Polymerase II into productive elongation. Bartman et al. used murine cells with an Estrogen receptor tagged to the erythroid transcription factor Gata1 (G1-ER4) to demonstrate that both RNA Polymerase II recruitment and subsequent pause-release occur after burst initiation (Bartman et al., 2019). Importantly, they showed that in early erythroid differentiation (4 h after estradiol stimulation) transcription burst frequency increases, which suggests that RNA Polymerase II recruitment rates are higher at erythroid genes during early stages of erythropoiesis. At later "peak" erythroid differentiation (13 h after stimulation) both burst frequency per cell and burst intensity was higher, and this increased burst intensity specifically correlated with high nascent mRNA production and changes in RNA Polymerase II traveling ratio suggesting an increased rate of transcription elongation at erythroid genes during erythroid maturation. Therefore, these studies continue to support a model where RNA Polymerase II recruitment and pausing at erythroid genes occurs during the period of erythroid commitment, which transitions to pause-release and elongation during later stages of terminal erythroid maturation to increase the transcriptional output of erythroid genes.

5. Control of erythroid transcription elongation requires extensive interplay between erythroid lineage transcription factors and transcription co-factors

Most transcription factors have the ability to activate gene transcription by increasing the rates of both RNA Polymerase II recruitment and pausing (Larke et al., 2021), and subsequent pause-release. Likewise, erythroid lineage transcription factors can be involved in context-dependent interactions with various transcription co-factor and chromatin remodeling complexes to activate both transcription initiation and elongation. The erythroid transcription factor EKLF/KLF1 interacts with the mammalian transcription co-activator complexes SWI/SNF (Armstrong et al., 1998) and CBP/p300 (Zhang & Bieker, 1998). Both of these complexes mediate chromatin remodeling and histone modifications that directly recruit elongation factors. Further, the CBP-EKLF interaction also results in acetylation of EKLF at two amino acids – Lys288 in the transactivation domain and Lys302 in the Zn finger DNA binding domain, but only the Lys288 acetylation appears to be necessary for EKLF/KLF1 transactivation ability (Zhang et al., 2001). Acetylated EKLF/KLF1 also has a higher affinity for the SWI/SNF complex ATPase subunit BRG1 (SMARCA4), which is an important mediator of triggering transcription elongation (Subtil-Rodriguez & Reyes, 2010). Finally, the EKLF-CBP interaction and EKLF acetylation at Lys288 was necessary for CBP recruitment to β-globin enhancers (Sengupta et al., 2008). GATA1 also interacts with CBP and BRG1 and this interaction enhances its ability to activate transcription (Blobel et al., 1998) and increase promoter-enhancer contacts at mouse β-globin locus (Kim et al., 2009). As with EKLF,

GATA1 is also acetylated by CBP/p300 and this enhances GATA1 chromatin occupancy and histone acetylation at target sites such as globins (Hung et al., 1999; Lamonica et al., 2006; Letting et al., 2003). Acetylated GATA1 is bound by the bromodomain protein BRD3 and this interaction stabilizes chromatin occupancy of both GATA1 and BRD3 at erythroid genes (Lamonica et al., 2011). The transcription factor Ikaros (Ikzf1) also interacts with Gata1, Brg1 and the Cdk9 P-TEFb subunit in mouse erythroid cells to directly trigger RNA Polymerase II elongation at the mouse β-globin genes (Bottardi et al., 2011). Additionally, Cdk9 recruitment to the mouse β-globin regulatory regions is also dependent on the Ldb1 transcription factor (Song et al., 2010), that may be dependent on Gata1 directly interacting with P-TEFb (Elagib et al., 2008) and, together with other erythroid transcription factors, stabilizing the elongation competent RNA Polymerase II complex.

Genome-wide approaches expanded on these molecular studies to provide a global view of this coordination between transcription factors, co-factors and the elongation machinery to drive erythroid gene activation by increasing the rate of transcription elongation. *First*, the Blobel lab expanded their work on the elongation regulating BRD proteins to show that in G1-ER mouse cells upon estradiol stimulation, the global occupancy of Brd2, Brd3 and Brd4 increased with the highest increases noted for Brd3 and Brd4 at both promoters and enhancers (Stonestrom et al., 2015). Interestingly, Brd2 was preferentially enriched at the promoters upon G1-ER stimulation. The overall effect was increased occupancy of both Gata1 and Brd proteins at erythroid genes, and this was reversed by specific inhibition of acetyl-protein binding ability of BET proteins using pharmacological small molecules called BET inhibitors. *Second*, the Bieker lab used a mouse model of chronic anemia called Nan (Neonatal Anemia) caused by an EKLF mutation that leads to a globally neomorphic defects due to ectopic expression of non-erythroid genes (Heruth et al., 2010; Nebor et al., 2018; Planutis et al., 2017; Siatecka et al., 2010). They demonstrated that the EKLF-CBP interaction plays an important role in RNA Polymerase II elongation control (Mukherjee & Bieker, 2022). Upon assaying global RNA Polymerase II Ser5p and Ser2p using ChIP-seq and correlating this with chromatin accessibility, EKLF binding, CBP occupancy and H3K27ac they showed that many of the non-erythroid genes with low expression in the embryonic E13.5 mouse fetal liver possess paused RNA Polymerase II Ser5p at their TSS. However, elongation was low in normal erythropoiesis (wild type) embryos and this resulted in low or no expression of these genes. In Nan embryonic fetal livers, these ectopic genes had high expression and this correlated with mutant Nan-EKLF binding to nearby intergenic or intronic enhancers, increasing the occupancy of CBP and H3K27ac, and leading to higher levels of elongating RNA Polymerase II Ser2p at their gene bodies (Mukherjee & Bieker, 2022). These studies therefore show that erythroid lineage transcription factors such as GATA1 and EKLF/KLF1 can convert the paused RNA Polymerase II complex to an elongation competent complex partly through their interactions with BRD-type transcription co-factors such as BRD2/3/4 and CBP at gene enhancers (Fig. 6A).

Transcription cofactors, however, are global regulators of transcription and thus are able to regulate transcription elongation of all erythroid genes during erythropoiesis. An important elongation regulator is HEXIM1 that antagonizes P-TEFb by sequestering it in an inactive ribonucleoprotein complex with a small RNA called 7SK snRNA (Chen et al., 2004; Yik et al., 2003). Both HEXIM1 and 7SK are highly expressed in erythroid cells and 7SK expression increases during terminal maturation whereas both HEXIM1 and P-TEFb components CDK9 and CycT1 levels decrease (Murphy et al., 2021). This correlates with the pattern of erythroid genes having a high traveling ratio in later stages of maturation due to low elongation at most genes. To directly test this idea, Murphy et al. depleted HEXIM1 using shRNA knockdown in HUDEP2 immortalized erythroid progenitor cells and found that knockdown reduced proliferation and severely affected viability during maturation, while HEXIM1-null cells were non-viable suggesting that HEXIM1 is necessary for erythroid cell survival. Conversely, HEXIM1 overexpression (OE) increased cellular proliferation during both expansion and erythroid maturation phases but these maturing cells were larger with bigger nuclei suggesting that maturation proceeds with loss of control over cell size and nuclear condensation (Lv et al., 2023). OE of a HEXIM1 phosphorylation site mutant that is unable to effectively release P-TEFb rescued the HEXIM1 OE phenotypes indicating that HEXIM1 functions in erythropoiesis are linked to its ability to antagonize P-TEFb function. HEXIM1 perturbations also affect global levels of elongation competent RNA Polymerase II and have variable effects on the RNA Polymerase II traveling ratio at erythroid genes depending on the timing of their expression (Murphy et al., 2021). Further exploration of HEXIM1′s ability to control transcription elongation of erythroid genes showed that HEXIM1′s function is largely context dependent. *First,* HEXIM1 OE leads to increased RNA Polymerase II

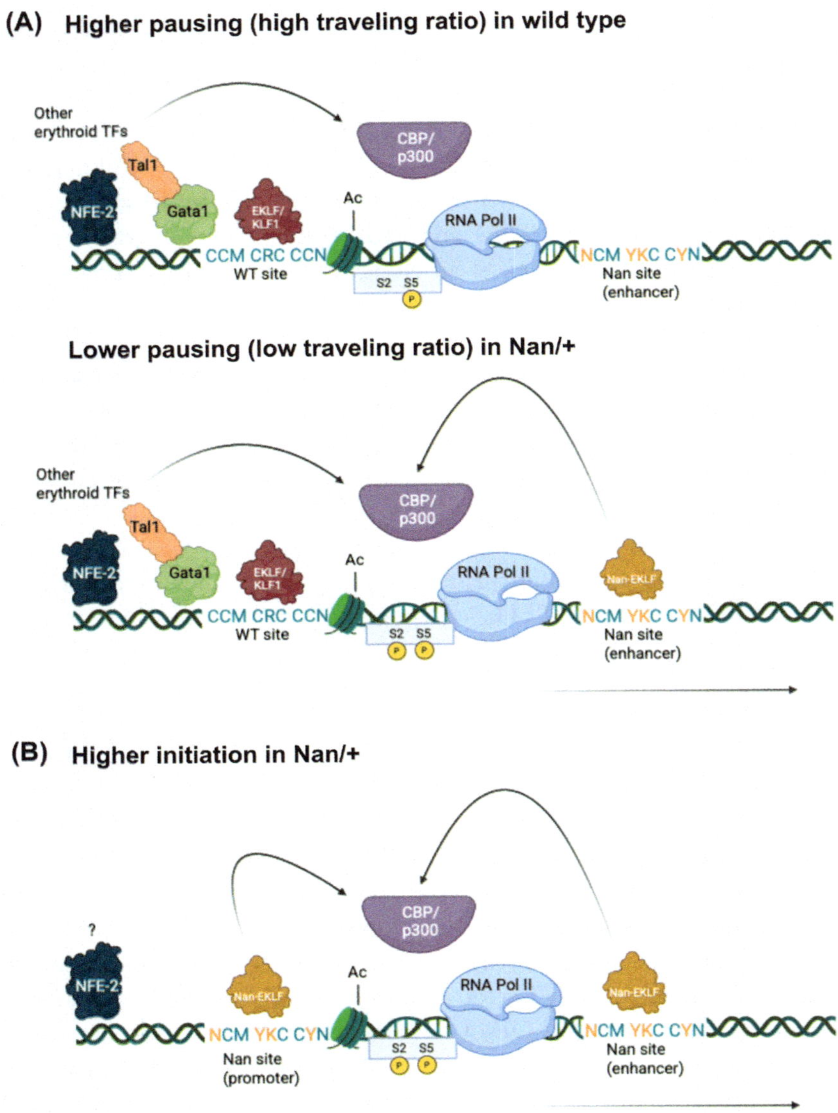

Fig. 6 Interplay of erythroid transcription factors and co-factors to drive ectopic RNA Polymerase II elongation in disordered erythropoiesis. (A). Top: At a subset of ectopically expressed genes in the Nan/+ mouse E13.5 fetal liver, in healthy WT embryos EKLF/KLF1 is bound to the promoter and EKLF binding sites are flanked by sites for binding of Gata1, NFE-2 and other erythroid transcription factors. These genes have paused RNA Polymerase II in WT cells with little to no elongating RNA Polymerase II and gene expression. Bottom: In Nan/+ cells, de the ectopic binding of mutant Nan-EKLF to a nearby enhancer site triggers release of RNA polymerase II and ectopic expression of these genes. (B). At other ectopic genes distinct from the subset in (A), Nan-EKLF drives higher initiation and elongation of ectopic genes, which aggravates the anemia.

occupancy at GATA1 target genes but reduced RNA Polymerase II occupancy at GATA1+ KLF1 target genes (Lv et al., 2023). *Second,* HEXIM1 OE caused genes to be both upregulated (heme synthesis genes) and downregulated (non-erythroid genes) (Murphy et al., 2021), but there was a global increase in HEXIM1 chromatin occupancy at both these gene sets dependent on its phosphorylation (Lv et al., 2023). This increase in HEXIM1 occupancy also correlated with higher RNA Polymerase II occupancy at both these gene sets but they partitioned into genes with higher traveling ratio that were downregulated and genes with lower traveling ratio that were upregulated upon HEXIM1 OE. The genes with increased RNA Polymerase II occupancy and lower traveling ratio (high elongation) were also enriched for GATA1 motifs and were involved in cell cycle regulation. Thus, a pattern emerges where at GATA1 targets, HEXIM1 is able to trigger release of RNA Polymerase II into productive elongation while at other sites it increases RNA Polymerase II recruitment and pausing but does not affect elongation (Lv et al., 2023).

Another important elongation regulator studied in the context of erythropoiesis is the elongation factor, SPT5 (DSIF subunit) encoded by the gene *SUPTH5* (Martell et al., 2023). Martell et al. investigated SPT5 function during erythroid differentiation of human HSPCs by creating heterozygous null SPT5 (*SUPTH5 +/−*) primary human HSPCs using Crispr-Cas9. They showed that SPT5+/− causes downregulation of cell cycle genes such as CDK1, and the erythroid transcription factors GATA1 and KLF1 due to an increase in their RNA Polymerase II traveling ratio (decreased pause-release and elongation) during the transition from erythroid progenitor (early stage) to precursor (intermediate stage). The decrease in GATA1 expression is reminiscent of earlier studies in *Zebrafish* showing that *spt5/foggy* knockdown decreases *gata1* expression and also that of its targets (Taneda et al., 2011). Consequently, this delayed erythroid maturation and these cells experienced a shorter S-phase due to cell cycle disruptions, but this delay was transient suggesting that SPT5 controls timing of erythroid maturation and cell cycle progression during erythropoiesis. There were no differences in maturation characteristics between healthy and SPT5+/− mature erythroid cells at later stages of differentiation, suggesting that either other mechanisms compensate for SPT5 haploinsufficiency or SPT5 haploinsufficiency is not enough to cause a markedly severe erythroid maturation phenotype.

6. Mechanisms that control transcription elongation may contribute significantly to the pathogenesis of erythroid disorders

Two examples of disrupted transcription elongation control that causes erythroid disorders were already alluded to in the previous sections. *First*, our work on the Neonatal anemia mouse model showed that the BRD protein CBP was ectopically deposited at non-erythroid genes in fetal liver erythroid cells, triggering ectopic RNA Polymerase II elongation and causing systemic anemia (Fig. 6A and B) (Mukherjee & Bieker, 2022). Single cell analysis of the Nan mouse bone marrow also showed ectopic expression in the erythroid populations (Arif et al., 2025) revealing that disrupted elongation control due to the EKLF/KLF1 Nan mutation persists till adulthood, thus explaining the lifelong anemia in these mice. The Nan mutation is highly relevant to human erythroid disorders since the same amino acid in the EKLF/KLF1 zinc finger (Glutamate 339 in mouse = Glutamate 325 in humans) is mutated in humans to Lysine causing Congenital Dyserythropoietic Anemia Type IV (CDA IV) ((Arnaud et al., 2010; Jaffray et al., 2013; Ravindranath et al., 2018; Singleton et al., 2011); reviewed in (Bieker & Philipsen, 2024)). Studies using CDA IV patient cells and cellular models revealed similar transcriptomic and proteomic changes as in the Nan mouse such as downregulation of erythroid genes and cell cycle genes, and ectopic expression of non-erythroid genes (Ferrer-Vicens et al., 2023; Ilsley et al., 2019; Kohara et al., 2019; Kulczynska et al., 2020; Varricchio et al., 2019) suggesting that disrupted transcription elongation control may also contribute to CDA IV pathogenesis.

Second, the study from Martell et. al. on SPT5 (Martell et al., 2023) initially sought to discover novel genetic mutations associated with the globin chain imbalance in individuals with β-thalassemia trait and found multiple individuals with various types of mutations in *SUPTH5,* which ultimately leads to loss-of-function causing microcytic hypochromic anemia. SPT5+/− erythroid cultures recapitulated the globin imbalance phenotype of *SUPTH5* mutant β-thalassemia patients. Finally, treatment of cells with the SPT5 inhibitors at calibrated doses that mimicked SPT5 haploinsufficiency, also recapitulated the β-thalassemia phenotype and showed a global increase in RNA Polymerase II traveling ratio (higher pausing). Thus, transcription elongation control likely plays an important role in β-thalassemia pathology and possibly involves additional elongation regulators apart from SPT5.

The observations linking GATA1 to transcription elongation control suggest that unexplored mechanisms likely play a role in erythroid disorders triggered by mutations in GATA1 such as Diamond Blackfan Anemia (Ling et al., 2019), caused by an N-terminal truncation in GATA1 (GATA1s) (Sankaran et al., 2012). Regarding clinical applications, it is reasonable to anticipate that inhibiting a crucial mechanism of erythroid gene expression control is highly likely to lead to anemia, requiring care in dose/delivery. Indeed, it is not surprising that the clinical use of BET inhibitors to treat various conditions such as hematological malignancies have shown that one of the predominant side-effects in these patients is anemia (Amorim et al., 2016; Sun et al., 2020).

7. Conclusions and open questions

We have provided an overview of the essential aspects of transcription elongation control mechanisms as they pertain to erythropoiesis. However, basic mechanistic questions continue to arise from these observations, particularly in the context of epigenetics. For example, given the importance of chromatin accessibility and transcription factor function in the control of initiation/pausing/elongation, what is the role of deacetylase (HDAC) proteins in maintaining the right balance of acetylated histone and non-histone proteins with acetylases such as CBP, and how are these regulated in erythropoiesis? The same can be asked for histone methylases and demethylases that regulate levels (for example, of H3K79me2) during elongation: are their relative levels coordinated during erythroid expansion/differentiation? Regarding histone H3, is there a differential requirement for the replication-independent H3.3 variant (Sengupta et al., 2008) compared to the canonical H3 in the process of pause/release, particularly as it relates the role of K36me3 modification (Shah et al., 2026; Wen et al., 2014)? How do histone H3 chaperones (such as HIRA, DAXX, ATRX (Lewis et al., 2010; Soni et al., 2014; Voon et al., 2015)) play into these details, both in normal and perhaps in aberrant erythroid expansion?

Acknowledgements

Studies in the Bieker lab are supported by NIH R01 DK046865 and R21 DK139543.

Conflicts of interest

The authors declare that they have no conflicts of interet.

References

Adelman, K., & Lis, J. T. (2012). Promoter-proximal pausing of RNA polymerase II: Emerging roles in metazoans. *Nature Reviews Genetics, 13*, 720–731.

Amorim, S., Stathis, A., Gleeson, M., Iyengar, S., Magarotto, V., Leleu, X., ... Rezai, K., et al. (2016). Bromodomain inhibitor OTX015 in patients with lymphoma or multiple myeloma: A dose-escalation, open-label, pharmacokinetic, phase 1 study. *The Lancet Haematology, 3*, e196–e204.

Arif, T., Mukherjee, K., Xue, L., Catrow, J. L., Cox, J., & Bieker, J. J. (2025). STING and VDAC inhibitors attenuate inflammation and ineffective erythropoiesis caused by an altered metabolome in the Nan (EKLF/E339D) mouse model of neonatal anemia. *bioRxiv*, https://doi.org/10.64898/2025.12.11.693792.

Armstrong, J. A., Bieker, J. J., & Emerson, B. M. (1998). A SWI/SNF-related chromatin remodeling complex, E-RC1, is required for tissue-specific transcriptional regulation by EKLF in vitro. *Cell, 95*, 93–104.

Arnaud, L., Saison, C., Helias, V., Lucien, N., Steschenko, D., Giarratana, M. C., ... de Brevern, A. G. (2010). A dominant mutation in the gene encoding the erythroid transcription factor KLF1 causes a congenital dyserythropoietic anemia. *American Journal of Human Genetics, 87*, 721–727.

Bahar Halpern, K., Tanami, S., Landen, S., Chapal, M., Szlak, L., Hutzler, A., ... Itzkovitz, S. (2015). Bursty gene expression in the intact mammalian liver. *Molecular Cell, 58*, 147–156.

Bai, X., Kim, J., Yang, Z., Jurynec, M. J., Akie, T. E., Lee, J., et al. (2010). TIF1gamma controls erythroid cell fate by regulating transcription elongation. *Cell, 142*, 133–143.

Bai, X., Trowbridge, J. J., Riley, E., Lee, J. A., DiBiase, A., Kaartinen, V. M., ... Zon, L. I. (2013). TiF1-gamma plays an essential role in murine hematopoiesis and regulates transcriptional elongation of erythroid genes. *Developmental Biology, 373*, 422–430.

Bartman, C. R., Hamagami, N., Keller, C. A., Giardine, B., Hardison, R. C., Blobel, G. A., & Raj, A. (2019). Transcriptional burst initiation and polymerase pause release are key control points of transcriptional regulation. *Molecular Cell, 73*, 519–532.e514.

Bartman, C. R., Hsu, S. C., Hsiung, C. C., Raj, A., & Blobel, G. A. (2016). Enhancer regulation of transcriptional bursting parameters revealed by forced chromatin looping. *Molecular Cell, 62*, 237–247.

Bender, M. A., Ragoczy, T., Lee, J., Byron, R., Telling, A., Dean, A., & Groudine, M. (2012). The hypersensitive sites of the murine beta-globin locus control region act independently to affect nuclear localization and transcriptional elongation. *Blood, 119*, 3820–3827.

Bentley, D. L., & Groudine, M. (1986). Novel promoter upstream of the human c-myc gene and regulation of c-myc expression in B-cell lymphomas. *Molecular and Cellular Biology, 6*, 3481–3489.

Bentley, D. L., & Groudine, M. (1988). Sequence requirements for premature termination of transcription in the human c-myc gene. *Cell, 53*, 245–256.

Bieker, J. J., & Philipsen, S. (2024). Erythroid kruppel-like factor (KLF1): A surprisingly versatile regulator of erythroid differentiation. *Advances in Experimental Medicine and Biology, 1459*, 217–242.

Blobel, G. A., Nakajima, T., Eckner, R., Montminy, M., & Orkin, S. H. (1998). CREB-binding protein cooperates with transcription factor GATA-1 and is required for erythroid differentiation. *Proceedings of the National Academy of Sciences of the United States of America, 95*, 2061–2066.

Bottardi, S., Zmiri, F. A., Bourgoin, V., Ross, J., Mavoungou, L., & Milot, E. (2011). Ikaros interacts with P-TEFb and cooperates with GATA-1 to enhance transcription elongation. *Nucleic Acids Research, 39*, 3505–3519.

Burgess, R. R., Travers, A. A., Dunn, J. J., & Bautz, E. K. (1969). Factor stimulating transcription by RNA polymerase. *Nature (London), 221*, 43–46.
Chamberlin, M., & Berg, P. (1962). Deoxyribo ucleic acid-directed synthesis of ribonucleic acid by an enzyme from Escherichia coli. *Proceedings of the National Academy of Sciences of the United States of America, 48*, 81–94.
Chen, F. X., Smith, E. R., & Shilatifard, A. (2018). Born to run: Control of transcription elongation by RNA polymerase II. *Nature Reviews. Molecular Cell Biology, 19*, 464–478.
Chen, R., Yang, Z., & Zhou, Q. (2004). Phosphorylated positive transcription elongation factor b (P-TEFb) is tagged for inhibition through association with 7SK snRNA. *The Journal of Biological Chemistry, 279*, 4153–4160.
Chen, X., Shen, Y., Draper, W., Buenrostro, J. D., Litzenburger, U., Cho, S. W., ... East-Seletsky, A. (2016). ATAC-see reveals the accessible genome by transposase-mediated imaging and sequencing. *Nature Methods, 13*, 1013–1020.
Chubb, J. R., Trcek, T., Shenoy, S. M., & Singer, R. H. (2006). Transcriptional pulsing of a developmental gene. *Current Biology: CB, 16*, 1018–1025.
Core, L., & Adelman, K. (2019). Promoter-proximal pausing of RNA polymerase II: A nexus of gene regulation. *Genes & Development, 33*, 960–982.
Core, L. J., & Lis, J. T. (2008). Transcription regulation through promoter-proximal pausing of RNA polymerase II. *Science (New York, N. Y.), 319*, 1791–1792.
Crick, F. (1970). Central dogma of molecular biology. *Nature (London), 227*, 561–563.
Dar, R. D., Razooky, B. S., Singh, A., Trimeloni, T. V., McCollum, J. M., Cox, C. D., ... Weinberger, L. S. (2012). Transcriptional burst frequency and burst size are equally modulated across the human genome. *Proceedings of the National Academy of Sciences of the United States of America, 109*, 17454–17459.
Dignam, J. D., Martin, P. L., Shastry, B. S., & Roeder, R. G. (1983). Eukaryotic gene transcription with purified components. *Methods in Enzymology, 101*, 587–597.
Elagib, K. E., Mihaylov, I. S., Delehanty, L. L., Bullock, G. C., Ouma, K. D., Caronia, J. F., ... Goldfarb, A. N. (2008). Cross-talk of GATA-1 and P-TEFb in megakaryocyte differentiation. *Blood, 112*, 4884–4894.
Ferrer-Vicens, I., Ferguson, D. C. J., Wilson, M. C., Heesom, K. J., Bieker, J. J., & Frayne, J. (2023). A novel human cellular model of CDA IV enables comprehensive analysis revealing the molecular basis of the disease phenotype. *Blood, 141*, 3039–3054.
Fraser, N. W., Sehgal, P. B., & Darnell, J. E. (1978). DRB-induced premature termination of late adenovirus transcription. *Nature (London), 272*, 590–593.
Fujinaga, K., Irwin, D., Huang, Y., Taube, R., Kurosu, T., & Peterlin, B. M. (2004). Dynamics of human immunodeficiency virus transcription: P-TEFb phosphorylates RD and dissociates negative effectors from the transactivation response element. *Molecular and Cellular Biology, 24*, 787–795.
Furth, J. J., Hurwitz, J., & Anders, M. (1962). The role of deoxyribonucleic acid in ribonucleic acid synthesis. I. The purification and properties of ribonucleic acid polymerase. *The Journal of Biological Chemistry, 237*, 2611–2619.
Gariglio, P., Bellard, M., & Chambon, P. (1981). Clustering of RNA polymerase B molecules in the 5′ moiety of the adult beta-globin gene of hen erythrocytes. *Nucleic Acids Research, 9*, 2589–2598.
Ghamari, A., van de Corput, M. P., Thongjuea, S., van Cappellen, W. A., van Ijcken, W., van Haren, J., ... Grosveld, F. G. (2013). In vivo live imaging of RNA polymerase II transcription factories in primary cells. *Genes & Development, 27*, 767–777.
Gilmour, D. S., & Lis, J. T. (1986). RNA polymerase II interacts with the promoter region of the noninduced hsp70 gene in Drosophila melanogaster cells. *Molecular and Cellular Biology, 6*, 3984–3989.

Grummt, I. (1981). Specific transcription of mouse ribosomal DNA in a cell-free system that mimics control in vivo. *Proceedings of the National Academy of Sciences of the United States of America, 78*, 727–731.

Guenther, M. G., Levine, S. S., Boyer, L. A., Jaenisch, R., & Young, R. A. (2007). A chromatin landmark and transcription initiation at most promoters in human cells. *Cell, 130*, 77–88.

Hargreaves, D. C., Horng, T., & Medzhitov, R. (2009). Control of inducible gene expression by signal-dependent transcriptional elongation. *Cell, 138*, 129–145.

Hawley, D. K., & Roeder, R. G. (1985). Separation and partial characterization of three functional steps in transcription initiation by human RNA polymerase II. *The Journal of Biological Chemistry, 260*, 8163–8172.

Hawley, D. K., & Roeder, R. G. (1987). Functional steps in transcription initiation and reinitiation from the major late promoter in a HeLa nuclear extract. *The Journal of Biological Chemistry, 262*, 3452–3461.

Heruth, D. P., Hawkins, T., Logsdon, D. P., Gibson, M. I., Sokolovsky, I. V., Nsumu, N. N., ... Lewing, K. B. (2010). Mutation in erythroid specific transcription factor KLF1 causes Hereditary Spherocytosis in the Nan hemolytic anemia mouse model. *Genomics, 96*, 303–307.

Hung, H. L., Lau, J., Kim, A. Y., Weiss, M. J., & Blobel, G. A. (1999). CREB-binding protein acetylates hematopoietic transcription factor GATA- 1 at functionally important sites. *Molecular and Cellular Biology, 19*, 3496–3505.

Ilsley, M. D., Huang, S., Magor, G. W., Landsberg, M. J., Gillinder, K. R., & Perkins, A. C. (2019). Corrupted DNA-binding specificity and ectopic transcription underpin dominant neomorphic mutations in KLF/SP transcription factors. *BMC Genomics, 20*, 417.

Jacob, F., & Monod, J. (1961). On the regulation of gene activity. *Cold Spring Harbor Symposia on Quantitative Biology, 26*, 193–211.

Jaffray, J. A., Mitchell, W. B., Gnanapragasam, M. N., Seshan, S. V., Guo, X., Westhoff, C. M., ... Manwani, D. (2013). Erythroid transcription factor EKLF/KLF1 mutation causing congenital dyserythropoietic anemia type IV in a patient of Taiwanese origin: Review of all reported cases and development of a clinical diagnostic paradigm. *Blood Cells, Molecules & Diseases, 51*, 71–75.

Johnson, K. D., Christensen, H. M., Zhao, B., & Bresnick, E. H. (2001). Distinct mechanisms control RNA polymerase II recruitment to a tissue-specific locus control region and a downstream promoter. *Molecular Cell, 8*, 465–471.

Johnson, K. D., Grass, J. A., Park, C., Im, H., Choi, K., & Bresnick, E. H. (2003). Highly restricted localization of RNA Polymerase II within a locus control region of a tissue-specific chromatin domain. *Molecular and Cellular Biology, 23*, 6484–6493.

Kaya-Okur, H. S., Wu, S. J., Codomo, C. A., Pledger, E. S., Bryson, T. D., Henikoff, J. G., ... Henikoff, S. (2019). CUT&Tag for efficient epigenomic profiling of small samples and single cells. *Nature Communications, 10*, 1930.

Kim, J. B., & Sharp, P. A. (2001). Positive transcription elongation factor B phosphorylates hSPT5 and RNA polymerase II carboxyl-terminal domain independently of cyclin-dependent kinase-activating kinase. *The Journal of Biological Chemistry, 276*, 12317–12323.

Kim, S. I., Bresnick, E. H., & Bultman, S. J. (2009). BRG1 directly regulates nucleosome structure and chromatin looping of the alpha globin locus to activate transcription. *Nucleic Acids Research, 37*, 6019–6027.

Kininis, M., Isaacs, G. D., Core, L. J., Hah, N., & Kraus, W. L. (2009). Postrecruitment regulation of RNA polymerase II directs rapid signaling responses at the promoters of estrogen target genes. *Molecular and Cellular Biology, 29*, 1123–1133.

Kohara, H., Utsugisawa, T., Sakamoto, C., Hirose, L., Ogawa, Y., Ogura, H., ... Iwasaki, T. (2019). KLF1 mutation E325K induces cell cycle arrest in erythroid cells differentiated from congenital dyserythropoietic anemia patient-specific induced pluripotent stem cells. *Experimental Hematology, 73*, 25–37.e28.

Kornberg, A. (1969). Active center of DNA polymerase. *Science (New York, N. Y.), 163*, 1410–1418.

Kulczynska, K., Bieker, J. J., & Siatecka, M. (2020). A Kruppel-like factor 1 (KLF1) mutation associated with severe congenital dyserythropoietic anemia alters its DNA-binding specificity. *Molecular and Cellular Biology, 40*.

Lamonica, J. M., Deng, W., Kadauke, S., Campbell, A. E., Gamsjaeger, R., Wang, H., ... Mackay, J. P. (2011). Bromodomain protein Brd3 associates with acetylated GATA1 to promote its chromatin occupancy at erythroid target genes. *Proceedings of the National Academy of Sciences of the United States of America, 108*, E159–E168.

Lamonica, J. M., Vakoc, C. R., & Blobel, G. A. (2006). Acetylation of GATA-1 is required for chromatin occupancy. *Blood, 108*, 3736–3738.

Larke, M. S. C., Schwessinger, R., Nojima, T., Telenius, J., Beagrie, R. A., Downes, D. J., ... Bender, M. A. (2021). Enhancers predominantly regulate gene expression during differentiation via transcription initiation. *Molecular Cell, 81*, 983–997.e987.

Larsson, A. J. M., Johnsson, P., Hagemann-Jensen, M., Hartmanis, L., Faridani, O. R., Reinius, B., ... Sandberg, R. (2019). Genomic encoding of transcriptional burst kinetics. *Nature (London), 565*, 251–254.

Letting, D. L., Rakowski, C., Weiss, M. J., & Blobel, G. A. (2003). Formation of a tissue-specific histone acetylation pattern by the hematopoietic transcription factor GATA-1. *Molecular and Cellular Biology, 23*, 1334–1340.

Lewis, P. W., Elsaesser, S. J., Noh, K. M., Stadler, S. C., & Allis, C. D. (2010). Daxx is an H3.3-specific histone chaperone and cooperates with ATRX in replication-independent chromatin assembly at telomeres. *Proceedings of the National Academy of Sciences of the United States of America, 107*, 14075–14080.

Ling, T., Birger, Y., Stankiewicz, M. J., Ben-Haim, N., Kalisky, T., Rein, A., ... Zhang, K. (2019). Chromatin occupancy and epigenetic analysis reveal new insights into the function of the GATA1 N terminus in erythropoiesis. *Blood, 134*, 1619–1631.

Lis, J. T. (2019). A 50 year history of technologies that drove discovery in eukaryotic transcription regulation. *Nature Structural & Molecular Biology, 26*, 777–782.

Lv, X., Murphy, K., Murphy, Z., Getman, M., Rahman, N., Nakamura, Y., ... Mohandas, N. (2023). HEXIM1 is an essential transcription regulator during human erythropoiesis. *Blood, 142*, 2198–2215.

Martell, D. J., Merens, H. E., Caulier, A., Fiorini, C., Ulirsch, J. C., Ietswaart, R., ... Cappellini, M. D. (2023). RNA polymerase II pausing temporally coordinates cell cycle progression and erythroid differentiation. *Developmental Cell, 58*, 2112–2127.e2114.

Matsui, T., Segall, J., Weil, P. A., & Roeder, R. G. (1980). Multiple factors required for accurate initiation of transcription by purified RNA Polymerase II. *The Journal of Biological Chemistry, 255*, 11992–11996.

Mayer, A., di Iulio, J., Maleri, S., Eser, U., Vierstra, J., Reynolds, A., ... Churchman, L. S. (2015). Native elongating transcript sequencing reveals human transcriptional activity at nucleotide resolution. *Cell, 161*, 541–554.

Mukherjee, K., & Bieker, J. J. (2022). EKLF/Klf1 regulates erythroid transcription by its pioneering activity and selective control of RNA Pol II pause-release. *Cell Reports, 41*, 111830.

Murphy, Z. C., Murphy, K., Myers, J., Getman, M., Couch, T., Schulz, V. P., ... Mohandas, N. (2021). Regulation of RNA Polymerase II activity is essential for terminal erythroid maturation. *Blood, 138*, 1740–1756.

Muse, G. W., Gilchrist, D. A., Nechaev, S., Shah, R., Parker, J. S., Grissom, S. F., ... Adelman, K. (2007). RNA polymerase is poised for activation across the genome. *Nature Genetics, 39*, 1507–1511.

Nebor, D., Graber, J. H., Ciciotte, S. L., Robledo, R. F., Papoin, J., Hartman, E., ... Blanc, L. (2018). Mutant KLF1 in adult anemic nan mice leads to profound transcriptome changes and disordered erythropoiesis. *Scientific Reports, 8*, 12793.

Ng, S. Y., Parker, C. S., & Roeder, R. G. (1979). Transcription of cloned Xenopus 5S RNA genes by X. laevis RNA Polymerase III in reconstituted systems. *Proceedings of the National Academy of Sciences of the United States of America, 76*, 136–140.

Nicolas, D., Zoller, B., Suter, D. M., & Naef, F. (2018). Modulation of transcriptional burst frequency by histone acetylation. *Proceedings of the National Academy of Sciences of the United States of America, 115*, 7153–7158.

Osman, S., & Cramer, P. (2020). Structural biology of RNA Polymerase II transcription: 20 years on. *Annual Review of Cell and Developmental Biology, 36*, 1–34.

Penagos-Puig, A., Claudio-Galeana, S., Stephenson-Gussinye, A., Jacome-Lopez, K., Aguilar-Lomas, A., Chen, X., ... Furlan-Magaril, M. (2023). RNA polymerase II pausing regulates chromatin organization in erythrocytes. *Nature Structural & Molecular Biology, 30*, 1092–1104.

Planutis, A., Xue, L., Trainor, C. D., Dangeti, M., Gillinder, K., Siatecka, M., ... Bieker, J. J. (2017). Neomorphic effects of the neonatal anemia (Nan-Eklf) mutation contribute to deficits throughout development. *Development (Cambridge, England), 144*, 430–440.

Rahl, P. B., Lin, C. Y., Seila, A. C., Flynn, R. A., McCuine, S., Burge, C. B., ... Young, R. A. (2010). c-Myc regulates transcriptional pause release. *Cell, 141*, 432–445.

Ransom, D. G., Bahary, N., Niss, K., Traver, D., Burns, C., Trede, N. S., ... Hersey, C. (2004). The zebrafish moonshine gene encodes transcriptional intermediary factor 1gamma, an essential regulator of hematopoiesis. *PLoS Biology, 2*, E237.

Ravindranath, Y., Johnson, R. M., Goyette, G., Buck, S., Gadgeel, M., & Gallagher, P. G. (2018). KLF1 E325K-associated congenital dyserythropoietic anemia type IV: Insights into the variable clinical severity. *Journal of Pediatric Hematology/Oncology: Official Journal of the American Society of Pediatric Hematology/Oncology, 40*, e405–e409.

Rodriguez, J., & Larson, D. R. (2020). Transcription in living cells: Molecular mechanisms of bursting. *Annual Review of Biochemistry, 89*, 189–212.

Roeder, R. G. (2005). Transcriptional regulation and the role of diverse coactivators in animal cells. *FEBS Letters, 579*, 909–915.

Roeder, R. G., & Rutter, W. J. (1969). Multiple forms of DNA-dependent RNA polymerase in eukaryotic organisms. *Nature (London), 224*, 234–237.

Rougvie, A. E., & Lis, J. T. (1988). The RNA polymerase II molecule at the 5′ end of the uninduced hsp70 gene of D. melanogaster is transcriptionally engaged. *Cell, 54*, 795–804.

Sankaran, V. G., Ghazvinian, R., Do, R., Thiru, P., Vergilio, J. A., Beggs, A. H., ... Lander, E. S. (2012). Exome sequencing identifies GATA1 mutations resulting in Diamond-Blackfan anemia. *The Journal of Clinical Investigation, 122*, 2439–2443.

Sawado, T., Halow, J., Bender, M. A., & Groudine, M. (2003). The beta -globin locus control region (LCR) functions primarily by enhancing the transition from transcription initiation to elongation. *Genes & Development, 17*, 1009–1018.

Schubeler, D., Groudine, M., & Bender, M. A. (2001). The murine beta-globin locus control region regulates the rate of transcription but not the hyperacetylation of histones at the active genes. *Proceedings of the National Academy of Sciences of the United States of America, 98*, 11432–11437.

Sengupta, T., Chen, K., Milot, E., & Bieker, J. J. (2008). Acetylation of EKLF is essential for epigenetic modification and transcriptional activation of the beta-globin locus. *Molecular and Cellular Biology, 28*, 6160–6170.

Shah, A., Katiyar, S., Chaturvedi, C. P., Rai, B., Rahman, K., Tripathy, N., ... Gupta, A. (2026). Histone methyltransferase NSD3 orchestrates early erythropoiesis by regulating erythroid progenitor cell differentiation and survival. *Biochimica et Biophysica Acta (BBA) - Molecular Cell Research, 1873*, 120095.

Siatecka, M., Sahr, K. E., Andersen, S. G., Mezei, M., Bieker, J. J., & Peters, L. L. (2010). Severe anemia in the Nan mutant mouse caused by sequence-selective disruption of erythroid Kruppel-like factor. *Proceedings of the National Academy of Sciences of the United States of America, 107*, 15151–15156.

Singleton, B. K., Lau, W., Fairweather, V. S., Burton, N. M., Wilson, M. C., Parsons, S. F., ... Anstee, D. J. (2011). Mutations in the second zinc finger of human EKLF reduce promoter affinity but give rise to benign and disease phenotypes. *Blood, 118*, 3137–3145.

Song, S. H., Kim, A., Ragoczy, T., Bender, M. A., Groudine, M., & Dean, A. (2010). Multiple functions of Ldb1 required for beta-globin activation during erythroid differentiation. *Blood, 116*, 2356–2364.

Soni, S., Pchelintsev, N., Adams, P. D., & Bieker, J. J. (2014). Transcription factor EKLF (KLF1) recruitment of the histone chaperone HIRA is essential for beta-globin gene expression. *Proceedings of the National Academy of Sciences of the United States of America, 111*, 13337–13342.

Stadhouders, R., Thongjuea, S., Andrieu-Soler, C., Palstra, R. J., Bryne, J. C., van den Heuvel, A., ... van der Sloot, A. (2012). Dynamic long-range chromatin interactions control Myb proto-oncogene transcription during erythroid development. *The EMBO Journal, 31*, 986–999.

Stonestrom, A. J., Hsu, S. C., Jahn, K. S., Huang, P., Keller, C. A., Giardine, B. M., ... Hardison, R. C. (2015). Functions of BET proteins in erythroid gene expression. *Blood, 125*, 2825–2834.

Struhl, K. (1999). Fundamentally different logic of gene regulation in eukaryotes and prokaryotes. *Cell, 98*, 1–4.

Subtil-Rodriguez, A., & Reyes, J. C. (2010). BRG1 helps RNA polymerase II to overcome a nucleosomal barrier during elongation, in vivo. *EMBO Reports, 11*, 751–757.

Sun, J., Pan, H., Lei, C., Yuan, B., Nair, S. J., April, C., ... Ruan, J. (2011). Genetic and genomic analyses of RNA polymerase II-pausing factor in regulation of mammalian transcription and cell growth. *The Journal of Biological Chemistry, 286*, 36248–36257.

Sun, Y., Han, J., Wang, Z., Li, X., Sun, Y., & Hu, Z. (2020). Safety and Efficacy of bromodomain and extra-terminal inhibitors for the treatment of hematological malignancies and solid tumors: A systematic study of clinical trials. *Frontiers in Pharmacology, 11*, 621093.

Suter, D. M., Molina, N., Gatfield, D., Schneider, K., Schibler, U., & Naef, F. (2011). Mammalian genes are transcribed with widely different bursting kinetics. *Science (New York, N. Y.), 332*, 472–474.

Tamm, I., & Kikuchi, T. (1979). Early termination of heterogeneous nuclear RNA transcripts in mammalian cells: Accentuation by 5,6-dichloro 1-beta-D-ribofuranosylbenzimidazole. *Proceedings of the National Academy of Sciences of the United States of America, 76*, 5750–5754.

Taneda, T., Zhu, W., Cao, Q., Watanabe, H., Yamaguchi, Y., Handa, H., & Wada, T. (2011). Erythropoiesis is regulated by the transcription elongation factor Foggy/Spt5 through gata1 gene regulation. *Genes to Cells: Devoted to Molecular & Cellular Mechanisms, 16*, 231–242.

Varricchio, L., Planutis, A., Manwani, D., Jaffray, J., Mitchell, W. B., Migliaccio, A. R., & Bieker, J. J. (2019). Genetic disarray follows mutant KLF1-E325K expression in a congenital dyserythropoietic anemia patient. *Haematologica*.

Voon, H. P., Hughes, J. R., Rode, C., De La Rosa-Velazquez, I. A., Jenuwein, T., Feil, R., ... Gibbons, R. J. (2015). ATRX plays a key role in maintaining silencing at interstitial heterochromatic loci and imprinted genes. *Cell Reports, 11*, 405–418.

Weil, P. A., Luse, D. S., Segall, J., & Roeder, R. G. (1979). Selective and accurate initiation of transcription at the Ad2 major late promotor in a soluble system dependent on purified RNA polymerase II and DNA. *Cell, 18*, 469–484.

Weiss, S. B. (1960). Enzymatic incorporation of ribonucleoside triphosphates into the interpolynucleotide linkages of ribonucleic acid. *Proceedings of the National Academy of Sciences of the United States of America, 46*, 1020–1030.

Welboren, W. J., van Driel, M. A., Janssen-Megens, E. M., van Heeringen, S. J., Sweep, F. C., Span, P. N., & Stunnenberg, H. G. (2009). ChIP-Seq of ERalpha and RNA polymerase II defines genes differentially responding to ligands. *The EMBO Journal, 28*, 1418–1428.

Wen, H., Li, Y., Xi, Y., Jiang, S., Stratton, S., Peng, D., ... Wu, J. (2014). ZMYND11 links histone H3.3K36me3 to transcription elongation and tumour suppression. *Nature (London), 508*, 263–268.

Wong, P., Hattangadi, S. M., Cheng, A. W., Frampton, G. M., Young, R. A., & Lodish, H. F. (2011). Gene induction and repression during terminal erythropoiesis are mediated by distinct epigenetic changes. *Blood, 118*, e128–e138.

Yamada, T., Yamaguchi, Y., Inukai, N., Okamoto, S., Mura, T., & Handa, H. (2006). P-TEFb-mediated phosphorylation of hSpt5C-terminal repeats is critical for processive transcription elongation. *Molecular Cell, 21*, 227–237.

Yik, J. H., Chen, R., Nishimura, R., Jennings, J. L., Link, A. J., & Zhou, Q. (2003). Inhibition of P-TEFb (CDK9/Cyclin T) kinase and RNA polymerase II transcription by the coordinated actions of HEXIM1 and 7SK snRNA. *Molecular Cell, 12*, 971–982.

Zeitlinger, J., Stark, A., Kellis, M., Hong, J. W., Nechaev, S., Adelman, K., ... Young, R. A. (2007). RNA polymerase stalling at developmental control genes in the Drosophila melanogaster embryo. *Nature Genetics, 39*, 1512–1516.

Zhang, W., & Bieker, J. J. (1998). Acetylation and modulation of erythroid Kruppel-like factor (EKLF) activity by interaction with histone acetyltransferases. *Proceedings of the National Academy of Sciences of the United States of America, 95*, 9855–9860.

Zhang, W., Kadam, S., Emerson, B. M., & Bieker, J. J. (2001). Site-specific acetylation by p300 or CREB binding protein regulates erythroid Kruppel-like factor transcriptional activity via its interaction with the SWI-SNF complex. *Molecular and Cellular Biology, 21*, 2413–2422.

Printed and bound by CPI Group (UK) Ltd, Croydon, CR0 4YY
24/09/2026
02222378-0004